ICIAM '87: Proceedings of the First International Conference on Industrial and Applied Mathematics

Proceedings of the First International Conference on Industrial and Applied Mathematics, la Cité des Sciences et de l'Industrie, Paris, June 29–July 3, 1987.

The Conference was co-sponsored by Gesellschaft für Angewandte Mathematik und Mechanik (GAMM), Institute of Mathematics and its Applications (IMA), Society for Industrial and Applied Mathematics (SIAM), and Société de Mathématiques Appliquées et Industrielle (SMAI). The Conference was organized by l'Institute National de Recherche en Informatique et en Automatique (INRIA).

ICIAM '87: Proceedings of the First International Conference on Industrial and Applied Mathematics

EDITED BY
James McKenna
AT&T Bell Laboratories
and
Roger Temam
Université Paris-Sud

Philadelphia 1988

Library of Congress Catalog Card Number 88-61320
ISBN 0-89871-224-6

Preface

These Proceedings of ICIAM '87, the First International Conference on Industrial and Applied Mathematics, record several important events. On the surface it is a permanent historical record of the conference, recording the talks of the invited speakers, the abstracts of the minisymposia, and listing the titles and authors of the contributed talks. In addition, it lists many other people who labored to put on this important conference.

ICIAM '87, held in Paris at the Cite des Sciences et de l'Industrie from June 29 through July 3, 1987, was attended by approximately 1800 applied mathematicians from over 50 countries in 7 continents. There were 16 invited speakers and 69 minisymposia, and approximately 1500 contributed papers were submitted. The conference was co-organized by Gesellschaft fur Angewandte Mathematik und Mechanik (GAMM), Institute of Mathematics and its Applications (IMA), Society for Industrial and Applied Mathematics (SIAM), and Societe de Mathematiques Appliquees et Industrielles (SMAI). L'Institut National de Recherche en Informatique et Automatique (INRIA) handled the logistics of the meeting. You will find this information carefully recorded in these Proceedings.

At a deeper level, however, these Proceedings record the size, vigor, and explosive growth of applied mathematics. Mathematics is a live science; real world phenomena provide its inspiration and nourishment. It then leaves this world for a formalization of the intrinsic properties of these phenomena, becomes aesthetical, and then cannot make further progress unless again exposed to reality. Mathematics needs a compass otherwise it loses its essence. All these steps are necessary and there is no hierarchy among them.

In the first half of this century a majority of the mathematicians began to accept the view that mathematics should be studied for its own sake, without regard for its roots or applications in the sciences and the world at large; this was the beginning of a period of abstraction. In the second half of this century the incredible growth of high technology in the industrialized world generated an increased need for applied mathematics--and an accompanying array of fascinating mathematical problems--and the tide shifted. More and more mathematicians and students of mathematics are realizing the beauty and excitement of applied mathematics. Needless to say, the advent of the digital computer, which has transformed applied mathematics in the second half of the twentieth century, has greatly magnified this development.

This new vigor has been reflected in the formation of new societies and the revitalization of older societies devoted to applied mathematics. SIAM, SMAI, IMA, a new society just formed in Italy, and the affiliation of a Scandinavian group with SIAM are products fo the rebirth of applied mathematics which has been taking place during the last forty years. Societies such as GAMM, which was founded much earlier, have taken on new importance.

SMAI, the host organization of the first ICIAM, was founded in France in 1983. Immediately this new society wanted to establish ties with similar societies in other countries. The contacts were made, the response was overwhelming, and the plans for a first ICIAM in Paris were underway.

These Proceedings illustrate the international character of applied mathematics today. The number of countries represented does not tell the whole story. A careful examination of the abstracts of the minisymposia reveals the current high level of international collaboration. In the majority of the minisymposia, at least two countries are represented; in many, three or more countries are included. This representation reflects not just shared knowledge, but actual collaboration. This is remarkable given the important role applied mathematics plays in the highly competitive and highly technological world marketplace.

Finally, these Proceedings exhibit the breadth and vigor of the field. Some of the topics prominently represented at this meeting--fluid flow, for example--would have been prominent topics at a similar meeting held twenty-five or thirty years ago. Other topics, such as discrete mathematics and combinatorics, are relatively new in applied mathematics. Underlying almost everything is the pervasive influence of digital computers. Often the computer is the direct subject of investigation. In many of the remaining cases, the subject of research is only of interest because of calculations that would be impossible without computers.

In summary, these Proceedings are more than just the record of an exciting event. They celebrate the vigor, growth, and worldwide importance of applied mathematics in the second half of the twentieth century.

The editors of this volume were deeply involved in the preparation of ICIAM '87, and very much enjoyed that experience as well as working with the numerous participants. They are happy to complete their work by editing this volume and they wish, once more, to thank all those who helped.

James McKenna
AT&T Bell Laboratories

Roger Temam
Université Paris-Sud

ICIAM '87 Committees

Honorary Presidents

The late Professor Peter Henrici,
Eidgenossische Technische Hochschule
Sir James Lighthill, Cambridge University
Professor Joseph B. Keller,
Stanford University
Professor Jacques-Louis Lions,
Centre National d'Etudes Spatiales

Program Committee

James McKenna, AT&T Bell Laboratories
Alain Bensoussan, INRIA
R. F. Churchhouse,
University of Cardiff
Albert M. Erisman,
Boeing Computer Services Company
K. P. Hadeler, Universitat Tubingen
J.C.R. Hunt, Cambridge University
M. Metivier, Ecole Polytechnique
R. Rannacher, Universität des Saarlandes
H. J. Stetter, Technische Universitat Wien
Roger Temam, Université Paris-Sud
Robert Voigt, ICASE, National
Aeronautics and Space Administration

Steering Committee

Roger Temam, Universite Paris-Sud
Francoise Chatelin, IBM Corporation
N. Clarke, Institute of Mathematics
and Its Applications
Gene H. Golub, Stanford University
Ettore Infante, University of Minnesota
Egon Krause, Technische Hochschule
Aachen
Patrick Lascaux, Centre d' Etudes de Limeil
Helmut Neunzert, Universität Kaiserslautern
Jacques Periaux,
Avions Marcel Dassault/Bréguet Aviation
C. Richard, Institute of Mathematics and
Its Applications
J. Stoer, Universität Wurzburg
J. M. Varah, University of British Columbia

Local Organizing Committee

Patrick Lascaux, Centre d' Etudes de Limeil
I. Edward Block, SIAM
J. P. Boujot, Informatique Internationale
T. Bricheteau, INRIA
S. Gosset, INRIA
C. Jouron, Université Paris-Sud
Gerard Meurant, Centre d' Etudes de Limeil
M. F. Neumann, INRIA
Jacques Periaux,
Avions Marcel Dassault/Bréguet Aviation
F. Thomasset, INRIA

Coordination/Organization

T. Bricheteau, INRIA

Contents

Invited Speakers

Wolfgang Alt

Karl Johan Åström

Michael Atiyah

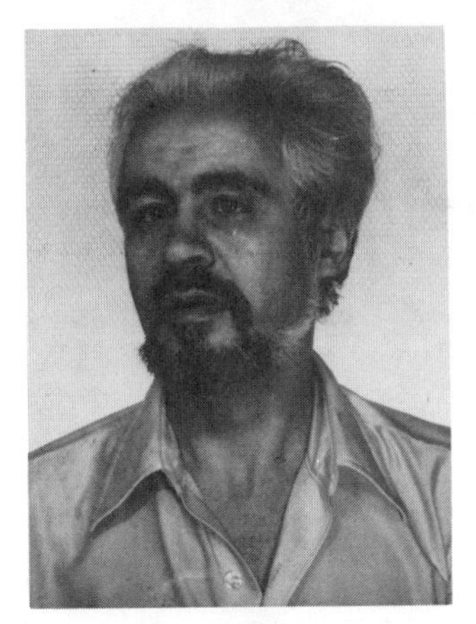

Robert Azencott

D. G. Crighton

Philippe G. Ciarlet

Carl de Boor

Yves Genin

John Hopcroft

Peter D. Lax

L. Lovász

Andrew Majda

F. Natterer

M. J. D. Powell

Please note: Photos unavailable for W. Hackbusch and P. Perrier.
Peter Lax photo is courtesy of NYU/L. Pellettieri.

Scenes from the Conference

Computing Vortex
Illustrate with

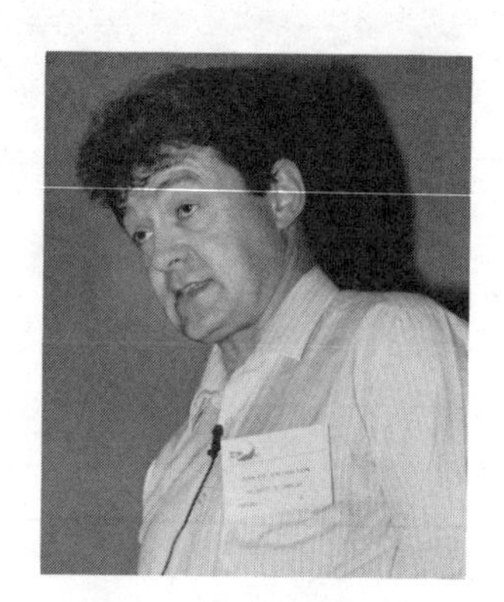

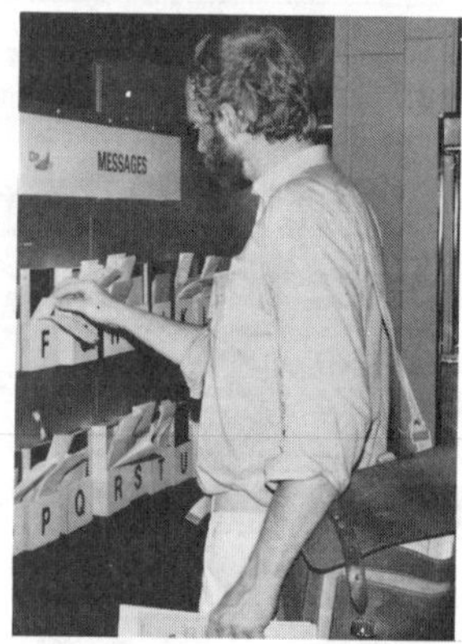
MESSAGES

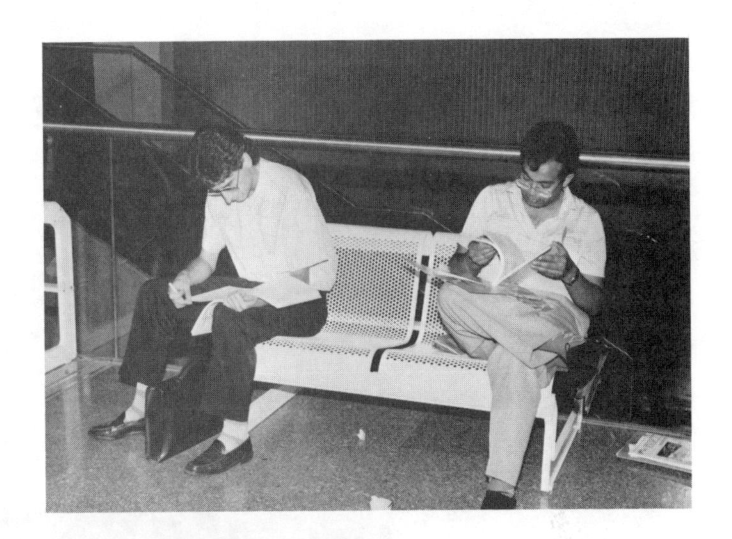

All photographs in this section taken by I. Edward Block.

PART I: WELCOMING AND OPENING PRESENTATIONS

Welcome to ICIAM '87

ROGER TEMAM, SMAI*

You have come in large numbers to attend the First International Conference on Industrial and Applied Mathematics. Some of you have traveled far to be here this morning. I am very happy to welcome all of you to Paris.

As this congress opens, I would like to remind you with sadness of the absence of Professor Henrici, who was one of the honorary presidents of this conference and who passed away this year. Professor Walter will present a eulogy of Professor Henrici and his work in a few minutes, but we all know of the immense influence he has had on the development of applied mathematics through his books and his work.

As you know, ICIAM '87 has been co-organized by: GAMM (Gesellschaft fur Angewandte Mathematik und Mechanik), IMA (Institute of Mathematics and its Applications), SIAM (Society for Industrial and Applied Mathematics), and SMAI (Societe de Mathematiques Appliquees et Industrielles).

In addition, INRIA (l'Institut National de Recherche en Informatique et Automatique) has also taken part in the organization of this conference.

The preparation of an event of such importance could not have been achieved without the efforts of many people. I would like to thank all those who participated and gave us their encouragement.

Monsieur le President de la Republique and Monsieur le Ministre Delegue Charge de la Recherche et de L'Enseignement Superieur have given their patronage to this conference. I thank them deeply for having done so.

I am happy to thank Professor Paul Germain, perpetual secretary of the French Academy of Sciences, who has honored us with his presence this morning and who will welcome you in a few minutes on behalf of the Academy of Sciences.

I would like to thank the honorary presidents of ICIAM, who by agreeing to endorse the conference lend their own renown to the endeavor. In addition to the late Professor Henrici, the honorary presidents are Professor Joe Keller, Sir James Lighthill, and Professor Jacques-Louis Lions. I am especially happy, on behalf of SMAI, to pay tribute to Professor Lions, who is the author of a very important mathematical oeuvre and who has played a considerable role in the development of applied mathematics in France.

Several groups participated in the organization of this conference. I would especially like to thank my colleagues from the international steering committee; the program

---

*Roger Temam is a professor of mathematics at the Université Paris-Sud and president of SMAI.

committee, chaired by Jim McKenna; the organizing committee, chaired by Patrick Lascaux; the exhibition committee, chaired by Jean Paul Boujot; the public relations office of INRIA, directed by Mademoiselle Bricheteau; the group from the society SCOIR; and the representatives of GAMM, IMA, SIAM, and SMAI.

I would also like to thank the many organizations and societies that subsidized this conference; a list of their names is printed in the conference program.

Finally, it is also your presence here today that has made this conference possible. Without trying to greet each one of you individually, I would nevertheless like to mention the presence of Professor Lax, who is returning from Jerusalem, where he was awarded the Wolf Prize in Mathematics.

The idea for this conference was conceived three years ago, when SMAI had just been founded in France. From the beginning, our society wished to make itself known in the international community and to establish ties with other societies working toward similar goals.

Applied mathematics has made great strides in recent years, stimulated by the enormous computational power made available by computers as well as by the influx of new problems arising from new technologies. The time seemed right for the organization of a congress that would bring together researchers with similar interests whose interdisciplinary work involves the interaction of mathematics, computers, and real-world problems.

When we began to plan this conference two years ago, we had no previous experience, no idea of the number of participants we could expect, and our conservative estimates led us to envision a total of 600 to 800 participants. You have more than doubled those estimates, testifying both to your interest in this congress and to the dynamic and vital nature of our discipline.

If the congress is new, perhaps the first of a new series, the buildings housing us are also new. The conference center in which we are meeting has been open for only three months; a year ago, when the scientific program of the conference was already quite well established, none of what you see here today existed. In addition to the conference center, the Cite des Sciences et de l'Industrie at La Villette contains a science and technology museum within this same building, which has been open for only a year, and also the spherical building, la Geode, that you have seen in photographs and perhaps already glimpsed. La Geode is a movie theater with a spherical screen, and, on close inspection, the specialists in finite elements among you will notice that the outer shell of the building is composed of triangular elements.

The organizers of ICIAM did not spare their efforts and did everything possible to make the conference a success. I hope that in spite of the imperfections and the inconveniences caused by the novelty of the endeavor, this congress will be profitable and pleasant for everyone. On behalf of the International Steering Committee of ICIAM '87, I declare the First International Conference of Industrial and Applied Mathematics open.

Greetings from SIAM

C. WILLIAM GEAR*

On behalf of SIAM, I would like first to thank our hosts and the organizers of this conference--the Societe de Mathematiques Appliquees et Industrielles and its president, Roger Temam, who chaired the steering committee; the Institut National de Recherche en Informatique et en Automatique, and particularly Mademoiselle Therese Bricheteau, who worked with the local organizing committee chaired by Patrick Lascaux to coordinate the local arrangements; and the program committee, chaired by Jim McKenna. Many others--representatives of the four sponsoring societies, of other supporting societies, and of assisting organizations--deserve credit for what I am sure is going to be a superb first international meeting, but I particularly want to recognize the efforts of Gene Golub, who has done as much as anyone to encourage international cooperation in applied mathematics and who played a major role in the conception of this meeting. Gene, the immediate past president of SIAM, has been the driving force behind many successful SIAM ventures and is to be particularly congratulated on the success of this one.

The most important part of any meeting is the participants. For ICIAM '87, you have submitted a record number of papers and registered in record numbers. The success of the meeting will be determined by the way we interact and share information during the next five days. Let me thank you for your participation and welcome all of you to Paris.

---

*C. William Gear is a professor of computer science at the University of Illinois at Urbana-Champaign and president of SIAM.

Greetings from GAMM

WOLFGANG WALTER*

In recent years mathematics has attained new dimensions. It has expanded into new fields of natural science and into almost every domain of technology. Is it really true that we are witnessing a historic change in the relation between mathematics and its applications, that mathematics has become an essential part of technological development?

The Industrial Revolution of the first half of the last century was made possible by progress in physics and mechanics, but mathematics was not essential to its development, and in the great technical inventions of the 19th century, mathematics played only a marginal role.

In the first quarter of this century, however, a gradual change took place. After the First World War, the German engineering society Verein Deutscher Ingenieure (VDI) established committees for technical mechanics and physics. Their purpose was twofold:

--To make the results of scientific research accessible to the engineer.

--To make the problems of engineering amenable to scientific treatment.

This description, which is taken from a contemporary VDI source, addresses clearly two essential aspects of applied mathematics: When mathematicians engage in applications, they are amply rewarded with new ideas and deep problems.

In 1920, to further these two goals, VDI founded a new journal--Zeitschrift fur Angewandte Mathematik und Mechanik (ZAMM). Two years later the Gesellschaft fur Angewandte Mathematik und Mechanik (GAMM) was founded by VDI and the Deutsche Mathematiker-Vereinigung (DMV). The following activities--mentioned, among others, in the VDI report for 1919-20--might give some idea of the engineering problems of that time:

--mechanical vibrating systems as models for wireless telegraphic systems (they had no computers, so they relied on mechanical analogies, i.e., special-purpose analog computers);

---

*Wolfgang Walter is a professor of mathematics at the University of Karlsruhe and president of GAMM.

--longitudinal vibrations of airplanes;

--the influence of eigentensions on stability experiments;

--lubrication problems;

--the vibration of electrical train locomotives;

--and the theory of the lift of airplane wings.

In particular, the technical problems encountered in the construction of airplanes could be mastered only with mathematics. Ludwig Prandtl, the eminent scientist of aerodynamics, invented boundary layer theory and brought aerodynamics within reach of numerical treatment. Prandtl was the first president of GAMM.

Today, specialists from many disciplines and all parts of the world have gathered here. Your contributions have provided a program that is impressive for its quality, quantity, and diversity. The program will provide insight into current technological developments and find time to contemplate their impact on human society.

With sadness, I have to report the death of Peter Henrici, who was to have been one of the honorary presidents of this conference. He died in Zurich on March 13 of this year after a nine-month illness.

Peter Henrici was born in 1923 in Basel, Switzerland; he attended school in Basel and then studied law at the University of Basel from 1942 to 1944. After the Second World War he studied electrical engineering and mathematics at ETH in Zurich, receiving diplomas in both fields and a doctorate in mathematics.

In 1951, after working at ETH in applied mathematics and aerodynamics, he went to the United States, where he remained for 11 years. He spent six years at UCLA, first as an associate professor and later as a full professor of mathematics. In 1962 he returned to his alma mater, ETH in Zurich, where he remained for the rest of his life. He held a variety of visiting professorships at Harvard, Stanford, and other American universities. From 1985 to his death, he was William Rand Kenan Distinguished Professor of Mathematics at the University of North Carolina in Chapel Hill.

Peter Henrici was an eminent scientist and one of the pioneers of numerical mathematics. His mathematical opus consists of about 80 research articles and 11 books. His first book, Discrete Variable Methods in Ordinary Differential Equations, was published in 1962; it won international recognition and soon became a standard text on the subject. Much of his later work is concerned with applied and numerical complex analysis. His monumental three-volume treatise, Applied and Computational Complex Analysis, which appeared between 1974 and 1986, is the highlight of his scientific work.

Peter Henrici was SIAM's 1978 John von Neumann lecturer, and he served as president of GAMM from 1977 to 1980. He was the editor of Zeitschrift fur Angewandte Mathematik und Physik and an associate editor of no fewer than 12 other journals.

This brief summary provides only an incomplete picture of Peter Henrici the scientist, of his energy and his immense diligence. Peter Henrici the man--his personality, his warmth and humor and modesty--will remain in our memories, as will for some of us his passion for music, the experience of listening to or playing with an extraordinarily gifted pianist. He is no longer among us, but his work is here to stay.

I want to thank Professor Temam and all those who have worked hard and continue to do so to prepare and organize this conference. I welcome you all and wish you a successful and rewarding week, and hope that you will return home with new insights and ideas.

Greetings from IMA

M.J.D. POWELL*

I am sorry that the president of the Institute of Mathematics and its Applications is not here to address you on this important occasion. My name is Michael Powell, and I am one of the fellows of the IMA. It is a great pleasure to speak on behalf of the institute at this opening ceremony. Last year the IMA cosponsored a meeting with the Society for Industrial and Applied Mathematics, but I believe that this is the first time it has cooperated in this way with the Gesellschaft fur Angewandte Mathematik und Mechanik (GAMM) and the Societe de Mathematiques Appliquees et Industrielles (SMAI). It is delightful to collaborate with all these societies. The number of participants who have registered for ICIAM shows clearly the demand for meetings that bring together students and practitioners from a wide range of mathematics, and it shows also the attraction of Paris.

*Michael Powell is a Fellow of the IMA.

Mathematics and Mechanics: Two Very Close Neighbors

PAUL GERMAIN*

I am honored and pleased to be here this morning to welcome you to the First International Conference on Industrial and Applied Mathematics. I would like to greet you, first of all, on behalf of the French Academy of Sciences, an institution that aims to represent the entire French scientific community. But since my young colleagues have been kind enough to invite me to deliver an opening adress at this meeting, I would prefer to speak to you this morning as one of your former colleagues. Indeed, I dare say that I could call myself a precursor of the true applied mathematicians that you are, because I have tried to keep the mathematical ideal alive in all my work that is devoted to mechanics.

In limiting my remarks to that branch of applied mathematics, I have no intention of neglecting the numerous other applications of mathematics or of diminishing their importance. I simply prefer to talk about what is familiar to me.

Let me begin with three personal recollections. In 1948, having completed my training in mathematics at the Ecole Normale and then a three-month visit at the National Physical Laboratory, I attended my first International Congress of Mechanics in London. A lunch with K.O. Friedrichs, whom I met for the first time during the congress, was an unforgettable episode. This wonderful and very kind mathematician was very encouraging to me, then a young scientist, explaining that his experience at New York University with Richard Courant and James Stoker had clearly demonstrated that it took longer to become an applied mathematician--a concept practically unknown at that time in France--than to become a pure mathematician. He also pointed out the main differences between the objectives of the two types of scientists and between the types of satisfaction they could hope to derive from their work.

The second of my recollections concerns a talk given by William Prager in November 1953 at a meeting held at Columbia University to celebrate the tenth anniversary of the creation of applied mathematics departments in American universities or, perhaps more correctly, of the momentum imparted to applied mathematics by President Roosevelt during the war. Prager likened the concepts of "pure" and "applied" in mathematics to "right" and "left" in politics: The terms are relative--the same person may be described as a pure mathematician by one person and as an applied mathematician by another.

---

*Paul Germain, president of the French Academy of Sciences, delivered the opening address at the First International Conference on Industrial and Applied Mathematics. The following version of the address was edited from an English translation.

At that time, applied mathematics departments often included mechanics, fluid mechanics, and sometimes even applied mechanics. Two of the most well known examples are the departments at Manchester, headed by Sydney Goldstein, and at Brown University, headed by William Prager. It is interesting that the International Congress of Applied Mechanics, the institution that later gave rise to the International Union of Theoretical and Applied Mechanics, was founded a few years after the end of World War I by Theodore von Karman and T. Levi-Civita, the former representing theoretical mechanics and the latter mathematics.

My third experience was a series of illuminating discussions with Walter Noll and a circle of his friends during a IUTAM symposium on Cosserat continua held in Freudenstadt in 1967. The topic was the relation between mathematics--or more precisely applied mathematics--and mechanics.

To summarize briefly the main conclusions of those discussions, one has to distinguish two main objectives when applying mathematics to mechanics. The first, and the most obvious, is to solve problems, to formulate and prove theorems, and to find new properties of the solutions and new ways to compute them, at least approximately. The second is to use one's mathematical insight to investigate mechanical phenomena, to build the right concepts, to derive new formulations, to look at the mechanical world from the viewpoint of a mathematician who is trying to discover new ways, new methods, new statements. Walter Noll emphasized that in this respect, the theorist in continuum mechanics is somewhat like a psychologist: When faced with engineers, he tries to discover what lies behind their words.

These three personal recollections convinced me of the continuity of the spectrum that extends from pure mathematics to applied mechanics, and helped me appreciate the variety of talents and motivations of the scientists working in these disciplines.

It is difficult, however, to describe in everyday language the psychological and cultural factors that motivate a scientist. Accordingly, I prefer to remain on safer ground by choosing a few examples from my own experience in mechanics to illustrate the potential benefits of the interaction between mathematics and a field of application.

During the few months that I spent at NPL in 1945, I was able to read and study the preprint of the famous book Supersonic Flow and Shock Waves, by Richard Courant and K.O. Friedrichs. It is remarkable that a topic like nonlinear hyperbolic systems of PDEs, which dates back to the 19th century, was so completely refreshed and broadened when it was studied in light of gas dynamics. As far as I know, many questions remain open, despite the number of papers that have been written and the deep and significant results that have been obtained on this topic.

For a system of conservation laws, a clear and simple definition of a discontinuous solution may be given within the framework of the theory of Schwartz distributions. But the uniqueness theorem for the Cauchy problem requires additional information; shock solutions are a subset of all generalized solutions. The answer may be provided in gas dynamics by the entropy inequality or, in a more abstract context, either by requiring continuity with respect to the data or by considering the perfect fluid case as a limit of a slightly dissipative fluid.

The application--in this case gas dynamics--led to the discovery of interesting mathematical concepts; conversely, the mathematical analysis enhanced our understanding of a physical system. For instance, it is tempting to think that the entropy inequality is merely a consequence of the privileged direction of time: Knowing the present, one can predict the future, but one cannot return to the past. The entropy statement is a necessary condition that has to be fulfilled, but it is not a sufficient condition, as shown by the study of shock waves in magneto-fluid dynamics: The stability criterion with respect to perturbations of the data at the initial time or with the type of dissipative effects schematized by the discontinuity is the convenient requirement for selecting the appropriate physically meaningful solutions. Time does not permit me to comment here on the interesting interactions between transonic flows and PDEs of the mixed type.

My second example concerns nonuniformly valid asymptotic approximations, singular perturbation theory, and multiple scale expansions. The boundary layer concept, introduced by Prandtl in 1904, was immediately recognized by fluid dynamicists as the breakthrough they had been seeking for nearly a century to correct the deficiencies of a perfect fluid flow approach. It provided a good description of the effects of viscosity and a satisfactory way to predict approximate values of the principal characteristics of the flow. A mathematical approach to the field, to my knowledge, was not undertaken until nearly 50 years later! Time does not permit me to describe here the various concepts and techniques that have been applied in the efforts to discover the useful asymptotic expansions and the hierarchy that rules their domain of validity, the matching conditions, and so on. I will remark only that today one would probably not have to wait for decades from the emergence of a new idea and the recognition of its mathematical significance to the development of its implications and applications in various other fields.

My third example is so well known that I will only mention it briefly here. The fantastic advances made in functional and numerical analysis have been applied to a large number of problems in continuum mechanics. The interactions are so obvious that there is no need to elaborate on them. It is certain that continuum mechanics, which emerged during the 1950s, would not have experienced such rapid growth without the amazing efficiency of the mathematical methods that were so well adapted to the problems formulated in the context of theoretical mechanics. I would venture to say that the variety of the problems arising from continuum mechanics was probably a stimulating challenge for applied mathematicians.

But, as I said earlier, mathematics, when applied in a particular field--whether mechanics, physics, biology, or economics--may be extremely helpful in revealing the mathematical structures of complex phenomena and consequently in providing better and deeper understanding of those phenomena. As my last example, I would like to mention the crucial role of constitutive equations. It is not certain that nature obeys simple mathematical schemes. But it is not forbidden to design convenient, simple mathematical schemes for approaching some real-world situations. Continuum thermodynamics, a generalization of the thermodynamics of irreversible processes introduced by mathematicians into theoretical mechanics, makes possible the development of an extensive theory of the structures of elastoplastic or viscoplastic materials, including stability, crack motions, bifurcations, and rupture. More recently, homogenization theories have led to the development of realistic macroscopic constitutive equations by a micro-macro analysis that permits one to capture more of the physics and to deal, for example, with composites. Homogenization has given rise to some new and highly efficient mathematical concepts.

With these few examples taken from my own experience, I have tried to show the fantastic and rapidly expanding domains of applied mathematics. If mathematics is to be applied, it must be useful. Thus, applied mathematicians must be connected in some way with the applications. They may solve problems; they may also delve deeply into a particular field, finding new ways to look at it and, consequently, improving our understanding of it. It is at least my conviction that mathematicians will find stimulating new challenges and ideas in this exercise.

I wish you a good First International Conference on Industrial and Applied Mathematics, good work, and a good time in Paris. I hope that this conference will bring all of you what you are expecting and that it will not be the last one.

PART II: INVITED PRESENTATIONS

Modelling of Motility in Biological Systems

WOLFGANG ALT*

Abstract. Locomotion of animals, organisms or single cells is provided by endogeneous energy supply inducing moving paths, which are different from the Brownian motion of particles and depend on the physiological control of their motile machineries (as of flagellated bacteria or crawling slime molds). Several questions concerning dispersal rates, velocity auto-correlations or searching times can be answered by modelling the individual locomotion with stochastic processes and by extracting relevant parameters as ''persistence index'' or ''motility coefficient''. Model extensions are important which also consider direct or indirect interactions between moving individuals as e.g. in swarming behavior.

The general problem consists in relating experimental data of the individual behavior to quantities at the population level. On the theoretical side this corresponds to different simulation models ranging from Monte Carlo simulations to nonlinear parabolic differential equations.

For several typical examples (leukocytes, gametes and slime bacteria) this wide spectrum of modelling and analyzing approaches is reviewed. Particular attention is drawn to the auto-correlation functions for the moving directions and to continuum equations and their numerical simulations in one-dimensional configurations.

1. Concepts and examples — motile machineries. The physical concept ''mobility of particles'' is closely connected to the phenomenon of Brownian motion, which was detected in 1827 by the botanist Robert BROWN, when he observed pollen grains fluctuating in water. The forces driving such a particle against its drag and inducing a continued movement were in question. It turned out that pollen grains do not move by themselves due to some biological activity, but are pushed by other particles in a stochastic manner. Thus, the *external* ''thermal energy'' known as heat, leads to a sustained *passive motion* of physical particles.

Comparing with the motion of bacteria shown in Fig. 1, for example, we get an obvious suggestion how to define the contrasting biological concept ''motility of individuals'': it is a

* Abteilung Theoretische Biologie, Universität Bonn, Kirschallee 1, D-5300 Bonn 1, F.R.G.

self-sustained *active locomotion* driven by *internal* energy production. Moreover, it does not only depend on exogeneous conditions as temperature, pressure, ionic strength etc., but on some *endogeneous physiological control* "apparatus".

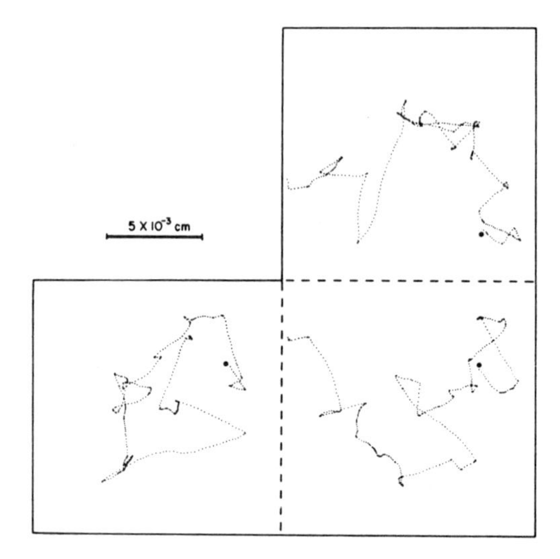

FIG. 1. Projections of 3-dimensional tracks of E.coli bacteria recorded by Berg and Brown (5).

Examples of such types of biological locomotion are:

→ *jumping* of fleas, grass- or sandhoppers, kangaroos
→ *walking* of beetles, spiders, ants, humans
→ *creeping* of worms, slime molds, leukocytes, tumor cells
→ *swimming* of bacteria, ciliates, sperms, fishes
→ *flying* of midges, bees, birds.

Within each particular species different aspects of motility can be investigated as shown in the following scheme:

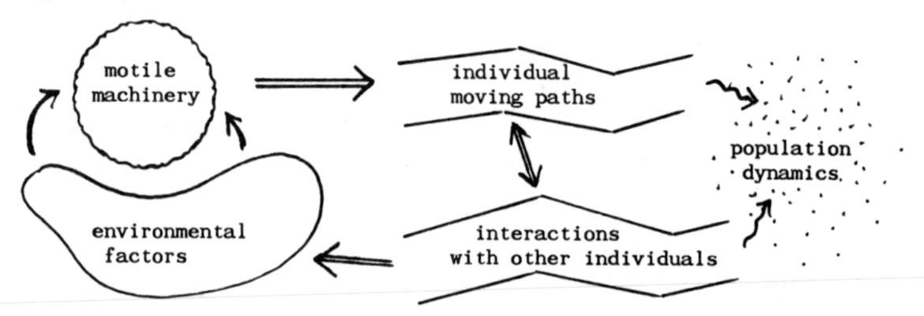

Experimental studies and theoretical models of these aspects of motility have been performed for many species. Trying to review mathematical modelling involved in all different fields would be an enormous task. This already becomes apparent when we look at the diversity of *motile machineries* in different species.

For example, there are successful attempts to describe the function of the *flagellar "rotor"* in bacteria, which is driven by a proton flux across the cell membrane (6). The locomotion of a bacterium is regulated by the switch between clockwise and counter-clockwise rotation mode of several helical flagella leading to subsequent "tumble" and "moving" periods. Experiments with one tethered flagellum show that both rotation times are exponentially distributed, suggesting an underlying *Poisson switching process* (7), Models for the dynamics of this "tumble generator" try to combine additional information about the biochemical control pathway, usually obtained from genetic mutation experiments.

The other important type of flagellar motion is that of eukaryontic cells as protozoa, sperms or gametes (see Fig. 2). Given the *bending wave* profile of the driving flagellum, which is flexible and usually beats in form of a travelling wave from base to tip, hydrodynamical slender body theory at low Reynolds number can be used to compute the forward thrust of the cell. These models, going back to Gray and Hancock in the 50's, are only one particular topic within the large biophysical area "Mechanics of swimming and flying", see e.g. the book by Childress (11).

However, these hydrodynamical considerations do not solve the problem, how the *bending movement* of the flagellum is *produced and regulated.* According to the well-accepted "sliding filament" theory the driving force is produced by dynein-bridges acting similar to the myosin-bridges in muscle contraction. Several groups have tried to incorporate this into mathematical models simulating the bending of one single flagellum (8,10,38,40). All models have the common feature of writing the bending moment, change of local curvature κ per unit time, as the sum of three contributions

$$\partial_t \kappa = M_{\text{active}} + M_{\text{elastic}} + M_{\text{viscous}} .$$

Here the first and the last term involve global integrals of κ, whereas the elastic moment creates a simple "parabolic" term of the kind $-\alpha \cdot \kappa + \beta \cdot \partial_s^2 \kappa$. However, realistic computer simulations require crucial conditions on the way of activation of sliding, which are not yet completely understood (10).

A further well-known example of motility is the "*amoeboid motion*" of single cells as dictyostelium (slime mold cells), leukocytes (white blood cells) or fibroblasts (tissue cells). Contractile filament systems within the cell plasma induce protoplasmic streaming, deformations of the plasma membrane and force application to specific adhesion sites on the substratum along which the cells crawl. Biochemical and biophysical details of these processes are on the way to be known, and therefore mathematical modelling has recently been started (14,35). The idea of these models is to consider the filament-meshwork as a highly viscous (or

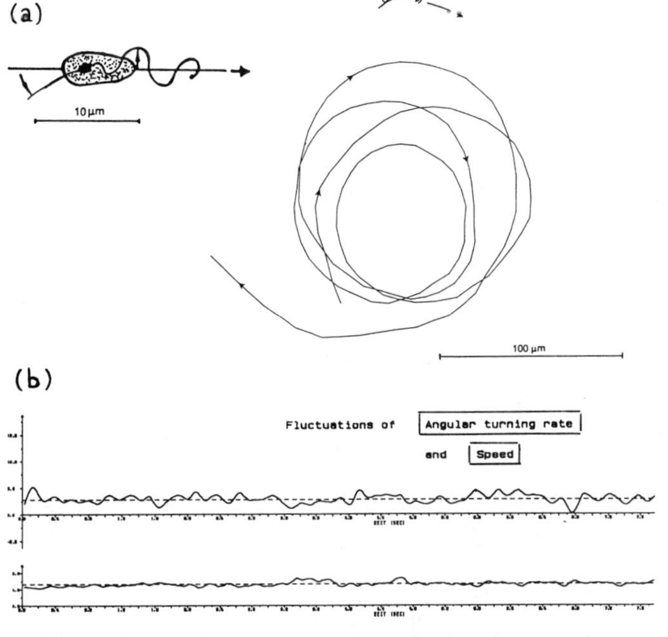

FIG.2. Typical loop-like 2-dimensional path performed by a male gamete of the brown algae Ectocarpus siliculosus (a), with considerable fluctuations of the angular turning rate, but with small variations in speed (b). From (2).

visco-elastic) fluid, in which the constituent elements can be assembled or degraded — according to a simple chemical kinetics $f(u)$, so that the *mass balance equation* for the mean density $u(t,x)$ and velocity $v(t,x)$ becomes (in space dimension one):

$$\partial_t u + \partial_x (u \cdot v) = f(u). \tag{1.1}$$

Further, from a more general two-phase fluid model (13) one can derive a *force balance equation*, which equilibrates friction, shear forces, contractile and swelling forces:

$$\phi \cdot v = \partial_x [\mu u \partial_x v + \psi/2 \cdot u^2 + \sigma \cdot \ln(1-u/u_{max})]. \tag{1.2}$$

These simple one-dimensional model equations together with suitable boundary conditions constitute a well-posed hyperbolic-elliptic system, namely the generalized compressible Stokes equations, and they already show important qualitative phenomena as phase separation between low-density "structural gels" and high-density "contracted gels" and other observed types of dynamical cell plasma motions, for more details see (1). Two-dimensional events of cytoplasmic motility are even more dynamical and dramatic, as can be seen from movies of demembranated amoeboid cells (25: Fig. 6.6) and the corresponding two-dimensional model simulations by Dembo (12).

There exist other types of motions which are far from being understood theoretically, as for example the *gliding* of slime-producing myxobacteria. Alternative models have been proposed (22,26) and are waiting for verification.

The examples above were intended to demonstrate the problems and diversities in modelling biological motility already at the basic level of explaining the motile machineries. Although the mentioned models help to understand "How an individual moves", the crucial theoretical question remains: "Why does it move?"

In the moving living world there can be found several, partially antagonistic goals or functions of movement:

→ *feeding*: grazing cattle, bees searching for flowers, bacteria searching for nutrients,
→ *protection*: swarming and homing of insects or bids, escaping from predators, hiding or purely resting, white blood cells encountering bacteria or toxic material
→ *reproduction*: gametes and sperms searching for sexual partners, parasites searching for hosts, slug formation of slime-molds and -bacteria.

The locomotion pattern of each biological species in different situations or life periods might have evolved according to some of the above mentioned goals or functions. Therefore, theoretical biology has to address the questions:

(1) Is the specific locomotion performed in some "optimal" way?
(2) How is locomotion controlled internally resp. externally?

Experimental investigations and mathematical models analyzing such problems have been developed along two main lines, by recording the *moving paths of individuals* and by description on the *population level*.

2. Stochastic models for individual locomotion. Under homogeneous external conditions many motile species swimming in three dimensions or migrating in two dimensions maintain a surprisingly uniform speed with minor fluctuations around a constant positive value s. This uniformity, which obviously is caused by the self-sustained internal energy supply, seems to be disturbed only when the individual performs a sharper turn or when it rests. Examples herefore are the

(a) 3-d swimming paths of E.coli-bacteria (5), see Fig. 1
(b) 2-d swimming paths of Ectocarpus-gametes (2), see Fig. 2
(c) 2-d migrating paths of leukocytes (18), see Fig. 3
(d) and even 2-d hopping paths of thrushes (39).

Stochastic models for these kinds of moving paths have been developed and analyzed previously, see e.g. (37) or the reviews by Berg (4) and Nossal (32). Under the above assumptions the location X_t of an individual and its velocity V_t at time t satisfy the following equations

$$dX_t = V_t \; dt \tag{2.1}$$

with

$$V_t = s \cdot \theta_t \tag{2.2}$$

where the moving direction $\theta_t \in S^{n-1} \cup \{0\}$ is a stochastic process living on the unit sphere in n dimensions, or in zero during resting periods. When applying this model to longer observation periods under uniform conditions we can assume the *stationarity* of the process θ_t, so that the corresponding *auto-correlation function*

$$G(\tau) = \langle \theta_t \cdot \theta_{t+\tau} \rangle \tag{2.3}$$

is independent of t and an even function of the correlation time τ. On the other hand, the expected dispersal of individuals after a time t can be measured by the *mean-squared displacement*

$$D^2(t) = \langle (X_t - X_0)^2 \rangle \, . \tag{2.4}$$

From (2.1)-(2.3) we obtain the formula

$$D^2(t) = \int_0^t \int_0^t \langle V_{t_1} \cdot V_{t_2} \rangle \, dt_1 \, dt_2$$

$$= 2 s^2 \int_0^t \int_0^{t_1} G(\tau) \, d\tau \, dt_1 \, . \tag{2.5}$$

Thus, the "curvature" of the mean-squared displacement curve, as given by the second derivative, is directly proportional to the auto-correlation function for the directional process θ_t. Let us consider some specific modelling cases and compare them with experimental data.

2.1 Pure random walks — velocity jump processes. We assume that the moving paths are piecewise straight and that resting periods are negligibly short compared to the mean moving time T, as it is approximately the case in locomotion of bacteria (see Fig. 1) or also of leukocytes (see Fig. 3). Moreover, the subsequent changes in moving direction (velocity jumps) should be independent of each other. Then θ_t is just a stationary Markov jump process on S^{n-1} determined by

a *"moving time" distribution* $f(\tau)$ for $\tau \geq 0$,

with $T = \int_0^\infty \tau \cdot f(\tau) \, d\tau$ being the *mean moving time*,

and a *jump kernel* $K(\eta, \theta)$ describing the probability to change the moving direction from θ to η. If K is *symmetric* then we have the relation

$$\int \eta \cdot K(\eta, \theta) \, d\eta = \psi \cdot \theta \tag{2.6}$$

where the uniquely determined factor ψ with $-1 \leq \psi \leq 1$ is called the *directional persistence index.*

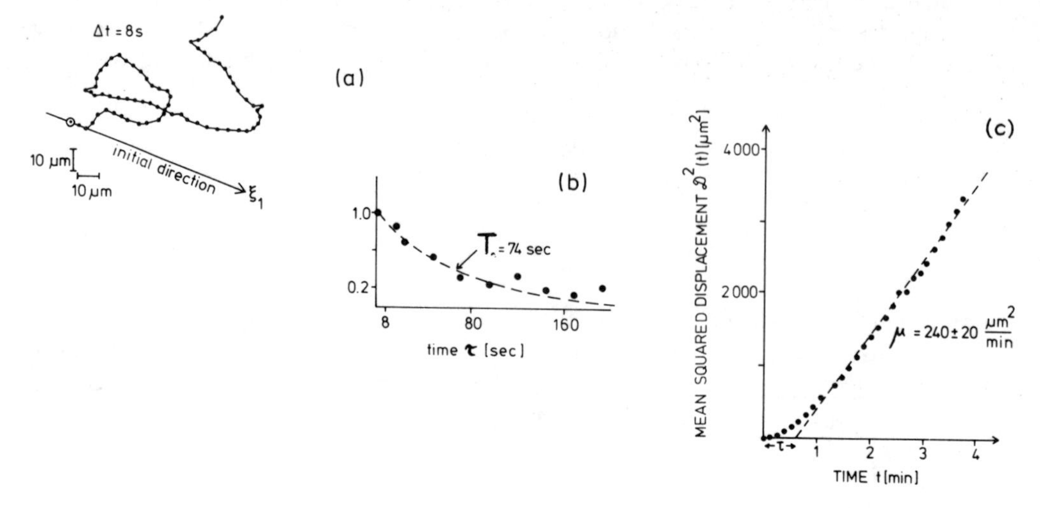

FIG. 3. Leukocytes migrating on a surface (a) analyzed by Gruler and Bültmann (18). The persistence time T_o can be estimated from the directional auto-correlation function (b), whereas the motility coefficient $\mu = s^2 T_o/2$ can be determined from the asymptotic slope in (c).

Then the following formula for the Laplace-transform of the auto-correlation function $G(\tau)$ can be derived, using the master equations for the above process:

$$\hat{G}(z) = \frac{1}{z} - \frac{1-\psi}{T \cdot z^2} \cdot \frac{1-\hat{f}(z)}{1-\psi \cdot \hat{f}(z)} \tag{2.7}$$

where $\hat{f}$ is the Laplace-transform of the moving time distribution f.

In particular, for a simple Poisson jump process, where f is an *exponential distribution*, we obtain $\hat{f}(z) = \frac{1}{1+z/T}$ and $\hat{G}(z) = \frac{1}{z+1/T_o}$, thus the auto-correlation function is the well-known exponential

$$G(\tau) = e^{-\tau/T_o} \tag{2.8}$$

where $T_o = T/(1-\psi)$ is the so-called *persistence time*. From (2.5) we get the mean-squared displacement formula

$$D^2(t) = 2\, s^2 T_o \cdot \left[t - T_o \cdot (1-e^{-t/T_o}) \right]. \tag{2.9}$$

These two theoretical curves can be compared with different experimental measurements, see Fig. 3.

Similar computations for the first *Gamma-distribution* f in (2.7) yield the auto-correlation function

$$G(\tau) = \frac{1}{4\sqrt{\psi}} \cdot \left[(1+\sqrt{\psi})^2 \cdot e^{-2\tau/T_-} - (1-\sqrt{\psi})^2 \cdot e^{-2\tau/T_+} \right] \tag{2.10a}$$

with $T_\pm = T/(1 \pm \sqrt{\psi})$, if ψ is positive, and

$$G(\tau) = \left[(1-\psi)/T \cdot \omega \sin \omega t = \cos \omega t \right] \cdot e^{-2\tau/T} \tag{2.10b}$$

with $\omega = 2\sqrt{|\psi|}/T$, if ψ is negative.

In Fig. 4(a) and (b) these functions are shown for two typical values of the persistence index, namely $\psi = 0.3$ for leukocytes and $\psi = -0.9$ for the motion of spirochetes which perform almost strict reversals after stopping, see (25).

2.2 Continuous directional fluctuations — angular turning processes. In many situations the moving paths of crawling cells can hardly be idealized as piecewise straight. In these cases we have to consider small stochastic changes of the directional process, which in two space dimensions can be written as $\theta_t = (\cos \phi_t , \sin \phi_t)$. The angle ϕ_t satisfies

$$d \phi_t = B_t \, dt \tag{2.11}$$

where B_t is the *angular turning rate* of an individual moving with constant speed s. Then the auto-correlation function in (3) becomes

$$G(\tau) = \langle \theta_t \cdot \theta_{t+\tau} \rangle = \langle \cos (\phi_{t+\tau} - \phi_t) \rangle . \tag{2.12}$$

Let us choose B_t to be a simple Gauss process, namely an Ornstein-Uhlenbeck process defined by

$$dB_t = - \lambda \cdot (B_t - b_o) \, dt + \sqrt{\beta} \, dW_t \tag{2.13}$$

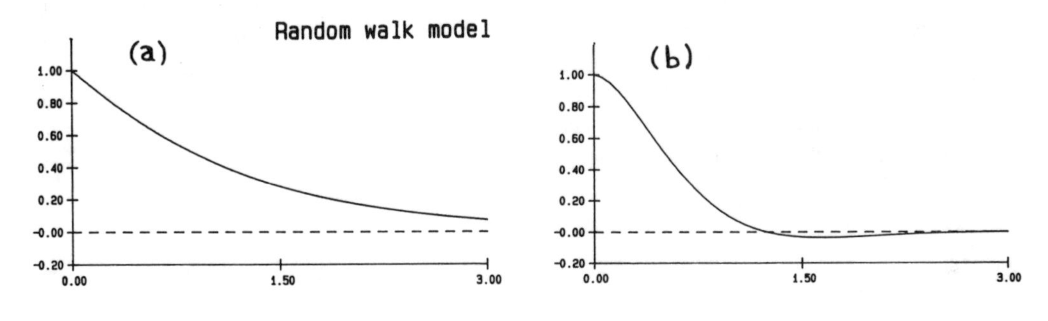

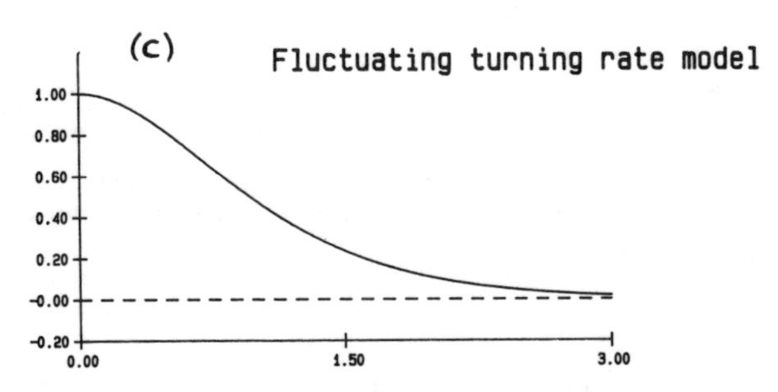

FIG. 4. Theoretical auto-correlation functions $G(\tau)$ of the moving direction for a *random walk model* with Gamma distributed moving times ($T = 1$) and persistence index $\psi = 0.3$ (a) resp. $\psi = -0.9$ (b). The same for a *fluctuating turning model* as in (2.13) with $\lambda = 1$, $\sigma = 1$ (c).

with the standard Wiener process W_t. This means that in a stationary situation the angular turning rate fluctuates around a mean value b_o and is normally distributed with variance $\sigma^2 = \beta / 2\lambda$.

Taking first $b_o = 0$ then, with the aid of characteristic functionals, one derives the following auto-correlation formula, see (23: chap. IV, sect. 3):

$$G(\tau) = \exp\left[-\sigma^2 T\left[\tau - T\cdot(1-e^{-\tau/T})\right]\right] \tag{2.14}$$

where $T = 1/\lambda$ is the mean relaxation time. Looking at the plot in Fig. 4(c) we notice that in contrast to the functions in (2.8) and (2.10) this $G(\tau)$ has zero slope at $\tau = 0$.

Obviously the experimental auto-correlation data for leukocytes in Fig. 3(b) much more resemble the theoretical function in Fig. 4(a) for a random walk model than the one in Fig. 4(c) for the fluctuating curvature model, suggesting that leukocytes migrate in a more "structural" way, possibly alternating between periods of effective displacement and more stationary periods of reorientation.

If we would include such resting periods, as they can often be observed for crawling leukocytes (31) and even more clearly for migrating insects or hopping birds (39), then for exponentially distributed moving and resting times we obtain a mean-squared displacement curve (36) and a corresponding auto-correlation function, which contains two exponential terms similar to (2.10). As we would expect, the auto-correlation function alone does not uniquely determine the underlying stochastic process. Nevertheless, together with additional data about the mean moving and resting times as well as the moving speed, for example, the theoretical expressions could help to discriminate between alternative models for the directional process θ_t.

However, more biological and physiological information is necessary to investigate the "true" reason for the stochasticity of the tracks. Let us take, for example, the loop-like movement of Ectocarpus gametes as shown in Fig. 2, where the fluctuating curvature obviously induces an intense area search. Looking at the flagellar beating pattern of these gametes as recorded by Geller and Müller (16) one observes that the flagellar bending waves have varying amplitude, represented by a stationary process A_t. Using a corresponding "alternating stroke" process for the angular turning rate, namely

$$B_{t_{i+1}} - B_{t_i} = \beta(-1)^i \cdot A_t$$

we were able to perform "realistic" Monte Carlo simulations of the gametes tracks. Clearly, for small beating periods $2 \cdot | t_{i+1} - t_i |$ the process B_t above can be approximated by an Ornstein-Uhlenbeck process as in (2.13).

Similarly, Tranquillo and Lauffenburger (41) made a modelling approach to reduce the stochasticity of the leukocyte motion to *fluctuations in the occupancy of certain cell surface receptors* $R_t^{\pm}$ on the left (+) and right (–) of a moving cell. In the classical diffusion limit, cp. (15), $R_t^{\pm}$ satisfy a system of stochastic differential equations generalizing (2.13). The concentrations $M_t^{\pm}$ of certain intracellular diffusible messengers, produced by bound receptors, then obey a system of the kind

$$dM_t^{\pm} = \pm D \cdot (M_t^{-} - M_t^{+}) - \alpha M_t^{\pm} + \beta R_t^{\pm} \tag{2.15}$$

so that finally the angular turning rate is determined by measuring the difference between left and right messenger concentrations

$$B_t = \kappa \cdot (M_t^{+} - M_t^{-})\,. \tag{2.16}$$

This extended fluctuating curvature model is also capable of describing a biased turning behavior of leukocytes in a spatial gradient of the stimulating ligand, the well-known phenomenon of "chemotaxis". Theoretical analyses and Monte Carlo simulations of the system reveal similar, but slightly different characteristics of the resulting movement paths when compared to the pure random walk model, see Fig. 5 below.

Let us briefly mention the attempts to understand the locomotive behavior of protozoa and higher organisms during *feeding*. It is obvious that many individuals have to stop their motion in order to uptake their food, as e.g. thrushes do it (39), or that they adapt their moving speed resp. their turning behavior to the presence or to the concentration of food. For a recent mathematical modelling approach for zooplankton see (28).

So far, direct interactions between moving individuals have been ignored. This assumption is often justified, e.g. when swarming is indirectly induced by chemotaxis (see section 3). The situation is quite different for other swarms as, for example, swarms of midges, see (34). Since the full "N-body problem" with mutual interactions, as studied in analogous physical situations, is too difficult, Okubo and coworkers developed a simplifying model for the individual distances r_i from a fixed "swarm center". The linearized version

$$r_i'' + k \cdot r_i' + \omega^2 \cdot r_i = F_t \tag{2.17}$$

constitutes a pendulum equation, which is perturbed by fluctuating forces F_t, representing the averaged contribution of other swarming individuals. F_t is assumed as a stationary process, similar to B_t above. In the limit of vanishing time correlations for F_t one obtains a formula for the velocity auto-correlation function which is oscillating around zero and can be compared to experimental data, see Fig. 6 below. In spite of the simplifying nature of this "random oscillator" swarming model, simulations of three-dimensional moving paths (20) show a surprising similarity with observed paths.

3. Diffusion limits and population models. There are many situations where the "microscopic" individual motion cannot be directly observed, for example when leukocytes in the so-called Boyden-chamber migrate into a micropore filter. However, the evolution of the "macroscopic" population density profile within the filter can be recorded from time to time (25) and approximate diffusion equations are used to describe these profiles. A further example is the indirect measurement of bacterial swimming velocities by laser light scattering applied to whole populations, see e.g. (9).

In other situations, as in the sexual search behavior of male Ectocarpus gametes, one can make use of "macroscopic" space and time scales in order to answer questions like

→ How long does it take for a randomly locomoting male to meet a female partner?
→ How much is he better off, if the female is able to attract him by a sexual pheromone?
→ How many males will be searching near the female?

Since a "swarm" of attracted Ectocarpus males really looks like a "diffusive cloud" with the female in its center, see e.g. (29), the so-called *diffusion limit* is justified: For large speed s and short mean runtime T with finite *motility coefficient*

$$\mu = s^2 \cdot T_o / n \qquad (n : \text{space dimension}) \tag{3.1}$$

the above mentioned random walk processes converge to a diffusion process, whose probability density $p(t, x)$ satisfies the well-known *diffusion-drift-equation*

$$\partial_t p = \nabla_x \left[\mu \nabla_x p - p\, v \right] \tag{3.2}$$

where v is a mean moving drift. For example, if a female produces a diffusing pheromone of concentration $\sigma(t, x)$ which leads to a biased angular turning of the male in gradient direction, then v will be proportional to that gradient

$$v = \chi \cdot \nabla_x \sigma . \tag{3.3}$$

These *"chemotaxis" equations* have been widely used to model aggregative behavior of different kinds, see (21, 3) and the book of Okubo (33). For the loop-like motion of Ectocarpus gametes the situation is more complicated, since there μ and χ turn out to be

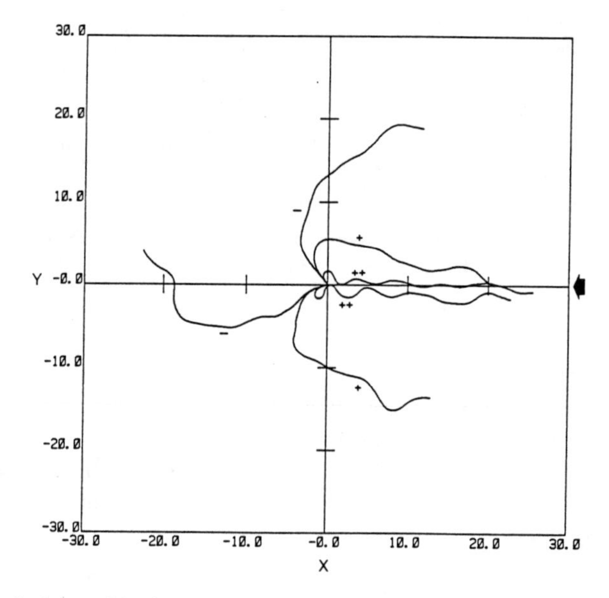

FIG. 5. Simulations of leukocyte moving paths in homogeneous conditions and in chemotactic gradients, according to the receptor fluctuation model. From (41).

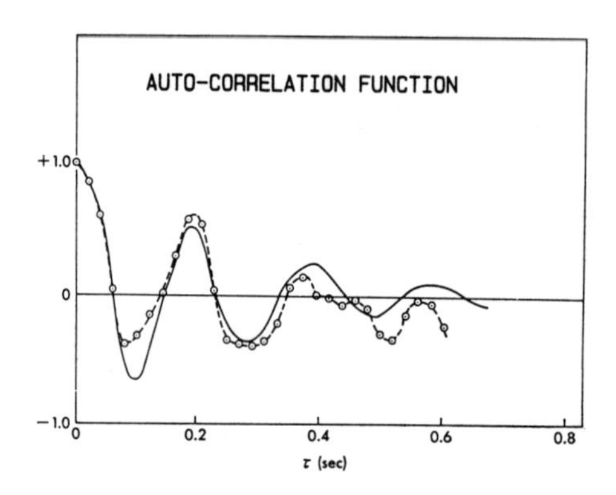

FIG. 6. Velocity auto-correlation function of swarming midges. --●-- data, —— theory. From (34).

2×2-matrices (2). However, in a radially symmetric situation with $\sigma = \sigma(r)$, $\chi = \chi(r)$ and $\mu = \mu(r)$ we can use equation (3.2) and obtain for the transformed function

$$z(t,r) = p(t,r) \cdot e^{\beta(r)}$$

the differential equation

$$\partial_t z = \frac{1}{\alpha} \partial_r \left[\alpha \cdot \mu \cdot \partial_r z \right] \tag{3.4}$$

with

$$\alpha = r \cdot e^{\beta(r)} \quad \text{and} \quad \beta(r) = \int_{r_o}^{r} \sigma' \cdot \chi / \mu .$$

Assuming that the motion is restricted to a disk with radius R around the female and that a male gamete starts at distance $r_o \leq R$, then the *mean searching time* during which the male reaches a contact circle with radius a can be computed as

$$T_a(r_o) = \int_a^{r_o} f_R(r)\, dr , \tag{3.5}$$

where the *searching effort density* per unit distance (during approaching $r = a$) is

$$f_R(r) = \frac{1}{\mu(r) \cdot \alpha(r)} \int_r^R \alpha(r_1)\, dr_1 .$$

For other formulas of this kind cp. (15). The plot of this function in Fig. 7 shows that the introduction of a chemotactic drift drastically reduces the mean searching time.

However, we can question whether it is the mean searching time that is minimized by the gametes, or some other quantity as e.g. the mating probability within a given time period. As an example for this latter optimization principle we can look at the "homing" behavior of some desert isopods, whose cumulative probability distribution reaches the value 1 much faster than a Brownian search would predict, see the simulations by Hoffmann (19). The reason for the improved random search efficiency might be that these animals are meandering, i.e. their migrating directions have negative auto-correlations. Model equations are desired, which describe this kind of movement on a "microscopic" level and lead to an estimation of "macroscopic" quantities as e.g. the "homing probability".

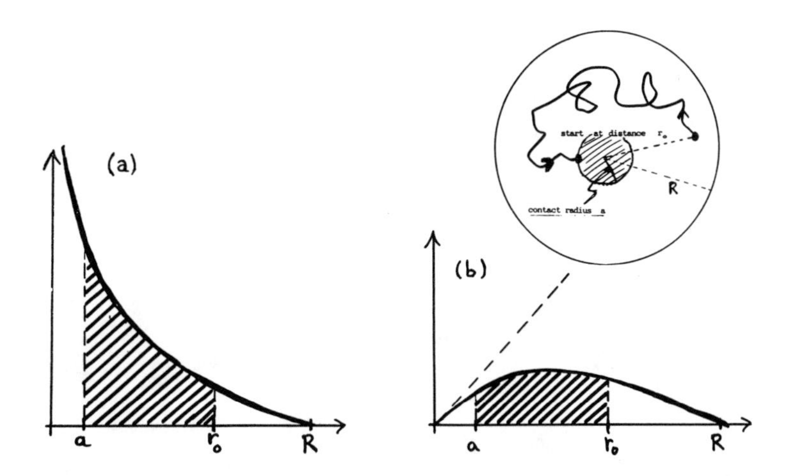

FIG. 7 The searching effort to reach the a-circle from a starting distance r_o is given by the area under the function f_R, sketched in (a) for $\beta = 0$, no chemotaxis, and in (b) for a strong chemotactic gradient.

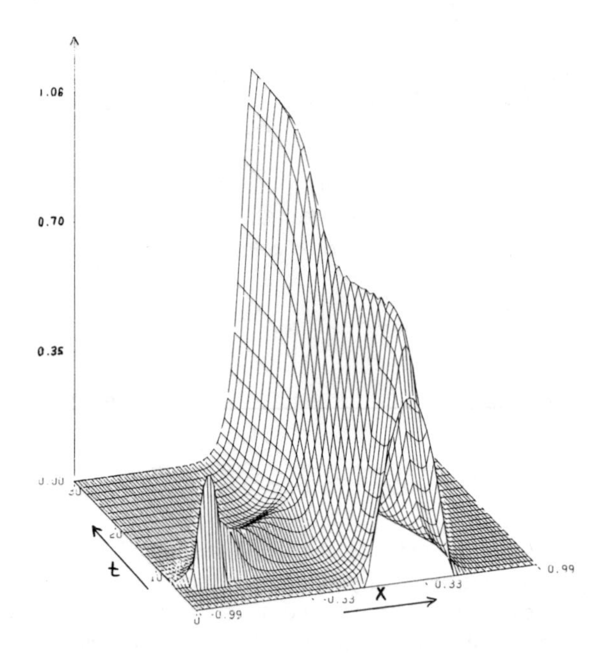

FIG. 8. Simulation of a swarm attracting a second smaller one, using a modified Keller-Segel model for chemotaxis. From (17).

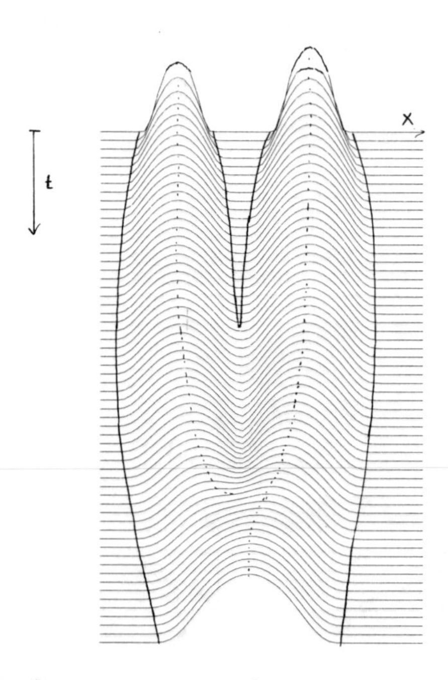

FIG. 9. Simulation of two swarms attracting each other only after their borders come into contact. Solution obtained by numerical integration of system (3.7), (3.8).

Similar to the well-known slime mold aggregation is the phenomenon of *swarming* in the much smaller slime bacteria. It has been observed (24) that such swarms keep their shape for longer times and are even able to attract each other. The famous Keller-Segel system (21) has been used by Lauffenburger and Keller (27) to support the idea that these phenomena are induced through production of a chemotactic substance by the bacteria themselves. A numerical simulation of a modified pseudo-stationary system is shown in Fig. 8, where the larger swarm absorbs the smaller one in a uniform manner. Compare also the theoretical and numerical results of similar aggregation models by Nagai and Mimura (30).

The model of Lauffenburger and Keller does not rely on direct interactions of the cells. However, when observing movies with high resolution (24) one frequently detects cells gliding along each other, thereby obviously enhancing locomotion and inducing a characteristic *"cooperative" movement* of smaller cell groups. These are not fixed, but change their consistency in a stochastic manner. Also, the degree of motility crucially depends on the amount of slime produced by the cells. In later stages the myxobacterial swarms resemble a highly viscous, since slowly moving, but dramatically pulsating fluid. These observations suggested a model approach which we already used earlier for the description of cytoplasma motion, namely a *"viscous and contractive fluid" analogon*, see equations (1.1)-(1.2), but now including active, random motility of the living "fluid elements". Considering only one-dimensional situations and specific non-linearities, the model equations for density $u = u(t,x)$ and mean velocity $v = v(t,x)$ of cells become:

$$\partial_t u = \partial_x \left[u (\mu \cdot \partial_x u - v) \right] \tag{3.7}$$

$$\phi \cdot v = \partial_x \left[u \cdot \partial_x v + \psi \cdot u^2 \cdot (1 - p \cdot u) \right] . \tag{3.8}$$

It can be shown that steady state solutions (with compact support) exist, provided the total cell mass M is above some critical value

$$M_{\min} = 2\pi \cdot \phi \cdot (\mu/\psi)^{3/2} .$$

This means, a swarm with subcritical mass is going to diverge and spread out. If then two subcritical swarms eventually meet at the edges, they relatively fast attract themselves towards each other and finally join into one larger swarm which now can have supercritical cell mass and thus form a stable aggregation. A computer simulation of this is shown in Fig. 9.

Both model descriptions of myxobacteria swarming are hypothetical and wait for experimental verification. However, the crucial question whether *chemotaxis* or direct *cooperative gliding* is responsible for swarming, probably will not be answered by such simple models on the population level. There is a need for more refined procedures and models able to analyze and characterize the interactions of individuals within a swarm.

Thus, as a general conclusion from the foregoing review of various examples in modelling of biological motility, we could formulate a slight warning, namely, not to insist on one specific approach but rather combine different theoretical viewpoints and mathematical techniques in order to describe the complex phenomena of living and moving nature.

REFERENCES

[1] W. ALT, *Mathematical models of actin-myosin interaction*, In: Nature and Function of Cytoskeletal Proteins in Mobility and Transport, Progress in Zoology, **34** (1987) pp. 1-22.

[2] W. ALT, T. EISELE, AND R. SCHAAF, *Chemotaxis of gametes: A diffusion approximation*, J. Math. Appl. in Med. and Biol., **2** (1985) pp. 109-129.

[3] W. ALT AND D. A. LAUFFENBURGER, *Transient behavior of a chemotaxis system modelling certain types of tissue inflammation*, J. Math. Biol., **24** (1987) pp. 691-722.

[4] H. C. BERG, *Random walks in biology*, Princeton Univ. Press 1983.

[5] H. C. BERG AND D. A. BROWN, *Chemotaxis in E. coli analyzed by three-dimensional tracking*, In: Antibiotics and Chemotherapy, Vol. 19, Karger, Basel 1974, pp. 55-78.

[6] H. C. BERG AND S. KHAN, *A model for the flagellar rotary motor*, In: Mobility and Recognition in Cell Biology, H. Sund, C. Veeger eds., de Gruyter, Berlin 1983, pp. 485-497.

[7] S. M. BLOCK, J. E. SEGALL AND H. C. BERG, *Adaptation kinetics in bacterial chemotaxis*, J. Bacteriol., **154** (1983) pp. 312-323.

[8] J. J. BLUM AND J. LUBLINER, *Analysis of wave propagation in cilia and flagella*, Lect. Notes in Biomath., **2** (1974) pp. 8-28.

[9] J. P. BOON, *Motility of living cells and micro-organisms*, In: The Application of Laser Light Scattering to the Study of Biological Motion, J. C. Earnshaw, M. W. Steer eds., Plenum Press, New York 1983.

[10] C. J. BRAKAW, *Models for oscillation and bend propagation by flagella*, Sympos. Soc. Exper. Biol., **35** (1982) pp. 313-338.

[11] S. CHILDRESS, *Mechanics of swimming and flying*, Cambridge Univ. Press, London 1981.

[12] M. DEMBO, *The mechanics of motility in dissociated cytoplasma*, Biophys. J., **50** (1986) pp. 165-1183.

[13] M. DEMBO AND F. HARLOW, *Cell motion, contractile networks, and the physics of interpenetrating reactive flow*, Biophys. J., **50** (1986) pp. 109-121.

[14] M. DEMBO, F. HARLOW AND W. ALT, *The biophysics of cell surface motility*, In: Cell Surface Dynamics, A. S. Perelson, Ch.DeLisi, F. W. Wiegel eds., M. Dekker, New York 1984, pp. 495-542.

[15] C. W. GARDINER, *Handbook of stochastic methods for physics, chemistry and the natural sciences*, Springer, Berlin 1985.

[16] A. GELLER AND D. G. MÜLLER, *Analysis of the flagellar beat pattern of male Ectocarpus siliculosus gametes in relation to chemotactic stimulation by female cells*, J. Exp. Biol., **92** (1981) pp. 53-66.

[17] J. GREENBERG AND W. ALT, *Stability results for a diffusion equation with functional drift approximating a chemotaxis model*, Trans. AMS, **300** (1987) pp. 235-258.

[18] H. GRULER AND B. D. BÜLTMANN, *Analysis of cell movement*, Blood Cells, **10** (1984) pp. 61-77.

[19] G. HOFFMANN, *The random elements in the systematic search behavior of the desert isopod Hemilepistus reaumuri*, Behav. Ecol. Sociobiol., **13** (1983) pp. 81-92.

[20] M. H. JANSEN, Thesis, Purdue Univ., Lafayette (Indiana) 1982.

[21] E. F. KELLER AND L. A. SEGEL, *Model for chemotaxis*, J. Theor. Biol., **30** (1971) pp. 225-234.

[22] K. H. KELLER, M. GRADY AND M. DWORKIN, *Surface tension gradients: Feasible model for gliding motility of Myxococcus xanthus*, J. Bacteriol., **155** (1983) pp. 1358-1366.

[23] N. G. van KAMPEN, *Stochastic processes in physics and chemistry*, North-Holland, Amsterdam 1981.

[24] H. KÜHLWEIN AND H. REICHENBACH, Schwarmverhalten und Morphogenese bei Myxobacterien (Film), IWF C 839, Göttingen 1965.

[25] J. M. LACKIE, *Cell movement and cell behaviour*, Allen & Unwin, London 1986.

[26] I. R. LAPIDUS AND H. C. BERG, *Gliding motility of Cytophaga sp. Strain V67*, J. Bacteriol., **151** (1982) pp. 384-398.

[27] D. A. LAUFFENBURGER, M. GRADY AND K. H. KELLER, *A hypothesis for approaching swarms of myxobacteria*, J. Theor. Biol., **110** (1984) pp. 257-274.

[28] M. LEVANDOWSKY, J. KLAFTER AND B. S. WHITE, *Feeding and swimming behavior in grazing microzooplankton*, J. Protozool., to appear.

[29] D. G. MÜLLER, *Locomotive responses of male gametes to the specific sex attractant in Ectocarpus siliculosus*, Arch. Protistenkunde, **120** (1978) pp. 371-377.

[30] T. NAGAI AND M. MIMURA, *Asymptotic behavior for a nonlinear degenerate diffusion equation in population dynamics*, SIAM J. Appl. Math., **43** (1983) pp. 449-464.

[31] P. B. NOBLE AND M. D. LEVINE, *Computer-assisted analyses of cell locomotion and chemotaxis*, CRC Press, Boca Raton (Florida) 1986.

[32] R. NOSSAL, *Stochastic aspects of biological locomotion*, J. Statist. Physics, **30** (1983) pp. 391-400.

[33] A. OKUBO, *Diffusion and ecological problems: Mathematical models*, Springer, New York 1980.

[34] A. OKUBO, *Dynamical aspects of animal grouping: Swarms, schools, flocks, and herds*, Adv. Biophys., **22** (1986) pp. 1-94.

[35] G. F. OSTER AND G. M. ODELL, *Mechanics of cytogels I: Oscillations in Physarum*, Cell Motility, **4** (1984) pp. 469-503.

[36] H. G. OTHMER, S. R. DUNBAR AND W. ALT, *Models of dispersal in biological systems*, J. Math. Biol. (1988) to appear.

[37] C. S. PATLAK, *Random walk with persistence and external bias*, Bull. Math. Biophys., **15** (1953) pp. 311-338.

[38] R. RIKMENSPOEL, *Ciliary contractile model applied in sperm flagellar motion*, J. Theor. Biol., **96** (1982) pp. 617-645.

[39] J. N. M. SMITH, *The food searching behavior of two European thrushes*, Behaviour, **48** (1974) pp. 276-302.

[40] K. SUGINO AND Y. NAITOH, *Computer simulations of the movement of Paramecium*, Zool. Mag. (Tokyo), **89** (1980) pp. 197.

[41] R. T. TRANQUILLO AND D. A. LAUFFENBURGER, *Stochastic models of chemosensory cell movement*, J. Math. Biol., **25** (1987) pp. 229-262.

Stochastic Control Theory

KARL JOHAN ÅSTRÖM*

Abstract. Uncertainties in the form of disturbances and model uncertainties are essential elements of feedback control. Without uncertainties there will be no distinction between open loop control and feedback control. To develop a theory which can capture the essence of feedback it is thus essential to model uncertainties mathematically. One possibility is to describe disturbances as random processes. This leads to stochastic control theory which combines ideas of stochastic processes with calculus of variations. This lecture presents some of these ideas in particular, system identification, adaptive and dual control.

1. Introduction

One of the goals of applied mathematics has been to use mathematics to get a deeper insight into natural phenomena. The field of automatic control introduces another aspect because it deals with man made devices. Mathematics will then also serve as an inspiration to construct better devices. The field of automatic control has developed rapidly over the past 50 years and control theory is now an established subspeciality. An illustration of this is SIAM Journal on Control (Now SIAM Journal of Control and Optimization) which first appeared about 25 years ago. Control theory spans a wide range of mathematics from algebra, analysis and topology to probability theory and numerics and computer science. The field is still in a state where it is possible to combine theory with experiments. Most control systems are well instrumented with computers which simplifies experimentation. There is also a long tradition in the field to do experimental mathematics based on simulation.

Uncertainties in the form of unknown disturbances or imprecise knowledge are key features of automatic control systems. Feedback was in fact developed to deal with these issues. This lecture deals with the special branch of control theory that is called stochastic control for the reason that stochastic processes are used to model uncertainties. The selection of topics is based on purely personal bias.

Before going into technicalities it should also be emphasized that even today most industrial processes are controlled by simple regulators of the PID type. Such a regulator

* Department of Automatic Control, Lund Institute of Technology, Box 118, S-221 00 Lund, SWEDEN.

is ideally described by the equation:

$$u(t) = k\left(e(t) + \frac{1}{T_i}\int^{T} e(s)ds + T_d\frac{de(t)}{dt}\right) \tag{1}$$

where u is the control variable and e is the error, i.e., the difference between the reference value and the controlled signal. The terms in Equation 1 are called proportional, integral and derivative actions. The derivative action is not often used. The integral term ensures that there is no steady state error and the derivative term predicts the future outputs. The regulator described by 1 has three parameters k, T_i, and T_d, which have to be chosen appropriately. The PID regulator works very well for a wide range of processes provided that the regulator parameters are chosen with care and that the requirements on the control system are not too high.

There are two different ways to find suitable values of the parameters of a regulator. One possibility is to use heuristic adjustment rules. Another method is to go through the steps of mathematical modeling supported by plant experiments and application of systematic design methods. The first method is not very reliable. The second approach requires a substantial engineering effort. It does however also give additional insights into the problem e.g., if there are benefits in using a more complex control law than (1). For this reasons it has been a long standing goal for control engineers to devise schemes to find regulator parameters automatically (auto-tuners) and techniques to continuously update regulator parameters (adaptive control).

The purpose of this paper is to describe some methods for obtaining mathematical models of a control system and for automatic tuning and adaptive control. The paper is organized as follows. Methods for determining mathematical models from experimental data are discussed in Section 2. Adaptive control is discussed in Section 3. Section 4 deals with automatic tuning of PID regulators. The notion of dual control is developed in Section 5. The paper ends with some conclusions and suggestions for further reading.

2. System Identification

The stochastic control problem is well understood for linear system, see e.g., Åström (1970). The problem can be formulated as follows. Consider a system described by the linear stochastic differential equation

$$\begin{aligned} dx &= Axdt + Budt + dv \\ dy &= Cxdt + de \end{aligned} \tag{2}$$

where $x \in R^n$ is the state variable, $u \in R^r$ the control variable and $y \in R^p$ the measured signal. The initial state $x(0)$ is a gaussian random variable and $v \in R^n$ and $e \in R^p$ are Wiener processes. The state vector x represents the dynamics of the system and the disturbances. Find a control law that minimizes the quadratic criterion

$$J = E\left\{x'(T)Q_0x(T) + \int_0^T \left(x'(t)Q_1x(t) + u'(t)Q_2u(t)\right)dt\right\} \tag{3}$$

The control law is given by

$$u = -L\hat{x} \tag{4}$$

where $\hat{x}(t)$ is the conditional mean value of the state given all observed data up to time t. This is given by the Kalman filter

$$d\hat{x} = A\hat{x}\,dt + Bu\,dt + K(dy - C\hat{x}\,dt) \tag{5}$$

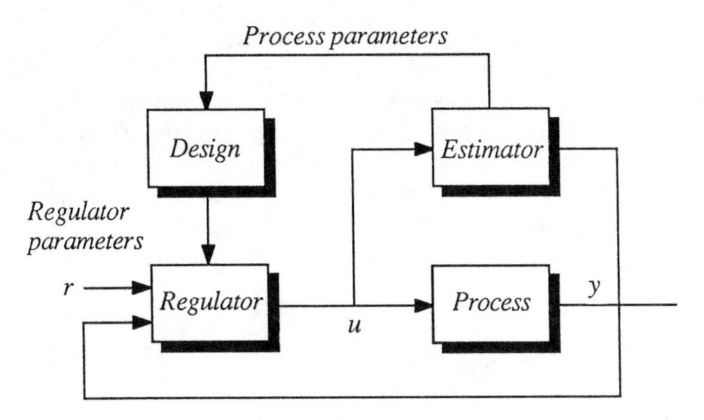

Figure 1. Block diagram of an adaptive regulator, which automatically performs estimation and control design.

The matrices K and L are obtained as solutions of Riccati equations, see Åström and Wittenmark (1985). The result is of considerable interest because it gives a structure of the controller, which tells that the regulator contains a mathematical model of the process. The complexity of the process (2) is thus directly reflected in the complexity in the regulator. For simple models the control law given by (4) and (5) reduces to (1).

When trying to apply the result to a practical problem we are immediately faced with the problem of obtaining the mathematical model (2) of the process and the disturbances. In some cases the model can be derived from first principles. In many cases this may, however, be extremely difficult. There will at least be some parameters θ in the model (2) that are unknown. The matrices A, B, C, and the incremental covariances of the Wiener processes v and e in the model, will thus depend on the unknown parameter.

To determine the parameter an experiment is performed where the input signal u is perturbed and the output is observed. This results in the data

$$\mathcal{D} = \{u(t_k), y(t_k), k = 1, 2, \ldots, N\}$$

The problem is thus to determine the unknown parameters given this data. This problem is the statistical problem for the dynamical system which in the control literature is known as the identification problem. It is a straightforward generalization of static statistics problem. the problem includes the steps of experimental planning, selection of model structure, parameter estimation and validation. A significant research effort has been devoted to this problem. The likelihood estimates have been derived, see Åström (1980). Consistency efficiency and asymptotic properties of the estimates have been explored, see Ljung (1987). Applications are found in Åström and Källström (1976) and Källström and Åström (1981). Recursive methods are discussed in Ljung and Söderström (1983).

3. Adaptive Control

The procedure of system identification and control design described in Section 2 is quite time consuming. Several attempts have been made to automate it. Figure 1 shows a block diagram of an adaptive regulator which attempts this. The adaptive regulator can be thought of as composed of two loops. The inner loop consists of the process and an ordinary linear feedback regulator. The parameters of the regulator are adjusted by the outer loop, which performs recursive parameter estimation and control design calculations. To obtain good estimates it may also be necessary to introduce perturbation signals. Notice that the system automatically performs the tasks of modeling and control design that is normally carried out by an engineer.

The diagram shown in Figure 1 is quite general. Many different design methods and many different parameter estimation schemes can be used. There are adaptive regulators based on phase- and amplitude margin design methods, pole-placement, minimum variance control, linear quadratic gaussian control and optimization methods. Many different parameter estimation schemes have also been used, for example stochastic approximation, least squares, extended and generalized least squares, instrumental variables, extended Kalman filtering and the maximum likelihood method. Further details are given in Goodwin and Sin (1984), Gupta (1986), Åström (1987), and Åström and Wittenmark (1988). An example illustrates a typical case.

EXAMPLE 1
Estimate the parameters of the second order model

$$y(t) + a_1 y(t-h) + a_2 y(t-2h) = b_1 u(t-h) + b_2 u(t-2h)$$

recursively. Let $\hat{a}_i$ and $\hat{b}_i$ denote the parameter estimates. The control law

$$u(t) = t_0 r(t) - s_0 y(t) - s_1 y(t-h) - r_1 u(t-h)$$

where

$$\begin{aligned}
t_0 &= \frac{1 + p_1 + p_2}{\hat{b}_1 + \hat{b}_2} \\
r_1 &= \frac{(p_1 - \hat{a}_1)\hat{b}_2^2 - (p_2 - \hat{a}_2)\hat{b}_1\hat{b}_2}{N} \\
s_0 &= \frac{(p_1 - \hat{a}_1)(\hat{a}_2\hat{b}_1 - \hat{a}_1\hat{b}_2) + (p_2 - \hat{a}_2)\hat{b}_2}{N} \\
s_1 &= -\frac{\hat{a}_2 r_1}{\hat{b}_2} \\
N &= \hat{b}_2^2 - \hat{a}_1\hat{b}_1\hat{b}_2 + \hat{a}_2\hat{b}_1^2
\end{aligned}$$

gives a closed loop system whose pulse transfer function from the command signal to the output is given by

$$H_m(z) = \frac{1 + p_1 + p_2}{b_1 + b_2} \cdot \frac{b_1 z + b_2}{z^2 + p_1 z + p_2}$$

where

$$\begin{aligned}
p_1 &= -2e^{\zeta\omega h} \cos \omega h \sqrt{1 - \zeta^2} \\
p_2 &= e^{-2\zeta\omega h}
\end{aligned}$$

The closed loop system thus attempts to make the closed loop poles correspond to a sampled second order system with bandwidth ω and relative damping ζ. □

The adaptive regulator shown in Figure 1 is called an indirect scheme because the regulator parameters are obtained indirectly by first estimating process parameters and solving a design problem. It is sometimes possible to reparameterize the process so that it can be expressed in terms of the regulator parameters. This gives a significant simplification of the algorithm because the design calculations are eliminated. In terms of Figure 1 the block labelled design calculations disappears and the regulator parameters are updated directly.

Mathematical Problems

There are many interesting mathematical problems in adaptive control. Analysis of specific algorithms is one broad category of problems, creation of new adaptive control laws is another. Among the problems of analysis are questions like stability, convergence and

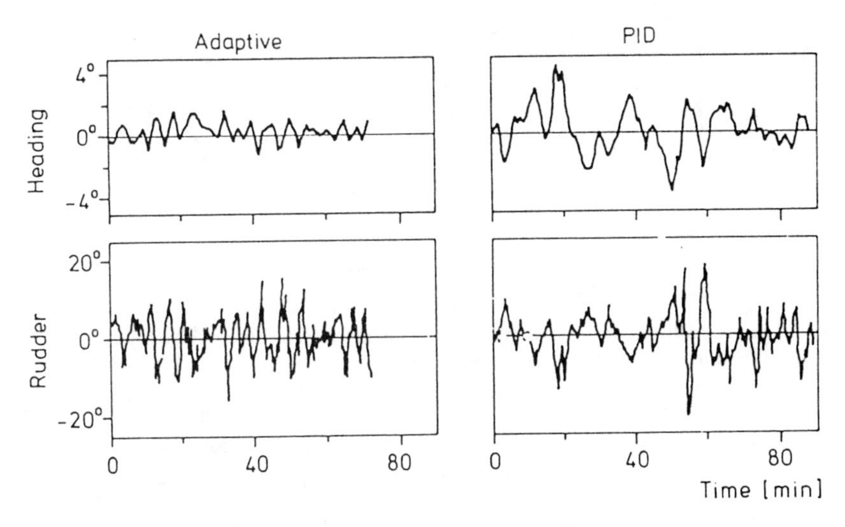

Figure 2. Heading and rudder angles for a conventional and an adaptive autopilot.

robustness. Progress in theory has been slow and painstaking. Much work remains to be done before a reasonably complete coherent theory is available.

Because of the complex behavior of adaptive systems it is necessary to consider them from several points of view. Theories of nonlinear systems, stability, system identification, recursive estimation, convergence of stochastic algorithms and optimal stochastic control all contribute to the understanding of adaptive systems.

Many adaptive algorithms are motivated by the assumption that the parameters change slower than the state variables of the system. We can make sure that the parameters change slowly by choosing a small adaptation gain. The variables describing the adaptive system can then be separated into two groups which change at different rates. The adjustable parameters are the slow parameters and the state of the controlled dynamical system are the fast variables. It is possible to derive approximations to that the fast and the slow variables can be treated separately. The method of averaging originally developed by Bogoliobov and Krylov for analysis of nonlinear oscillations is thus a natural tool, see Anderson *et al.* (1986). Adaptive systems can also exhibit chaotic behavior when the adaptation gains are too high.

An Example

An example illustrates use and performance of adaptive control. See Källström *et al.* (1979). A ship operates in an environment that changes with wind, waves and currents. The dynamics of a ship depend on trim, loading, ship speed and water depth. A conventional autopilot for a ship is based on the PID algorithm given by Equation (1). An adaptive autopilot based on recursive least squares estimation and linear quadratic control theory gives a control law which is more complicated than a PID regulator.

Figure 2 shows results of steering experiments with conventional and adaptive control. The experiments are performed under similar conditions. The figure shows clearly the superior performance of the adaptive system. The heading variations are considerably smaller while the rudder motions have a similar magnitude. A closer inspection shows however that there are more high frequencies in the control signal for the adaptive autopilot. The curves shown in Figure 2 can be translated to fuel consumption. The adaptive autopilot requires about 2.7% less fuel consumption. The reason why the adaptive regulator performs much better is that it is more complex. It has an internal model which describes the dynamics of the ship and of wind, waves and currents. If the parameters were frozen it would not be much change of performance in the short run. The performance would,

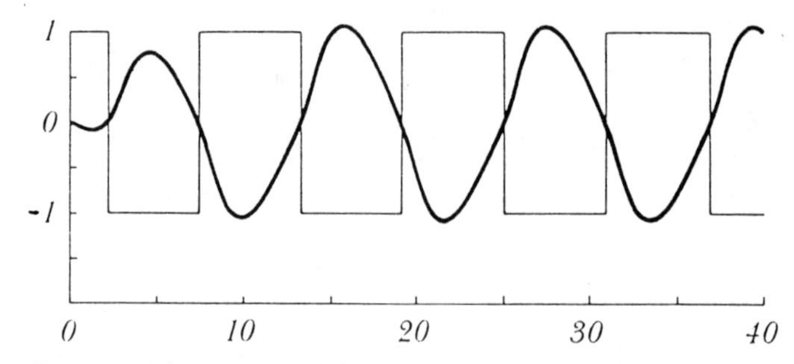

Figure 3. Input and output signals for a system under relay feedback. The linear system has the transfer function $G(s) = 0.5(1-s)/s(s+1)^2$.

however, deteriorate when conditions change. There are altogether 8 parameters that are estimated on line. It is quite difficult and tedious to change so many parameters manually. Adaptation is thus a necessity for using a regulator of this complexity.

4. Automatic Tuning

The adaptive system described in Section 3 can be viewed as a local gradient algorithm. It works quite well provided that enough *a priori* information is available so that the system can be started close to the correct operating point. A key problem is that it is essential to know the order of magnitude of the time scales of the systems. It is therefore of significant interest to also have cruder methods that require less prior information. Such a method is described in Åström and Hägglund (1984, 1988). It was motivated by the need for a simple method to automatically find the parameters of the PID controller (1). The method is based on a special technique for system identification which automatically generates an appropriate test signal and a variation of a classical method for determining the parameters of a PID regulator.

The Basic Idea

The parameters of a PID regulator can be determined from knowledge of the point where the Nyquist curve of the open loop system intersects the negative real axis. It is traditionally described in terms of the ultimate gain k_u and the ultimate period T_u. In the original scheme, described in Ziegler and Nichols (1943), the ultimate gain and the ultimate period are determined in the following way: A proportional regulator is connected to the system. The gain is gradually increased until an oscillation is obtained. The gain k_u when this occurs is the ultimate gain and the oscillation has the ultimate period. It is difficult to perform this experiment automatically in such a way that the amplitude of the oscillation is kept under control.

The auto-tuner is based on the idea that the ultimate gain and the ultimate period can be determined by introducing relay feedback. A periodic oscillation is then obtained. The ultimate period T_u is simply the period of the oscillation and the critical gain can be determined from the relay amplitude and the amplitude of the oscillation, see Figure 3.

If the process attenuates high frequencies so that the first harmonic component dominates the response it follows that the input and the output are out of phase. Furthermore if the relay amplitude is d it follows from a Fourier series expansion that the first harmonic of the input is $4d/\pi$. If the amplitude of the output is a the process gain is $\pi a/4d$ at the critical frequency and the ultimate gain becomes

$$k_c = \frac{4d}{\pi a}$$

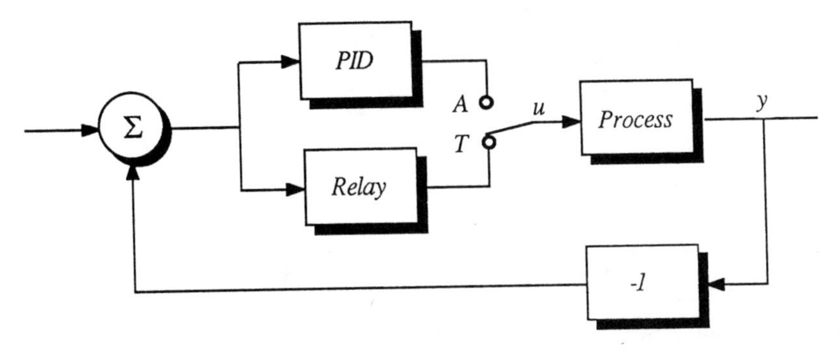

Figure 4. Block diagram of an auto-tuner. The system operates as a relay controller in the tuning mode (T) and as an ordinary PID regulator in the automatic control mode (A).

Control Design

When the ultimate gain k_u and the ultimate period are known, the parameters of a PID regulator (1) can be determined by the Ziegler-Nichols rule which can be expressed as

$$k = \frac{k_u}{2}, \qquad T_i = \frac{T_u}{2}, \qquad T_d = \frac{T_u}{8} \tag{6}$$

This rule gives a closed loop system which is sometimes too poorly damped. There are therefore many modifications of the Ziegler-Nichols rule which give improved performance.

Exact analyses of relay oscillations are also available, see Åström and Hägglund (1984). The period of an oscillation can be determined by measuring the times between zero-crossings. The amplitude may be determined from the peak-to-peak values of the output. These estimation methods are easy to implement because they are based on counting and comparison only. The sensitivity to disturbances can be reduced significantly by filtering the signals adaptively and introducing hysteresis in the relay.

The Auto-tuner

A block diagram of the auto-tuner is shown in Figure 4. The tuner is very easy to use. The process is simply brought to an equilibrium by setting a constant control signal in manual mode. The tuning is then activated by pushing the tuning switch. The regulator is automatically switched to automatic mode when the tuning is complete.

A major advantage of the auto-tuner is that it requires little prior information. Only two parameters—the relay amplitude and the hysteresis width of the relay—are required. These parameters can determined automatically in the auto-tuner. The relay amplitude is initially set to fixed proportion of the output range. The amplitude is adjusted after one half period to give an output oscillation of specified amplitude. The modified relay amplitude is stored for the next tuning. The hysteresis width is set automatically based on measurements of the measurement noise.

An Example

The properties of the auto-tuner are illustrated in Figure 5, which shows an application to temperature control in a distillation column. A PI regulator was used originally. The plant personnel had great difficulties to get the control loop to function well using conventional tuning rules. The loop was oscillating when the experiment started as is shown in Figure 5. The regulator parameters are also shown in the figure. Notice in particular the absence of derivative action. To perform automatic tuning the regulator was first switched to manual control at time 11.00. The auto-tuning mode was activated at time 14.00. At time 20.00 the tuning is complete and the regulator switches to automatic control mode. Notice the

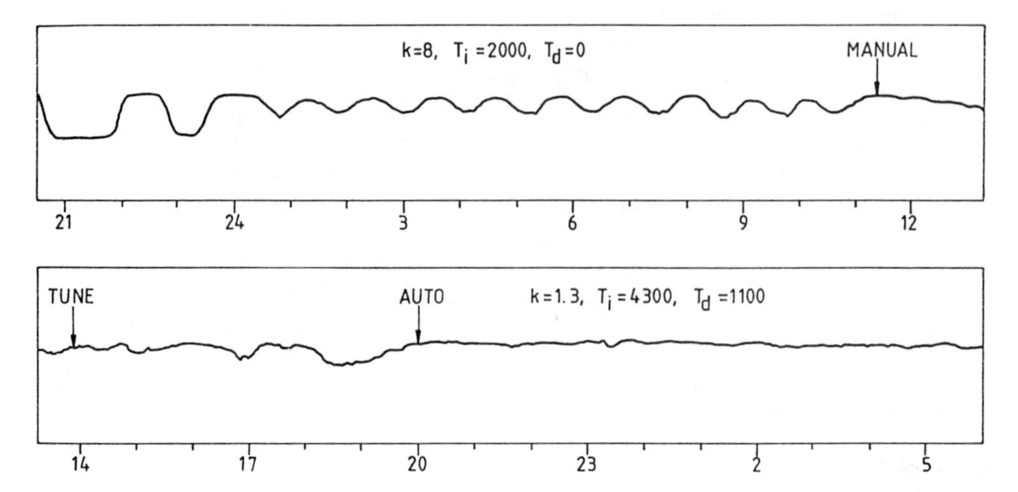

Figure 5. Results obtained applying an auto-tuner to temperature control of a distillation column. The figure is a copy of a strip chart recorder, which explains why time increases from right to left.

drastic improvement in the behavior of the closed loop. Also notice that the regulator parameters have been changed significantly.

There are several interesting observations that can be drawn from the experiment. First, notice that tuning of a slow process takes a considerable time. It is not likely that an operator has the time and patience to tune the regulator manually. Secondly, if the tuning is done manually by heuristic methods it is necessary to make several tuning experiments which will increase the time substantially. These are probably the reasons why the loop was poorly tuned to start with. It is also worth observing that there are significant disturbance during the tuning phase, from time 14.00 to 20.00. The adaptive filtering in the tuner can, however, handle the disturbances very well. After one half period of the oscillation a crude estimate of the period is obtained. This value can be used to set the bandwidth of the filters.

Theoretical Issues

Analysis of limit cycles in linear systems under relay feedback is a key issue for the auto-tuner. A central problem is to determine the class of all linear systems which give a unique stable limit cycle under relay feedback. This problem is not completely solved although considerable effort has been devoted to relay oscillations, see Tsypkin (1984).

5. Dual Control

The adaptive schemes discussed in Section 3 are based on purely heuristic arguments. It would be appealing to obtain adaptive systems from a unified theoretical framework. This can be done using nonlinear stochastic control theory where the system and its environment are described by a stochastic model. To do so the parameters are introduced as state variables and the parameter uncertainty is modeled by stochastic models. An unknown constant is thus modeled by the differential equation

$$\frac{d\theta}{dt} = 0$$

with an initial distribution that reflects the parameter uncertainty. This corresponds to a Bayesian approach where unknown parameters are viewed as random variables. Parameter

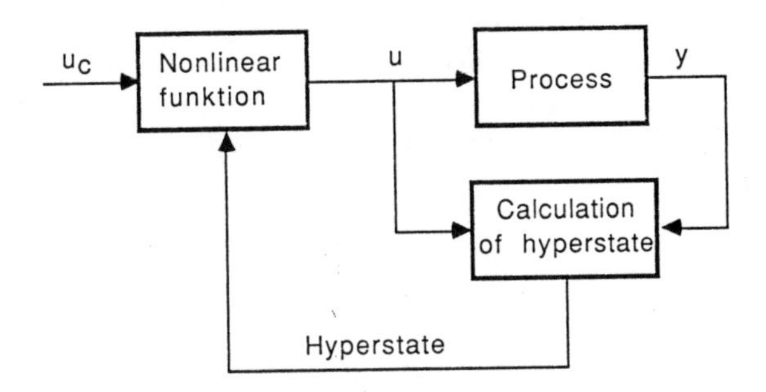

Figure 6. Block diagram of an adaptive regulator obtained from stochastic control theory.

drift is captured by adding random variables to the right hand sides of the equations. A criterion is formulated as to minimize the expected value of a loss function, which is a scalar function of states and controls.

The problem of finding a control, which minimizes the expected loss function, is difficult. Under the assumption that a solution exists, a functional equation for the optimal loss function can be derived using dynamic programming, see Bellman (1957, 1961), and Bertsekas (1976). The functional equation, called the *Bellman equation*, is a generalization of the Hamilton-Jacobi equation in calculus of variations. It can be solved numerically only in very simple cases. The structure of the optimal regulator obtained is shown in Figure 6. The controller can be thought of as composed of two parts: a nonlinear estimator and a feedback regulator. The estimator generates the conditional probability distribution of the state from the measurements. This distribution is called the *hyperstate* of the problem. The feedback regulator is a nonlinear function, which maps the hyperstate into the space of control variables. This function can be computed off-line. The hyperstate must, however, be updated on-line. The structural simplicity of the solution is obtained at the price of introducing the hyperstate, which is a quantity of very high dimension. Updating of the hyperstate requires in general solution of a complicated nonlinear filtering problem. Notice that there is no distinction between the parameters and the other state variables in Figure 6. This means that the regulator can handle very rapid parameter variations. Notice, however, that it is necessary to have prior information about the stochastic properties of the variations of states and parameters.

The optimal control law has interesting properties which have been found by solving a number of specific problems. The control attempts to drive the output to its desired value, but it will also introduce perturbations (probing) when the parameters are uncertain. This improves the quality of the estimates and the future controls. The optimal control gives the correct balance between maintaining good control and small estimation errors. The name *dual control* was coined by Feldbaum (1965) to express this property.

It is interesting to compare the regulator in Figure 6 with the self-tuning regulator in Figure 1. In the STR the states are separated into two groups, the ordinary state variables of the underlying constant parameter model and the parameters which are assumed to vary slowly. In the optimal stochastic regulator there is no such distinction. There is no feedback from the variance of the estimate in the STR although this information is available in the estimator. In the optimal stochastic regulator there is feedback from the full conditional distribution of parameters and states. The design calculations in the STR are made in the same way as if the parameters were known exactly, there are no attempts to modify the control law when the estimates are uncertain. In the optimal stochastic regulator the control law is calculated based on the hyperstate which takes full account of uncertainties. This also introduces perturbations when estimates are poor. The comparison indicates that it may be useful to add parameter uncertainties and probing to the STR.

A simple example illustrates the dual control law and some approximations.

EXAMPLE 2 — From Åström and Helmersson (1986)
Consider a discrete time system described by

$$y(t+1) = y(t) + bu(t) + e(t) \tag{7}$$

where u is the control, y the output and e normal $(0, \sigma_e)$ white noise. Let the criterion be to minimize the mean square deviation of the output y. When the parameters are known the optimal control law is given by

$$u(t) = -\frac{y(t)}{b} \tag{8}$$

If the parameter b is assumed to be a random variable with a Gaussian prior distribution, the conditional distribution of b, given inputs and outputs up to time t, is Gaussian with mean $\hat{b}(t)$ and standard deviation $\sigma(t)$. The hyperstate can then be characterized by the triple $(y(t), \hat{b}(t), \sigma(t))$. The equations for updating the hyperstate are the same as the ordinary Kalman filtering equations, see Åström (1970) and Åström (1978).

Introduce the loss function

$$V_N = \min_u E \frac{1}{\sigma^2} \left\{ \sum_{k=t+1}^{t+N} y^2(k) \;\middle|\; Y_t \right\} \tag{9}$$

where Y_t denotes the data available at time t, i.e., $\{y(t), y(t-1), \ldots\}$. The loss function is thus the conditional mean square error of the control errors N steps ahead. By introducing the normalized variables

$$\eta = \frac{y}{\sigma_e}, \qquad \beta = \frac{\hat{b}}{\sigma}, \qquad \mu = -\frac{u\hat{b}}{y} \tag{10}$$

it can be shown that V_N depends on η and β only. The Bellman equation for the problem can be written as

$$V_T(\eta, \beta) = \min U_T(\eta, \beta, \mu) \tag{11}$$

where

$$V_0(\eta, \beta) = 0$$

and

$$U_T(\eta, \beta, \mu) = 1 + \eta^2 \left(1 - \mu^2\right) + \frac{\mu^2 \eta^2}{\beta^2} + \int_{-\infty}^{\infty} V_{T-1}(y, b) \varphi(\epsilon) d\epsilon \tag{12}$$

where φ is the normal probability density and

$$y = \eta + \beta\mu + \epsilon\sqrt{1+\mu^2}$$
$$b = \epsilon\mu + \beta\sqrt{1+\mu^2}$$

see Åström (1978). When the minimization is performed the control law is obtained as

$$\mu_T(\eta, \beta) = \arg\min U_T(\eta, \beta, \mu) \tag{13}$$

The minimization can be done analytically for $T = 1$. We get

$$\mu_1(\eta, \beta) = \arg\min \left((1 - \beta\mu)^2 + 1 + \mu^2\right) = \frac{\beta^2}{1+\beta^2}$$

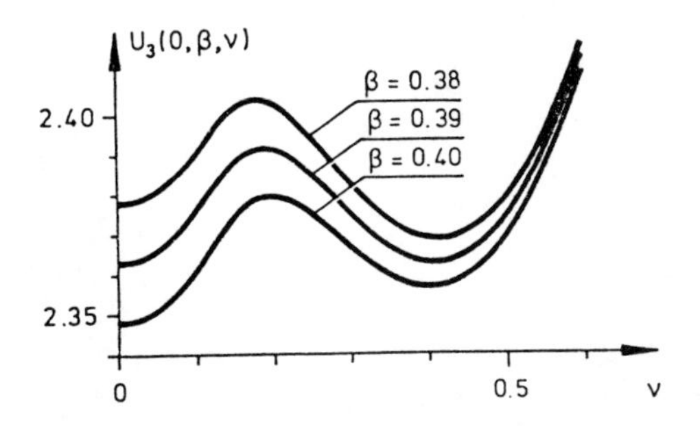

Figure 7. Graph of the function $U_3(0, \beta, \mu)$.

Transforming back to the original variables we get

$$u(t) = -\frac{1}{\hat{b}(t)} \cdot \frac{\hat{b}^2(t)}{\hat{b}^2(t) + \sigma^2(t)} y(t) \tag{14}$$

This control law is called *one-step* control or *myopic* control because the loss function V_1 only looks one step ahead. It is also called *cautious* control because in comparison with the certainty equivalence control it hedges by decreasing the gain when the estimate of b is uncertain.

For $T > 1$ the optimization can no longer be made analytically. Instead we have to resort to numerical calculations. The solution has some interesting properties. Figure 7 shows the function $U_3(0, \beta, \mu)$ for different values of β. Notice that the function has several local minima with respect to μ. For $\beta = 0.40$ the minimum at $\mu = 0$ is the smallest one but for $\beta = 0.38$ the minima at $\mu = 0.42$ is the smallest one. The control $\mu_3(\eta, \beta)$ is thus discontinous in β. For $\eta = 0$ the control signal is zero if β is sufficiently large, i.e., the estimates are reasonable accurate. When the estimates become sufficiently poor $\beta < 0.39$ the control signal μ jumps to about 0.4. The discontinuity of the control law corresponds to the situation that a probing signal is introduced to improve the estimates. Figure 8 shows the dual control laws obtained for different time horizons N.

Some approximations to the optimal control law will also be discussed. The *certainty equivalence* control

$$u(t) = -\frac{y(t)}{\hat{b}} \tag{15}$$

is obtained simply by taking the control law (8) for known parameters and substituting the parameters by their estimates. The self-tuning regulator can be interpreted as a certainty equivalence control. Using normalized variables the certainty equivalence control law becomes

$$\mu = 1$$

Using normalized variables the cautious control law can be expressed as

$$\mu = \frac{\beta^2}{1 + \beta^2}$$

Notice that all control laws are the same for large β, i.e., if the estimate is accurate. The optimal control law is close to the cautious control for large control errors. For estimates with poor precision and moderate control errors the dual control gives larger control actions than the other control laws. A graphical representation of the control laws are given in Figure 8.

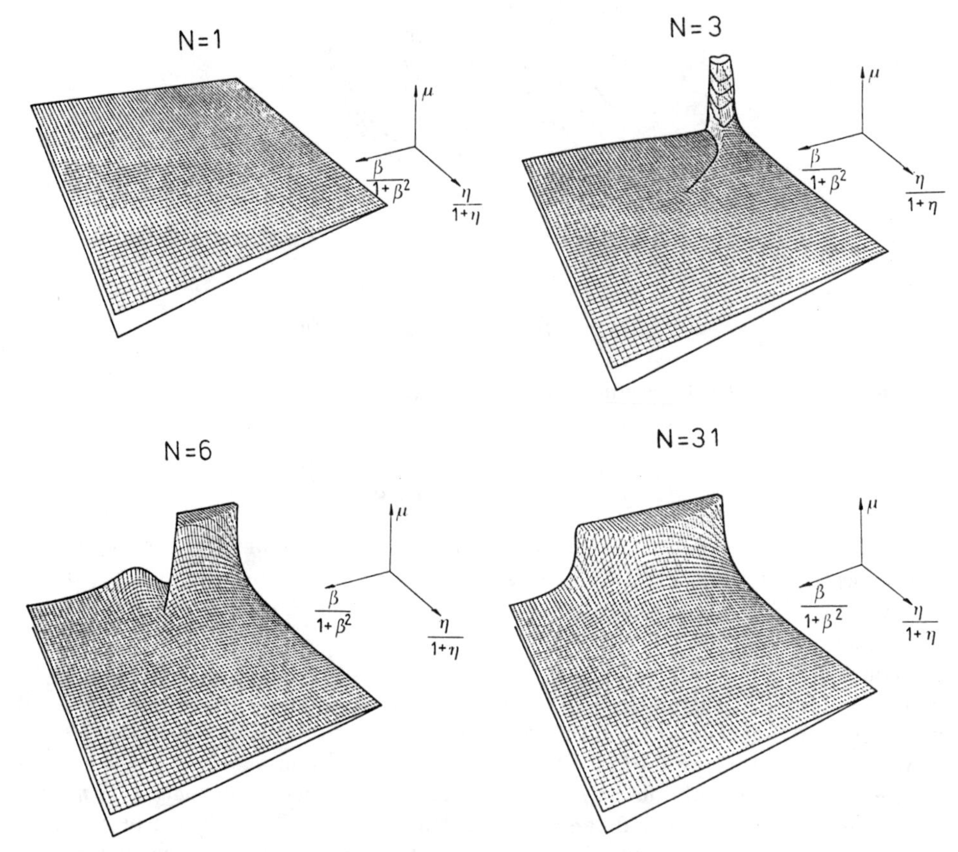

Figure 8. Illustration of the cautious control μ_1 and the dual control laws μ_3, μ_6, and μ_31. The graphs show the normalized control variable μ as a function of the normalized parameter uncertainty $\beta^2/(1+\beta^2)$ and the normalized control error $\eta/(1+\eta)$.

6. Conclusions

Linear stochastic control theory is well understood. It is a standard tool of control engineering. The theory is well established. The computations required to determine the control law in specific cases are extensions of standard numerical linear algebra. These problems are now starting to attract the attention of numerical analysts. The statistical problems associated with system identification are also well understood at least asymptotically for large samples.

Adaptive control systems are much more complicated because they are inherently nonlinear. Some progress has been made in unification and analysis of adaptive schemes. The empirical knowledge of adaptive systems is also growing rapidly because of their emerging industrial use. Much theoretical work remain, however. It is my belief that the understanding of adaptive systems can benefit significantly from different branches of applied mathematics. The auto-tuning problems is quite special. An improved understanding of relay oscillations could however be very beneficial to determine the classes of systems where auto-tuning can be applied.

Dual control is difficult both conceptually and computationally. It is challenging to develop existence theory for solutions to the Bellman equation for certain classes of problems. The numerical solution of the Bellman equation is also difficult. The solution of selected problems do however give significant insight into the nature of adaptive control.

In conclusion I hope that this survey of a special branch of control theory has shown you some of the challenges in this branch of applied mathematics.

7. References

Anderson, B. D. O., R. R. Bitmead, C. R. Johnson, Jr, P. V. Kokotovic, R. L. Kosut, I. M. Y. Mareels, L. Praly, and B. D. Riedle (1986): *Stability of Adaptive Systems—Passivity and Averaging Analysis*, MIT Press, Cambridge, MA.

Åström, K. J. (1970): *Introduction to Stochastic Control Theory*, Academic Press, New York.

Åström, K. J. (1978): "Stochastic control problems," in W. A. Coppel (Ed.): *Mathematical Control Theory. Lecture Notes in Mathematics*, Springer-Verlag, Berlin, FRG.

Åström, K. J. (1980): "Maximum likelihood and prediction error methods," *Automatica*, **16**, 551–574.

Åström, K. J. (1987): "Adaptive feedback control," *Proc. of the IEEE*, **75**, 185–217.

Åström, K. J., and A. Helmersson (1986): "Dual control of an integrator with unknown gain," *Comp. & Maths. with Appls.*, **12A**, 653–662.

Åström, K. J., and T. Hägglund (1984): "Automatic tuning of simple regulators with specifications on phase and amplitude margins.," *Automatica*, **20**, 645–651.

Åström, K. J., and T. Hägglund (1988): *Automatic Tuning of PID Regulators*, Instrument Society of America, Research Triangle Park, NC.

Åström, K. J., and C. G. Källström (1976): "Application of system identification techniques to determination of ship dynamics," *Automatica*, **12**, 9–22.

Åström, K. J., and B. Wittenmark (1985): *Computer Controlled Systems—Theory and Design*, Prentice Hall, Englewood Cliffs, NJ.

Åström, K. J., and B. Wittenmark (1988): *Adaptive Control*, Addison-Wesley, MA, To be published in the fall of 1988.

Bellman, R. (1957): *Dynamic Programming*, Princeton Univ. Press, Princeton, NJ.

Bellman, R. (1961): *Adaptive Processes—A Guided Tour*, Princeton Univ. Press, Princeton, NJ.

Bertsekas, D. P. (1976): *Dynamic Programming and Stochastic Control*, Academic Press, New York, NY.

Egardt, B. (1979): *Stability of Adaptive Controllers. Lecture Notes in Control and Information Sciences*, vol. 20, Springer-Verlag, Berlin, FRG.

Feldbaum, A. A. (1965): *Optimal Control System*, Academic Press, New York, NY.

Goodwin, G. C., and K. S. Sin (1984): *Adaptive Filtering, Prediction and Control*, Prentice Hall, Englewood Cliffs, NJ.

Gupta, M. M. (Ed.) (1986): *Adaptive Methods for Control System Design*, IEEE Press, The Institute of Electrical and Electronics Engineers, Inc., New York.

Källström, C. G., and K. J. Åström (1981): "Experiences of system identification applied to ship steering.," *Automatica*, **17**, 187–198.

Källström, C. G., K. J. Åström, N. E. Thorell, J. Eriksson, and L. Sten (1979): "Adaptive autopilots for tankers," *Automatica*, **15**, 241–254.

Ljung, L., and T. Söderström (1983): *Theory and Practice of Recursive Identification*, MIT Press, Cambridge, MA.

Ljung, L. (1987): *System Identification: Theory for the user*, Prentice Hall, Englewood Cliffs, NJ.

Tsypkin, Y. A. (1984): *Relay Control Systems*, Cambridge University Press, Cambridge, UK.

Ziegler, J. G., and N. B. Nichols (1943): "Optimum settings for automatic controllers," *Trans. ASME*, **65**, 433–444.

Topology and Differential Equations

MICHAEL ATIYAH*

Abstract. In many areas of differential equations topological methods can provide important qualitative information. This particularly applies to complicated problems where the number of variables is large. Examples will be given illustrating the scope and power of topological ideas.

For constant coefficient linear hyperbolic systems the fundamental solution is supported in the forward "light-cone". For general systems this cone has many compartments and, in some of these (called lacunas) the fundamental solution may vanish. The determination of lacunas is a topological problem (solved originally by Petrovsky) involving the topology of multi-dimensional contour integrals in the complex domain.

For elliptic boundary value problems of Cauchy-Riemann or Dirac type the index, measuring the difference between the number of solutions of the problem and its adjoint, is a topological invariant which can be computed from the geometrical data. This has important applications to models in elementary particles physics.

For a periodic family of self-adjoint elliptic operators the spectral flow, counting the net number of eigenvalues which change sign over a period, is a topological invariant. This can be computed from the geometrical data and can be used to derive sharp bounds on gaps in the spectrum.

In non-linear PDE, "solitons" may have a topological origin, and soliton-interaction can be related to underlying topological features. This is illustrated by models in 2 and 3 spatial dimensions, following original ideas of Skyrme.

1. Introduction. Since I am a firm believer in the over-all unity of mathematics, I welcome this opportunity of addressing the first International Congress of Applied and Industrial Mathematics. Perhaps I may begin by describing my own view on the general relation between pure and applied mathematics. Probably the most important thing to emphasize is the different time scales on which the two operate. Pure mathematicians are not in a hurry, they look to the long-term and build essentially for the future. By contrast applied mathematicians

* Mathematical Institute, University of Oxford, 24-29 St. Giles, Oxford OX1 3LB, England, U.K.

are interested in getting fairly immediate, albeit imperfect answers. For this reason applied mathematicians who take their problems to a pure colleague are usually disappointed: only rarely are they given useful help. On the other hand, if we take the long view, measuring progress over decades rather than days, pure mathematical ideas eventually percolate through and make their impact.

The relation between pure and applied is of course two sided. Many of the most important developments in pure mathematics have arisen from the stimulus of physical problems, and progress has frequently come from finding the right mathematical formulation of physical concepts.

The topic I have chosen for this talk is the application of topology to differential equations. This seems a good illustration of the bridge between modern pure mathematics and the physical world, where most phenomena are governed ultimately by some differential equation. Topology is one of the most recently developed branches of pure mathematics and it can appear very abstract and far-removed from reality. My aim here is to show that, on the contrary, topological ideas are very important in a wide variety of different problems concerning differential equations.

I will of course have to be selective in the choice of illustrations I give, and I have picked out topics which I have been involved with personally. On these I can speak with more confidence but there are other examples, just as important, which another speaker might have selected.

Let me begin with some general remarks. Topology should be viewed as an *additional tool* to be used, not as a substitute for the more conventional tools of classical analysis or computer calculation. Moreover, it is essentially concerned with information which is *qualitative* and *stable.* Finally it is especially significant for complicated situations where the number of dependent or independent variables is large (i.e. not 1 or 2), or where the situation is non-linear in some way.

In describing my examples I will in each case focus on two basic questions. Where does the topology enter? What good does it do? I hope to persuade you that topology enters naturally in a variety of ways and that it can help to answer interesting problems about differential equations.

2. Huygens Principle. The simplest P.D.E. are linear constant coefficient equations, and these are classically solved by Fourier transform methods. Thus to solve

$$P(D)u = f\ , \quad D_j = -i\,\frac{\partial}{\partial x_j}, \quad (P \ \text{ a polynomial})$$

we have formally

$$u(x) = \int_{R^n} \frac{e^{ix\xi}\hat{f}(\xi)}{P(\xi)}\,dx$$

where $\hat{f}$ is the Fourier transform of f. The difficulty with this expression for $u(x)$ is of course that $P(\xi)$ can have zeros. The way round this is to "push into the complex plane", i.e. to deform the contour R^n inside C^n so as to avoid the zeros. Immediately we see that the topology of "multi-dimensional contour integration" has appeared. For $n = 1$, the deformation of contours is familiar and standard in connection with Cauchy's theorem. For higher n the topology becomes more serious and has to be treated with respect.

Suppose now we concentrate on *hyperbolic* P.D.E., and consider solutions of the Cauchy problem, i.e. we specify initial data for time zero and look at wave propagation. The fundamental solution is then supported in a forward *cone.* For the standard wave equation for light propagation this cone is quadratic and the fundamental solution is actually supported on the surface of the light-cone, i.e. it vanishes both outside and inside. This is the classical

Huygens principle and holds for the wave equation in odd space dimensions ≥3, but not for even space dimensions.

What happens if the polynomial P is of higher degree than 2, so that the cone is no longer quadratic? One might first think this an academic question since physically interesting equations tend to involve only one or two derivatives. However, if we deal with systems, rather than scalars, then the relevant polynomial is given by the *determinant* of the matrix system and will have a high degree. Classical examples arise in crystal optics.

In 2 space dimensions we can draw pictures of cones by considering a plane slicing the cones in curves. For example when the degree is 4, given by a product of two quadratics, we have the picture in ξ-space of the *characteristic* cone

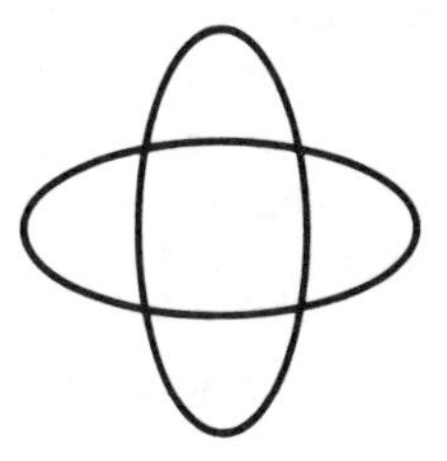

and the dual picture in x-space of the *wave-front* cone.

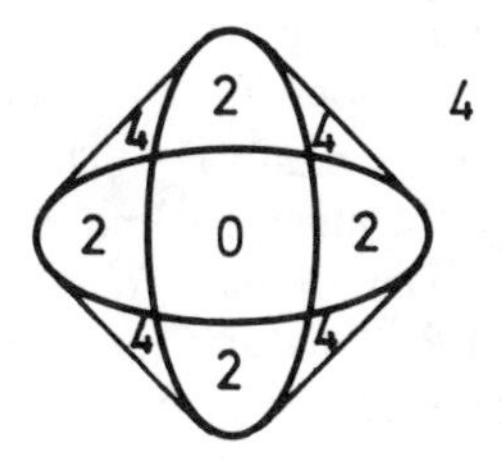

Note that the 4 double points in the first picture produce the 4 double tangents drawn in the second. The fundamental solution is supported inside the wave-front cone but may actually vanish in some of the interior regions. These are called lacunas and were studied in general by Petrovsky [7]. He gave the following simple criterion for a lacuna. This is a region in x-space for which the line $x \cdot \xi = 0$ in ξ-space has all its intersections with the characteristic surface *real*. Thus in the example this number is 0, 2 or 4 and regions 4 are lacunas.

In 3-space dimensions the pictures become surfaces, $x \cdot \xi = 0$ is a plane and the intersection with the characteristic surface is a real *curve*. Petrovsky showed that we get a lacuna when this real curve, viewed as lying on the complex curve (Riemann surface) can be shrunk to a point.

In higher dimensions these topological criteria for lacunas involve more serious topology and were studied in [2].

3. Elliptic boundary-value problems. Turning from constant to variable coefficients, and from hyperbolic to elliptic let us consider the classical type of elliptic boundary-value problem, such as the Dirichlet problem. Here we look for solutions of the Laplace equation $\Delta f = 0$ in some domain with f equal to a prescribed function g on the boundary. Such problems are self-adjoint and we usually have existence and uniqueness of solutions. If we have zero eigenvalues then existence and uniqueness just fail, but can be restored by a small perturbation of the equation (add a constant).

There are however important elliptic problems which are not self-adjoint. The prototype are the Cauchy-Riemann equations whose solutions give holomorphic functions, while solutions of the adjoint equation give anti-homomorphic functions. These may be studied with, for example, periodic boundary conditions (with multipliers) and this leads to the classical theory of elliptic functions.

For such non-self-adjoint problems the equations $Pf = 0$ and its adjoint $P^*g = 0$ may have a different number of linearly independent solutions, say p and q respectively. The *index* of p is then defined by

$$\text{index } P = p - q$$

and it is a fundamental analytical fact that this is constant under continuous deformations. If index $P > 0$ we may expect $q = 0$ generically so that index $P = p$ gives the number of solutions of $Pf = 0$ (if index $P < 0$ switch the roles of P and P^*). We may therefore think of the index as the generic number of linear parameters on which a solution of $Pf = 0$ will depend. It is clearly an important number to know and there are very general topological formulas to compute this index [5].

We should now ask: where does the topology come from? There are in fact three different sources of topology:

(i) boundary conditions,
(ii) the differential operator,
(iii) topology of the domain.

A boundary condition may involve a multiplier, which should be non-zero and, for systems, a *non-singular matrix*. The space of non-singular matrices has some interesting topology (generalizing C^*, the non-zero complex numbers). This space again enters via the differential operator. Recall that for ellipticity we must have no real characteristics and this is expressed by the non-vanishing of a determinant. Finally our domain may have holes or may be a constrained surface (e.g. a sphere in R^3), and this produces topology.

In recent years index problems of this type have become very important in elementary particle physics, where the fundamental operator is the Dirac operator. As Dirac discovered this has positive and negative energy levels (associated with positrons and electrons). Moreover reversing the orientation of 3-space (the "parity" operator) interchanges plus and minus (in close analogy with the behaviour of the Cauchy-Riemann equations). We now know that the physical world is not ambidextrous, but contains phenomena which distinguish preferentially between right and left handedness. The mathematics which underlies this involves the index of the Dirac operator, and in searching for the correct mathematical model of elementary particle physics the index provides a very useful tool.

Another physical problem of current interest which involves index notions is that of the Quantum Hall Effect, concerned with the behaviour of electrons on a 2-dimensional surface, when subjected to a magnetic field.

4. Dependence on Parameters Differential operators frequently depend on continuous parameters and this dependence can contain interesting topological information. As an example consider a 1-parameter family $P(t)$ of elliptic self-adjoint operators which are *periodic* in t in the sense that

$$P(t+1) = U^{-1}P(t)U \tag{4.1}$$

where U is a unitary operator. This implies that the spectrum of $P(t)$ is periodic in t. Assuming that we are in a good situation (e.g. arising from a boundary-value problem) so that the spectrum is discrete and that the t-dependence is analytic, it follows that the eigenvalues $\lambda_n(t)$ are analytic in t and satisfy

$$\lambda_n(t+1) = \lambda_{\sigma(n)}(t) \tag{4.2}$$

for some *permutation* σ. That this permuation may be non-trivial is already shown by the following very elementary example.

Example: Take $P(t) = -i\frac{d}{dx} + kt$ where x is an angular parameter on the circle S^1 and k is an integer. Equation (4.1) is satisfied with U being multiplication by e^{ikx}. The eigenvalues are $\lambda_n(t) = n + kt$, so that the permutation σ of (4.2) is just the k-shift

$$\sigma(n) = n + k .$$

Note that k is the degree or "winding number" of the map $x \to e^{ikx}$ defined on the circle.

The integer k in this example has a generalization as the "spectral flow" of any periodic family $P(t)$. We consider the eigenvalues $\lambda_n(t)$ for $0 \le t \le 1$ and count the total number p of times an eigenvalue changes sign from $-$ to $+$ and the number q of times we get a change from $+$ to $-$. The spectral flow k is the net gain

$$k = p - q . \tag{4.3}$$

It is not hard to see that this is a topological invariant of the periodic family $P(t)$. Note that it is only of interest (i.e. non-zero) for operators with infinitely many positive and negative eigenvalues (e.g. of Dirac type).

In our example, as we noted, the integer k is also the degree of the map $x \to e^{ikx}$ defined by U. Similarly in the general case if U is implemented by a matrix multiplier $U(x)$ then the spectral flow k can be computed topologically [4] from a suitable "degree" of $U(x)$. In fact this result is closely related to the index formulae of §3.

A few years ago two physicists, C. Vafa and E. Witten, applied these ideas to a situation of the following type. They considered $P(t) = P_0 + V(t)$ with $V(t)$ a bounded family of potentials and P_0 a fixed first-order elliptic system of Dirac type. Assuming that the spectral shift $k \neq 0$ (say $k = 1$), and this can be arranged purely topologically, they argued roughly as follows. As t increases from 0 to 1, $\lambda_n \to \lambda_{n+1}$ with uniformly bounded speed. Hence we have a uniform bound on the *gaps* $|\lambda_{n+1} - \lambda_n|$.

I find this a remarkable result because it uses *topological* methods to prove an inequality with quite precise bounds. Moreover the results turn out to be very sharp. For a simplified account of this work see [1].

5. Non-linear Equations. So far I have discussed only *linear* equations. I deliberately began with these in order to emphasize that topological notions are important even here. However, when we turn to non-linear equations topological aspects become even more fundamental. I will concentrate on one such aspect which has attracted increasing attention over the past twenty years, namely the phenomenon of *solitons*.

The term soliton has been coined to describe certain solutions of non-linear P.D.E. which are *approximately localized* and retain their identity after interaction. The real pioneer in this field, a man far ahead of his time, was Skyrme [8] who was seeking soliton models for elementary particles. The prototype equation (in 1-space dimension) is the sine-Gordon equation

$$u_{xx} - u_{tt} = \sin u . \tag{5.1}$$

This has a static "kink" solution with graph as indicated.

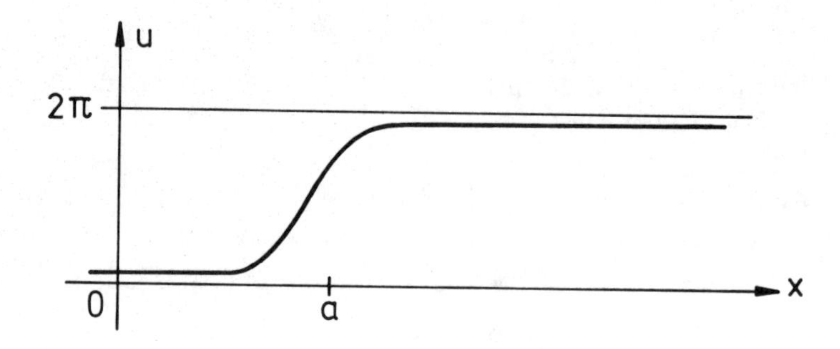

The energy density u_x^2 is approximately localized at $x = a$, the "location" of the soliton. Since

$$\lim_{x \to -\infty} u(x) = 0$$

$$\lim_{x \to +\infty} u(x) = 2\pi$$

the map $x \to \exp iu(x)$ of $R \cup \infty \to S^1$ has winding number 1. In this way the soliton may be seen to have a topological origin.

If we increase our space dimensions from 1 to 2 and for simplicity restrict to the static case we can consider functions $u(x,y)$ where $u \in R^3$ with $|u| = 1$ represents a point *constrained* to lie on the unit sphere. As energy we take the standard expression

$$E(u) = \tfrac{1}{2} \int_{R^2} |\operatorname{grad} u|^2 \, dxdy . \tag{5.2}$$

Here we have not added any explicit non-linear term (like $\sin u$ in (5.1)). However, the constraint $|u| = 1$ essentially generates the non-linearity because the 2-sphere (unlike the circle) is *not flat*. Moreover functions u of finite energy must decay at ∞ in the (x,y)-plane and so u defines a map

$$R^2 \cup \infty \to S^2$$

which will have a *degree* k.

For this problem it turns out that, for $k \geq 0$, the functions of *minimum energy* are precisely the *rational functions* of degree k in the complex variable $z = x + iy$, where S^2 is identified with the Riemann sphere $C \cup \infty$. Thus the general solution has the form

$$u(z) = \sum_{i=1}^{k} \frac{b_i}{z - a_i} . \tag{5.3}$$

If the a_i are far apart this looks like a superposition of k separate "particles" or solitons, located at the points a_i (with complex weights b_i).

Note: The form (5.3) shows that our minimum energy solutions depend on $4k$ real parameters. This can be derived from the index theory of §3 applied to the linearized elliptic problem.

One of the most interesting aspects of soliton theory is the behaviour of two "interacting solitons". What happens when solitons collide? For the sine-Gordon equation and others of the same type in 1-space dimension there is an extensive theory showing that solitons are essentially unaffected by collision (except for a phase-shift). For higher space dimensions the situation is more complicated and more interesting. For our rational function solutions (5.3) the situation is easy to describe: we only have to investigate the effect of multiple poles. For $k = 2$ consider the rational function

$$u(z) = \frac{1}{z^2 - t}$$

with t as a real parameter. For $|t| \to \infty$ this represents 2 particles located at $z = \pm\sqrt{t}$. Letting t increase from $-\infty$ to $+\infty$ this gives a picture of 2 particles coming in on the *imaginary* axis and emerging, after "collision" at $t = 0$, along the *real* axis. This 90° scattering indicates an interesting "interaction". For $t = 0$, the "collision state", the energy density vanishes at the centre ($z = 0$) (because grad u vanishes at a branch point) but has a maximum on a *ring* round the origin. Schematically the collision process can therefore be depicted as follows

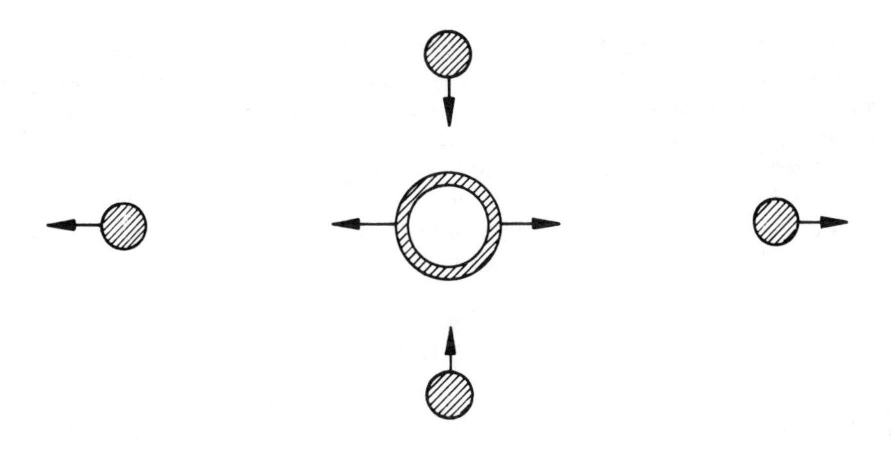

Moving up from 2 to 3 space-dimensions we can consider analogously a function $u(x, y, z) \in R^4$, $|u| = 1$. As energy function we take not only the term $|\text{grad } u|^2$ as in (5.2), measuring distortion of length, but in addition a term measuring distortion of area. This gives the original *Skyrme model* for nucleons. The corresponding solitons, now called *Skyrmions*, are once more in favour for nucleons. Again there is an integer degree k and $k = 1$ has a minimal energy solution which is spherically symmetric and represents a single proton or neutron. Special interest attaches to the $k = 2$ solution which represents a deuteron.

There are no explicit formulae for these Skyrmion solutions but numerical evidence has led Manton [6] to suggest that the deuteron should be axially symmetric, with pictures that resemble the 2-dimensional case described earlier.

Another 3-dimensional model, describing magnetic monopoles, exhibiting similar features can be analyzed in detail and is described fully in [3].

In all these cases it seems that the topology of the function space plays a major role in determining the type of soliton interaction. However, this is probably only the tip of the iceberg and we have much to learn yet about topological effects in such non-linear problems.

6. Summary. It may be helpful to summarize our examples in the following table

TOPOLOGY	DIFFERENTIAL EQUATIONS	
	INPUT	OUTPUT
Complex Algebraic Geometry	Characteristics	Lacunas of hyperbolic equations
Matrix Groups	Boundary conditions Ellipticity Domain	Index Theory Spectral information
Function Spaces	Non-linear functional	Solitons Interaction

The topics of the first column constitute a large part of modern pure mathematics. Understanding the role of topology in these areas has been a major preoccupation of mathematicians for much of this century. The second column indicates where, in the theory of differential equations, these topics enter, while the last column indicates what kind of information about differential equations finally emerges.

Perhaps I can end with a word of advice for any applied mathematician who has been persuaded by my talk and is keen to apply topology in his work. My advice is: don't try to read the books (they are usually unreadable); find a friendly topologist!

REFERENCES

[1] M. F. ATIYAH, *Eigenvalues of the Dirac Operator*, in Lecture Notes in Mathematics, 1111 (Proc. of 25th Arbeitstagung, Bonn 1984) Springer-Verlag (1985), pp. 251-260.

[2] M. F. ATIYAH, R. BOTT AND L. GARDING, *Lacunas for hyperbolic differential operators with constant coefficients I & II*, Acta Math. **124** (1970), pp. 109-189, and Acta Math. 131 (1973), pp. 145-206.

[3] M. F. ATIYAH AND N. J. HITCHIN, *The Geometry and Dynamics of Magnetic Monopoles*, Princeton Univ. Press (1988).

[4] M. F. ATIYAH, V. K. PATODI AND I. M. SINGER, *Spectral Asymmetry and Riemannian Geometry*, Bull. London. Math. Soc. **5** (1973), pp. 229-234.

[5] M. F. ATIYAH AND I. M. SINGER, *The Index of Elliptic Operators on Compact Manifolds*, Bull. Amer. Math. Soc. **69** (1963) pp. 422-433.

[6] N. MANTON, *Is the $B = 2$ Skyrmion Axially Symmetric?* Cambridge preprint DAMTP/87-5.

[7] I. G. PETROVSKY, *On the Diffusion of Waves and the Lacunas for Hyperbolic Equations*, Mat. Sb., 17 (59) (1945), pp. 289-370.

[8] T. H. R. SKYRME, *A Non-linear Field Theory*, Proc. Roy. Soc. **A(260)** (1961) pp. 127-138.

Image Analysis and Markov Fields

ROBERT AZENCOTT*

1. Introduction. Gibbs fields (or Markov fields) have been standard probabilistic tools in statistical mechanics for a long time. The link between Gibb's laws and two dimensional Markov properties dates back to a celebrated result of Hammersley. However applications of these notions to images considered as two-dimensional arrays is much more recent, and a major step in that direction was D. and S. Geman's first and exciting paper on image restoration.

Research is very active in this–young–domain, and has been linked with stochastic optimization methods such as simulated annealing as well as a number of low-level image analysis tasks. This paper intends to give a synthetic view of the subject and to sketch a few interesting results reached at University Paris-Sud.

2. Low-Level Tasks of Image Analysis. We shall list here several basic low-level tasks which we feel can be correctly handled by *sophisticated* Markov field approaches. Here ''low-level'' is a traditional terminology for tasks which often are preliminary to ''image understanding''.

2.1 Compression. Starting with an image x, one wants to find a new image y as close as possible to x, but described at a much smaller cost (for instance with cruder intensity discretisation).

2.2 Restoration. Starting with an *unknown* image x, one actually observes a *noisy version* y of x (for instance $y = x + w$ where w is a random noise). The goal is to approximately ''recover'' x from y.

2.3 Contours extraction. Find *piecewise smooth boundaries* separating zones where the intensity $s \to x_s$ varies ''smoothly''.

2.4 Texture segmentation. Partition the image into homogeneous zones where the homogeneity refers to *local statistical characteristics* of the intensity map $s \to x_s$.

* Professor, Université Paris-Sud (Maths), F-91405 ORSAY Cedex, (Lab. Statistiques Appliquées) and Ecole Normale Superieure, 45 rue d'Uln, 75005 Paris, France.

2.5 Speed detection. In the analysis of sequences of images, construct *a piecewise smooth field of velocities* linking one image to the next.

2.6 Tomography. Reconstruct a 2-dimensional "image" from a finite set of 1-dimensional "projections", where the intensity at any pixel of a projection is obtained by adding up all the intensities of the original image along the corresponding projection ray.

This list is of course not limitative. Let us now describe briefly the mathematical set up associated to these tasks and the Markov field approach.

3. Images. Call S a rectangular lattice of sites (the so-called pixels = picture elements) and let Λ be a finite set. An image $x = (x_s)_{s \in S}$ will be any element of $\Omega = \Lambda^S$. Typically, $x_s \in \Lambda$ represents the light intensity for the pixel $s \in S$. In standard black and white images, one may have card $(S) = 512 \times 512$ and card $\Lambda = 256$, so that card Ω is a huge number.

4. Edges.

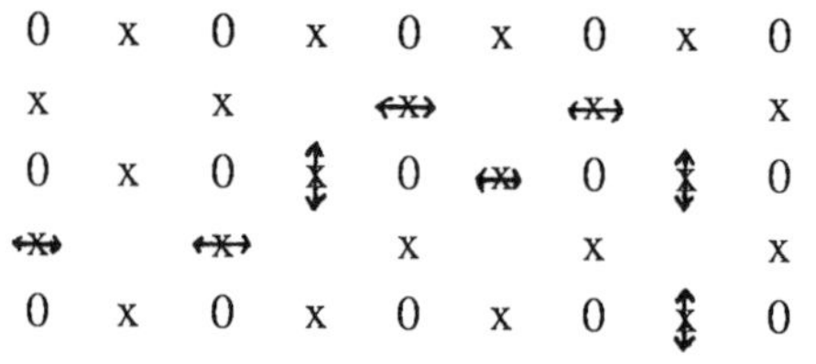

```
0   x   0   x   0   x   0   x   0
x       x      ↔x      ↔x       x
0   x   0   ↕   0   ↔   0   ↕   0
↔       ↔       x       x       x
0   x   0   x   0   x   0   ↕   0
```

FIG. 4.1. pixel sites and edge sites (0 and x)

Consider the array Σ formed by all the midpoints of *consecutive* pairs of sites belonging to S . Following (4), we call Σ the set of "edge sites" and define an *edge configuration* $k = (k_\sigma)_{\sigma \in \Sigma}$ by letting $k_\sigma = 1$ if an "elementary edge" is present at point σ (for a given image $x \in \Omega$), and $k_\sigma = 0$ if this is not the case.

Intuitively "elementary edges" are contact elements tangent to a "contour curve", and such contour curves for the image x are the locations of highly marked "discontinuities" for the map $s \to x_x$. The above description can be much refined, but we'll stick to this simple outline here, and call Ω_E the set of all pairs (x , k) where x is an intensity configuration on S and k an edge configuration on Σ.

5. Labels. In many concrete tasks of image analysis, one seeks a partition of the array S of pixels into r "homogeneous" zones $Z_1 \cdots Z_r$, where homogeneity refers to a criterion such as *colour, texture,* etc. Defining a *label configuration* $\eta = (\eta_s)_{s \in S}$ by $\{\eta_s = j\}$ whenever the pixel s belongs to Z_j , we thus introduce the set Ω_L of labelled images with typical element $(x , \eta) \in \Omega \times \{1 \cdots r\}^S$.

Note that labelling can be more complex: for instance η_s could be a *speed vector* in $\mathbf{R}^2$, attached to the site s in the analysis of sequences of images.

6. Markov Fields. They are a specific class of probability laws on configuration sets such as Ω, Ω_E , Ω_L (introduced in §3, 4, 5 above). Their main characteristic is that they are *completely determined* by *local statistical characteristics.*

Namely, one attaches to each site $s \in S$ a neighborhood $V_s \subset S$. Then a probability law P on the configuration space Ω is said to be a Markov field (with respect to the chosen family of neighborhoods) if one has for all $s \in S$, all $\lambda \in \Lambda$

$$P(x_s = \lambda \mid x_t , t \in V_s - s) = P(x_s = \lambda \mid x_t , t \in S - s) .$$

In other words, local prediction of x_s is just as efficient when one knows only the restriction of

x to $V_s - s$ as when one knows all of the image except x_s.

Typically local neighborhoods V_s are taken to be invariant for translation ($V_{s+t} \equiv s + V_t$) but *this is not required*, and they often consist of the 4 or 8 nearest neighbors in the lattice.

We now recall, in the following paragraph, how one can build explicitly such Markov fields.

7. Gibbs Laws. An elegant theorem of Hammersley characterizes Markov fields P on Ω, relative to a given local neighborhood family V_s, $s \in S$, as given by

$$P(x) = \frac{1}{Z} \exp[-H(x)], \text{ for all } x \in \Omega,$$

where $Z = \sum_{y \in \Omega} \exp[-H(y)]$, and the so-called *energy function* H on the configuration space Ω has the specific structure

$$H(x) = \sum_C J_C(x_C)$$

In the last formula, C runs over the set of all *cliques* in S (a clique is any set of sites which are all pairwise mutual neighbors) and $J_C(x)$ are "interaction functionals" depending only on the restriction x_C of x to the clique C.

Of course, such laws P have long been used in statistical mechanics under the name of *Gibbs measures* or *Gibbs fields.*

8. Actual Markov Field Models for Images. Given a family F of images on a lattice S, local statistics on F such as the actual best predictors of a pixel intensity x_s knowing only its immediate neighbors, etc. are often meaningful characteristics of F. The natural mathematical object which *synthesises consistently* all such local statistics looks essentially like a Markov field P on Ω, and hence an energy function H on Ω will characterize P.

The essential problem here is to actually exhibit the *best* energy function H modelizing F as far as local statistics are concerned.

The theoretical and practical difficulties are serious for such a task, but can be "decently" overcome. In my opinion, strongly based on a 3 years research program which I have led at University of Paris-Sud, the crucial conclusions are the following:

8.1. The structure of the energy function H must be *task-oriented*: for instance good models for edge detection or for velocities analysis will be quite distinct.

8.2. It is highly recommended to use energies H which are *not spatially homogeneous*: local statistics of an image in the North-East corner can *widely* differ from local statistics in the South-West corner.

8.3. *Good sophisticated estimation algorithms* for the actual numerical parameters of the energy function H are essential: within a given *parametric* family H_θ of energies, the best estimates of θ may well outperform spectacularly "rough" or "educated" guesses.

8.4. Simulated annealing (introduced in this context by D. and S. Geman) is useful in *suitably accelerated* versions, and that for certain *specific tasks only*. However, *stochastic simulations of P by so-called Gibbs samplers* turn out to be *a fundamental computing tool*, of much wider scope.

8.5. *Computations can and must remain highly local* for all usual low level tasks of image-analysis.

8.6. *Adaptative energies H with limited parameter flexibility*, automatically *tuned to* incoming images, are often feasible and quite desirable.

8.7. *Suitable structured forms* for H to achieve a given task are far from unique, but quite relevant. *They may be analysed qualitatively in a systematic fashion.*

9. Example: Cleaning Up NMR Images. Nuclear Magnetic Resonance images obtained with magnetic fields of low intensity (Sauzade et Altri) turn out to be quite noisy. In the spirit of Geman's approach to image restoration, Chalmond has implemented a Markov field modelisation and an algorithm achieving the cleaning up of such images.

The energy initially used was on the space Ω_E of (pixel, edges) configuration and given by

$$H(x,k) = Q(x) + R(k) + \delta \sum_{\substack{s,t \\ \text{neighbors}}} (x_s - x_t)^2 \, k_{\frac{s+t}{2}} ,$$

where $Q(x)$ is a positive quadratic form in the intensities, $R(k)$ a polynomial of degree 4 in the (boolean) variables k recording presence or absence of edges, and δ a negative parameter governing interaction between edges and pixels. Actually, 8 parameters were used to specify H and were estimated *directly* and *locally* by statistical techniques.

The performances were quite satisfactory and required about 35 mn of VAX 750 to clean up — and estimate H — for each 512×512 NMR image. Since then, the energy function is being *revamped* to remove a few undesirable artifacts, improve edge detection, accelerate computations (Azencott, Chalmont, Younes).

10. Example: Edge and Texture Detection. An interesting sophisticated Markov model has been proposed and implemented by D. and S. Geman and C. Graffigne. Due to lack of space, we refer to their paper, but point out that their energy function involves simultaneously *3 configurations*: intensities x, edges k, texture labels η.

Their goal was *segmentation*, and the interactions between those 3 configurations were skillfully handled by a fairly intricate energy function.

11. Probabilistic Principle for Solving Low-Level Image Analysis Tasks. We place ourselves in the situation where two (often large) sets of variables have been defined: the *visible variables* y and the *hidden variables* η.

For instance in *edge detection*, y will be the configuration x of pixel intensities and η the edge configuration k. In *image restoration*, η will be the unknown clean image intensities x, possibly coupled with unknown edge configuration k, while y will be the observed noisy image intensities $(x+w)$.

In *texture analysis*, η will be the label configuration assigning texture type to site while y will be the pixel configuration x.

In *image sequences analysis*, η will be the speed velocity field while y will be a sequence of two consecutive images.

Assume that a satisfactory probabilistic *joint distribution* $P(y,\eta)$ has been obtained (a crucial step which will be reexamined below).

Then the probabilistic principle to solve the task at hand is

$$\text{Find the ``best'' estimator } \hat{\eta} \text{ of } \eta \text{ given } y . \qquad (11.1)$$

Statisticians have been handling such problems for almost a century, so that there is no lack of available techniques. Of course, the computational problems are crucial here; however they are not the only relevant considerations, as pointed out by Maroquin, Besag and many others.

Let us sketch the main solutions actually proposed — so far — for problem 11.1 *in the image context.*

12. Maximum Likelihood a Posteriori Estimators. As in the preceding paragraph, we have visible variables y, hidden variables η, and are looking for an estimator $\hat{\eta}$ of η of the form $\hat{\eta} = F(y)$ *where the deterministic function F is to be constructed* skillfully.

The *a posteriori maximum likelihood estimator* $\hat{\eta}$ is classically (for statisticians) given by solving

$$P(\hat{\eta} \mid y) = \max_{\eta} P(\eta \mid y) \tag{12.1}$$

where $P(\eta \mid y)$ is the conditional probability of η given that y is observed.

When one uses a Markov field model with energy function H

$$P(y,\eta) = \frac{1}{Z} \exp[-H(y,\eta)],$$

then $\hat{\eta}$ is given by

$$H(y,\hat{\eta}) = \min_{\eta} H(y,\eta). \tag{12.2}$$

This is the so-called *M.A.P. estimator* used for instance by Geman, Chalmond in restoration tasks.

13. Marginal M.A.P. Estimators. As pointed out by Maroquin and Besag, M.A.P. estimators may sometimes be too costly or outperformed by more readily accessible estimators. Maroquin's analysis, in terms of various loss functions, is couched in a quite familiar set-up for any seasoned statistician, but is nevertheless quite relevant, and justifies neatly the introduction of the following estimator.

Coming back to our visible variables y and hidden variables x, we could for instance "compute" for each site s and each possible value λ of η_s the conditional probability p

$$p_s(\lambda,y) = P(\eta_s = \lambda \mid y)$$

and then obtain an estimator $\hat{\eta}_s$ of η_s by

$$p_s(\hat{\eta}_s, y) = \max_{\lambda} p_s(\lambda,y). \tag{13.1}$$

These *marginal MAP estimators* are tempting when η_s takes only a small number of values, but (see section 14 below) the actual computation of $p_s(\lambda,y)$ is still costly.

In the context of image restoration, Besag has suggested a faster algorithm (the I.C.M. estimator, for Iterated Conditional Mode), which provides a configuration $\bar{\eta}$ realizing a *local* maximum in η for $P(\eta \mid y)$. In the more general set-up presented here, $\bar{\eta}$ would be a solution of

$$q_s(\bar{\eta}_s, y, \bar{\eta}) = \max_{\lambda} q_s(\lambda, y, \bar{\eta}) \tag{13.2}$$

where $q_s(\lambda,y,\eta) = P(\eta_s = \lambda \mid y, \eta_{S-s})$ and (13.2) should be solved iteratively.

14. Computing Estimators. For a standard 512×512 image, the hidden variable η may easily involve more than 260,000 variables so that the optimisation problems (12.1) or (13.1) are heavy computational tasks.

In the context of image restoration (13.2) is much more tractable at least for labels η_s taking *small* number of values. However by construction the I.C.M. approach will strongly depend on a good initialisation for the estimator $\bar{\eta}$.

I still have an open mind on the choice between (12.1) (13.1) (13.2) or other variants, *which in any case will be task dependent*, but I feel that if only to be able to probe these questions a bit deeper, one needs to tackle seriously the approaches (12.1) and (13.1).

Computationally (13.1) looks simpler than (12.1); however note that in (12.1) the energy function H is used directly, while problem (13.1) requires the marginal conditional distributions $P(\eta_s = \lambda \mid y)$; and in general their actual computation is as costly as obtaining partition functions such as $Z = \sum_{x \in \Omega} e^{-H(x)}$.

Finally, as we shall see below, the effective Markov field modelisation of a given problem relies on the computation of actual parameters θ for a given parametrized family of energies H_θ.

Such estimations involve the computation of expected values $E[f(x)]$ for given known functionals $f(x)$ of the configuration x, and these expectations again have no simple closed analytic forms.

For these three computational problems, two remarkable stochastic algorithms have been introduced:

1. the *Gibbs sampler* well known in statistical mechanics since Metropolis, Glauber and many others;

2. the *simulated annealing* technique introduced by Kirkpatrick, Gelatt, Vecchi, and linked to image restoration by D. and S. Geman.

15. Gibbs Sampler. Given a set Ω of configurations x (indexed by a lattice S) and an energy function $H(x)$, one wants to *simulate* the Markov field $P(x)$ associated to $H(x)$, and to compute expected values $E[f(x)]$ where $f : \Omega \to \mathbf{R}$ is given.

The Gibbs sampler creates *iteratively* a *stochastic* sequence $x^n \in \Omega$ of configurations starting with an *arbitrary* x^o. It has the two following essential properties

$$\lim_{n \to \infty} \text{Law}(x^n) = P$$

$$\lim_{n \to \infty} \frac{1}{n}[f(x^1) + f(x^2) + \cdots + f(x^n)] = E[f(x)].$$

Thus for n large, x^n *looks like* a configuration drawn at random with law P.

Many variants of the Gibbs sampler have been used; for instance one may construct x^{n+1} from x^n by modifying the configuration x^n at a *single site* s^n, where (s^n) is a preassigned periodic sequence of sites visiting the whole lattice S. The choice of the new value $x_{s^n}^{n+1}$ is made *at random* and takes the value λ with probability

$$p(\lambda) = q(s^n, x^n, \lambda),$$

where

$$q(s, x, \lambda) = P(x_s = \lambda \mid x_{V_s - s}).$$

16. Simulated Annealing. The problem is here to minimize a given energy function $H(x)$ on a given space of configurations Ω. One selects a sequence T_n of numbers *decreasing to zero* as $n \to \infty$, the so-called *temperatures*; then one builds *iteratively* a *stochastic* sequence x^n where the passage from x^n to x^{n+1} is governed by a Gibbs sampler using the energy function $H_n(x) = \frac{1}{T_n} H(x)$.

Let Ω^* be the set of minimizing configurations

$$\Omega^* = \{z \in \Omega \mid H(z) = \min_{x \in \Omega} H(x)\}$$

and call Q the *uniform* probability distribution on Ω^*.

One can show (Geman, Hajek) that if $\lim_{n \to \infty} T_n \log n \geq K > 0$ with K large enough, then as $n \to \infty$, the probability law of x^n converges to Q.

This "universal" optimization algorithm is now widely used outside of image analysis but practical applications almost always involve speeds of decrease for T_n much faster than the prescribed logarithmic rate. This was the case for instance in Chalmond's restoration algorithm where he used $T_n = c\ \alpha^n$ with $0 < \alpha < 1$.

Of course, for very fast decreasing speed of the T_n, simulated annealing becomes quite close to gradient descent, with all the standard problems of spurious local minima.

From a practical point of view, I feel that logarithmic annealing is too costly in the context of actual image analysis and should thus mainly be used as a probing tool, while "fast" annealing is much more relevant for building efficient algorithms.

17. Parameter Estimation for Markov Fields. For a given family of images, Markov field modelization requires first the *choice* of a particular *structure* for the energy function, a task which no technical result has actually "solved", but for which specialist are beginning to get a good intuition (see §18 below). This leads to a parametrized family $H_\theta(x)$ of energy functions where $\theta \in \mathbf{R}^k$ is an unknown vector of parameters.

At this point, the problem is *formally* a standard one in statistical analysis: find *the best estimate* $\hat{\theta}$ of θ based on a single typical image x, or on a finite set $x^1 \cdots x^r$ of images. A classical solution for this, say in the case of a single image x, is the maximum likelihood estimator $\hat{\theta} = G(x)$ given by solving (*in* $\theta \in R^k$) the system of non linear equations, *where* x *is fixed and known*,

$$\text{grad}_\theta\,[P_\theta(x)] = 0 \tag{17.1}$$

where $P_\theta(x) = \frac{1}{Z_\theta} \exp[-H_\theta(x)]$

$$Z_\theta = \sum_{y \in \Omega} \exp[-H_\theta(y)]\,.$$

Due to the absence of simple closed forms for Z_θ or $\text{grad}_\theta Z_\theta$, the system (17.1) is most unwieldy.

Approximations of maximum likelihood methods, such as the pseudomaximum likelihood method (studied for instance by Guyon, Kuntsch ...) or codage methods (Besag) have been used. Several devious schemes have been suggested such as Gidas' renormalization approach for instance.

However, it is feasible to solve efficiently and neatly system (17.1) using an interesting algorithm developed and thoroughly studied as Université Paris-Sud in Younès' thesis.

The basic idea is to solve system (17.1) by a *stochastic* algorithm which looks like "Newton" descent with decreasing steps, while the gradient value necessary to establish which direction to use at step n is based on a *simultaneous Gibbs sampling scheme*, for which the value of the parameter θ is *refreshed* at each step.

Younès proved the convergence of this algorithm, and checked that in most cases it can even be used *concurrently* with a simulated annealing scheme, so that one can in principle *minimize an energy* $K_\theta(x)$ *in* x *while simultaneously estimating* θ.

Detailed asymptotics for the behaviour of Younès' algorithm have been obtained and used to construct optimal stopping rules for actual implementation.

18. The Method of Qualitative Boxes When trying to discover a good structure for the energy function $H(x)$ in a given framework, one may use several (additive) building blocks $B_1(x) \cdots B_k(x)$ each of which reinforces or penalizes a given type of local pattern, ending thus with an energy function

$$H_\theta(x) = \theta_1 B_1(x) + \cdots + \theta_k B_k(x) = \langle\theta, T(x)\rangle ,$$

where $T : \Omega \to \mathbf{R}^k$ is a fixed functional and $\theta \in \mathbf{R}^k$. To make rough guesses about the qualitative behaviour of the corresponding Markov field, one can (easily) compute

$$p_s(\lambda, z) = P\{x_s = \lambda \mid x_{V_s - s} = z\} , \tag{18.1}$$

where V_s is the set of neighbors for the site s.

Now one may impose a finite set of "natural constraints" of the following type (18.2) For the "interesting" specific local pattern $x_{V_s - s} = z$, the value λ_1 for x_s is α-times more likely than the value λ_2, with $\alpha_1 \le \alpha \le \alpha_2$.

Each such rule (with specific α_1, α_2, λ_1, λ_2 and z) is easily seen to be equivalent to an inequality of the form

$$\gamma_1 \le \langle\theta, u\rangle \le \gamma_2 \tag{18.3}$$

where $u \in \mathbf{R}^k$, $\gamma_1 \in \mathbf{R}$, $\gamma_2 \in \mathbf{R}$ are readily computed from ($\alpha_1\ \alpha_2\ \lambda_1\lambda_2 z$).

Thus the finite set of natural empirical constraints (18.2) translates, modulo (18.3) into a single *polyhedral constrain*

$$\theta \in \Sigma \tag{18.4}$$

where Σ is a (possibly empty!) polyhedron in $\mathbf{R}^k$, which we call a *qualitative box for* θ.

I have used several times this simple but powerful approach for *rough guesses* allowing *a fast evaluation* of energy functions. I have gathered, from casual conversations with D. and S. Geman, that they had indulged in similar practices.

19. State of the Art. I have proposed in Section 8 above a set of seven important *technical conclusions* concerning the use of Markov fields in image analysis.

I would like to conclude this paper with a few remarks on the "state of the art" in this area.

19.1. All low-level tasks of image analysis mentioned in Section 2 seem to be *tractable* by the Markov field approach. *Sophisticated modelization* can make an important *qualitative difference.*

19.2. Actual experiments yield results of *excellent quality* whenever serious technical (and that includes statistical) tools are used.

19.3. *Computation time* on sequential machines is a problem, but not much more than for standard relaxation methods, for instance. Moreover, the possibility of *massive parallelisation* is natural in that context.

19.4. From a mathematical point of view, more readily usable asymptotics are still sorely needed for theoretical studies, but from a practical point of view, many of the tools needed are now at hand, and quite a few experimental algorithms are being developed.

PARTIAL BIBLIOGRAPHY

[1] J. BESAG, *Spatial interaction and statistical analysis of lattice systems*, J. Roy. Stat. Soc., vol. B-36, p. 192-236, 1974.

[2] J. BESAG, *Statistical analysis of dirty pictures*, J. Roy. Stat. Soc., vol. B-148, 1986.

[3] B. CHALMOND, *Image restoration using an estimated Markov model*, Preprint Université Paris-Sud, to appear.

[4] D. AND S. GEMAN, *Stochastic relaxation, Gibbs distributions, and Bayesian restoration of images*, IEEE Trans. on Pattern Ann. Mach. Intel. vol. PAMI-6, p. 721-741, 1984.

[5] D. S. GEMAN AND C. GRAFFIGNE, *Locating texture and object boundary*, preprint Brown University.

[6] S. GEMAN AND D. E. MAC CLURE, *Bayesian image analysis: an application to single photon emission tomography*, preprint Brown University, 1985.

[7] B. GIDAS, *Non-stationary Markov chains and convergence of the annealing algorithm*, J. Stat. Phys. **39**, p. 73-131, 1985.

[8] B. GIDAS, *Renormalization methods*, preprint Brown University, 1987.

[9] C. GRAFFIGNE, Ph. D. thesis, Brown University, USA, 1987.

[10] X. GUYON, *Pseudo maximum de vraisemblance et champs markoviens*, preprint Université Paris-Sud, 1985.

[11] X. GUYON, *Estimation d'un champ de Gibbs*, preprint Université Paris-Sud, 1986.

[12] B. HAJEK, *Tutorial survey of theory and applications of simulated annealing*, Proc. 24th Conf. on Decision and Control, Fort Lauderdale, p. 755-760, 1985.

[13] S. KIRKPATRICK, C. GELATT AND M. VECCHI, *Optimisation by simulated annealing*, Science, **220**, p. 671-680, 1983.

[14] J.-L. MAROQUIN, *Optimal Bayesian estimation for image segmentation and surface reconstruction*, M.I.T. Artif. Int. Memo 839, 1985.

[15] N. METROPOLIS, A. AND M. ROSENBLUTH AND A. AND E. TELLER, *Equation of State calculations*, J. Chem. Physics **21**, p. 1087-1092, 1953.

[16] M. SAUZADE, L. DARASSE AND H. SAINT-JAMES, *An efficient low cost NMR imaging machine*, Proceeding 3rd Annual Meeting, San Diego, California, 1985.

[17] L. YOUNES, *Estimation and annealing for Gibbs fields*, preprint 1986 Université Paris-Sud, to appear in Ann. Inst. Henri Poincaré, 1988.

[18] L. YOUNES, *Estimation for Gibbs fields: applications and numerical results*, preprint Université Paris-Sud, 1987.

Modeling and Numerical Analysis of Junctions Between Elastic Structures

PHILIPPE G. CIARLET*

Abstract. We consider a problem in three-dimensional linearized elasticity, posed over a domain consisting of a plate with thickness 2ϵ, inserted into a solid whose Lamé constants are independent of ϵ. If the Lamé constants of the material constituting the plate vary as ϵ^{-3}, the solution of the three-dimensional problem converges, as ϵ approaches zero, to the solution of a coupled, "pluri-dimensional" problem of a new type, posed simultaneously over a three-dimensional open set with a slit and a two-dimensional open set. Other problems are also amenable to the same method, such as junctions between plates and rods, and folded plates.

1. Introduction. *The modeling of junctions between three-dimensional, two-dimensional (plates), and one-dimensional (rods)*, elastic structures is a problem of outstanding practical importance: Consider for instance the structure formed by an H-shaped beam (Fig. 1), or a satellite (Fig. 2). Nevertheless, this problem does not seem to have been so far investigated from a mathematical viewpoint. By contrast, the mathematical theory of *three-dimensional elastic structures* is well established; see e.g., Ciarlet [8], Duvaut & Lions [23], Fichera [24], Germain [25], Gurtin [27], Marsden & Hugues [34], Wang & Truesdell [40]; likewise, the modeling of *elastic plates and rods* have been thoroughly analyzed, and (more recently) justified, by *asymptotic methods* applied in the spirit of Lions [31]: See Caillerie [4], Ciarlet & Destuynder [10, 11], Ciarlet [5, 6, 9], Destuynder [20, 22], Raoult [35, 36] for plates; see Aganovič & Tutek [1], Bermudez & Viaño [2], Cimetière, Geymonat, Le Dret, Raoult & Tutek [17], Rigolot [37,38], Trabucho & Viaño [39], for rods.

We present here an approach that has been successfully used for modeling the junctions between such structures, by Ciarlet, Le Dret & Nzengwa [14] for junctions between three-dimensional and two-dimensional structures, by Le Dret [28, 29, 30] *for junctions between two-dimensional and two-dimensional structures (folded plates, possibly with corners), and by Ciarlet [7] for junctions between two-dimensional and one-dimensional structures (junctions between plates and rods).*

* Laboratoire d'Analyse Numérique, Université Pierre et Marie Curie, 4, place Jussieu, 75005 Paris, France.

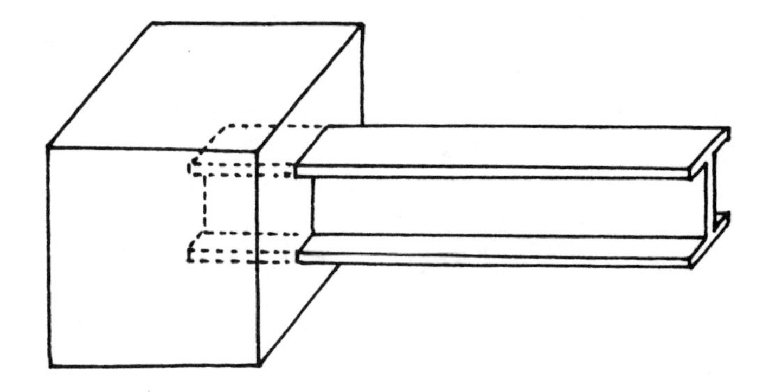

FIG. 1. An H-shaped beam inserted into a block of concrete.

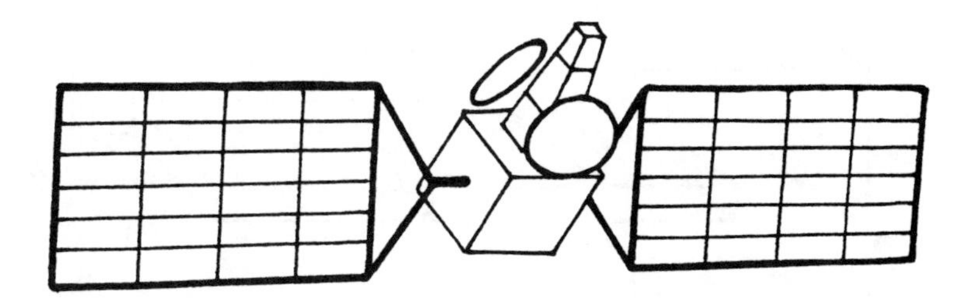

FIG. 2. The solar panels of a satellite are two-dimensional structures which are held together, and connected to the central structure, by one-dimensional structures (rods). This is a drawing of the satellite TDF-1, reproduced here by courtesy of the Centre National d'Etudes Spatiales.

In each instance, one, or several, portions of the whole three-dimensional structure has a "small" thickness, or diameter, denoted ϵ. If the various data (Lamé constants, applied body or surface force densities) behave as specific powers of ϵ as $\epsilon \to 0$, one can establish the *H^1-convergence of the appropriately scaled components of the displacements towards the solution of a "limit" variational problem of a new type*, posed simultaneously over an open subset of $\mathbb{R}^m$ and an open subset of $\mathbb{R}^n$, with $1 \leq m, n \leq 3$.

The crucial idea for treating the junction seems to be of wide applicability; It consists in *scaling the different parts of the full structure independently of each other, each part* being scaled as is usually done in "single plate" or "single rod" theories, but *counting the junction twice, once in each portion that it connects*. The scaled components of the displacement, which are defined in this fashion on two separate domains, thus contain the information about the junction twice. That they correspond to the same displacement of the whole structure then yield the "*junction conditions*" that the solution of the limit problem must satisfy.

2. The three-dimensional problem. Latin indices take their values in the set $\{1, 2, 3\}$ and Greek indices take their values in the set $\{1, 2\}$; the repeated index convention for summation is systematically used in conjunction with the above rules.

We are given constants $a_1, b_1, a_2, a_3, b_3, \beta$ which are all > 0, and which satisfy $\beta < b_1$. For each $\epsilon > 0$, we let (cf. Fig. 3):

$$\omega = \{(x_1, x_2) \in \mathbb{R}^2;\ 0 < x_1 < b_1,\ |x_2| < a_2\},\quad \Omega^\epsilon = \omega \times]-\epsilon, \epsilon[,$$

$$\gamma_0 = \{(b_1, x_2) \in \mathbb{R}^2;\ |x_2| \leq a_2\},\quad \Gamma_0^\epsilon = \gamma_0 \times]-\epsilon, \epsilon[,$$

$$\omega_\beta = \{(x_1, x_2) \in \mathbb{R}^2;\ 0 < x_1 < \beta,\ |x_2| < a_2\},\quad \Omega_\beta^\epsilon = \omega_\beta \times]-\epsilon, \epsilon[,$$

$$O = \{(x_1, x_2, x_3) \in \mathbb{R}^3;\ -a_1 < x_1 < \beta,\ |x_2| < a_2,\ -a_3 < x_3 < b_3\},$$

$$O_\beta^\epsilon = 0 - \overline{\Omega}_\beta^\epsilon,\quad S^\epsilon = O \cup \Omega^\epsilon.$$

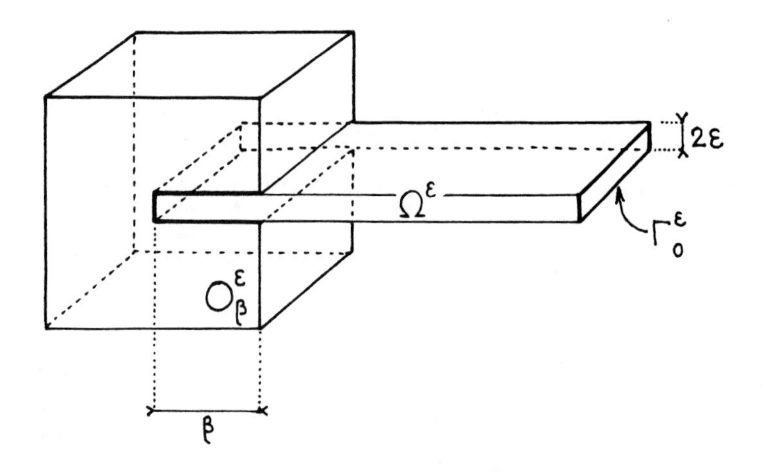

FIG. 3. The three-dimensional elastic structure.

Remark. Since ϵ is a *dimensionless parameter*, the thickness of the "thin" structure should be written as $2\epsilon h$, for some fixed length $h > 0$. For notational simplicity, we however avoided introducing an additional letter. ■

Remark. The results presented here are by no means restricted to structures with faces orthogonal to each other; but such shapes render the drawings considerably easier! ■

The set $\overline{O}_\beta^\epsilon$ is the reference configuration of a linearly elastic material whose *Lamé constants* $\tilde{\lambda}$, $\tilde{\mu}$ are *assumed* to be independent of ϵ; the set $\overline{\Omega}^\epsilon$ is the reference configuration of a linearly elastic material whose *Lamé constants* λ^ϵ, μ^ϵ are *assumed* to be of the form

$$\lambda^\epsilon = \epsilon^{-3}\lambda,\quad \mu^\epsilon = \epsilon^{-3}\mu, \tag{1}$$

where $\tilde{\lambda}$ and λ are two constants *independent* of ϵ. The unknown is the *displacement vector field* $u^\epsilon = (u_i^\epsilon) : S^\epsilon \to \mathbb{R}^3$, which is assumed to satisfy a *boundary condition of place* $u^\epsilon = 0$ on Γ_0^ϵ. In linearized elasticity, u^ϵ is solution of the following variational equations:

$$\int_{O_\beta^\epsilon} \{\tilde{\lambda} e_{pp}(u^\epsilon) e_{qq}(v^\epsilon) + 2\tilde{\mu}\, e_{ij}(u^\epsilon) e_{ij}(v^\epsilon)\} dx^\epsilon - \int_{O_\beta^\epsilon} f_i^\epsilon v_i^\epsilon \, dx^\epsilon$$

$$+ \int_{\Omega^\epsilon} \{\lambda^\epsilon e_{pp}(u^\epsilon) e_{qq}(v^\epsilon) + 2\mu^\epsilon e_{ij}(u^\epsilon) e_{ij}(v^\epsilon)\} dx^\epsilon - \int_{\Omega^\epsilon} f_i^\epsilon v_i^\epsilon \, dx^\epsilon = 0 \quad \text{for all} \quad v^\epsilon \in V^\epsilon, \tag{2}$$

where

$$V^\epsilon = \{v^\epsilon \in H^1(S^\epsilon)\,;\ v^\epsilon = 0 \ \text{ on } \ \Gamma_0^\epsilon\}, \tag{3}$$

where $e_{ij} = 1/2(\partial_i u_j + \partial_j u_i)$ denote the components of the linearized strain tensor, and where the vector field $(f_i^\epsilon) : S^\epsilon \to \mathbb{R}^3$ represents the applied body force density. This is a coercive variational problem (since Korn's inequality applies), whose solution satisfies, at least formally, a classical "transmission problem" of three-dimensional elasticity (cf. Dautray & Lions [18, p. 1245]).

3. Equivalent formulation of the three-dimensional problem over two open sets independent of ϵ. With the sets $\bar{\Omega}^\epsilon$ and $\bar{O}$, *which overlap over the "inserted" part* $\bar{\Omega}_\beta^\epsilon$ *of the "thin" part* $\bar{\Omega}^\epsilon$, we associate *two disjoint sets* $\bar{\Omega}$ and $[\tilde{\Omega}]^-$, as follows: First, as in the case of a single plate (cf. Ciarlet & Destuynder [10]), we let $\Omega = \omega \times]-1, 1[$; with each point $x^\epsilon = (x_1, x_2, x_3^\epsilon) \in \bar{\Omega}^\epsilon$, we associate the point $x = (x_1, x_2, \epsilon^{-1} x_3^\epsilon) \in \bar{\Omega}$ (cf. Fig. 4); finally, with the restriction (still denoted) $u^\epsilon = (u_i^\epsilon) : \bar{\Omega}^\epsilon \to \mathbb{R}^\epsilon$ of the unknown u^ϵ to the set $\bar{\Omega}^\epsilon$, we associate the function $u(\epsilon) = (u_i(\epsilon)) : \bar{\Omega} \to \mathbb{R}^3$ defined by

$$u_\alpha^\epsilon(x^\epsilon) = \epsilon^2 u_\alpha(\epsilon)(x) \ \text{ for all } \ x^\epsilon \in \bar{\Omega}^\epsilon, \tag{4}$$

$$u_3^\epsilon(x^\epsilon) = \epsilon\, u_3(\epsilon)(x) \ \text{ for all } \ x^\epsilon \in \bar{\Omega}^\epsilon, \tag{5}$$

Secondly, we define the translated set $\tilde{\Omega} = \Omega + t$, the vector t being such that $\{\tilde{\Omega}\}^- \cap \bar{\Omega} = \varnothing$; with each point $x^\epsilon \in \bar{O}$, we associate the translated point $\tilde{x} = (x^\epsilon + t) \in \{\tilde{\Omega}\}^-$ (Fig. 4); finally, with the restriction (still denoted) $u^\epsilon = (u_i^\epsilon) : \bar{O} \to \mathbb{R}^3$ of the unknown u^ϵ to the set $\bar{O}$, we associate the function $\tilde{u}(\epsilon) = (\tilde{u}_i(\epsilon)) : \{\tilde{\Omega}\}^- \to \mathbb{R}^3$ defined by

$$u_i^\epsilon(x^\epsilon) = \epsilon\, \tilde{u}_i(\epsilon)(\tilde{x}) \quad \text{for all} \quad x^\epsilon \in \bar{O}. \tag{6}$$

The function $u^\epsilon \in V^\epsilon$, where V^ϵ is the space defined in (3), is thus mapped through relations (4), (5), (6), into a pair $(\tilde{u}(\epsilon), u(\epsilon))$ which belongs to the space $H^1(\tilde{\Omega}) \times H^1(\Omega)$, which satisfies the boundary condition $u(\epsilon) = 0$ on $\Gamma_0 = \gamma_0 \times]-1, 1[$, and which satisfies the *compatibility conditions*

$$\tilde{u}_\alpha(\epsilon)(\tilde{x}) = \epsilon\, u_\alpha(\epsilon)(x), \tag{7}$$

$$\tilde{u}_3(\epsilon)(\tilde{x}) = u_3(\epsilon)(x), \tag{8}$$

for each point $\tilde{x} \in \tilde{\Omega}_\beta^\epsilon = \Omega_\beta^\epsilon + t$ and each point $x \in \Omega_\beta = \omega_\beta \times]-1, 1[$ that correspond to the *same* point $x^\epsilon \in \Omega_\beta^\epsilon$ (Fig. 4).

Finally, we *assume* that there exist functions $f_i : \Omega \to \mathbb{R}$ and $\tilde{f}_i : \tilde{\Omega} \to \mathbb{R}$ such that

$$f_\alpha^\epsilon(x^\epsilon) = \epsilon^{-1} f_\alpha(x) \qquad \text{for all } \ x^\epsilon \in \Omega^\epsilon, \tag{9}$$

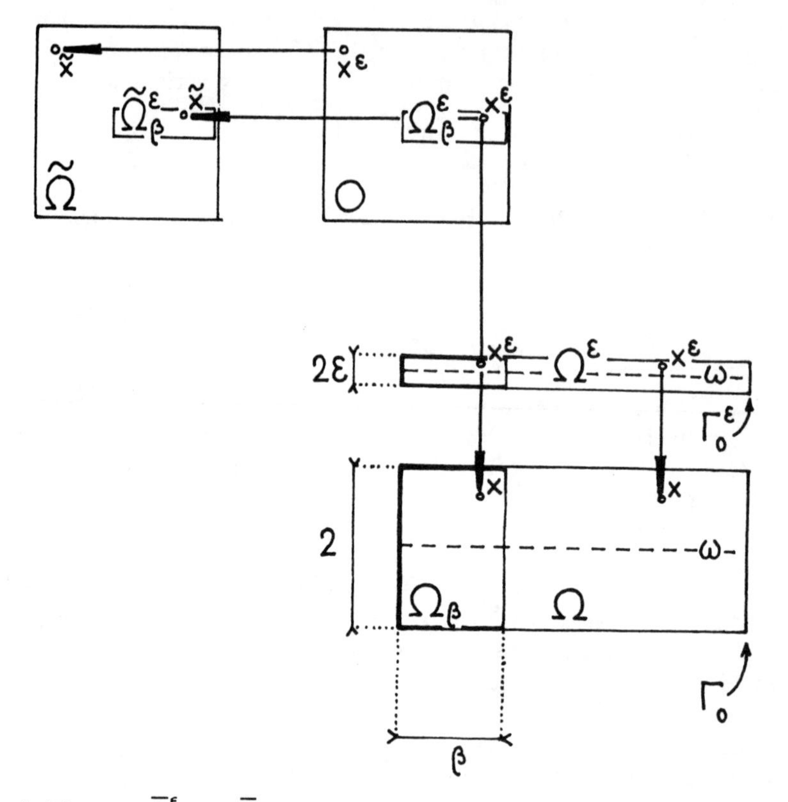

FIG. 4. The sets $\overline{\Omega}^\epsilon$ and $\overline{O}$, which are respectively occupied by the "thin" part and the "three-dimensional" part of the elastic structure, are mapped into two disjoint sets $\overline{\Omega}$ and $[\tilde{\Omega}]^-$. The "inserted" part $\overline{\Omega}_\beta^\epsilon$ of the thin part $\overline{\Omega}^\epsilon$ is thus mapped twice, once onto $\overline{\Omega}_\beta \subset \overline{\Omega}$ and once onto $\{\tilde{\Omega}_\beta^\epsilon\}^- \subset \{\tilde{\Omega}\}^-$.

$$f_3^\epsilon(x^\epsilon) = f_3(x) \qquad \text{for all } x^\epsilon \in \Omega^\epsilon , \tag{10}$$

$$f_i^\epsilon(x^\epsilon) = \epsilon \tilde{f}_i(\tilde{x}) \qquad \text{for all } x^\epsilon \in O_\beta^\epsilon . \tag{11}$$

Using the assumptions (1) and (9)-(11) made on the data, and the transformation formulas (4)-(6), we can thus re-formulate the variational problem (2) in the following *equivalent* form: *The pair* $(\tilde{u}(\tilde{u}(\epsilon), u(\epsilon))$, *constructed in* (4)-(6), *belongs to the space*

$$V(\epsilon) = \{(\tilde{v}, v) \in H^1(\tilde{\Omega}) \times H^1(\Omega);\ v = 0 \text{ on } \Gamma_0 ,$$

$$\tilde{v}_\alpha(\tilde{x}) = \epsilon v_\alpha(x) ,\ \tilde{v}_3(\tilde{x}) = v_3(x) \text{ at all}$$

$$\text{corresponding points } \tilde{x} \in \tilde{\Omega}_\beta^\epsilon \text{ and } x \in \Omega_\beta\} , \tag{12}$$

and it satisfies the following variational equations:

$$\int_{\tilde\Omega} \chi(\tilde O_\beta)\{\tilde\lambda e_{pp}(\tilde u(\epsilon))e_{qq}(\tilde v) + 2\tilde\mu e_{ij}(\tilde u(\epsilon))e_{ij}(\tilde v)\}d\tilde x - \int_{\tilde\Omega} \chi(\tilde O^\epsilon_\beta)\tilde f_i \tilde v_i\, d\tilde x$$

$$+ \int_\Omega \{\lambda e_{\alpha\alpha}(u(\epsilon))e_{\beta\beta}(v) + 2\mu e_{\alpha\beta}(u(\epsilon))e_{\alpha\beta}(v)\}dx - \int_\Omega f_i v_i\, dx$$

$$+ \epsilon^{-2} \int_\Omega \{\lambda[e_{\alpha\alpha}(u(\epsilon))e_{33}(v) + e_{33}(u(\epsilon))e_{\beta\beta}(v)] + 4\mu e_{\alpha 3}(u(\epsilon))e_{\alpha 3}(v)\}dx$$

$$+ \epsilon^{-4} \int_\Omega \{(\lambda+2\mu)e_{33}(u(\epsilon))e_{33}(v)dx = 0 \text{ for all } (\tilde v, v) \in V(\epsilon), \tag{13}$$

where $\chi(A)$ denotes in general the characteristic function of a set A, and $\tilde O^\epsilon_\beta = O^\epsilon_\beta + \tau$.

4. Convergence of $(\tilde u(\epsilon))$ as $\epsilon \to 0$; identification of the limit problem. The detailed proofs of the following theorems are found in Ciarlet, Le Dret and Nzengwa [14]. In what follows, $\tilde\omega_\beta$ denotes the translated set $(\omega_\beta + t)$; $\phi_{|A}$ denotes the restriction of a function ϕ to a set A; in (16), the equality $\tilde v_{3|\tilde\omega_\beta} = \eta_{3|\omega_\beta}$ is to be understood up to a translation by the vector t; ∂_τ denotes the outer normal derivative operator along $\partial\omega$.

THEOREM 1. *As $\epsilon \to 0$, the family $(\tilde u(\epsilon), u(\epsilon))_{\epsilon>0}$ converges strongly in the space $H^1(\tilde\Omega) \times H^1(\Omega)$ towards an element $(\tilde u, u)$ that satisfies the following relations:*

(a) *The limit $u \in H^1(\Omega)$ vanishes on $\Gamma_0 = \gamma_0 \times]-1, 1[$ and is a Kirchhoff-Love field in Ω, i.e., there exist functions $\zeta_\alpha \in H^1(\omega)$ and $\zeta_3 \in H^2(\omega)$ satisfying $\zeta_i = \partial_\nu \zeta_3 = 0$ on γ_0 such that*

$$u_3(x_1, x_2, x_3) = \zeta_3(x_1, x_2) \text{ for all } (x_1, x_2, x_3) \in \Omega, \tag{14}$$

$$u_\alpha(x_1, x_2, x_3) = \zeta_\alpha(x_1, x_2) - x_3\, \partial_\alpha \zeta_3(x_1, x_2) \text{ for all } (x_1, x_2, x_3) \in \Omega. \tag{15}$$

(b) *The pair $(\tilde u, \zeta_3)$ belongs to the space*

$$[H^1(\tilde\Omega) \times H^2(\omega)]_\beta \overset{\text{def}}{=} \{(\tilde v, \eta_3) \in H^1(\tilde\Omega) \times H^2(\omega);\ \eta_3 = \partial_\nu \eta_3 = 0 \text{ on } \gamma_0,$$

$$\tilde v_{3|\tilde\omega_\beta} = \eta_{3|\omega_\beta},\ \tilde v_{\alpha|\tilde\omega_\beta} = 0], \tag{16}$$

and solves the variational equations:

$$\int_{\tilde\Omega} \{\tilde\lambda e_{pp}(\tilde u)e_{qq}(\tilde v) + 2\tilde\mu e_{ij}(\tilde u)e_{ij}(\tilde v)\}d\tilde x - \int_{\tilde\Omega} \tilde f_i \tilde v_i d\tilde x$$

$$- \int_\omega m_{\alpha\beta}(\zeta_3)\partial_{\alpha\beta}\eta_3\, d\omega - \left(\int_\omega \left\{ \int_{-1}^{1} f_3\, dx_3 \right\} \eta_3 d\omega + \int_\omega \left\{ \int_{-1}^{1} x_3 f_\alpha\, dx_3 \right\} \partial_\alpha \eta_3 d\omega \right) = 0$$

$$\text{for all } (\tilde v, \eta_3) \in [H^1(\tilde\Omega) \times H^2(\omega)]_\beta, \tag{17}$$

where

$$m_{\alpha\beta}(\zeta_3) \stackrel{\text{def}}{=} \frac{2E}{3(1-\nu^2)} \{(1-\nu)\partial_{\alpha\beta}\zeta_3 + \nu\Delta\zeta_3\,\delta_{\alpha\beta}\}. \tag{18}$$

(c) *The pair* (ζ_1, ζ_2) *belongs to the space*

$$H(\omega) \stackrel{\text{def}}{=} \{(\eta_1, \eta_2) \in H^1(\omega) \times H^1(\omega);\ \eta_\alpha = 0 \text{ on } \gamma_0\}, \tag{19}$$

and solves the variational equations:

$$\int_\omega n_{\alpha\beta}(\zeta_1, \zeta_2)\partial_\alpha \eta_\beta \, d\omega - \int_\omega \left\{ \int_{-1}^{1} f_\alpha \, dx_3 \right\} \eta_\alpha \, d\omega = 0 \quad \text{for all } (\eta_1, \eta_2) \in H(\omega), \tag{20}$$

where

$$n_{\alpha\beta}(\zeta_1, \zeta_2) \stackrel{\text{def}}{=} \frac{2E}{(1-\nu^2)} \left\{ (1-\nu) \left(\frac{\partial_\alpha \zeta_\beta + \partial_\beta \zeta_\alpha}{2} \right) + \partial_\gamma \zeta_\gamma \, \delta_{\alpha\beta} \right\}. \tag{21}$$

Sketch of proof: Using the boundary condition $v = 0$ on Γ_0 and the *compatibility conditions* $\tilde{v}_\alpha(\tilde{x}) = \epsilon v_\alpha(x)$ and $\tilde{v}_3(\tilde{x}) = v_3(x)$ satisfied at all corresponding points $\tilde{x} \in \tilde{\Omega}_\beta^\epsilon$ and $x \in \Omega_\beta$ by the functions $(\tilde{v}, v) \in V(\epsilon)$ (cf. (12)), one first shows that *the family* $(\tilde{u}(\epsilon), u(\epsilon))_{\epsilon>0}$ *is bounded in the space* $H^1(\tilde{\Omega}) \times H^1(\Omega)$ (for this part of the proof, it is convenient to introduce the quadratic energy that is naturally associated with the variational equations (13)). Hence there exists a subsequence that weakly converges to a pair $(\tilde{u}, u) \in H^1(\tilde{\Omega}) \times H^1(\Omega)$.

Adapting techniques that are now standard in linearized plate theory (cf. Destuynder [20,22], Ciarlet & Kesavan [12]), one then shows that the limit u is a Kirchhoff-Love field (in the sense that relations (14), (15) are satisfied); passing to the limit in the compatibility conditions (7), (8), one establishes next that the limit $(\tilde{u}, u)$ satisfies the *junction conditions*

$$\tilde{u}_{3|\omega_\beta} = \zeta_{3|\omega_\beta} \text{ and } \tilde{u}_{\alpha|\tilde{\omega}_\beta} = 0, \tag{22}$$

used in the definition (16) of the space $[H^1(\tilde{\Omega}) \times H^2(\omega)]_\beta$.

The main difficulty consists in passing to the limit in the variational equations (13), in order to eventually obtain the variational formulations (17) and (20) of the limit problems, which are posed respectively over the *limit spaces* $[H^1(\tilde{\Omega}) \times H^2(\omega)]_\beta$ and $H(\omega)$ of (16) and (19). Since the aforementioned subsequence of $(\tilde{u}(\epsilon), u(\epsilon))_{\epsilon>0}$ is only weakly convergent, *arbitrary test functions of either limit spaces must be strongly approximated with respect to the* H^1*-norm by functions in the space* $V(\epsilon)$ of (12); the construction of such approximating functions is rather delicate.

Finally, one shows that the solution of each limit problem exists and is unique, and that the whole family $(\tilde{u}(\epsilon), u(\epsilon))_{\epsilon>0}$ strongly converges in the space $H^1(\tilde{\Omega}) \times H^1(\Omega)$ to the limit $(\tilde{u}, u)$. ∎

It remains *to describe the boundary value problems that are, at least formally, associated with the variational equations* (17) *and* (20). In what follows, we define the open set

$$\tilde{\Omega}_\beta = \tilde{\Omega} - [\tilde{\omega}_\beta]^-, \tag{23}$$

which is *thus a three-dimensional open set with a two-dimensional slit* (Fig. 5), we let $\tilde{\omega}_\beta^+$ and $\tilde{\omega}_\beta^-$ *denote the upper and lower faces of the slit.* When viewed as sets, these faces are fictitiously distinguished since they coincide with the set $\tilde{\omega}_\beta$; on the other hand, the introduction of different notations allows for a convenient distinction between the traces "from above" or "from below" of a function defined over the set $\tilde{\Omega}_\beta$, as in eqn. (27). In what follows $\tilde{n} = (\tilde{n}_i)$ denotes the unit outer normal vector along the set $\partial\tilde{\Omega}_\beta - \{\tilde{\omega}^+_\beta \cup \tilde{\omega}^-_\beta\}$; (ν_α) and (τ_α) denote the unit outer normal and tangential vectors along $\partial\omega$; ∂_τ denote the tangential derivative operator along $\partial\omega$.

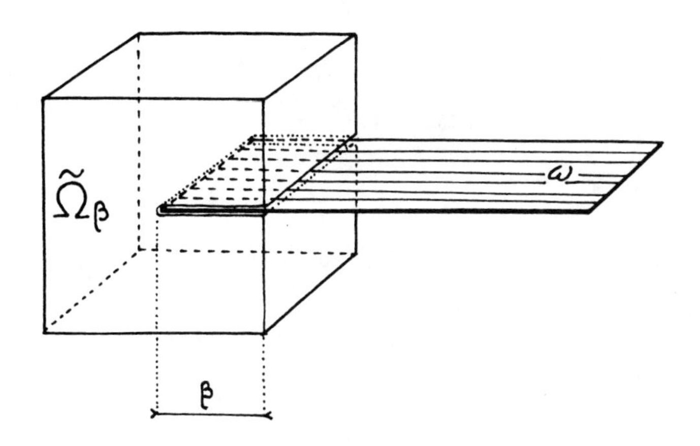

FIG. 5. The limit problem is a coupled, pluri-dimensional, problem posed over the three-dimensional open set $\tilde{\Omega}_\beta$ and a two-dimensional open set ω. The three-dimensional set has a slit into which the two-dimensional set is inserted.

THEOREM 2. (a) *The function* $\tilde{u}$ *solves*

$$-\tilde{\partial}_j\, \sigma_{ij}(\tilde{u}) = \tilde{f}_i \quad in \quad \tilde{\Omega}_\beta\,, \tag{24}$$

$$\sigma_{ij}(\tilde{u})\tilde{n}_j = 0 \quad on \quad \partial\tilde{\Omega}_\beta - \{\tilde{\omega}^+_\beta \cup \tilde{\omega}^-_\beta\}, \tag{25}$$

where $\tilde{\partial}_i = \partial/\partial\tilde{x}_j$, *and*

$$\sigma_{ij}(\tilde{u}) = \tilde{\lambda} e_{pp}(\tilde{u})\delta_{ij} + 2\tilde{\mu} e_{ij}(\tilde{u})\,. \tag{26}$$

(b) *The function* ζ_3 *solves*

$$\frac{2E}{3(1-\nu^2)}\,\Delta^2\zeta_3 = \int_{-1}^{1} f_3\,dx_3 - \int_{-1}^{1} x_3\,\partial_\alpha f_\alpha\,dx_3$$

$$+ \chi(\omega_\beta)\{\sigma_{33}(\tilde{u})_{|\tilde{\omega}^+_\beta} - \sigma_{33}(\tilde{u})_{|\tilde{\omega}^-_\beta}\} \quad in \quad \omega\,, \tag{27}$$

$$\zeta_3 = \partial_\nu\zeta_3 = 0 \quad on \quad \gamma_0\,, \tag{28}$$

$$m_{\alpha\beta}(\zeta_3)\nu_\alpha\nu_\beta = 0 \quad on \quad (\partial\omega - \gamma_0)\,, \tag{29}$$

$$- \partial_\tau\{m_{\alpha\beta}(\zeta_3)\nu_\alpha \tau_\beta\} - \partial_\alpha m_{\alpha\beta}(\zeta_3)\nu_\beta = \left\{ \int_{-1}^{1} x_3 f_\alpha \, dx_3 \right\} \nu_\alpha \quad on \quad \partial\omega - \gamma_0 , \tag{30}$$

where $m_{\alpha\beta}(\zeta_3)$ is defined as in (18).

(c) *The pair $(\tilde{u}, \zeta_3)$ satisfies the "junction conditions"*

$$\tilde{u}_{3|\tilde{\omega}_\beta^+} = \tilde{u}_{3|\tilde{\omega}_\beta^-} = \zeta_{3|\omega_\beta} , \tag{31}$$

$$\tilde{u}_{\alpha|\tilde{\omega}_\beta^+} = \tilde{u}_{\alpha|\tilde{\omega}_\beta^-} = 0 . \tag{32}$$

(d) *The pair (ζ_1, ζ_2) solves*

$$- \partial_\beta n_{\alpha\beta}(\zeta_1, \zeta_2) = \int_{-1}^{1} f_\alpha \, dx_3 \quad in \quad \omega , \tag{33}$$

$$\zeta_\alpha = 0 \quad on \quad \gamma_0 , \tag{34}$$

$$n_{\alpha\beta}(\zeta_1, \zeta_2)\nu_\alpha = 0 \quad on \quad (\partial\omega - \gamma_0) , \tag{35}$$

where $n_{\alpha\beta}(\zeta_1, \zeta_2)$ is defined as in (21). ■

5. Comments. We now list various comments and conclusions about the results of Theorems 1 and 2:

(i) The main conclusion is that the pair $(\tilde{u}, \zeta_3)$ solves a *coupled, pluri-dimensional, variational problem of a new type* (cf. eqns. (17) and (19)), posed over a subspace of $H^1(\tilde{\Omega}) \times H^2(\omega)$, whose functions satisfy the *junction conditions* (31)-(32).

(ii) The function

$$h_3 \overset{\text{def}}{=} \chi(\omega_\beta)\{\sigma_{33}(\tilde{u})_{|\tilde{\omega}_\beta^+} - \sigma_{33}(\tilde{u})_{|\tilde{\omega}_\beta^-}\} \tag{36}$$

appearing in the right-hand side of the otherwise familiar two-dimensional plate equation (27) is nothing but the *Lagrange multiplier* associated with the equality constraints (in the sense of optimization theory) imposed to the "trial" function $(\tilde{v}, \eta_3)$ in the form of the *junction conditions* (31)-(32).

(iii) *The mechanical interpretation of the limit problem* is straightforward: The function $\tilde{u}$ solves the standard equations (24)-(26) of three-dimensional linearized elasticity, while the functions ζ_3 and (ζ_1, ζ_2) solve the standard equations (27)-(30) and (33)-(35) satisfied in two-dimensional plate theory by the transverse and in-plane displacements respectively (cf. Germain [26, Sect. VIII.7]). The added function h_3 of (36) in the right-hand side of the biharmonic equation (27) balances the vertical component of the forces that act on the three-dimensional part $\tilde{\Omega}_\beta$, since one can prove that

$$\int_\omega h_3 \, d\omega = \int_{\tilde{\Omega}} \tilde{f}_3 \, d\tilde{x}_3 \tag{37}$$

(iv) While the junction conditions $\tilde{u}_{3|\omega_\beta^+} = \tilde{u}_{3|\tilde{\omega}_\beta} = \zeta_{3|\omega_\beta}$ express the *continuity of the vertical displacement at the junction*, the other junction conditions $\tilde{u}_{\alpha|\tilde{\omega}_\beta^+} = \tilde{u}_{\alpha|\tilde{\omega}_\beta^-} = 0$

do not involve the functions $\zeta_{\alpha|\omega_\beta}$, which do not vanish in general (they do vanish if $f_\alpha = 0$ in Ω, however). This is only an apparent paradox, for the convergence result of Theorem 1 implies that $(u_\alpha^\epsilon|\Omega^\epsilon)|_{\omega_\beta} = \epsilon^2 \int_\alpha + o(\epsilon^2)$ in $H^{1/2}(\omega_\beta)$; hence the first-order term (with respect to ϵ) of the horizontal components of the displacement of the three-dimensional structure should also vanish in $H^{1/2}(\omega_\beta)$; but this is precisely what is implied by conditions (32) and the scalings (6).

(v) While the "full" three-dimensional problem is well defined for any $\epsilon > 0$ if $\beta = 0$, it is not possible to define a "coupled" limit problem in this case: Even if an appropriate boundary condition of place is satisfied along the boundary of the three-dimensional part (in order to "fix" this part), the limit problem consists of two completely unrelated problems, i.e., *there are no longer any junction conditions in the limit problem when* $\beta = 0$.

(vi) Relations (1) express that the rigidity of the material constituting the thin portion of the structure must increase as ϵ^{-3} when $\epsilon \to 0$. *That such asymptotic orders, as well as the assumed asymptotic orders (9)-(11) on the applied forces, are needed in order that a limit problem exist* has already been observed by Caillerie [4] and Ciarlet [5] in the case of a "single" plate. The reader is referred to Ciarlet [5,6,9] for more detailed discussions of the meaning of such asymptotic orders.

(vii) In the same spirit, assume that a boundary condition of place is satisfied along the boundary of the three-dimensional part, and that its Lamé constants converge in an appropriate manner to $+\infty$ as $\epsilon \to 0$ (while here they were independent of ϵ); in order words, "the three-dimensional part becomes rigid in the limit". Then in this case the junction conditions of the limit problem represent the *genuine boundary conditions that a clamped plate should satisfy at the junction with its three-dimensional support* (in this case, no boundary condition of place is needed along the boundary of the thin part); cf. Ciarlet et Le Dret [13].

(viii) *Numerical results* obtained with the limit model are satisfactory (cf. Fig. 6).

6. Extensions. We now list a *"programme" of various applications and possible extensions* of the present approach (some of these are already completed; others are currently investigated):

(i) *Identification and mathematical analysis of the "limit" problem for nonlinearly elastic materials*, in the spirit of Ciarlet & Destuynder [11], Ciarlet [5], Blanchard & Ciarlet [3], Ciarlet & Rabier [16], Davet [19];

(ii) *Computation of the stresses, especially in the neighborhood of the junction*, in the spirit of the "(σ, u)-approach" of Ciarlet & Destuynder [10,11];

(iii) *Identification of the "limit" eigenvalue problem*, in the spirit of Ciarlet & Kesavan [12];

(iv) *Identification of the "limit" dynamical problem*, in the spirit of Raoult [35,36];

(v) *Controllability of the "limit" problem*, by the Hilbert uniqueness method of Lions [32,33];

(vi) *Behavior as the depth* β *of the insertion also approaches 0 as* $\epsilon \to 0$ (there are two "small" parameters in this case);

(vii) *"Curved" lower-dimensional structures*, in the spirit of Destuynder [21] or Ciarlet & Paumier [15];

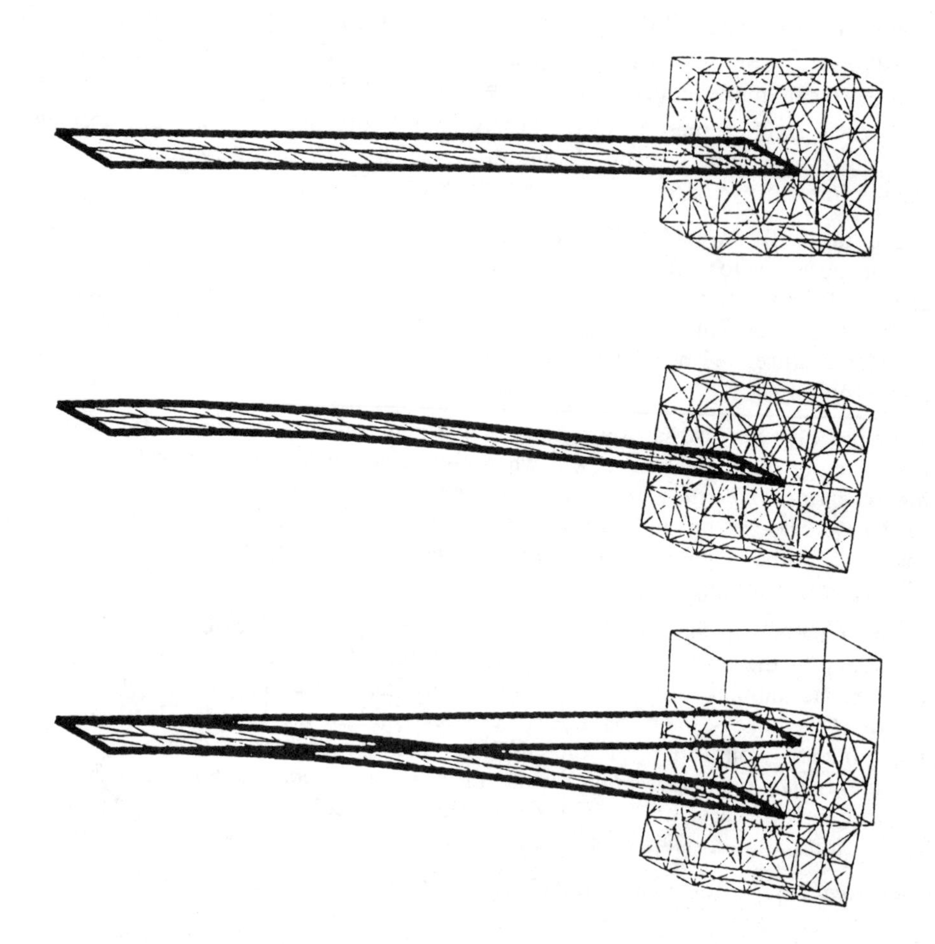

FIG. 6. Computation of the displacement vector field of a linearly elastic structure which comprises a "thin" part (plate) inserted into a "three-dimensional" part. The "limit" problem found in Theorem 1 has been used in this computation, which has been performed by Martial Aufranc, at the Institut National de la Recherche en Informatique et Automatique (I.N.R.I.A.), using the MODULEF Code. These results are reproduced by courtesy of Michel Bernadou, director of the MODULEF project at the I.N.R.I.A.

(viii) *Folded plates, possibly with corners* (cf. Le Dret [28,29,30];

(ix) *Junctions between plates and rods* (cf. Ciarlet [7]).

REFERENCES

[1] I. AGANOVIČ AND Z. TUTEK, A justification of the one-dimensional model of elastic beam, *Math. Methods in Applied Sci.* **8**, 1-14.

[2] A. BERMUDEZ AND J. M. VIAÑO, Une justification des équations de la thermoélasticité des poutres à section variable par des méthodes asymptotiques,

RAIRO Analyse Numérique **18**, 1984, 347-376.

[3] D. BLANCHARD AND P. G. CIARLET, A remark on the von Kármán equations, *Comput. Methods Appl. Mech. Engrg.* **37**, 79-92.

[4] D. CAILLERIE, The effect of a thin inclusion of high rigidity in an elastic body, *Math. Methods in Applied Sci.* **2**, 1980, 251-270.

[5] P. G. CIARLET, A justification of the von Kármán equations, *Arch. Rational Mech. Anal.* **73**, 1980, 349-389.

[6] P. G. CIARLET, Recent progresses in the two-dimensional approximation of three-dimensional plate models in nonlinear elasticity, in *Numerical Approximation of Partial Differential Equations* (E. L. Ortiz, Editor), pp. 3-19, North-Holland, Amsterdam, 1987.

[7] P. G. CIARLET, Junctions between plates and rods, to appear.

[8] P. G. CIARLET, *Mathematical Elasticity, Vol. I: Three-Dimensional Elasticity*, North-Holland, Amsterdam, 1988.

[9] P. G. CIARLET, *Mathematical Elasticity, Vol. II: Lower-Dimensional Theories of Plates and Rods*, North-Holland, Amsterdam, to appear.

[10] P. G. CIARLET AND P. DESTUYNDER, A justification of the two-dimensional linear plate model, *J. Mécanique* **18**, 1979, 315-344.

[11] P. G. CIARLET AND P. DESTUYNDER, A justification of a nonlinear model in plate theory, *Comput. Methods Appl. Mech. Engrg.* **17/18**, 1979, 227-258.

[12] P. G. CIARLET AND S. KESAVAN, Two-dimensional approximation of three-dimensional eigenvalue problems in plate theory, *Comput. Methods Appl. Mech. Engrg.* **26**, 1980, 149-172.

[13] P. G. CIARLET AND H. LE DRET, Justification of the boundary conditions of a clamped plate by an asymptotic analysis, to appear.

[14] P. G. CIARLET, H. LE DRET AND R. NZENGWA, Junctions between three-dimensional and two-dimensional linearly elastic structures, *J. Math Pure Appl.* to appear.

[15] P. G. CIARLET AND J. C. PAUMIER, A justification of the Marguerre-von Kármán equations, *Computational Mechanics* **1**, 177-202.

[16] P. G. CIARLET AND P. RABIER, *Les Equations de von Kármán*, Lecture Notes in Mathematics, Vol. 826, Springer-Verlag, Berlin.

[17] A. CIMETIÈRE, G. GEYMONAT, H. LE DRET, A. RAOULT AND Z. TUTEK, Asymptotic theory and analysis for displacements and stress distribution in nonlinear elastic straight slender rods, *J. Elasticity*, **19**, 1988, 111-161.

[18] R. DAUTRAY AND J.-L. LIONS, *Analsyse Mathématique et Calcul Numérique pour les Sciences et les Techniques*, Vol. 1, Masson, Paris, 1984.

[19] J.-L. DAVET, Justification de modèles de plaques non linéaires pour des lois de comportement générales, *Modélisation Mathématique et Analyse Numérique* **20**, 1986, 225-249.

[20] P. DESTUYNDER, Comparaison entre les modeles tridimensionnels et bidimensionnels de plaques en élasticité, *RAIRO Analyse Numérique* **15**, 1981, 331-369.

[21] P. DESTUYNDER, A classification of thin shell theories, *Acta Applicandae Mathematicae* **4**, 1985, 15-63.

[22] P. DESTUYNDER, *Une Théorie Asymptotique des Plaques Minces en Elasticité Linéaire*, Masson, Paris, 1986.

[23] G. DUVAUT AND J.-L. LIONS, *Les Inéquations en Mécanique et en Physique*, Dunod, Paris, 1972.

[24] G. FICHERA, Existence theorems in elasticity, in *Handbuch der Physik, VIa/2*, pp. 347-389, Springer-Verlag, Berlin.

[25] P. GERMAIN, *Mécanique des Milieux Continus*, Tome 1, Masson, Paris, 1972.

[26] P. GERMAIN, *Mécanique, Tome II, Ecole Polytechnique*, Ellipses, Paris, 1986.

[27] M. GURTIN, *Introduction to Continum Mechanics*, Academic Press, New York, 1981.

[28] H. LE DRET, Modélisation d'une plaque pliée, *C. R. Acad. Sci. Paris, Sér. I*, **304**, 1987, 571-573.

[29] H. LE DRET, Modeling of a folded plate, to appear.

[30] H. LE DRET, Folded plates revisited, to appear.

[31] J.-L. LIONS, *Perturbations Singulières dans les Problèmes aux Limites et en Contrôle Optimal*, Lecture Notes in Mathematics, Vol. 323, Springer-Verlag, Berlin, 1973.

[32] J.-L. LIONS, Exact controllability, stabilization and perturbations for distributed systems, John von Neumann Lecture, Boston, 1986, *SIAM Review*, **30**, 1988, 1-68.

[33] J.-L. LIONS, Controlabilité exacte et perturbations singuliéres (II). La méthode de daulité, in *Applications of Multiple Scaling in Mechanics* (P. G. Ciarlet and E. Sanchez-Palencia, Editors), pp. 223-237, Masson, Paris, 1987.

[34] J. E. MARSDEN AND T. J. R. HUGHES, *Mathematical Foundations of Elasticity*, Prentice-Hall, Englewood Cliffs, 1983.

[35] A. RAOULT, Construction d'un modèle d'évolution de plaques avec terme d'inertie de rotation, *Annali di Matematica Pura ed Applicata* **139**, 1985, 361-400.

[36] A. RAOULT, Asymptotic theory of nonlinearly elastic dynamic plates, to appear.

[37] A. RIGOLOT, *Sur une Théorie Asymptotique des Poutres*, Thèse de Doctorat d'Etat, Université Pierre et Marie Curie, Paris, 1976.

[38] A. RIGOLOT, Sur une théorie asymptotique de poutres, *J. Mécanique* **11**, 1972, 673-703.

[39] L. TRABUCHO AND J. M. VIAÑO, Derivation of generalized models for linear elastic beams by asymptotic expansion methods, in *Applications of Multiple Scaling in Mechanics* (P. G. Ciarlet and E. Sanchez-Palencia, Editors), pp. 302-315, Masson, Paris, 1987.

[40] C.-C. WANG AND C. TRUESDELL, *Introduction to Rational Elasticity*, Noordhoff, Groningen, 1973.

Aeronautical Acoustics: Mathematics Applied to a Major Industrial Problem

D. G. CRIGHTON*

Abstract. This paper gives a commentary on aeronautical acoustics as a scientific field, and on those features of it that make it a field worthy of attention by applied mathematicians. After an overview of the issues involved in the practical problem of aircraft noise, the "acoustic analogy" is outlined, typical scaling laws following from it are given, and recent progress in its detailed application to simple vortex flows is described. The re-introduction of propeller-driven commercial aircraft now raises the question of the noise of advanced propellers, and the paper sketches some of the principal aspects of modern propeller noise theory and emphasises the vital role that can be played in the formulation of a scientifically-based prediction scheme by asymptotic techniques allied to appropriate modelling of mechanisms.

1. Introduction. Modern aeronautical acoustics will be put forward in these notes (and in the lecture for which they were prepared) as a topic worthy of continuing study by mathematicians for the following reasons:-

* It provides an excellent example of the applied mathematical modelling discipline, involving repeated refinements of the model, and repeated exchanges between the mathematical model and the underlying scientific model (fluid mechanics) with the added complication of subjective human response as a strong factor.
* It is a field not susceptible to straightforward engineering approaches.

---

* Department of Applied Mathematics and Theoretical Physics, University of Cambridge, Cambridge CB3 9EW, U.K.

* It is a field where direct large-scale computation is not yet practicable, where experiments produce vast amounts of data needing assimilation and interpretation, and where analytical (asymptotic) modelling and solution remain essential.
* It is a field in which the long sequence of scientific problems and increasing social and economic pressures combine to repay prolonged interaction between academic applied mathematicians and industry.
* It is a field of critical importance (and increasingly so) to commercial aircraft development.

The subject of aeroacoustics (and more recently hydroacoustics for underwater applications of similar ideas) has developed as a scientific discipline in its own right since its creation by M.J. Lighthill in 1952 to address the then-impending and now still vital problem of commercial jet aircraft noise. Conventional acoustics, of the nineteenth and early twentieth centuries, envisages sound generation, in air or water, as accomplished primarily by the controlled motion of boundaries of a region of essentially static or uniformly moving fluid. Aero- and hydroacoustics deal with the generation of sound by vastly more powerful mechanisms, and primarily with sound generation by highly unsteady *turbulent* fluid flow.

Significant differences from conventional acoustics are:-

* Very intense aeroacoustic near field;

$$\frac{\text{near field energy flux}}{\text{radiated acoustic flux}} >>>> 1$$

(by a factor 10^4 in jet engines, 10^{12} or more in underwater boundary layers or wakes).
* Very broad frequency spectrum,

$$100\text{Hz} - 10\text{kHz} \qquad \text{aircraft noise}$$

$$10^{-1}\text{Hz} - 50\text{kHz} \qquad \text{underwater applications .}$$

* An essentially large-scale process is involved – a full-scale aircraft in flight in the atmosphere – which has proved almost impossible to simulate at any smaller scale.
* Many complex interacting source and propagation mechanisms are involved in producing an acoustic field which – even in the most acoustically efficient case of a large rocket motor – never accounts for as much as 1% of the total energy budget.
* Subjective human aural response is very important. There is particular sensitivity to sound in the 2-4 kHz band, and special sensitivity to narrow-band tones there – whereas the bulk of what little acoustic power there is lies broadband and at much lower frequencies.

Implications of these features are that computation will be difficult, making approximations will be dangerous, experiments will suffer from scaling and interaction problems, fluid dynamics knowledge will generally be unavailable or irrelevant, and particular care will have to be paid to aspects which in conventional steady or unsteady fluid dynamics would be of no significance whatever.

This all implies that the role of aeroacoustic theory must be to
- isolate mechanisms
- identify significant characteristics
- suggest correlations of experimental data
- predict gross trends
- suggest critical experiments.

Outright quantitative noise prediction will only rarely be feasible (only possible for very simple model flows, quite impossible for noise from fully turbulent jets in air, or boundary layers and wakes underwater).

Recent developments have, however, significantly changed this picture, as we shall see in §3. Proposals (now advanced to the stage at which flight tests have been made on a commercial aircraft) for high speed passenger transport aircraft propulsion by advanced propellers ("propfans") from about 1993 on imply that (1) sound sources in propfans will be largely deterministic and susceptible to detailed modelling, (2) outright noise prediction with reasonable accuracy should be possible (as indicated already in §3 below). Applied mathematical modelling thus has an enhanced role – but again one in which asymptotic analysis (complemented by numerical work) remains crucial both for understanding and for quantitative prediction!

2. Lighthill aeroacoustic analogy. The main tool of aeroacoustic theory is the "acoustic analogy". This has been extensively discussed in the review articles cited here. Our aim is to simply comment on a few aspects of importance for the status of the theory and for the applied mathematician.

Imagine an unsteady fluid flow, possibly containing solid bodies and material inhomogeneities, all confined to a finite region of space V, outside which the fluid is uniform and at rest, with sound speed c_o. The exact equations of motion (highly nonlinear, with a variety of possible dissipative mechanisms) are, in an acoustic analogy, written as a forced wave equation for density fluctuations ρ,

$$\left(\frac{\partial^2}{\partial t^2} - c_o^2\nabla^2\right)\rho = Q \tag{2.1}$$

with the hope that if the "source" Q, which must essentially vanish outside V, can be measured or estimated independently of the sound field it generates, then that sound field ρ can be calculated by procedures of classical acoustics.

Lighthill's expression for Q rightly emphasises its *quadrupole* structure (in the absence of bodies), advertised by a double space divergence,

$$Q = \frac{\partial^2 T_{ij}}{\partial x_i \partial x_j} \tag{2.2}$$

where (exactly)

$$T_{ij} = p_{ij} - c_o^2\rho\delta_{ij} + \rho u_i u_j \tag{2.3}$$

with p_{ij} the compressive stress tensor, u_i the fluid velocity. For low-speed flows with small temperature variations, Lighthill argued that $T_{ij} \simeq \rho_o v_i v_j$ where ρ_o is the mean density and $\underline{v}$ an incompressible approximation to $\underline{u}$ – which might be measured, or calculated analytically for simple flows (like two vortex rings moving under their mutual induction), or calculated numerically for more "realistic" flows.

It is impossible to carry out the detailed programme for a turbulent flow. For the mean acoustic intensity one would need the fourth time derivative of an 81-component space-time Reynolds stress correlation tensor,

$$\frac{\partial^4}{\partial \tau^4} < v_i v_j(\underline{y}, \sigma) v_k v_\ell(\underline{y} + \underline{r}, \sigma + \tau) >, \tag{2.4}$$

which would then have to be integrated over the space separation $d^3\underline{r}$ and then over the physical volume V, $d^3\underline{y}$.Such a task goes far beyond current experimental capabilities, which hardly extend beyond the study of the longitudinal simultaneous velocity correlation $< v_1(x_1, x_2, x_3, t) v_1(x_1 + r_1, x_2, x_3, t) >$! However the acoustic analogy immediately predicts one critically important feature. If U is a typical flow velocity (say the jet exhaust velocity), then the Reynolds stress correlation tensor scales with U^4, and the operator $\partial/\partial\tau$ scales with U/L where L is a characteristic length (say the jet nozzle diameter). Thus the intensity of jet noise increases as U^8. This classic Lighthill scaling law (a prime example of the "prediction of gross trends" of §1) represents a triumph of theory over experiment; before the publication of U^8, most reports of measured jet noise data gave a U^4 variation, which was then quite quickly recognised, post U^8, as associated with noise sources (primarily combustion) *within* the engine itself, rather than with the jet exhaust turbulent mixing downstream of the engine. In fact, variation of intensity with U^8 is now generally accepted as *defining* jet mixing noise, and identifying the presence of other "excess" or "internal" noise source mechanisms.

If the flow is other than one for which $T_{ij} \simeq \rho_o v_i v_j$ holds, one must use the Navier-Stokes equations, the continuity equation, the equations of energy, state, phase, ... to manipulate the quadrupole strength T_{ij} into an appropriate form. Thus the game for the applied mathematician takes three parts:

(1) Modelling: isolate those elements of T_{ij} that correspond to a particular physical process and obtain an expression for them which can be unambiguously estimated. (A vast range of such processes has been so modelled, in aeronautical and underwater applications, including jet mixing, turbulent combustion, shock wave-turbulence interaction, multiphase flow effects (bubbles in water, dust and aerosol droplets in air), breaking of surface waves, impingement of spray droplets, formation of bubbles, scattering of sound by turbulence and vortices, high temperature effects.)

(2) Determine a Green's function for the wave operator (perhaps generalised to include mean flow effects) appropriate to the geometry of any boundaries on which conditions may be prescribed (for example, a model of an aircraft wing with leading edge slats and trailing edge flaps).

(3) Convolve the Green's function with the source and extract information on scaling laws, directional variations of sound intensity, etc., and make detailed quantitative predictions if the source can be adequately characterised.

Step (2) has actually led to considerable developments in classical acoustic theory – in particular through the introduction of *causality* and *receptivity* ideas to determine the response to acoustic excitation of fluid systems in which there are unstable shear layers (as when a jet emerges from a nozzle into static surrounding fluid), and through the development of integral transform methods (matrix Wiener-Hopf problems) to treat particular inhomogeneous geometries. A problem which involves all these issues is the "engine-over-the-wing" problem, schematically illustrated in Figure 1. The wing surface is hard on one side, and lined with absorbing material on the other, leading to a matrix W-H problem, and separates mean flows of different speeds which create an unstable vortex sheet downstream of the wing trailing edge. Finding the (causal) Green's function for such a configuration amounts to determining the receptivity of the unstable flow coupled to the wing with one-sided lining.

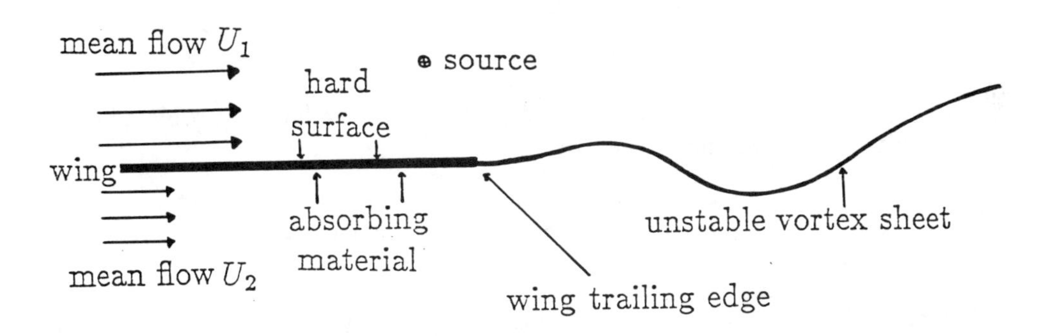

Figure 1. Illustration of the mathematical model for the "engine-over-the-wing" configuration, with flows of different speeds U_1, U_2 above and below the wing, acoustic lining on (say) the lower wing surface only, and an unstable shear layer downstream of the wing trailing edge.

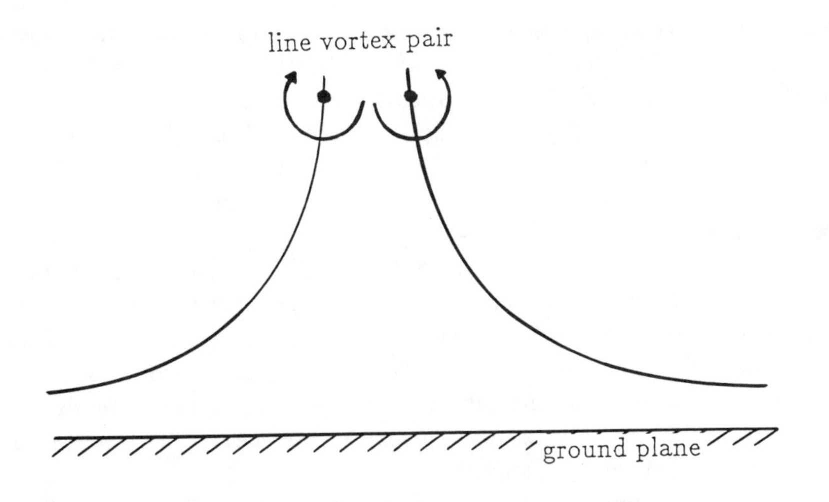

Figure 2. Sketch of the impact of a line vortex pair on a ground plane.

Detailed calculations following the steps (1)-(3) above can be used to illustrate the difficulties implied in §1 with regard to "noisy approximations". Consider the problem sketched in Figure 2, the impact on a ground plane $y = 0$ of a counter-rotating pair of line vortices. If the approach speed at infinity, U, is small compared with the sound speed c_o, the problem of acoustic radiation in the impact can be solved asymptotically as $M = U/c_o \to 0$, by a variety of methods. For example, one can calculate the vortex paths by elementary incompressible flow arguments, find the outer limit of the incompressible vortex impact problem, and match this (in the sense of matched asymptotic expansions) to an appropriate acoustic field. The problem is obviously one of the motion of four vortices in free space, the sound sources are pure Lighthill quadrupoles, and the total energy radiated over the whole motion satisfies $E \propto M^6$ (this corresponding exactly to the Lighthill U^8 prediction for the sound *power* in three dimensions). Alternatively, one can use the Powell-Howe analogy, in which the source Q in (2.1) is expressed as

$$Q = -\rho_o \mathrm{div}(\underline{\omega} \wedge \underline{u}) \tag{2.5}$$

where $\underline{\omega} = \mathrm{curl}\underline{u}$ is the vorticity. This is a convenient form for this problem because the vorticity is known, and concentrated,

$$\underline{\omega} = \sum_{j=1}^{4} (\omega_j \underline{\hat{k}})\, \delta(\underline{x} - \underline{x}_j(t)) \ , \quad \text{say} \ .$$

If, however, the $\underline{x}_j(t)$ are calculated *numerically* from incompressible flow theory, and Q evaluated and convolved with the free space Green's function, one finds, for a rigid ground plane $y = 0$, $E \propto (\Delta s)^2 M^4$, and for a flexible ground plane $E \propto (\Delta s)^2 M^2$, Δs being the step size for integration of the vortex paths. In underwater applications, with $M = 10^{-4} \to 10^{-3}$, these results are hopelessly out in level, and wrong in form. The reason is that the convenient Powell-Howe form for the source has the character of a dipole, with only one space divergence, the second divergence to comply with the Lighthill quadrupole structure being hidden in $\underline{\omega} \wedge \underline{u}$. The precise conditions for $\underline{\omega} \wedge \underline{u}$ to contain such a divergence are broken by the numerical integration, which makes the vortices travel in small linear segments and requires a small unsteady force to act on the fluid. This artificial numerically-induced force is more efficient acoustically than the quadrupoles equivalent to the actual vortex flow, and leads to dipole sound with the M^4 energy law. If the ground plane is not rigid, the numerical scheme gives a small uncancelled *monopole* source associated with the vibrations of the plane, and leading to the yet more intense field with $E \propto M^2$.

Thus the preservation of source structure is all-important – much more important than accurate tracking of the vortices, and approximations, whether analytical or numerical, which do not preserve that structure, are fatal. Approximations must be "good" in the sense of leading only to "quiet" errors. The vortex impact problem was actually a test case for a numerical program intended to predict the sound from helical cavitated ship propeller vortices - where it may also have grossly overpredicted the acoustic field.

There is thus a conflict between the use of an approximate form for Q which conveniently reflects some particular feature (e.g., the delta function concentration of vorticity) and the need not to allow this to introduce spurious sound sources more efficient than the actual flow itself. This has so far precluded reliable calculation of sound radiation from more realistic flows. For example, many studies have been made of the simulation of turbulent jet flows by clouds of axisymmetric vortex rings. This needs to be done numerically, the vortices being created periodically at some location, and being then allowed to interact through the standard induction law (the Biot-Savart law). Numerical round-off error is enough to make the periodic array of vortices interact and amalgamate in what looks like a fully "turbulent" way, and indeed the velocity field rapidly develops a chaotic turbulent form – although of course the whole field remains strictly axisymmetric. If long-time averages are calculated from the numerical output, one gets distributions of mean velocity and mean Reynolds stress in any azimuthal plane which are remarkably close to experimentally determined distributions (in which the individual realisations are far from axisymmetric). This ring vortex cloud method seems, then, to predict mean turbulent properties rather well. However, if the sound field is evaluated using the numerical output as input to Q in the form (2.5) again, one gets an overestimate by at least 30dB (factor of 10^3) in the non-dimensional intensity, as compared with measurements from carefully-controlled model jet rigs, and an incorrect distribution of the sound over frequency, especially at the subjectively important high frequencies where the agreement is even poorer.

All this means that *computational aeroacoustics* is a subject very much in its infancy. Many numerical calculations of unsteady low-speed fluid flows have been reported in the past few years, and a way of reliably calculating their associated sound fields is an important goal. There are several obvious possible lines to try:-

(i) compute the sound field directly, along with the flow field;

(ii) evaluate the sound field from some integral over a surface S which encloses all significant sources and within which the flow field can be computed;

(iii) use an acoustic analogy, according to which the unsteady flow acts as a volume source within S;

(iv) try analytical/numerical "matching" of the outer behaviour of the flow field to the inner behaviour of an acoustic wave field.

Of these, (i) looks hopeless, given the enormous disparity between near and far field levels, while at low speeds (ii), (iii) and (iv) all have similar subtle difficulties – though it is fair to say that these difficulties are really important only at low speeds. In some cases, however, "low-speed" implies that *fluctuation* velocities must be small compared with the sound speed, and this is always satisfied in ordinary engineering flows, while in other cases "low-speed" implies that the mean velocity must be small compared with c_o, and many engineering flows are "high-speed" in this sense and less sensitive to subtleties of acoustic source structure.

To understand the computational problems for low speed flows, what is needed now is a programme of complementary analytical and numerical study of simple model problems (potential flow past simple oscillating bodies is a good example) where the difficulties are fully understood analytically. The acoustics of realistic complex turbulent flows (involving, for example, vortex shedding and shear layer separation) cannot be attacked with current understanding and numerical capability.

These remarks may be unduly pessimistic, because great progress has in fact been made in the last ten years in the theory of "vortex sound" – sound generated by simple flows with concentrations of vorticity. This progress has led directly to integral expressions for the density fluctuations, the idea of an intervening acoustic analogy being somewhat by-passed. Moreover, the expressions are linear in the vorticity alone. For example, the sound of vorticity in free space is given by

$$\rho(\underline{x},t) = \frac{\rho_o}{12\pi c_o^4|\underline{x}|^3}\frac{\partial^3}{\partial t^3}\int(\underline{x}\cdot\underline{y})[\underline{y}\cdot\underline{\omega}\times\underline{x}]d^3\underline{y}\ , \tag{2.6}$$

where the observer listens at $\underline{x}$ at time t and the integral runs over all the vorticity elements $\underline{\omega}(\underline{y},\tau)d^3\underline{y}$ at the earlier times $t-|\underline{x}-\underline{y}|/c_o$. For the head-on collision of two vortex rings (the axisymmetric version of the vortex pair problem discussed earlier) this can be evaluated analytically, and the result compared very favourably with experiments (theory and experiment both due to T. Kambe, University of Tokyo, and his collaborators).

Equation (2.6) uses the free-space Green's function. Applications have also been made to the case where large or small solid bodies are present, and a simple vortex ring rapidly convected past the body, leading to a *scattering* of the hydrodynamic energy attached to the vortex into a radiating sound field. Figure 3 shows an experiment conducted by Kambe in which a vortex ring is blown past the sharp edge of a large flat plate, and Figure 4 shows the extraordinary agreement between the theoretical prediction for the magnitude and directional distribution of the sound and the experimental data. (The reader insufficiently impressed by the agreement should recall that all scales here are linear, in contrast to common acoustic practice which uses decibels (logarithmic) and where discrepancies of 3dB (a factor of 2) are generally thought acceptable (and difficult for the human ear to perceive) while discrepancies of even 20dB (factor of 100) are not always thought to invalidate a theory. Kambe's work has brought a new level of detail to aeroacoustics, making the decibel redundant, at least for the sound of simple vortical flows.)

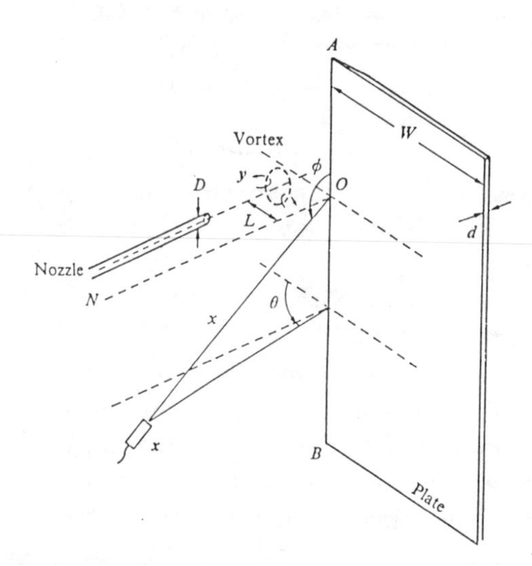

Figure 3. Experimental configuration for sound generation by convection of a vortex ring past the edge of a large flat plate.

While (2.6) is an expression which might one day be useful for jet exhaust noise studies, the situation of Figures 3 and 4 is of importance in a rather different context. In the approach of a large aircraft to landing, the engines operate at low power, and a significant noise heard on the ground is "airframe noise" – the noise made by turbulent flow interacting with wheels, landing gear, gear cavities, leading edge slats and trailing edge flaps on the wings, etc. The prediction represented by Figure 4, including the characteristic cardioid directivity pattern, is central to that part of all airframe noise prediction schemes which deals with the interaction of turbulence with the flap systems deployed from the wing trailing edges.

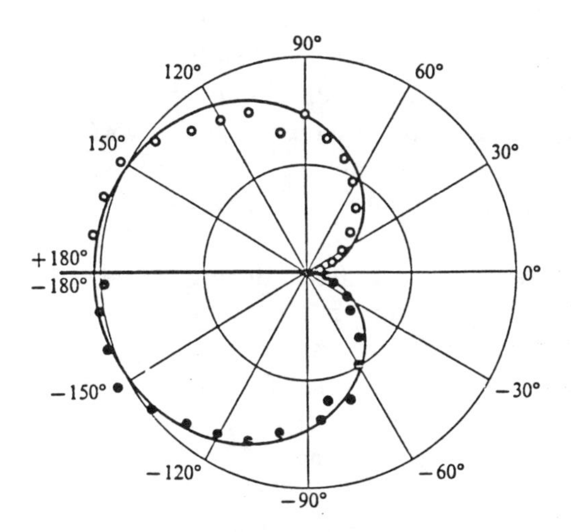

Figure 4. Sound generation by convection of a vortex ring past the edge of a large flat plate. The plate occupies $\theta = \pm 180^o$. Open and closed circles – measured data; solid line – theoretical prediction. (Figures 3 and 4 are from T. Kambe, Journal of Fluid Mechanics 1986, 173, 643-666, with permission.)

Despite these impressive advances in vortex sound, most of the best uses of aeroacoustic theory have been largely qualitative, and in line with the comments of §1. One spectacular example will suffice to close this section. The speed of sound in pure water is about $c_o = 1500ms^{-1}$, that in air about $c_1 = 340ms^{-1}$. If there is a volume concentration β of air bubbles in water one might expect the sound speed c_m in the bubbly mixture to steadily and monotonically decrease from 1500 to 340 as β increases from 0 to 1. An engineering approach might indeed try to fit a linear decrease! The truth is much more interesting; c_m rapidly decreases to small values, $c_m \sim 100ms^{-1}$ for $\beta = 0.04$, and reaches a (theoretical) minimum of about $20ms^{-1}$ at $\beta \simeq 0.4$, thereafter rising to $340ms^{-1}$ as β increases to 1. Experiments generally bear out the theory quite well, especially for small β, and have certainly shown values of c_m well below $100ms^{-1}$ for concentrations of 10-20%. How this arises can easily be seen by thinking of c_m^2 as the ratio of a bulk modulus of elasticity E to a density ρ; even small values of β suffice to give the mixture the elasticity of air, $E = E_1$, but the density of water, $\rho = \rho_o$, so that c_m is not only less than c_o, but less than c_1 also.

Now suppose our region V of turbulence to be occupied by bubbly water, the region outside V to be filled with pure water (V might be the wake of a ship, for example). If we use the Lighthill acoustic analogy , T_{ij} is no longer dominated by $\rho_o v_i v_j$ (of order $\rho_o U^2$ say) but by the isotropic quadrupole $q_{ij} = (p - c_o^2\rho)\delta_{ij}$, where p is the pressure. Inside V density fluctuations ρ are related to pressure fluctuations p by $\rho = p/c_m^2$, so $q_{ij} = (1 - c_o^2/c_m^2)p\delta_{ij} \simeq -(c_o^2/c_m^2)p\delta_{ij}$ because $c_m \ll c_o$. Now p is of order $\rho_o U^2$, so that while in pure water we have $T_{ij} \sim \rho_o U^2$, in bubbly water we have $T_{ij} \sim \rho_o U^2(c_o/c_m)^2$. The intensity of the radiated sound is therefore increased by $(c_o/c_m)^4$, a factor 5×10^4 when β is as small as 0.04 – and moreover, the sound field in the case when V is occupied by bubbly water is isotropic, uniform in direction, whatever the directionality properties may have been when no bubbles were present. Much more is needed (and has been done) to put these conclusions on a firm basis – but they are typical of the dramatic and subtle effects present in aeroacoustics and capable of being revealed by the Lighthill aeroacoustic theory and its developments.

Before moving on to propfan acoustics, it is worthwhile to summarise the achievements of thirty years of jet noise research with some impressive figures (all figures given here are approximate, but accurate enough to make the essential points correctly).

First generation jet transports (e.g., Boeing 707 of 1955)
Passengers 120 Range 3000 miles
4 engines, 4×14000 lb thrust
Noise at take-off = shouting power of whole world's
population (10^{10} people)

Current jet transports (e.g., Boeing 767 or Airbus A310 of 1985)
Passengers 240 Range 5000 miles
2 engines, 2×56000 lb thrust
Noise at take-off = shouting power of population of
a large city (10^7 people)

Engine thrust has been enormously increased (factor 4), range has been virtually doubled, passenger load at least doubled – but radiated noise energy has been reduced to 1/1000 of its level 30 years ago (30dB reduction). And it must be emphasized again that propulsive or aerodynamic efficiency has nothing to do with acoustic efficiency. If all the acoustic energy radiated in the 45-second take-off of a large jet transport were recovered, that energy would be about enough to fry one egg!

While it would be absurd to claim that the 30dB reduction has been achieved as a direct result of aeroacoustic theory, it is entirely fair to say that theory has often contributed greatly, to a very difficult engineering problem, in the ways suggested in §1, and has guided thinking about the problem along lines which are not part of conventional engineering approaches. Airframe noise provides a particularly good example, where all prediction schemes are firmly anchored in theory, and where theory is constantly used to discriminate, in experimental data, between different mechanisms.

3. Propfan acoustics. Advances in aerodynamics and manufacturing technology have now made it possible to contemplate propulsion of passenger transports over short and medium range by turbine engines driving propellers. Virtually the same speeds can be achieved as at present with jet propulsion, but the fuel burn is expected to be reduced by around 25-30%. Proposals of this kind were met with smiles of scepticism in the early 1970's, but no-one is laughing now, the prototype engine having been flown in 1986 and 1987 on Boeing 727 and MD-81 aircraft and introduction of a 150-passenger aircraft into service being likely in 1993. The actual design generally favoured involves a pusher-propeller mounted at the rear of the aircraft, with two counter-rotating blade rows, the second recovering the energy in the swirling flow created by the first.

"Propfans" of this kind deliver far greater power than any propellers of older design, and may have significant associated problems of community noise (heard by people on the ground) and cabin noise. The scientific problems associated with these two are rather different, and the biggest initial constraint is the legal one involved in meeting certification requirements for community noise. For the prediction of this we have (with some qualifications) essentially a problem of classical acoustics; the boundaries of the fluid region are propeller blades whose shape and location are known, and if the distribution over those blades of force between blades and fluid is known, then the acoustic field can be calculated from a classical integral formula. For single-rotation-propellers (SRP) with only one blade row, the force distribution is determined by the performance requirements, and one can assume that each blade is subjected to a known steady (in a rotating frame) force distribution specified by the aerodynamicist to the acoustician. For counter-rotating propellers (CRP) there is again a steady force distribution, but also an unsteady component as each row rotates through a nonuniform flow generated by the other; for example, the second row chops through the wakes of the first. The unsteady component is not given in advance – and indeed is not of much interest to the aerodynamicist; it must be regarded as the task of the acoustician first to calculate the unsteady aerodynamic interactions and then to use them to calculate the interaction noise as well as the steady force noise (and noise from the volume displacement effects of the blades).

The essentials of the force ("loading") noise are as follows. For an SRP with B uniformly spaced blades on a shaft rotating with angular velocity Ω, the classical expression for the acoustic field of a distributed force can be manipulated into the form

$$p \propto \frac{1}{R} \sum_{m=-\infty}^{\infty} \exp[imB\Omega(t - R/c_o)/(1 - M_x \cos\theta)] \times$$

$$\int_o^{\ell} dx \int_{z_h}^{z_o} dz \exp[i\phi(x,z)] J_{mB}(mBM_t z \sin\theta/(1 - M_x \cos\theta)) F(x,z) \,. \qquad (3.1)$$

Here the integrals run over the surface of one blade, that over x from the leading to trailing edge along the chord, that from z_h to z_o from the hub to the tip along the span, $F(x,z)$ being a force component (lift or drag) per unit area. The pressure fluctuations p decrease as the inverse of distance R between propeller shaft and observer, and comprise harmonics of the "blade passing frequency" $B\Omega$ only. The Bessel function is a "radiation efficiency factor" for a section of the blade at span z and for an observer at angle θ from the forward axis of the propeller, and $\exp[i\phi(x,z)]$ is a phase factor which describes interference effects between different parts of a given highly twisted and swept (scimitar-shaped) blade. Other parameters appearing in (3.1) are the axial Mach number M_x (forward speed of the propeller hub divided by sound speed c_o) and the tip rotational Mach number M_t.

Direct numerical evaluation of (3.1) is of course possible, and gives really very good agreement with experiment. However, (3.1) does not suggest any trends, or possibilities for control, and numerical evaluation for a range of configurations, for the number of harmonics m and the number of observer locations θ needed for a noise prediction, is very inefficient. Recent work I have been doing with A.B. Parry of Rolls-Royce seeks to complement accurate numerical studies by obtaining (for SRP and CRP) asymptotic expressions valid uniformly in all the parameters involved, and capable of not only providing insight into trends and mechanisms but also of providing quantitative estimates which are quite good enough for parametric studies and preliminary design. These asymptotics relate to the "many-bladed propeller", and the limit $B \to \infty$. For earlier propellers B was 2, 3 or 4 and $B \gg 1$ apparently seemed then inappropriate, although subsequent experience with asymptotics suggests that $B \gg 1$ will certainly cover $B = 4$ well, and probably $B = 2$ also. At any rate, current designs have $B = 8$, 10 or 12 and sketches for larger engines in the future have $B \sim 50$, and $B \gg 1$ must be an exploitable limit.

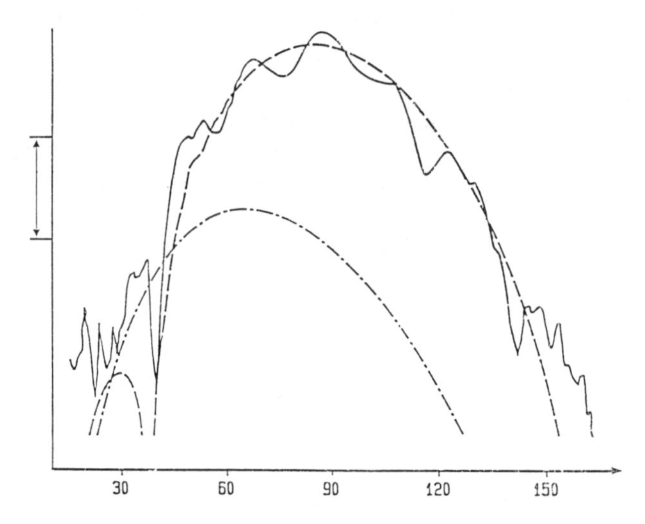

Figure 5. Sound pressure, as function of angle θ, of first harmonic ($m = 1$) of a 4-bladed propeller; — measurements on Gannet aircraft; - - - - predicted steady loading noise (numerical evaluation of (3.1)); -.-.-.-. predicted thickness noise. The marked segment of the ordinate corresponds to 10dB. Theoretical predictions by A.B. Parry (Rolls-Royce) and D.G. Crighton.

A most important result follows at once from $B \to \infty$. If the blade tip speed is less than the sound speed, the normalized tip radius $z_o < 1$ in (3.1), the argument of the Bessel function is less than the order for all θ and all z, and the Bessel function decreases exponentially as z decreases from z_o. Therefore (leaving aside the possible effects of the phase term ϕ) the integral is dominated by contributions from the tip – regardless of how the force is distributed elsewhere. This implies that SRP noise at subsonic tip speeds can be controlled by careful attention to the force distribution near the blade tips, which can be achieved without any loss of the total thrust delivered by the propeller. Similar remarks apply to the "thickness noise" generated by the displacement effect of the propeller. Detailed asymptotic calculations can be made, giving simple formulae, accurate to within 1 or 2dB when compared with numerical evaluation of (3.1), which yield the dependence of the amplitude and phase of each of the Fourier harmonics in (3.1) on harmonic number m, blade number B, observer location θ, the rotational and axial velocities of the propeller and the force and thickness variations near the tip. Figure 5 gives an example of the close agreement between the numerical evaluation of (3.1) and experiments conducted by Rolls-Royce on a Gannet aircraft; here $m = 1$, $B = 4$ and θ takes values from 30° to 150° (the harmonic amplitude varying by more than a factor of 10^3 over a 60° change in θ).

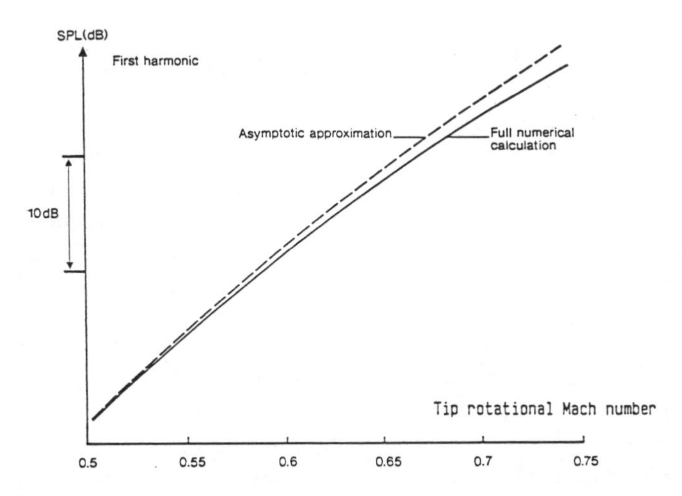

Figure 6. Comparison between asymptotic and numerical evaluations of (3.1) for the first harmonic ($m = 1$) of a 4-bladed propeller (A.B. Parry and D.G. Crighton).

The asymptotic prediction is indistinguishable from the numerical for these moderate speed conditions, as is clear from Figure 6 , and has the advantage of showing explicitly the functional dependence and the vital property of emphasising the importance of the tip region only.

For supersonic tip speeds, the maximum argument of the Bessel function exceeds the order , and for some angles θ the spanwise integration in (3.1) will run through the "sonic radius" z^*, defined by $z^* M_t \sin\theta/(1 - M_x \cos\theta) = 1$, where the Bessel function peaks and beyond which it oscillates rapidly. The sonic radius corresponds to that point along the span at which the blade element approaches the observer at exactly sonic speed, and the dominant contribution to (3.1) now comes from the vicinity of $z = z^*$, contributions from $z < z^*$ being suppressed exponentially, those from $z > z^*$ being suppressed by the rapid oscillations of the Bessel function. For given rotational and axial propeller speeds, the emphasis is now on control of the force and thickness at z^* for angles θ for which z^* exists, and at the tip where no z^* exists; for some conditions z^* may actually be close to the hub. Asymptotic expressions can be derived which again display all the parametric dependences explicitly, point out the essential mechanisms, and agree with full numerical computation to within one or two decibels.

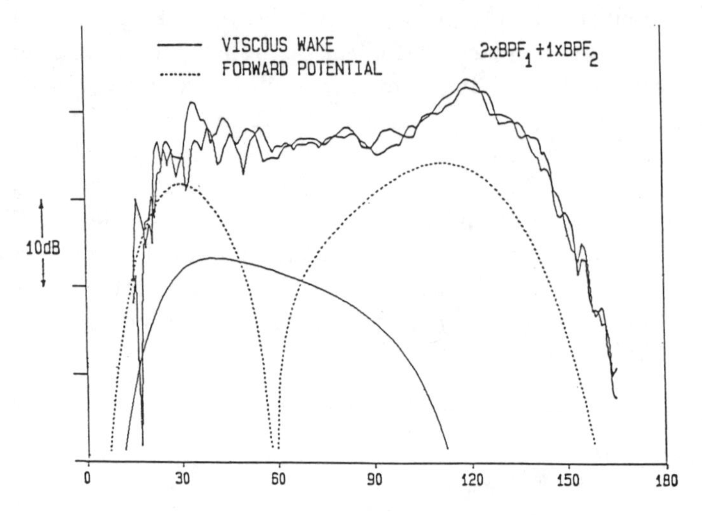

Figure 7. Comparison, with measured data from Gannet aircraft, of analytical prediction for the $m = 2$, $n = 1$ interaction harmonic of a 4×4 bladed CRP. The figure shows the fields generated by the cutting of the first row wakes by the second row blades, and by the interaction of the first row with the lifting potential field of the second. Other interactions, and compressibility effects, remain to be added. (Figures 5, 6 and 7 represent collaborative work between the author and Rolls-Royce.)

In the case of a counter-rotating propeller the situation is much more complex. The expression replacing (3.1) involves a double sum over integers m, n, representing the interaction between each Fourier component of the excitation generated by one row with each Fourier component of the response of the other. The excitation may take the form of a velocity nonuniformity of either a viscous kind (the wakes shed by the leading row) or a potential kind (the lifting and displacement fields associated with either row), and the interaction problem has to be solved to provide the necessary force distribution on each blade. Then an integral like that in (3.1) has to be performed, but with more complicated argument and order of the Bessel function. However, similar approximations can be employed ($B_1, B_2 \to \infty$), based upon uniform asymptotics for the Bessel function $J_\nu(\nu\varsigma)$ where $\nu \to \infty$ and ς is unrestricted. It is possible to complete the prediction scheme for CRP noise in purely analytical terms by using consistent asymptotic approaches for the solution of the aerodynamic interaction problems. We (the author, and Mr A.B. Parry of Rolls-Royce) have modelled the excitation and response aspects appropriately - using simple representations of the wakes, and concentrated horseshoe vortices for the lifting fields, and using semi-infinite flat plate models for the blade responses. The correctness of this turns on the idea that the generation of a particular interaction component of sound in the CRP always comes from an effective sonic radius, and this localisation of the sound source legitimises calculations which would globally (e.g. for calculation of the net unsteady force on one blade row) be quite inappropriate. Yet again, considerable advantages can be claimed for the asymptotic prediction scheme – especially in cases where up to 10,000 interactions may have to be calculated, following which numerical evaluation of an integral with many self-cancelling features has to be carried out for each geometrical and dynamic configuration and for each observer position. The asymptotic scheme is explicit, and based upon modifications of the Wiener-Hopf method plus asymptotics for the interactions, and on saddle point methods, with multiple nonuniformities, for the radiation integrals. Figure 7 gives a comparison with full-scale measurement of the analytically predicted sound pressure, as function of angle from the propeller axis, for the (2,1) interaction harmonic of a 4×4 bladed CRP. The generally good agreement with experiment of these incompressible flow calculations has been further improved by incorporation of other potential field interactions and of compressibility effects.

4. Concluding remarks. Aeronautical (and underwater) acoustics continues to offer prime opportunities for applied mathematicians. Understanding of the subtle near-cancellation effects which guarantee inefficient sound generation even by high speed flows and boundaries is all-important, and the accurate modelling of these effects must take precedence over accurate modelling of the flows in an overall sense. To put it another way, the sound-generating elements of a function $f(\underline{x},t)$ at any frequency ω are determined by the Fourier transform

$$\tilde{f}(\underline{k},\omega) = \int f(\underline{x},t)\exp(i\underline{k}\cdot\underline{x} + i\omega t)d^3\underline{x}dt$$

evaluated on the acoustic sphere $|\underline{k}| = \omega/c_o$. If f is a property of an unsteady flow with typical velocity U, then an accurate calculation of f means that $\tilde{f}(\underline{k},\omega)$ is well-determined in the spectral range $|\underline{k}| \sim \omega/U$, a region far from the acoustic sphere if $U/c_o \ll 1$. Nevertheless, developments of the Lighthill acoustic analogy have proved adequate to reveal the general features and typical magnitude of the acoustic fields radiated by particular types of flow, and very recently have proved adequate to predict all details of the sound from relatively simple vortex flows.

Propfan acoustics leads to many new opportunities for analytical modelling of mechanisms and the use of asymptotic methods for solution and efficient evaluation. Many refinements, potentially important for particular parameter ranges, remain to be carried out, but the ground has been laid for a prediction scheme for advanced CRP's which is soundly based in fluid mechanics and acoustics, which conveys the insight needed for understanding and for initial design, and which also is capable of rather accurate outright numerical prediction.

REFERENCES

M.J. Lighthill 1952 *On sound generated aerodynamically I. General theory*, Proc. R. Soc. Lond. A211, 564-587

J.E. Ffowcs Williams 1969 *Hydrodynamic noise*, Ann. Rev. Fluid Mech. 1, 197-222

D.G. Crighton 1975 *Basic principles of aerodynamic noise generation*, Prog. Aerosp. Sci. 16, 31-96

J.E. Ffowcs Williams 1977 *Aeroacoustics*, Ann. Rev. Fluid Mech. 9, 447-468

A.P. Dowling & J.E. Ffowcs Williams 1983 *Sound and sources of sound* , Ellis Horwood

M.E. Goldstein 1984 *Aeroacoustics of turbulent shear flows*, Ann. Rev. Fluid Mech. 16, 263-285

What Is a Multivariate Spline?

CARL DE BOOR*

Abstract. The various concepts and ideas that have contributed to univariate spline theory are considered with a view to finding a suitable definition of a multivariate spline. In this way, an overview of the existing more or less complete univariate spline theory is given along with a survey of some of the high points of the current research in multivariate splines.

My very first paper dealt with multivariate (well, bivariate) splines and I was then quite certain of what a multivariate spline, i.e., a spline function of many variables, might be. Now, many years and several answers later, I am not so sure any more and therefore consider the question worth a forty-minute talk.

It is a worthwhile question since univariate splines have been phenomenally successful and one would wish to have available a similarly useful tool for the approximation of functions of *several* variables. This raises the question of just which features of the univariate spline to generalize. My talk will therefore be in part a survey of the more or less complete univariate spline theory with the aim of deciding which parts to take along into the multivariate context.

But before embarking on that discussion, I want to point out that there is available one way of generalization that is specifically designed to require no thought, no new idea (if this construction is satisfactory for you, I have nothing further to tell you). This is the **tensor product** construct. Here one takes one's favorite univariate spline class $\$$ and fashion from it splines

$$\mathbb{R}^d \longrightarrow \mathbb{R} : (x, y, \ldots, z) \longmapsto f(x)g(y)\cdots h(z)$$

*CMS, University of Wisconsin, 610 Walnut St., Madison WI 53705.
This work was sponsored by the U.S. Army under Contract DAAG29-80-C-0041.

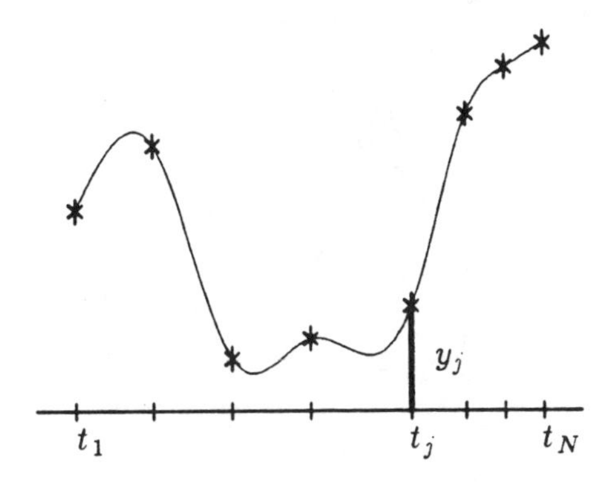

Figure 1. The 'natural' cubic spline interpolant. Are the two bottom bumps natural?

in the d variables $x, y, \ldots, z$ by taking (univariate) functions $f, g, \ldots, h$ in these variables from $ and multiplying them. One would take linear combinations of such functions, and the resulting approximation schemes are simply products of univariate schemes. This means that one can even use the univariate computer programs, and the resulting schemes are so efficient that it pays to force one's particular approximation problem into this form if one can do it. It does require that the data come in tensor product form, i.e., on a rectangular grid, and that raises questions. Is the proper multivariate version of an interval a (hyper)rectangle?Also, just how is one to deal with *scattered* data? This made me and others look for other ways of making up multivariate splines.

There are essentially two avenues to splines, the variational and the constructive. Although I have had the mathematical pleasure of writing papers using the variational approach, I am firmly in the constructive camp and so want to begin by doing a job on the variational approach.

The story is familiar since it is available wherever splines are sold, so I can be brief. If I am to fit data points $(t_j, y_j), j = 1, \ldots, N$, I ought to use the "natural" cubic spline interpolant, that is, the function which among all functions fitting the data has the smallest second derivative. This is a good thing, so the story goes, because in this way I am doing more or less what draftsmen have been doing even when they were still draughtsmen. More or less, because they would put a 'spline', i.e., a thin flexible rod, through the data, and this rod (if ideal) would take on the shape of that curve γ through the points which minimizes strain energy, i.e., the integral with respect to arclength of the squared curvature. Assuming now that curve γ to be a function, i.e., $\gamma = \{(t, f(t)) : a \leq t \leq b\}$, the integral being minimized can also be written

$$\int_\gamma \kappa^2 = \int_a^b \frac{(D^2 f)^2}{(1 + (Df)^2)^{3/2}},$$

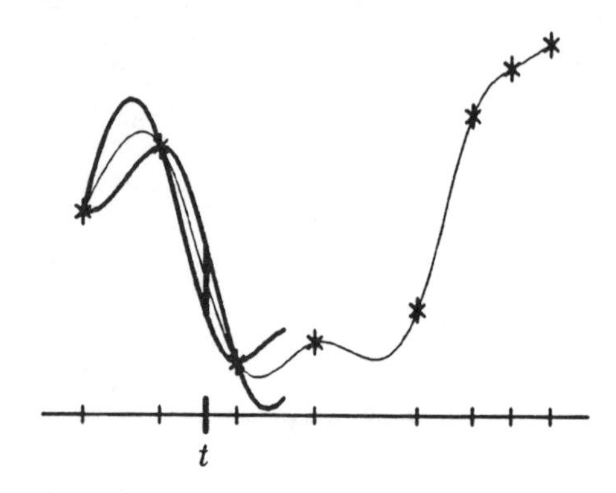

Figure 2. Part of the envelope in whose center the 'natural' cubic spline interpolant happens to lie.

and, for small Df, this is much like the integral

$$\int_a^b (D^2 f)^2$$

which is being minimized by the 'natural' cubic spline interpolant to the data.

You will discern several false notes in this story. For small Df, there is usually no call for any subtlety at all, a straight line or parabola will fit nicely. In any case, going from a curve to a function is a bit fishy. In fact, if we really believe in the draftsman's spline, then we should reject the cubic spline and compute the draftsman's spline instead. Of course, we will then run into some difficulties. For example, this minimization problem doesn't have a solution without further conditions, such as a bound on the length of the curve. Even with such a condition in place, the draftsman's spline (or *elastica*) is not easy to compute. This leads me to the conclusion that people use cubic splines, not because cubic splines provide them an automatic French curve, but because cubic splines are easy to compute.

There is a more serious variational approach to splines which these days goes under the name of *Optimal Recovery*. Here one starts with the worthwhile observation that, if we know nothing but the data points (t_j, y_j), then we can say nothing about the function between the data points. We need additional information. Suitable information could be a bound on some derivative. For example, to stay with our simple picture, we might also know that the L_2-norm of the second derivative $D^2 f$ is no bigger than some constant c. Then, for each t, the possible values of f at t form an interval, and we obtain in this way an envelope within which our function f must lie. Of course, this envelope depends on c. But, it so happens that, for each t, the midpoint of that interval lies, you guessed it, on our friend the 'natural' cubic spline interpolant, and this is so regardless of c. Thus, the cubic spline interpolant is rather central.

Yet I am not impressed, since all this depends on the decision to give a bound in terms of the L_2-norm and that decision seems arbitrary to me. Had we used, more reasonably to me, a bound on the maximum norm of $D^2 f$, we would again have found an envelope, but now the midpoint changes with c. It does converge, as $c \to \infty$, but not to the cubic spline interpolant, but to the broken line interpolant! Is that sufficient reason to reject the cubic spline in favor of the broken line?

In any case, if you look for the reason why splines occur as solutions to such extremal problems, you will find that it is so because they represent point evaluation with respect to bilinear forms involving some derivative, or, equivalently, they are sections of Green's functions (for D^4 or D^2 or whatever). In a multivariate variational approach, we would expect, correspondingly, to have sections of Green's functions of *partial* differential operators turn up. Such Green's functions are strongly domain dependent, i.e., the resulting multivariate 'splines' change in local detail as the domain of the minimization changes. This made me give up on this approach early on. It has recently been given a strong impetus by Duchon [D76] (see, e.g., [Me79]) who in effect declared that there is only one domain of interest, namely all of $\mathbb{R}^d$, and so created the **thin plate** splines which, for $d = 2$, are used in many places. They provide that interpolant f to given data points $(t_j, y_j), j = 1, \ldots, N$, which minimizes

$$\int_{\mathbb{R}^d} \sum_{i,j=1}^{d} (D_i D_j f)^2,$$

hence the name. But the resulting space of interpolants fails to have a *local* basis, hence the construction of the thin plate spline interpolant takes $O(N^3)$ effort, which is to be compared to the $O(N)$ effort required for the (univariate) spline interpolant.

I hasten to add to this diatribe that I am all for the variational approach in case the smoothness measure being minimized has some *a priori* justification. For example, in planning the path of the arm of a painting robot, one wants the acceleration to be as small as possible, hences its minimization subject to the constraints imposed by the painting job makes very good sense. As another example, we might eventually understand in a mathematical sense just what we mean by a 'good' shape, and it would then be very desirable to look for interpolants of best possible shape. But, given the computational history of the *elastica* or the thin plate spline, we are not likely to compute such a 'best' or 'shapeliest' interpolant exactly. Rather, we are likely to follow the example set by D. Terzopoulos and others and compute such 'splines' only approximately, by minimizing over a suitably flexible, fine-meshed space of piecewise polynomial functions with a local basis.

This brings me to the *constructive* approach to splines. In this approach, a spline is, most simply, a pp (:= piecewise polynomial) function of degree $\le r$ with breakpoint sequence $t = (t_j)$; in symbols:

$$\$ = \pi_{r,t},$$

or, perhaps,

$$\$ = \pi_{r,t}^{\rho} := \pi_{r,t} \cap C^{\rho}.$$

Correspondingly, a d-variate spline would be any element of

$$\pi_{r,\Delta}^{\rho} := \pi_{r,\Delta} \cap C^{\rho},$$

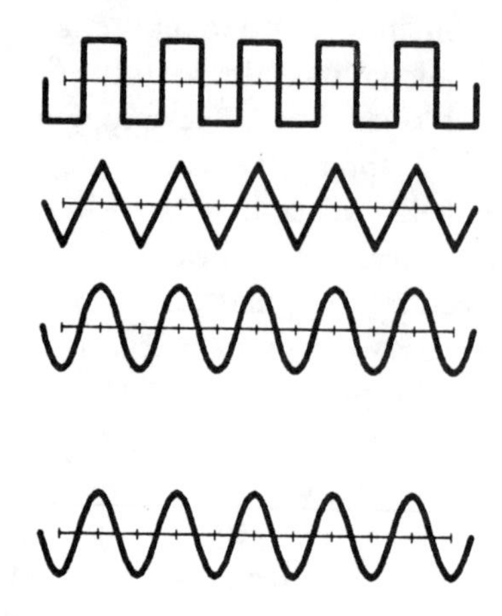

Figure 3. The first three Euler splines and their limit as their degree goes to infinity.

with $\pi_{r,\Delta}$ the collection of all functions which are pp of degree $\leq r$ with respect to some partition Δ. If this satisfies you, let me try to convince you that there is more to splines than that.

Already in the very early papers on splines ([E28], [QC38], [S46]), there is much more structure than that. These early papers are concerned with what we now call **cardinal** splines

$$\$ = \pi^{k-2}_{k-1,\mathbb{Z}},$$

i.e., smooth piecewise polynomials with uniformly spaced breakpoints, for example at the integers. Although cardinal spline theory did not quite develop this way, you will find that you can understand cardinal splines most simply if you think of them as smoothed-out step functions, i.e., as obtained from step functions by repeated convolution with the characteristic function

$$M_1 := \chi_{[0,1]}$$

of the unit interval.

For example, that most beautiful of cardinal splines, the **Euler** spline, is obtained in this way. Starting with the (shifted) cardinal step function which is alternately ± 1, a first averaging brings the alternating broken line, while a second averaging (followed by a shift and multiplication by 2) gives the alternating parabolic cardinal spline which is already hard to distinguish (see Figure 3) from the function reached after infinitely many such steps, viz. the cosine. Schoenberg [S73] called this spline function 'Euler spline' since it is made up of Euler polynomials. But it had been put to good use long before that baptism. It had appeared as the solution of various variational problems. For example, it provides [F37] Favard's best constant in the bound on the distance of a function from trigonometric polynomials in terms of that function's k-th derivative. It also occurs [K62] as the

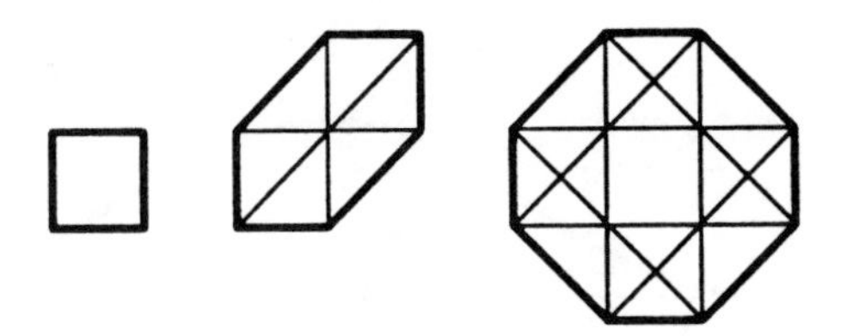

Figure 4. The support of the *bivariate* cardinal B-splines M_1, M_2, M_3.

simultaneous extremizer of the Landau-Kolmogorov inequalities in which the j-th derivative is bounded on $\mathbb{R}$ in terms of the zeroth and the kth.

Schoenberg's fundamental paper [S46] also introduced what became eventually the centerpiece of univariate spline theory, viz. the **B-spline**

$$M_k := \underbrace{M_1 * \cdots * M_1}_{k \text{ times}}.$$

Its integer translates provide a most suitable basis for the cardinal spline space of order k (i.e., of degree $k-1$).

This structure is easily generalized to d variables, as I will now illustrate for $d = 2$. We get bivariate cardinal B-splines by starting with the characteristic function of the unit square,

$$M_1 := \chi_{[0,1]^2}.$$

From it, by averaging, e.g., in the direction $(1,1)$, we obtain

$$M_2(x) := \int_{-1}^{0} M_1(x+vt)dt, \qquad v := (1,1),$$

the familiar piecewise linear pyramid function already used by Courant [C43]. By following up with another averaging, this time in the direction $(1,-1)$, we obtain the C^1-quadratic finite element

$$M_3(x) := \int_{-1}^{0} M_2(x+wt)dt \qquad w := (1,-1)$$

of Zwart and Powell (e.g., [PS77]).

The space

$$\$:= \{ \sum_{j\in\mathbb{Z}^d} M(\cdot - j)a(j) : a : \mathbb{Z}^d \mapsto \mathbb{R}\}$$

spanned by integer translates of such functions M is rightly thought to be a multivariate cardinal spline space, and there is a complete theory of its approximation power and use available, as developed by mathematicians working in Finite Elements around 1970 and given final form by Strang & Fix [SF73].

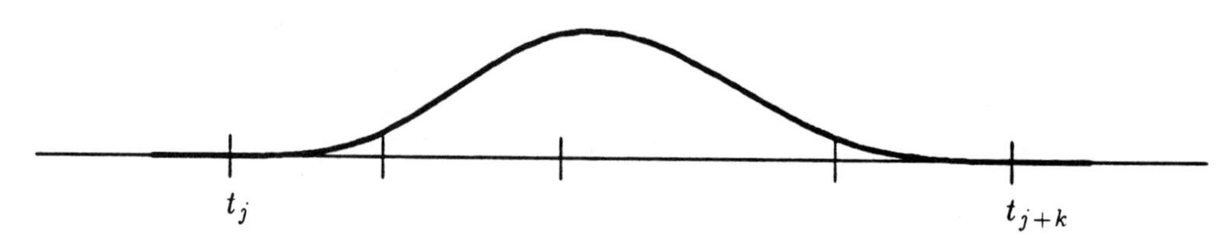

Figure 5. The B-spline $M(x|t_j,\ldots,t_{j+k})$.

If these splines of *uniform* structure do it for you, here is yet another point to quit listening (except that there will be more of this later on). But if you have to deal with scattered data or other nonuniform problems, you know that you need more than cardinal splines.

It was Schoenberg's colleague, the logician H. B. Curry, who pointed out in a review of Schoenberg's '46 paper that, with the aid of divided differences, such B-splines could be constructed for an arbitrary spacing of breakpoints as follows

$$N_j(x) := ((t_{j+k}-t_j)/k)M(x|t_j,\ldots,t_{j+k}) := (t_{j+k}-t_j)[t_j,\ldots,t_{j+k}](\cdot - x)_+^{k-1},$$

and that these points t_i, now called **knots**, could even be repeated to control precisely the smoothness across the knot. In this way, one obtains [CS66] a convenient basis for any space of piecewise polynomials of degree $< k$ and of specified smoothness across breakpoints.

The list of useful properties of the univariate B-spline is quite impressive. Here are some of the items on that list (cf., e.g., [B76] or [Sch81] for details and references).

- N_j depends continuously on its knots $t_j,\ldots,t_{j+k}$.
- N_j has **minimal support**, is **nonnegative**, and $\sum N_j = 1$, i.e., (N_j) provides a good and local **partition of unity**.
- (N_j) provides a **stable basis**, i.e, $d_k^{-1}\|a\|_\infty \le \|\sum N_j a_j\|_\infty \le \|a\|_\infty$ for all coefficient sequences a and some knot-independent (positive) constant d_k.
- **Good quasi-interpolants** are available, of the form $f \sim Qf := \sum N_j\lambda_j f$, with λ_j locally supported, uniformly bounded linear functionals.
- These quasi-interpolants provide **optimal approximation order,** i.e., $\|f - Qf\| \le \text{const}\, |t|^k \|D^k f\|$ (with $|t| := \sup \Delta t_j$).
- **Shape preserving** approximation schemes are available in the simple form $Vf := \sum N_j f(\tau_j)$.

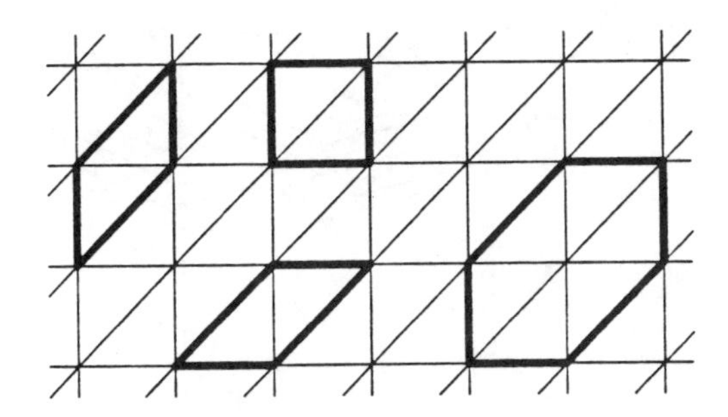

Figure 6. The minimally supported elements in $\pi^0_{2,\Delta}$ (on the left) have to be augmented by elements of far-from-minimal support (such as on the right) in order to obtain a basis for $\pi^0_{2,\Delta}$.

- Determination of the B-spline coefficients of a spline approximation leads to **banded systems**.
- Evaluation of the B-splines can be accomplished by a **stable recurrence**.
- ...

In fact, this list is so impressive that I have come to the conclusion that, in the univariate context, splines are, *by definition*, linear combinations of B-splines.

Once this is accepted, it is obvious what a multivariate spline is; it is a linear combination of multivariate B-splines. All that is now required is the construction of multivariate B-splines. This turned out to be a nontrivial task.

A generalization via divided differences turned out to be difficult since the divided difference $[t_j, \ldots, t_{j+k}]f$ is customarily defined as the leading coefficient of the polynomial of degree $\leq k$ which agrees with f at $t_j, \ldots, t_{j+k}$ and this definition becomes doubtful in the multivariate context because interpolating polynomials are only defined for certain pointsets and, even if defined, have several 'leading' coefficients.

While it is possible to develop the univariate B-splines entirely from their recurrence relation, there was no obvious way to extend these to a multivariate context. In fact, when multivariate B-splines were ultimately defined, it took two years of intense effort to find stable recurrence relations for them.

A very tempting approach was via the minimal support property. In this approach, one defines multivariate B-splines to be those functions in a given class of smooth piecewise polynomials whose support is as small as possible. Unfortunately, already very simple examples, such as C^0-parabolics on the 'three-direction mesh' (see Figure 6), show that the resulting functions may not be plentiful enough to staff a basis. (There is an alternative definition of minimal support in terms of the Bernstein-Bézier net for these pp's, but that idea has never been fully explored.)

The approach finally used in [B76] relied on yet another B-spline property, already found in [CS66] where it is shown that

$$M(y|t_j, \ldots, t_{j+k}) = \mathrm{vol}_{k-1}\sigma \cap P^{-1}y,$$

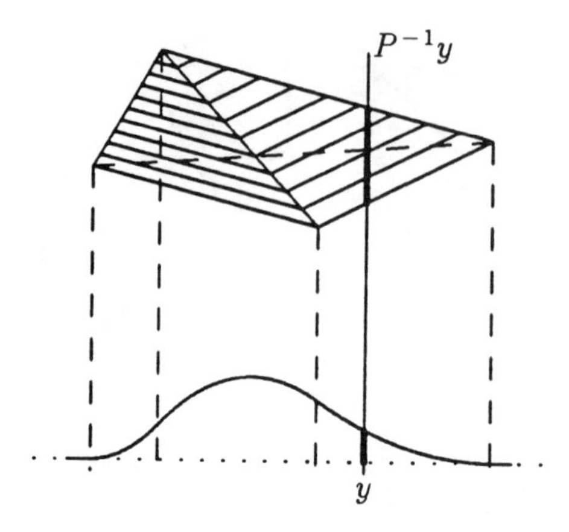

Figure 7. Curry-Schoenberg construction of the (univariate) B-spline as a simplex spline.

with P the canonical projector $P : x \mapsto x(1)$ on $\mathbb{R}^k$ to $\mathbb{R}$ and σ any appropriately scaled simplex $[v_0, \ldots, v_k]$ in $\mathbb{R}^k$ oriented in such a way that $Pv_i = t_{j+i}$, all i. This construction changes only in minor detail when P is taken to be the canonical projector onto $\mathbb{R}^d$ and then provides what is now called the d-variate **simplex** spline

$$M(y|\sigma) := \mathrm{vol}_{k-d}\, \sigma \cap P^{-1}y$$

(in order to distinguish it from other multivariate B-splines; see below).

It is obvious that $M(\cdot|\sigma)$ is a compactly supported nonnegative function. It is not hard to see that $M(\cdot|\sigma) \in \pi_{k-d,\Delta}$, with the partition Δ generated by the $(d-1)$-dimensional images under P of faces of σ. With somewhat more effort, one can establish that $M(\cdot|\sigma)$ is as smooth as possible , i.e., that $M(\cdot|\sigma) \in C^{k-d-1}$ in case σ is in general position. It is also easy to construct enough simplex splines to provide a (local) partition of unity: If the (essentially disjoint) simplices σ are so chosen that $\bigcup \sigma = \mathbb{R}^d \times G$, then $\sum_\sigma M(y|\sigma) = \mathrm{vol}_{k-d} G$ is constant.

But it took some time before Micchelli [M78] came up with stable recurrence relations for these simplex splines. For their proof, Micchelli described the simplex spline equivalently as the distribution which carries the smooth test function φ to the number $\int_\sigma \varphi(Px)dx$. This formulation made it easy to prove [BH82] similar results for the more general multivariate B-spline $M(\cdot|B, P)$ which is defined as the distribution on $\mathbb{R}^d$ which carries the test function φ to the number $\int_B \varphi(Px)dx$, with B, more generally, a (convex) polytope and P, more generally, some linear map on $\mathbb{R}^k$ to $\mathbb{R}^d$.

The most recent summary of material about multivariate B-splines is [H86]; see also [Ch88]. These results show that several of these multivariate B-splines have almost all the properties we listed earlier for the univariate B-spline, i.e., all the properties that we can expect them to have. (For example, we cannot hope for a 'shape-preserving' map to parallel Schoenberg's map V since there is as yet no satisfactory multivariate definition of 'shape preservation'.) Going down our list

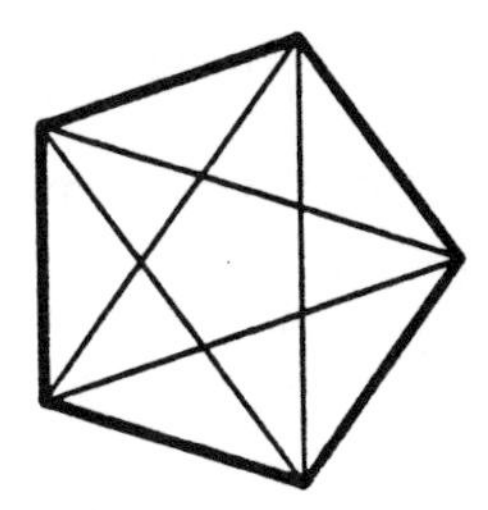

Figure 8. The partition for a simplex spline cannot be made to fit an arbitrary partition.

of good properties, we find stable recurrence relations and good quasi-interpolants. We also find much beautiful mathematics (see, e.g, Dahmen and Micchelli [DM84]), particularly when we use bodies other than simplices, and use projectors other than orthogonal projectors. For example, we obtain the so-called **box splines** [BH83] when we use the unit cube as the body. Such a box spline is, offhand, the distribution defined by

$$\int_{\mathbb{R}^d} M(x|V)\varphi(x)dx := \int_{[0,1]^V} \varphi(\sum_{v\in V} vt_v)dt$$

for some sequence V in $\mathbb{R}^d$, hence can be obtained recursively by

$$M(x|V) := \int_{-1}^{0} M(x+vt|V\backslash v)dt,$$

with $M(\cdot|V)$ the characteristic function of the convex hull of $0 \cup V$ in case $\#V = d$. This shows the multivariate cardinal B-splines introduced earlier to be box splines.

But the initial enthusiasm for these multivariate B-splines has somewhat abated for the simple reason that they are not entitled to the prefix 'B': they fail to be basic. Since they are obtained as shadows of polyhedra (or polytopes), their partition or mesh depends on the structure of those polyhedra. E.g., the 2-dimensional shadow of a simplex has any line connecting any two of the projected vertices as meshlines. This makes it in general impossible to suit multivariate B-splines to a given partition. Even if we restrict attention to partitions generated by such shadows, the collection of all pp functions of the appropriate degree and smoothness is usually larger than the span of all these B-splines. This puts into question the ultimate usefulness of these multivariate B-splines for **practical** work, except, perhaps, for the box splines (if a regular partition is satisfactory).

But it also puts into question that naive definition of a spline as a pp function of some degree and some smoothness on some partition. For this class can often be shown not to have a locally supported basis. I.e., even if the class contains locally supported elements, it also contains functions which cannot be represented by them. On the other hand, these non-local elements are usually not useful for approximation, i.e., it can often be shown that

$$\mathrm{dist}(f, \$) \sim \mathrm{dist}(f, \$_{\mathrm{loc}}),$$

with

$$\$_{\mathrm{loc}} := \mathrm{span}\{s \in \$: \mathrm{supp}\, s \text{ compact}\}.$$

If this leaves you a bit wondering what multivariate splines might be, I am pleased. For I don't know myself. I am coming to the realization, though, that it will be necessary to separate the various roles the univariate spline plays simultaneously. My guess is that, if there is ultimately a satisfactory definition of a multivariate spline as a tool for approximation, it will capture the best features of the univariate spline, i.e., it will refer to classes of functions probably pp of controllable smoothness which are spanned by a stable, locally supported basis which is not too hard to handle in computations. Such classes will also be used to provide suitable and entirely satisfactory approximations to 'splines' in the variational sense.

References

[B76] C. DE BOOR, *Splines as linear combinations of B-splines*, in *Approximation Theory II*, G. G. Lorentz, C. K. Chui and L. L. Schumaker, eds., Academic Press, (1976), 1–47.

[BH82] C. DE BOOR AND K. HÖLLIG, *Recurrence relations for multivariate B-splines*, Proc. Amer. Math. Soc. **85** (1982), 397–400.

[BH83] C. DE BOOR AND K. HÖLLIG, *B-splines from parallelepipeds*, J. d'Anal. Math. **42** (1982/83), 99–115.

[Ch88] C. CHUI, *Multivariate Splines. Theory and Applications*, CMBS-NSF Lectures, SIAM, Philadelphia PA 1988.

[C43] R. COURANT, *Variational methods for the solution of problems of equilibrium and vibrations*, Bull. Amer. Math. Soc. **49** (1943), 1-23.

[CS66] H. B. CURRY AND I. J. SCHOENBERG, *Pólya frequency functions.IV, The fundamental spline functions and their limits*, J. d'Anal. Math. **17** (1966), 71-107.

[DM84] W. DAHMEN AND C.A. MICCHELLI, *Some results on box splines*, Bull. Amer. Math. Soc. **11** (1984), 147–150.

[D76] J. DUCHON, *Interpolation des functions de deux variables suivant le principe de la flexion des plaques minces*, R.A.I.R.O. Analyse Numérique **10** (1976), 5-12.

[E28] A. EAGLE, *On the relation between Fourier constants of a periodic function and the coefficients determined by harmonic analysis*, Phil.Mag. **5** (1928), 113–132.

[F37] J. FAVARD, *Sur les meilleurs procédés d'approximation de certaines classes de fonctions par des polynômes trigonométriques*, Bull. Sci. Math. Sér. 2, **61** (1937), 209–224, 243–256.

[H86] K. HÖLLIG, *Multivariate splines*, in *Approximation Theory*, C. de Boor ed., Amer. Mathem. Soc., Providence RI 1986.

[K62] A. KOLMOGOROV, *On inequalities between upper bounds of the successive derivatives of an arbitrary function on an infinite interval*, Amer. Math. Soc. Translations, Ser.1,2 (1962), 233-243.

[Me79] J. MEINGUET, *Multivariate interpolation at arbitrary points made simple*, J. Appl. Math. Phys. (ZAMP) **30** (1979), 292-304.

[M78] C. A. MICCHELLI, *A constructive approach to Kergin interpolation in* $\mathbf{R}^k$*: Multivariate B-splines and Lagrange interpolation*, Rocky Mountain J. Math. **10** (1980), 485–497.

[PS77] M. J. POWELL AND M. A. SABIN, *Piecewise quadratic approximations on triangles*, ACM Trans. on Mathematical Software **3** (1977), 316–325.

[QC38] W. QUADE & L. COLLATZ, *Zur Interpolationstheorie der reellen periodischen Funktionen*, Akad.Wiss., Math.-Phys. Klasse **30** (1938), 383-429.

[S46] I. J. SCHOENBERG, *Contributions to the problem of approximation of equidistant data by analytic functions, Parts A & B*, Quarterly Appl.Math. **IV** (1946), 45-99, 112-141.

[S73] I. J. SCHOENBERG, *Cardinal Spline Interpolation*, SIAM, Philadelphia 1973.

[Sch81] L. L. SCHUMAKER, *Spline Functions: Basic Theory*, Wiley, New York 1981.

[SF73] G. STRANG AND G. FIX, *A Fourier analysis of the finite element variational method*, C.I.M.E., II Ciclo 1971, in *Constructive Aspects of Functional Analysis*, G. Geymonat, ed., (1973), 793–840.

On a Duality Relation in the Theory of Orthogonal Polynomials and its Application in Signal Processing

YVES GENIN*

Abstract. The aim of this paper is to put into light a simple algebraic duality existing between all families of real orthogonal polynomials on the unit circle and all families of odd/even orthogonal polynomials on the interval $(-1,+1)$. It turns out that these two types of orthogonal families can be mapped into each other by elementary transformations. This result is discussed in the context of the recurrence relations underlying the split Levinson algorithm, recently proposed in the digital signal processing literature to speed up the numerical solution of a Toeplitz system of linear equations. As an illustration of this duality, the family of orthogonal polynomials on the unit circle corresponding to the ultra-spherical polynomials is derived and shown to satisfy a remarkably simple recurrence formula.

1. Introduction. Let us consider a family of real monic polynomials $[P_k(x) = \sum_{i=0}^{k} P_{k,i}\, x^i;\ P_{k,k} = 1,\ k = 0,1,2,\cdots]$ *orthogonal on the interval (-1,+1) of the real axis*, with respect to a well-defined measure $d\mu(x)$ [1], [2]. With the notation $< P, Q >_\mu$ for the scalar product

$$< P, Q >_\mu = \int_{-1}^{+1} P(x)\, Q(x) d\mu(x) \tag{1}$$

defined for any pair of real polynomials $P(x)$, $Q(x)$, the members of the polynomial family $[P_k(x)]$ thus satisfy the orthogonal relations

$$< P_k, P_\ell >_\mu = \gamma_k\, \delta_{k,\ell} \tag{2}$$

where the γ_k are appropriate positive constants and with $\delta_{k,\ell}$ the Kronecker delta. Let us also introduce the $(k+1)\times(k+1)$ positive definite *Hankel matrix* $H_k = [s_{i+j};\ 0 \le i,j \le k]$ whose entries are the successive power moments relative to the measure $d\mu(x)$:

$$s_k = \int_{-1}^{+1} x^k\, d\mu(x), \quad k = 0,1,\cdots \tag{3}$$

*Philips Res. Lab. Brussels, 2, avenue Van Becelaere, Box 8, B-1170 Brussels, Belgium.

The coefficients $P_{k,i}$ of polynomial $P_k(x)$ are easily verified to be obtainable as the solution of the linear system of equations

$$[P_{k,0}, P_{k,1}, \cdots, P_{k,k}]H_k = [0, 0, \cdots, 0, \gamma_k] \tag{4}$$

The fast computation of the family of orthogonal polynomials $[P_k(x)]$ is classically performed by applying the recursive *Lanczos-Phillips algorithm* [3], [4] to the Hankel system (4).

Let us similarly consider a family of monic polynomials $[a_k(z) = \sum_{i=0}^{k} a_{k,i}\, z^i;\ a_{k,i}\, z^i;\ a_{k,k} = 1,\ k = 0, 1, \cdots]$ *orthogonal on the unit circle* $|z| = 1$ with respect to a well-defined measure $d\omega(\theta)$ [1], [2]. If

$$< p, q >_\omega = \int_0^{2\pi} p(e^{i\theta})\, \overline{q}(e^{i\theta}) d\omega(\theta) \tag{5}$$

stands for the scalar product of any pair of complex polynomials $p(z)$, $q(z)$, one has thus by assumption the orthogonality relations

$$< a_k, a_\ell >_\omega = \sigma_k\, \delta_{k,\ell} \tag{6}$$

for appropriate positive constants σ_k. The trigonometric moments $c_k = \overline{c}_{-k}$ relative to the measure $d\omega(\theta)$ are defined by

$$c_k = \int_0^{2\pi} e^{-ik\theta}\, d\omega(\theta), \quad k = \cdots, -1, 0, 1, \cdots \tag{7}$$

It turns out that the coefficients $a_{k,i}$ of polynomial $a_k(z)$ can be obtained by solving the linear system of equations

$$[a_{k,0}, a_{k,1}, \cdots, a_{k,k}]T_k = [0, 0, \cdots, 0, \sigma_k] \tag{8}$$

where $T_k = [c_{i-j};\ 0 \le i, j \le k]$ is the positive definite Hermitian *Toeplitz matrix* built from these trigonometric moments. The *Levinson algorithm* [3], [5] is the standard fast algorithm to achieve the solution of system (8), whence to construct the family of orthogonal polynomials $[a_k(z)]$.

Obviously, the theory of orthogonal polynomials on the interval (-1,+1) on the one hand and on the unit circle on the other hand can be approached within the same conceptual framework: their different algebraic properties simply reflect the difference in the definition of their respective scalar products (1) and (5). In view of this, it is a natural question to ask whether this conceptual analogy can be strengthened and possibly turned into a full algebraic equivalence with the help of some appropriate transformations. It turns out that this question can be answered by the affirmative under an additional symmetry hypothesis.

As a matter of fact, the interest in this problem is not new in the literature. For example, the following result is wellknown [1]. Let $\mathring{\mu}(x)$ be an arbitrary weight function on the interval (-1,+1) and $[P_k(x);\ k = 0, 1, \cdots]$, $[Q_k(x);\ k = 0, 1, \cdots]$ be the two families of orthogonal polynomials respectively associated with the measures $\mathring{\mu}(x)\, dx$ and $(1 - x^2)\, \mathring{\mu}(x)\, dx$. Let us furthermore introduce the family of *real* orthogonal polynomials on the unit circle $[a_k(z)\ ;\ k = 0, 1, \cdots]$ relative to the derived measure $|sin\, \theta|\ \mathring{\mu}(cos\, \theta) d\theta$. It can be proved that the pseudo-polynomials $z^{-k}\, a_{2k}(z)$ and $z^{1-k}\, a_{2k-1}(z)$ can be expressed as linear combinations of $P_k(x)$ and $Q_k(x)$ via the variable transformation $x = (z + z^{-1})/2$. These ideas have been recently applied by Cybenko in a digital processing context to solve Pisarenko's problem in an efficient manner [6].

As interesting as this result may be, it turns out that there exists a more direct relation for the subfamilies of *real* orthogonal polynomials on the unit circle; they can be put into a one-to-one correspondence with a well-defined subset of the families of orthogonal polynomials on the interval (-1,+1). Let us observe that a family of real orthogonal polynomials on the unit circle necessarily implies a symmetry relation of the underlying measure $d\omega(\theta)$: the trigonometric moments c_k must be all real; thus, one must have the symmetry relation $d\omega(\theta) = -d\omega(2\pi - \theta)$ in view of (7). As a result, it appears that a one-to-one mapping between the two sets of orthogonal polynomial families can only be possible if the corresponding measure $d\mu(x)$ on the interval (-1,+1) exhibits a similar symmetry property, i.e. $d\mu(x) = -d\mu(-x)$. It turns out that this symmetry condition is also sufficient to work out the algebraic equivalence in question. Note that under this assumption, the moments s_i of odd subscript are all zero due to (3); moreover, the numbers of independent entries of the Hankel matrix H_k and of the Toeplitz matrix T_k are readily verified to be identical.

Before proceeding further, let us briefly comment on the origin of the present new approach to the orthogonal polynomials equivalence problem. As a matter of fact, it has to be found in the applied mathematics literature, mainly pertaining to digital signal processing applications. Counting the zeros of a polynomial inside the unit disk $|z| < 1$ is a standard problem in that context; the Schur-Cohn criterion has been for decades the numerical procedure commonly used for that purpose [9]. A new approach to the same problem was proposed by Bistritz in 1983 [7], [8]; the resulting Bistritz criterion relies on polynomial recurrence relations of an unusual form, which turn out to be simpler and, at first glance, quite different from those underlying the Schur-Cohn test. They can be shown however to be intimately related, though in a rather disguised manner. It turns out that a recurrence relation of the same form constitutes the basis of a new algorithm to solve a Toeplitz system of linear equations [10]; the resulting split Levinson algorithm exhibits a reduced complexity with respect to the Levinson algorithm, the standard numerical tool in DSP applications to solve the problem above. The precise relationship between the Levinson and the split Levinson algorithm yields directly the one-to-one mapping between the families of orthogonal polynomials discussed in the present contribution.

In section 2, the Phillips version [4] of the Lanczos algorithm [3] to orthogonalize a Krylov space of vectors is first recalled to yield an efficient algorithm for the computation of a family of orthogonal polynomials on the real line; this algorithm actually consists of an implementation of the recurrence relations existing between three successive polynomials of such a family. The constraints imposed on the parameters of these recurrence relations when the measure underlying the family of orthogonal polynomials has a support restricted to the interval (-1,+1) are then discussed. Finally, the results above are specialized to the situation where this measure exhibits the required symmetry property, which makes the polynomials to enjoy the parity property, i.e. $P_k(x) = (-1)^k P_k(-x)$.

In section 3, the Levinson algorithm commonly used in DSP applications in the context of least squares estimation problems [5] is alternatively recalled to produce a fast numerical procedure for the actual calculation of a family of orthogonal polynomials on the unit circle. The special case where all the polynomials of the family are real is then considered and shown to imply a precise symmetry relation of the underlying measure.

In section 4, the split Levinson algorithm [10] recently proposed in the literature as an efficient substitute for the Levinson algorithm is first derived in the real case. It appears to implement polynomial recurrence relations of a non classicalform and to process appropriately transformed polynomials. The split Levinson algorithm involves the computation of a new set of parameters. The constraints on these parameters are then put into light so that conversely they yield a well-defined family of real orthogonal polynomials on the unit circle.

In section 5, the split Levinson algorithm (in the real case) is shown via elementary algebraic transformations to take the form of the Lanczos-Phillips algorithm particularized to the case of orthogonal polynomials on the interval (-1,+1) having the parity property. As the transformations above are invertible, this induces a one-to-one correspondence between all families of real orthogonal polynomials on the unit circle on the one hand and all families of odd/even orthogonal polynomials on the interval (-1,+1) on the other hand. This duality principle is illustrated by an example: the counterparts of the ultra-spherical polynomials, orthogonal on the unit circle, are derived and shown to satisfy a remarkably simple recurrence relation.

In the conclusion, it is finally pointed out that this algebraic correspondence does not extend beyond the restrictive assumption of symmetric measures; thus, it does not apply to *complex* orthogonal polynomials on the unit circle and to *general* orthogonal polynomials on the interval (-1,+1).

2. Orthogonal polynomials on the real line and Lanczos-Phillips algorithm. Any family $[P_k(x);\, k = 0, 1, \cdots]$ of monic orthogonal polynomials on the real line is well known to satisfy a three-term recurrence relation of the form [1], [2]

$$P_{k+1}(x) - (x + B_k)\, P_k(x) + C_k\, P_{k-1}(x) = 0 \tag{9}$$

for appropriate numbers B_k, C_k with $C_k > 0$ for all k.

The Lanczos algorithm [3] is a standard numerical procedure to orthogonalyze a Krylov space of vectors $(b, Ab, A^2b, \cdots)$ [11]. As shown by Phillips [4], the same algorithm can be used for the fast computation of polynomials $P_k(x)$; it actually corresponds to the special case where the operator A generating the Krylov space above is self-adjoint. Let us briefly recall how it can be derived in the later context.

Let us set the nested Hankel matrices

$$H_k = \begin{bmatrix} s_0 & s_1 & \cdot & s_k \\ s_1 & s_2 & \cdot & s_{k+1} \\ \cdot & \cdot & \cdot & \cdot \\ s_k & s_{k+1} & \cdot & s_{2k} \end{bmatrix}, \quad k = 0, 1, \cdots \tag{10}$$

built on the successive power moments (3) relative to the measure $d\mu(x)$ underlying the given family of orthogonal polynomials. One then easily verifies that the coefficients of three successive polynomials $P_{k-1}(x)$, $P_k(x)$, $P_{k+1}(x)$ satisfy the linear systems of equations

$$H_{k+1} \begin{bmatrix} P_{k+1,0} & P_{k,0} & 0 & P_{k-1,0} \\ P_{k+1,1} & P_{k,1} & P_{k,0} & P_{k-1,1} \\ \cdot & \cdot & \cdot & \cdot \\ \cdot & \cdot & \cdot & \cdot \\ P_{k+1,k-1} & P_{k,k-1} & P_{k,k-2} & 1 \\ P_{k+1,k} & 1 & P_{k,k-1} & 0 \\ 1 & 0 & 1 & 0 \end{bmatrix} = \begin{bmatrix} 0 & 0 & 0 & 0 \\ \cdot & \cdot & \cdot & \cdot \\ \cdot & \cdot & \cdot & \cdot \\ \cdot & \cdot & 0 & 0 \\ \cdot & 0 & \mu_k & \mu_{k-1} \\ 0 & \mu_k & \nu_k & \nu_{k-1} \\ \mu_{k+1} & \nu_k & \xi_k & \xi_{k-1} \end{bmatrix} \tag{11}$$

where the μ_k, ν_k, ξ_k are appropriate real numbers. Note, in particular, the equalities

$$\mu_k = \int_{-\infty}^{+\infty} P_k^2(x)\, du(x) > 0, \ \nu_k = \int_{-\infty}^{+\infty} P_{k-1}(x)\, P_k(x)\, d\mu(x) . \tag{12}$$

Moreover, a straightforward comparison of (9) with (11) yields the relations

$$C_k = \mu_k\, \mu_{k-1}^{-1}, \ \ B_k = \nu_{k-1}\, \mu_{k-1}^{-1} - \nu_k\, \mu_k^{-1} . \tag{13}$$

The Lanczos-Phillips algorithm to compute the orthogonal polynomials $P_k(x)$ then follows directly from (10), (11) and (13). It takes the form:

Lanczos – Phillips algorithm (general case)

a. set the initializations

$$P_0(x) = 1, \quad P_{-1}(x) = 0, \quad \mu_{-1}^{-1} = 0$$

b. compute for $k = 0, 1, \cdots$ (14)

$$\mu_k = \textstyle\sum_{i=0}^{k} P_{k,i}\, s_{k+i}, \ C_k = \mu_k\, \mu_{k-1}^{-1}$$

$$\nu_k = \textstyle\sum_{i=0}^{k} P_{k,i}\, s_{k+i+1}, \ B_k = -\nu_k\, \mu_{k-1}^{-1} - P_{k,k-1}$$

$$P_{k+1}(x) = (x + B_k)\, P_k(x) - C_k\, P_{k-1}(x)$$

Let us incidentally mention that a Cholesky factorization of H_k^{-1} is immediately available as a side result of the algorithm [3]. Indeed, one has $H_k^{-1} = U_k \Delta_k^{-1} U_k^T$ where the upper-triangular matrix U_k is made of the coefficient vectors of the successive polynomials $P_0(x), P_1(x), \cdots, P_k(x)$ while $\Delta_k = [\text{diag}\ \mu_i;\ i = 0, 1, \cdots, k]$.

In the framework of the equivalence problem discussed in this paper, we shall restrict our attention to these particular measures $d\mu(x)$ which are supported by the interval (-1,+1) on the one hand and which exhibit the symmetry property $d\mu(x) = -d\mu(-x)$ on the other hand. It turns out that the first condition is equivalent to the constraint that the zeros of the polynomials $P_k(x)$ lie on the interval (-1,+1) [1]. This polynomial zeros location property is easily verified to hold true if and only if one has the inequalities

$$P_k(1) > 0\,, \quad (-1)^k P_k(-1) > 0\,, \quad \text{all } k\,. \tag{15}$$

A straightforward calculation based on the recurrence relations (9) shows that these inequalities can alternatively be expressed in terms of the continued fraction expansions

$$\frac{P_{k+1}(1)}{P_k(1)} = 1 + B_k - \frac{C_k}{\lfloor 1}\!\rfloor + B_{k-1} - \cdots - \frac{C_1}{\lfloor 1}\!\rfloor + B_0 > 0\,, \tag{16}$$

$$-\frac{P_{k+1}(1)}{P_k(1)} = 1 - B_k - \frac{C_k}{\lfloor 1}\!\rfloor - B_{k-1} - \cdots - \frac{C_1}{\lfloor 1}\!\rfloor - B_0 > 0\,.$$

Let us now investigate the consequences of the symmetry relation $d\mu(x) = -d\mu(-x)$. As in this case the power moments of odd subscript are necessarily all zero, this clearly induces the identities

$$B_k = 0, \quad P_k(x) = (-1)^k P_k(-x), \quad \text{all } k\,. \tag{17}$$

As a result, the Lanczos-Phillips algorithm to compute the corresponding orthogonal polynomials $P_k(x)$ reduces to the simpler form:

Lanczos – Phillips algorithm $(d\mu(x) = -d\mu(-x))$

a. set the initializations

$$P_0(x) = 1, \quad P_{-1}(x) = 0, \quad \mu_{-1}^{-1} = 0$$

(18)

b. compute for $k = 0, 1, \cdots$

$$\mu_k = \sum_{i=0}^{\lfloor k/2 \rfloor} s_{2(k-i)}\, P_{k,k-2i}, \; C_k = \mu_k\, \mu_{k-1}^{-1}$$

$$P_{k+1}(x) = x\, P_k(x) - C_k\, P_{k-1}(x)$$

Furthermore, it appears in the present case that both expressions (16) coincide in view of (17) and simplify into the elementary continued fraction [9]

$$\frac{P_{k+1}(1)}{P_k(1)} = 1 - \frac{C_k|}{|1} - \frac{C_{k-1}|}{|1} - \cdots - \frac{C_1|}{|1} > 0 \tag{19}$$

3. Orthogonal polynomials on the unit circle and Levinson algorithm. A classical result of Szegö's theory of orthogonal polynomials on the unit circle [1] states that any such family of monic polynomials $[a_k(z);\ k = 0, 1, \cdots]$ satisfies a recurrence relation of the form

$$a_k(z) = z\, a_{k-1}(z) + \rho_k\, \hat{a}_{k-1}(z)\,, \tag{20}$$

where $\hat{a}_{k-1}(z)$ is the reciprocal of the polynomial $a_{k-1}(z)$, i.e. $\hat{a}_{k-1}(z) = z^{k-1}\, \overline{a}_{k-1}(1/\overline{z})$, while ρ_k is an appropriate complex number of modulus bounded to one ($|\rho_k| < 1$).

The Levinson algorithm [3] is a widely used algorithm in DSP applications in connection with a variety of least squares estimation and approximation problems [5]. In fact, it carries out the fast computation of the orthogonal polynomials $a_k(z)$ in a recursive manner. To see this, let us introduce the nested set of positive definite Hermitian Toeplitz matrices

$$T_k = \begin{bmatrix} c_0 & \overline{c}_1 & \cdot & \overline{c}_k \\ c_1 & c_0 & \cdot & \overline{c}_{k-1} \\ \cdot & \cdot & \cdot & \cdot \\ c_k & c_{k-1} & \cdot & c_0 \end{bmatrix}, \quad k = 0, 1, \cdots \tag{21}$$

associated with the trigonometric moments (7) relative to the measure $d\omega(\theta)$ underlying the considered family of orthogonal polynomials. The orthogonality relations (6) are easily verified to imply, in particular, the linear systems of equations

$$\begin{bmatrix} a_{k,0} & a_{k,1} & \cdot & a_{k,k-1} & 1 \\ 0 & a_{k-1,0} & \cdot & a_{k-1,k-2} & 1 \\ 1 & \overline{a}_{k-1,k-2} & \cdot & \overline{a}_{k-1,0} & 0 \end{bmatrix} T_k = \begin{bmatrix} 0 & \cdot\cdot\cdot & \sigma_k \\ -\rho_k\,\sigma_{k-1} & \cdot\cdot\cdot & \sigma_{k-1} \\ \sigma_{k-1} & \cdot\cdot\cdot & -\overline{\rho}_k\,\sigma_{k-1} \end{bmatrix} \quad (22)$$

for appropriate positive numbers σ_k and complex numbers ρ_k. Note in particular the resulting equalities

$$\sigma_k = \int_{-\pi}^{+\pi} |a_k(e^{i\theta})|^2 d\omega(\theta) > 0, \; -\rho_k\sigma_{k-1} = \int_{-\pi}^{+\pi} e^{i\theta}\,\overline{\hat{a}}_{k-1}(e^{i\theta}) a_{k-1}(e^{i\theta}) d\omega(\theta) \quad (23)$$

Furthermore, a comparison of (20) with (22) reveals that the parameters ρ_k are identical in both expressions, which thus yields the relations

$$-\rho_k\sigma_{k-1} = \sum_{i=0}^{k-1} c_{i+1} a_{k-1,i}, \; \sigma_k = \sigma_{k-1}(1 - |\rho_k|^2) = det\; T_k / det\; T_{k-1} \quad (24)$$

The Levinson algorithm to compute the orthogonal polynomials $a_k(z)$ consists of a recursive implementation of the relations involved in (20), (22) and (24). It takes the form

Levinson algorithm

a. set the initializations

$$a_0(x) = 1, \sigma_0 = c_0$$

b. compute for $k = 0, 1, \cdots$ (25)

$$-\rho_k\sigma_{k-1} = \textstyle\sum_{i=0}^{k-1} c_{i+1}\, a_{k-1,i}$$

$$\sigma_k = \sigma_{k-1}\,(1 - |\rho_k|^2)$$

$$a_k(z) = z\, a_{k-1}(z) + \rho_k\, \hat{a}_{k-1}(z)$$

The zeros of the orthogonal polynomials $a_k(z)$ are all located in the unit disk $|z| < 1$; this property can be shown [9] to be a direct consequence of the inequalities $|\rho_k| < 1$ resulting from (23) and (24). Let us also observe that, similarly to the Lanczos-Phillips algorithm, the Levinson algorithm provides, as a side result, a Cholesky factorization of T_k^{-1}; the relation $T_k^{-1} = V_k D_k^{-1} \tilde{V}_k$ is indeed easily established where V_k is the upper-triangular matrix made of the coefficient vectors of the successive polynomials $a_k(z)$ and with $D_k = [\text{diag}\; \sigma_i,\, i = 0, 1, \cdots, k]$.

In the remaining part of this paper, we will be exclusively concerned with the special case of *real* orthogonal polynomials on the unit circle. This contraint clearly implies the trigonometric moments $c_k = \overline{c}_{-k}$ to be all real, whence the symmetry relation $d\omega(\theta) = -d\omega(2\pi - \theta)$ of the underlying measure (7). In view of

(20) and (24), the parameters ρ_k are thus all real numbers satisfying the condition $-1 < \rho_k < 1$.

4. The split Levinson algorithm. The split Levinson algorithm, recently introduced in the literature [10], consists essentially of a parsimonious reformulation of the Levinson algorithm in terms of a new family of polynomials. The resulting algorithm requires roughly one-half the number of multiplications for the same cost in additions; moreover, the size of the vectors recursively processed by the algorithm appears to be reduced by a factor two.

Assuming the family of orthogonal polynomials $[a_k(z)\ ;\ k = 0, 1, \cdots]$ to be *real*, let us first note in view of (20) and its reciprocal version $\hat{a}_k(z) = \hat{a}_{k-1}(z) + \rho_k\, z\, a_{k-1}(z)$ that one has the identity

$$a_k(z) + \hat{a}_k(z) = (1 + \rho_k)[z\, a_{k-1}(z) + \hat{a}_{k-1}(z)] \tag{26}$$

Let us then substitute the family of polynomials $[p_k(z);\ k = 1, 2, \cdots]$ for the family $[a_k(z);\ k = 0, 1, \cdots]$, defined by

$$p_k(z) = z\, a_{k-1}(z) + \hat{a}_{k-1}(z) = (1 + \rho_k)^{-1}[a_k(z) + \hat{a}_k(z)]\,. \tag{27}$$

It is important to observe that the resulting polynomials are symmetric by definition, i.e.

$$\hat{p}_k(z) = p_k(z) \tag{28}$$

so that half the number of their coefficients need actually to be computed.

It turns out that the polynomial $a_k(z)$ and the parameter ρ_k can easily be recovered from $p_k(z)$ and $p_{k+1}(z)$. To see this, let us substract $(1 + \rho_k)p_k(z) = a_k(z) + \hat{a}_k(z)$ from $p_{k+1}(z) = z\, a_k(z) + \hat{a}_k(z)$ to obtain the relation $p_{k+1}(z) - (1 + \rho_k)p_k(z) = (z - 1)a_k(z)$. As a result, one deduces the required expressions from

$$1 + \rho_k = p_{k+1}(1)/p_k(1), \ a_k(z) = (z - 1)^{-1}[p_{k+1}(z) - (1 + \rho_k)p_k(z)]. \tag{29}$$

Let us now deal with the problem of translating the polynomial recurrence relations (20) in terms of the symmetric polynomials $p_k(z)$. To that aim, one first derived from (20) the identity $z\, a_k(z) + \hat{a}_k(z) = z^2 a_{k-1}(z) + \rho_k z[a_{k-1}(z) + \hat{a}_{k-1}(z)]$, which can be easily recast into $p_{k+1}(z) = (1 + z)p_k(z) - (1 + \rho_{k-1})z\, p_{k-1}(z) + (1 + \rho_{k-1})\rho_k z\, p_{k-1}(z)$ with the help of (27). Thus, the translated version of (20) assumes the form

$$p_{k+1}(z) - (1 + z)p_k(z) + \gamma_k\, z\, p_{k-1}(z) = 0 \tag{30}$$

provided the parameter γ_k is set equal to

$$\gamma_k = (1 + \rho_{k-1})(1 - \rho_k)\,. \tag{31}$$

To turn the recurrence relations (34) into an efficient algorithm for the computation of the family of symmetric polynomials $p_k(z)$, it remains to show how the new parameters γ_k can actually be determined. With $p_k(z) = \sum_{i=0}^{k} p_{k,i} z^i$, let us introduce the scalar product

$$\tau_k = \sum_{i=0}^{k} c_i\, p_{k,i}\,. \tag{32}$$

In view of the definition (27) of $p_k(z)$, it appears that τ_k can equivalently be expressed as $\tau_k = \sum_{i=0}^{k-1} [c_{i+1}a_{k-1,i} + c_i a_{k-1,k-i-1}]$ so that one deduces the equalities

$$\tau_k = \sigma_{k-1}(1-\rho_k) = \sigma_k(1+\rho_k)^{-1} \tag{33}$$

due to (22) and (24). Comparing (31) and (33) then yields the desired relation, i.e.

$$\gamma_k = \tau_k \tau_{k-1}^{-1} \tag{34}$$

The split Levinson algorithm [10] to compute the family of symmetric polynomials $p_k(z)$ follows directly from a recursive implementation of formulas (30), (32) and (34). For appropriate initializations, it takes the form

Split Levinson algorithm (real case)

a. set the initializations

$$p_0(z) = 1, \quad p_1(z) = (1+z), \quad \tau_0 = c_0/2$$

b. compute for $k = 0, 1, 2, \cdots$

$$\tau_k = \textstyle\sum_{i=0}^{k} c_i\, p_{k,i}, \quad \gamma_k = \tau_k \tau_{k-1}^{-1}$$

$$p_{k+1}(z) = (1+z)p_k(z) - \gamma_k\, z\, p_{k-1}(z)$$

(35)

Let us point out that the complexity reduction of the split Levinson algorithm with respect to the Levinson algoirthm is achieved from exploiting in a systematic way the symmetry relations $p_{k,i} = p_{k,k-i}$ implied by the definition (27) of the polynomials $p_k(z)$.

It is of obvious interest to address the issue of the constraints to be satisfied by the parameters γ_k so that the resulting polynomials $p_k(z)$ yield via (29) a set of real orthogonal polynomials on the unit circle. This property is clearly equivalent to impose the constraints $-1 < \rho_k < +1$ on the successive parameters ρ_k determined by (29). Noting that the recurrence relation (30) specializes for $z = 1$ into the identity

$$p_{k+1}(1)/p_k(1) = 2 - [p_k(1)/p_{k-1}(1)]^{-1}, \tag{36}$$

one deduces by induction via (23) that the constraints above will be satisfied if and only if one has

$$p_{k+1}(1)/p_k(1) > 1, \text{ all } k\,. \tag{37}$$

For further use, let us rewrite the necessary and sufficients conditions above in terms of a positivity condition imposed on all continued fractions

$$\frac{p_{k+1}(1)}{p_k(1)} = 1 - \frac{4^{-1}\gamma_k \rfloor}{\lceil 1} + B_{k-1} - \frac{4^{-1}\gamma_1 \rfloor}{\lceil 1} - \qquad \frac{4^{-1}\gamma_1 \rfloor}{\lceil 1} > 0\,, \tag{38}$$

which readily result from (37) via (29), (30) and (31).

Let us finally mention that there exist several versions of the split Levinson algorithm [10]. Furthermore, it can be generalized to accommodate the case of complex orthogonal polynomials on the unit circle [12], [13] as well as the case of their matrix extensions [14]. These interesting developments are however of no specific concern in the context of the algebraic equivalence between orthogonal polynomials on the unit circle and on the real interval (-1,+1) to be established in the next section of this paper.

5. Duality relation. It turns out that the recurrence relations underlying the split Levinson algorithm (35), (38) on the one hand and the special form of the Lanczos-Phillips algorithm (18), (19) on the other hand can be derived from each other by well-defined variable and polynomial transformations [10].

To see this, let us start from the polynomial recurrence relations $P_{k+1}(x) - x\,P_k(x) + C_k P_{k-1}(x) = 0$ valid for any family of odd/even orthogonal polynomials on the interval (-1,+1); thus, one has $P_k(x) = (-1)^k P_k(-x)$ for all k and the successive numbers $P_k(1)$ satisfy the constraints (19). Let us then introduce the transformations

$$x = (z^{1/2} + z^{-1/2})/2,\ p_k(z) = 2^k z^{k/2} p_k(x),\ \gamma_k = 4C_k\,. \tag{39}$$

It appears that the resulting functions $p_k(z)$ are well-defined polynomials obeying the split Levinson recurrence formulas $p_{k+1}(z) - (1+z)p_k(z) + \gamma_k z\,p_{k-1}(z) = 0$. Note that the initial polynomials $P_0(x) = 1$, $P_1(x) = x$ are adequately translated by (39) into the correct initializations $p_0(z) = 1$, $p_1(z) = 1 + z$. Furthermore, the support inequalities (19) are spontaneously transformed into the split Levinson algorithm inequalities (42). As a result, the family of polynomials $a_k(z)$ deduced from the $p_k(z)$ with the help of (29) is known to constitute a family of real orthogonal polynomials on the unit circle.

As the transformations (39) are clearly invertible, the algebraic equivalence just described induces a one-to-one mapping between all families $[a_k(z);\ k = 0,1,\cdots]$ of real orthogonal polynomials on the unit circle on the one hand and all families $[P_k(x) = (-1)^k P_k(-x);\ k = 0,1,\cdots]$ of odd/even orthogonal polynomials on the interval (-1,+1) on the other hand. Let us observe in particular that for $z = e^{i\theta}$, the change of variable (39) takes the form $x = \cos\theta/2$ and thus puts into one-to-one correspondence the interval $-1 < x \leq +1$ and the unit circle $z = e^{i\theta}$, $0 \leq \theta < 2\pi$.

It can be shown that the corresponding measures $d\omega(\theta)$ and $d\mu(x)$ are linked through the very simple relation

$$d\mu(x) = -d\omega(\theta) \tag{40}$$

while the entries in the corresponding Hankel and Toeplitz matrices are related by the identity

$$h_{2k} = 4^{-k} \sum_{i=0}^{2k} \binom{2k}{i} c_{i-k} \tag{41}$$

with the usual notation for binomial coefficients; the power moments thus appear to be well-defined linear combinations of the trigonometric moments and conversely [10].

Let us illustrate this duality relation by an example. Consider the parametric family of the ultra-spherical polynomials, classically defined [1] by the recurrence relations

$$P_{k+1}(x) - x\,P_k(x) + \frac{k(k+2\alpha-1)}{4(k+\alpha)(k+\alpha-1)} P_{k-1}(x) = 0 \tag{42}$$

initialized with $P_0(x) = 1$, $P_1(x) = x$. These polynomials are well known to constitute a family of odd/even orthogonal polynomials on the interval (-1,+1) for any choice of the real parameter α satisfying the condition $\alpha > -1/2$. The inverse of transformations (39) is easily verified to recast the recurrence formula (42) into the form

$$p_{k+1}(z) - (1+z)p_k(z) + \frac{k(k+2\alpha-1)}{(k+\alpha)(k+\alpha-1)} z\, p_{k-1}(z) = 0 \tag{43}$$

with the initializations $p_0(z) = 1$, $p_1(z) = (1+z)$. As the resulting parameter γ_k factorizes into

$$\gamma_k = \left(1 - \frac{\alpha}{k+\alpha}\right)\left(1 + \frac{\alpha}{k+\alpha-1}\right), \tag{44}$$

a comparison of (31) with (44) immediately reveals that the parameters ρ_k relative to the orthogonal polynomials on the unit circle associated with the ultra-spherical polynomials are given by

$$\rho_k = \alpha/(k+\alpha). \tag{45}$$

Thus, the ultra-spherical orthogonal polynomials on the unit circle appear to be defined by the simple recurrence formulas

$$a_k(z) = z\, a_{k-1}(z) + \frac{\alpha}{k+\alpha}\, \hat{a}_{k-1}(z), \quad k = 1, 2, \cdots \tag{46}$$

with $a_0(z) = 1$. Note incidentally that the constraint $\alpha > -1/2$ is precisely the necessary and sufficient condition such that the parameters ρ_k, as given by (45), satisfy the inequalities $-1 < \rho_k < +1$ for all $k \geq 1$.

The ultra-spherical orthogonal polynomials simplify into the Chebyshev orthogonal polynomials for the particular choice of the parameter $\alpha = 0$. In view of (45), the Chebyshev orthogonal polynomials on the unit circle thus simply consist of the successive powers of the variable z, i.e. $[z^k;\ k = 0, 1, \cdots]$

6. Conclusion. In this paper, the set of real orthogonal polynomials on the unit circle has been shown to be algebraically equivalent to the set of odd/even orthogonal polynomials on the interval (-1, +1). Thus, any result available in either one of these settings can be carried over to the other by elementary transformations.

It is obviously tempting to ask oneself whether this algebraic duality does not extend beyond the limit of the symmetry constraint imposed on the measures underlying the corresponding families of orthogonal polynomials. Actually, it does not seem to be the case. More precisely, the complex version of the split Levinson algorithm [12], [13] does not yield any generalized form of the algebraic equivalence discussed in this paper. In other words, there exists no known simple transformation which maps a family of *complex* orthogonal polynomials on the unit circle into a family of *general* orthogonal polynomials on the interval (-1,+1) and conversely.

Acknowledgement The material covered in this paper is part of some joint work with my colleague Philippe Delsarte at Philips Research Lab. Brussels.

References

[1] G. SZEGO, *Orthogonal Polynomials*, AMS Colloquium Publications, Vol. XXIII, New-York, 1959.

[2] N.I. AKHIEZER, *The Classical Moment Problem*, Oliver & Boyd, London, 1965.

[3] G. GOLUB, C. VAN LOAN, *Matrix Computations*, North Oxford Academic, Oxford, 1983.

[4] J.L. Phillips, *The Triangular Decomposition of Hankel Matrices*, Math. Comp., Vol. 25, pp. 509-602, 1971.

[5] N. LEVINSON, *The Wiener rms (root-mean-square) Error Criterion in Filter Design and Prediction*, J. Mathematical Physics, Vol. 25, pp. 261-278, 1947.

[6] G. CYBENKO, *Computing Pisarenko Frequency Estimates*, 1984, Proc. of the 1984 Conf. on Inf. Syst. and Sciences, Princeton Univ., March 14-16, 1984.

[7] Y. Bistritz, *A New Unit Circle Stability Criterion*, Proc. 1983 Intern. Symp. Mathem. Th. Networks & Systems, Beer-Sheva, Israël, pp. 69-87, 1983.

[8] Y. BISTRITZ, *Zero Location with Respect to the Unit Circle of Discrete-Time Linear System Polynomials*, IEEE Proc., vol. 72, pp. 1131-1142, 1984.

[9] P. HENRICI, *Applied and Computational Complex Analysis*, J. Wiley & Sons, New-York, 1974.

[10] P. DELSARTE, Y. GENIN, *The split Levinson algorithm*, IEEE Trans. on Acoust., Speech, Signal Processing, Vol. ASSP-34, pp. 470-478, 1986.

[11] G. CYBENKO, *Restrictions of Normal Operators, Padé Approximation and Autoregressive Time Series*, SIAM Journal on Math. Analysis, Vol. 15, pp. 753-767, 1984.

[12] P. DELSARTE, Y. GENIN, *The Tridiagonal Approach to Szegö Orthogonal Polynomials, Toeplitz Linear Systems and Related Interpolation Problems*, to appear in SIAM Journal on Math. Analysis.

[13] B. KRISHNA, S. MORGERA, H. KRISHNA, *Generalized Two-Term Recurrences and Fast Algorithms for Hermitian Toeplitz Matrices*, Proc. of the 1987 Int. Conf. on Acoustics, Speech and Signal Processing, Vol. 3, pp. 1839-1842, 1987.

[14] P. DELSARTE, Y. GENIN, *Multichannel Singular Predictor Polynomials*, Philips Res. Lab. Brussels, M-172, 1986.

A New Approach to Robust Multi-grid Solvers

W. HACKBUSCH*

Abstract. For a lot of problems special very efficient multi-grid methods can be constructed. On the other hand, one needs multi-grid software, which works for a class of problems which is as large as possible. Such a method would be called a robust method. Here, a new multi-grid approach with the following properties is presented. It is robust; it is applicable not only to elliptic problems; simple smoothers are sufficient; it allows parallel computations; further, it works for all dimensions. The new ingredient is a multiple coarse-grid correction with different coarse-grid problems.

1. THE GENERAL MULTI-GRID CONCEPT. Let

$$Lu = f \tag{1.1}$$

be a partial differential equation including the boundary conditions. Discretising Eq. (1.1) by a finite difference, finite volume or finite element method, we obtain a system

$$L_h u_h = f_h \tag{1.2}$$

of linear equations, where h denotes the discretisation parameter (mesh size etc.).

All traditional iterative methods try to solve the algebraic Eq. (1.2) without using the fact that the discrete solution u_h approaches the continuous solution u :

$$u_h \approx u \ . \tag{1.3}$$

It is a general property of all classical iterative methods, that in particular smooth errors e_h of the iterate $\tilde{u} = u_h + e_h$ are reduced very slowly. For the standard five-point discretisation of the Poisson equation

$$-\Delta u = f \ \text{ in } \Omega \, , \quad u = 0 \ \text{ on } \Gamma = \partial\Omega \, , \tag{1.4}$$

* Institut für Informatik und Praktische Mathematik, Christian-Albrechts-Universität zu Kiel, Olshausenstr. 40, D-2300 Kiel 1, Germany (FR).

the corresponding convergence speed is $1 - 0\,(h^2)$ for the Jacobi and Gauss-Seidel iteration. Hence, the smaller the step size the less efficient do these methods work. This statement is also true for accelerated methods as SOR, or optimised semi-iterative procedures, or preconditioned conjugate gradient methods.

When we apply the Gauss-Seidel iteration (or a damped Jacobi method) to the discretised Poisson problem (1.4), it can easily be observed that the iteration error $e_h^i := u_h^i - u_h$ (u_h^i : i-th iterate) or their difference $e_h^{i+1} - e_h^i = u_h^{i+1} - u_h^i$ decreases in the beginning until e_h^i is relatively smooth. Then the error changes very slowly [asymptotically as $e_h^{i+1} = e_h^i \cdot (1 - O\,(h^2))$]. This behaviour will be called the *smoothing effect* of the Gauss-Seidel (or damped Jacobi) method. An algebraic explanation of this property is as follows. The linear space V containing u_h can be represented as a sum of $V_1 + V_2$, where V_1 (V_2) corresponds to smooth (nonsmooth) grid functions. Although the iteration is slow in V_1, it is fast in V_2. Splitting the error e_h^i into $e_{h,1}^i + e_{h,2}^i$ ($e_{h,j}^i \,\varepsilon\, V_j$), we see that $e_{h,2}^i$ tends to zero much faster than $e_{h,1}^i$. Therefore, after few iterations the smooth part $e_{h,1}^i$ dominates the other part in e_h^i.

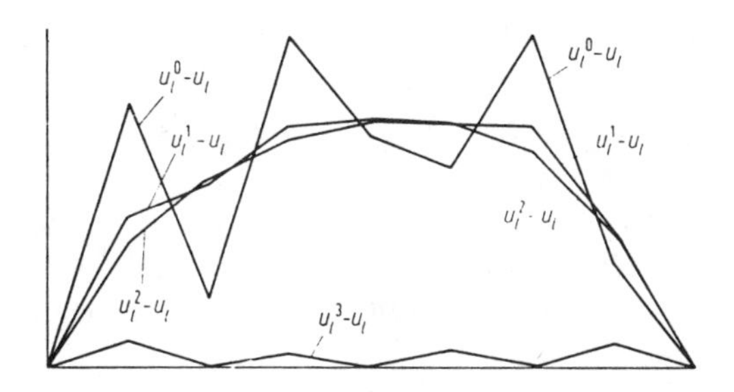

Fig. 1.1 Iteration error $u_h^i - u_h$
(i = 0,1,2) of the damped Jacobi iteration

Fig. 1.1 shows the errors e_h^i (i = 0, 1, 2) of the damped Jacobi iteration applied to the one-dimensional discrete Poisson equation. Here the space V consists of the grid functions $\{u_h(x) : x = h, 2h, \ldots, 1-h\}$ and V_1, V_2 are defined by

$$V_1 = \{s_h^\nu : \nu \,\varepsilon\, \mathbf{N},\ \nu h < 1/2\}\,,\quad V_2 = \{s_h^\nu : \nu \,\varepsilon\, \mathbf{N},\ 1/2 \le \nu h < 1\}\,,$$

where $s_h^\nu(x) = \sin(\nu\pi x)$ are the eigenfunctions of the iteration matrix. The corresponding eigenvalues λ_ν are $\cos^2(\nu\pi h/2)$. Since $\lambda_\nu \,\varepsilon\, (0, 1/2)$ for $\nu h < 1/2$, the damped Jacobi iteration has a convergence speed of 1/2 for all step sizes h with respect to V_2.

The foregoing discussion shows that the problem of constructing a fast iterative method reduces to the problem how to reduce efficiently errors from the "smooth" subspace V_1. Obviously, all classical methods do not have this property. For the solution of this problem we have to remember the relation (1.3): $u_h \approx u$. If

$$L_{2h}\, u_{2h} = f_{2h} \tag{1.2'}$$

is the discretisation of Eq. (1.1) w.r.t. a coarser grid of size $2h$, we have also

$$u_{2h} \approx u \tag{1.3'}$$

yielding

$$u_{2h} \approx u_h \ . \tag{1.5}$$

More precisely, (1.3), (1.3'), and (1.5) are only true, if the functions u, u_h, u_{2h} are smooth. Relation (1.5) shows that a smooth function u_h can be approximated by a coarse-grid function u_{2h}. Applying the above mentioned classical iterative methods we obtain an approximation $\tilde{u}_h$, the error of which is smooth:

$$e_h := \tilde{u}_h - u_h \text{ smooth} . \tag{1.6}$$

Multiplying by L_h and using (1.2) we see that the defect

$$d_h := L_h \tilde{u}_h - f_h \tag{1.7}$$

of $\tilde{u}_h$ satisfies

$$L_h e_h = d_h \ . \tag{1.8}$$

If we would be able to solve Eq. (1.8), we would correct $\tilde{u}_h$ to the exact discrete value $u_h = \tilde{u}_h - e_h$ (cf. (1.6)). Replacing u_h and f_h in (1.2) by e_h and d_h, we conclude from the smoothness of e_h that (1.5) holds in the form

$$e_h \approx e_{2h} \ , \text{ where } L_{2h} e_{2h} = d_{2h} \ , d_{2h} \approx d_h \ . \tag{1.9}$$

Here, $L_{2h} e_{2h} = d_{2h}$ is considered as a 'discretisation' of the discrete problem $L_h e_h = d_h$.

The value d_{2h} in (1.9) is obtained as a (weighted) restriction of d_h. Therefore, the linear mapping $r : d_h - d_{2h}$ is called a "restriction" (for precise definitions compare [4,§3.5]):

$$d_{2h} := r \ d_h \ . \tag{1.10a}$$

Since the solution e_{2h} is defined on the coarse grid, an extension onto the fine grid is needed:

$$\tilde{e}_h := p \ e_{2h} \ , \tag{1.10b}$$

where the "prolongation" p is a mapping from the coarse onto the fine grid by means of interpolation (cf. [4,§3.4]).

As $\tilde{e}_h$ is assumed to approximate e_h, Eq. (1.6) shows that

$$\begin{aligned} u_h^{i+1} &:= \tilde{u}_h - \tilde{e}_h = \tilde{u}_h - p e_{2h} = \tilde{u}_h - pL_{2h}^{-1} d_{2h} = \tilde{u}_h - pL_{2h}^{-1} r d_h = \\ &= \tilde{u}_h - p \ L_{2h}^{-1} \ r \ (L_h \tilde{u}_h - f_h) \end{aligned} \tag{1.11}$$

is a good candidate for the next iterate. The mapping $\tilde{u}_h \to u_h^{i+1}$ by (1.11) is called the "*coarse-grid correction*". Although (1.11) yields a good correction if the error $\tilde{u}_h - u_h$ is smooth, it can be shown that it is never a convergent process (cf. [4, p. 23]). Roughly speaking, the coarse-grid correction (1.11) is fast convergent on V_1, but not necessarily contracting on V_2. However, if we combine a classical method (as the Gauss-Seidel iteration) with (1.11), at least one of them reduces the errors from V_i ($i = 1,2$), and therefore, the combination — the "*two-grid method*" — is a fast convergent iteration.

The disadvantage of the two-grid method is the fact that we have still to solve a system $L_{2h} e_{2h} = d_{2h}$ exactly. But since this system is of the same form as the original problem $L_h u_h = f_h$, we can apply the same iteration; that means $L_{2h} e_{2h} = d_{2h}$ is solved by means of the next coarser grid of size $4h$ etc. This approach results in the multi-grid method (1.12), which requires an exact solution of a linear system only for the coarsest grid, where the number of unknowns is very small. In order to simplify the notation, we set $h = h_l$,

$2h = h_{l-1}$, $4h = h_{l-2}, \ldots, h_o = 2h_l$ coarsest grid size, $L_{h_l} = L_l$, $f_{h_l} = f_l$. Eq. (1.2) corresponds to the system $L_l u_l = f_l$ at the level l. The procedure (1.12) performs one step $u_l^i \to u_l^{i+1}$ of the multi-grid iteration at level l. The fixed number $\nu \geq 1$ (independent of h_l) is the number of smoothing iterations $u_l \to S_l(u_l, f_l)$ (e.g. of the Gauss-Seidel iteration). Suitable values of γ are $\gamma = 1$ ("V-cycle") or $\gamma = 2$ ("W-cycle", cf. [4, p. 33]).

procedure multigrid (l, u, f);
integer l; array u, f;
if $l = o$ *then* $u_o := L_o^{-1} f$ *else*
begin array d,v; integer j; (1.12)
  for $j := 1$ *step* 1 *until* ν *do* $u := S_l(u, f)$;
  $d := r(L_1 u - f)$;
  $v := o$;
  for $j := 1$ step 1 *until* γ *do* multigrid $(l-1, v, d)$;
  $u := u - pv$
end;

The computational work of one iteration (1.12) is proportional to the number of unknowns (cf. [4,§4.3]). On the other hand one can prove for a large class of elliptic boundary value problems, that the convergence speed of the iteration (1.12) is $\leq \zeta = \zeta(\nu) < 1$, where ζ depends on the problem but not of the step size $h = h_l$ (cf. [4,§§6,7]). Since for a suitable choice of the multi-grid components ν, S_l, p, r the value ζ is small, $\zeta \ll 1$, the iteration (1.12) is really efficient.

A further reduction of the computational work can be obtained from the *nested iteration*, where the starting value u_l^o at level l is chosen as interpolated value of the multi-grid result of level l -1 (cf. [4,§5]). Often, it is sufficient to perform the multi-grid iteration only once at all levels $l = 1,2, \ldots, l_{\max}$ in order to obtain approximations $\tilde{u}_l$ $(o \leq l \leq l_{\max})$ with an iteration error of the order of the discretisation error.

2. FIELDS OF APPLICATION. Multi-grid methods started in the field of scalar elliptic boundary value problems, where they entered into competition with many other direct, iterative and semi-iterative solvers. It turned out that different from other methods the multi-grid algorithm remains fast whether the problem has constant coefficients / is separable / is defined on a rectangle / is symmetric or not. The convergence speed reported in Table 3.1 or §3 for $\varepsilon = 1$ (Poisson equation) does not deteriorate remarkably, if the region, the kind of the boundary condition, or similar parameters are changed.

The multi-grid methods are not restricted to discrete problems obtained from difference schemes. Also finite element equations can be treated if there is a hierarchy of finite element subspaces. A very interesting multi-grid program including an adaptive construction of the finite element discretisation is due to R. Bank [1] (cf. also [4,§3.8.2] and [6]).

For elliptic systems (e.g. Stokes or Lamé equations) multi-grid algorithms can be constructed, too. They require special smoothing iterations, since the simple iterations of the scalar examples of §1 make no more sense. A general proposal is given in [4,§11.2]. Unfortunately, it is not robust in the sense of §3. Brandt – Dinar (cf. [5], [6], [4,§11.3]) propose a "distributed relaxation" as smoother, which is the Gauss-Seidel iteration applied to a transformed system. Using incomplete LU-decompositions of such a transformed system, Wittum [7] recently developed an interesting new smoothing process.

Nonlinear problems can be solved by two different multi-grid approaches. Either one applies the Newton method and solves the resulting linear systems by the (linear) multi-grid method or one applies a nonlinear multi-grid method, which has asymptotically the same rate of convergence as the linear multi-grid iteration applied to the linearised problem (cf. [4,§9]). A typical difficulty of nonlinear problems is the choice of the starting iterate. In this respect, the nested iteration (see end of §1) is very helpful, since the starting value is then needed for the coarsest grid only.

Eigenvalue problems are closely related to the solution of linear equations. Correspondingly, there are different multi-grid variants for the determination of eigenvalues and corresponding eigenfunctions (cf. [4,§12]).

In §3 we shall consider several singular perturbation problems, where the ellipticity vanishes with a parameter $\varepsilon \to o$. These are the most difficult problems, since the smoothing property of standard 'smoothing iterations' is lost with vanishing ellipticity. One has to note that the multi-grid approach also depends on the discretisation technique. If, for example, a sufficient quantity of artificial viscosity is added, the discrete problem is again elliptic enough to allow the application of simple multi-grid algorithms (cf. Brandt in [5], [6]).

Usually, parabolic equations are solved time step by time step. To allow reasonable time steps, implicit schemes are preferred, which require the solution of one system of equations per time step. This system is (discrete) elliptic. Therefore, multi-grid algorithms for elliptic (i.e. stationary) problems can be applied to this instationary problem, too. Besides, there are other variants allowing the approximation of several time steps at once. The new approach of §4 applies to parabolic problems also.

The solution of hyperbolic equations give different kinds of problems. For linear hyperbolic equations usual multi-grid iterations (differently from that of §4) do not work unless the discretisation contains enough artificial or numerical viscosity. For nonlinear hyperbolic problems in addition the shock discontinuities occur. For traditional multi-grid approaches to hyperbolic problems compare Jameson (in [6]) and Hemker (in [2] and [6]).

Another field, where the multi-grid approach is as direct as in the elliptic case, are the integral equations of the second kind and equations of the form $u = Ku + f$, if K behaves like an integral operator. Various applications of this "multi-grid method of the second kind" and its favourable convergence properties are reported in [4,§16]).

3. THE PROBLEM OF ROBUST MULTI-GRID SOLVERS. The components S_l (smoothing iteration), p (prolongation), r (restriction) of the multi-grid iteration have to be chosen according to the pde to be solved. The prolongation p is an interpolation the order of which depends mainly on the order of the differential equation. The restriction can be defined as the adjoint of p:

$$r := p^* . \tag{3.1}$$

The most sensible choice is the selection of the smoother S_l, as we shall see next. There is a further parameter to be chosen, if we start with the problem $L_l u_l = f_l$ at the finest grid size h_l at once. We need coarse-grid matrices $L_{l-1}, L_{l-2}, \ldots, L_o$. Among other choices the matrices may be defined recursively by

$$L_{l-1} := r \; L_l \; p \; . \tag{3.2}$$

The interesting fact about the choice (3.2) is

LEMMA 3.1. The coarse-grid correction (1.11) is a projection if (3.2) holds. In particular, all errors in the range of p are corrected exactly.

Even if one restricts oneself to problems $L_l u_l = f_l$ discretising scalar second order equations, it is not at all easy to develop a multi-grid program working efficiently for all these equations. This task has been started by P. Wesseling, who has introduced the incomplete LU decompositions as smoothing process (cf. [4,§ 10.1.3]). Another proposal for the choice of the smoothing process is the combination of x-line Gauss-Seidel with y-line Gauss-Seidel iterations (i.e. blockwise Gauss-Seidel with blocks being the $x-/y-$ lines of the grid; cf. Stüben-Trottenberg in [5]).

The problem can be exemplified by the anisotropic equation

$$-\varepsilon u_{xx} - u_{yy} = f \quad \text{in } \Omega = (0, 1) \times (0, 1)\,, \quad \varepsilon > 0\,, \tag{3.3}$$

with Dirichlet boundary conditions. The investigation of Eq. (3.3) is important, since the discretisation of $-\Delta u = f$ in a rectangular grid with step sizes $\Delta y = \sqrt{\varepsilon}\,\Delta x$ is equivalent to the discretisation of Eq. (3.3) in a standard square grid. Often, equations are discretised in body-fitted grids, which locally may have large and small ratios $\Delta y / \Delta x$ as well.

ε	10^3	100	10	2	1	0.8	0.5	0.4	0.2	0.1	.05	.01	.001
point-wise	.92	.89	.63	.18	.074	.09	.18	.24	.44	.63	.76	.89	.92
y -line	.92	.89	.63	.18	.058	.04	.02	.02	.03	.04	.04	.01	$5_{10}{-5}$

Table 3.1 Multi-grid convergence speed for the anisotropic problem (3.3) with pointwise and y -line Gauss-Seidel iteration

Applying a standard multi-grid iteration with pointwise Gauss-Seidel smoothing we obtain the convergence speeds shown in the first row of Table 3.1 (cf. [4,§10.1.1]). Obviously, the iteration converges for all ε, but it is inefficient as soon as $\min\{\varepsilon, 1/\varepsilon\} < 1/5$. For the special problem (3.3) there are two remedies. For $\varepsilon < 1$ ($\varepsilon > 1$) one may use a grid coarsening only w.r.t. the y -direction (x -direction). However, such a strategy fails in situations with variable coefficients, where $\varepsilon < 1$ and $\varepsilon > 1$ occurs.

The other remedy is the choice of suitable smoothing iterations. Candidates are y -line Gauss-Seidel for $\varepsilon < 1$ and x -line Gauss-Seidel for $\varepsilon > 1$ or ILU (incomplete LU decompositions) in both cases. The convergence speed of the multi-grid iteration with y -line Gauss-Seidel is shown in the 2nd row of Table 3.1.

The discretisation of Eq. (3.3) by the difference scheme

$$L_l = h_l^{-2} \begin{bmatrix} & -1 & \\ -\varepsilon & 2+2\varepsilon & -\varepsilon \\ & -1 & \end{bmatrix} \tag{3.4}$$

can be analysed very easily by Fourier analysis (cf. [4,§8]) using the fact that $\sin\nu\pi x \cdot \sin\mu\pi y$ are eigenfunctions for all frequencies $1 \le \nu, \mu \le n-1$ with $n = 1/h$. Low frequencies are those with $1 \le \nu, \mu \le n/2$ (region I in Fig. 3.1). Errors in these components should be reduced

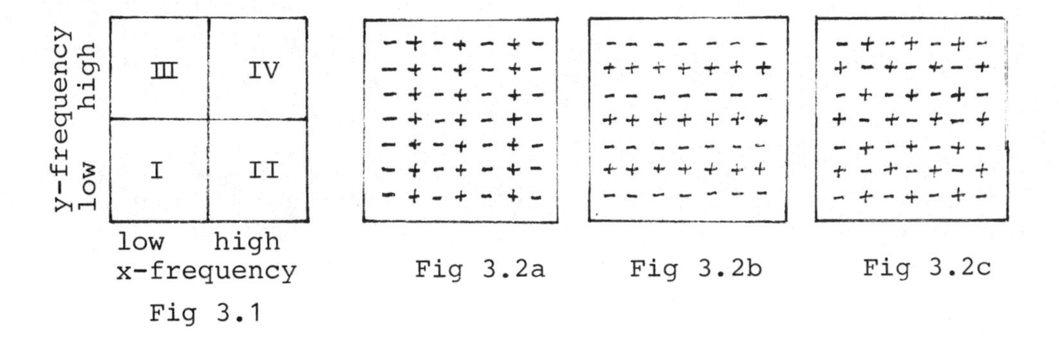

Fig 3.1 Fig 3.2a Fig 3.2b Fig 3.2c

by the coarse-grid correction, while the smoothing iteration must reduce all error components in the other regions II, III, IV ("high frequencies"). However, for $\varepsilon \ll 1$ the pointwise Gauss-Seidel iteration fails in region II of Fig. 3.1. A typical eigenfunction of region II is depicted in Fig. 3.2a: An error component smooth in y and highly oscillatory in x-direction is almost not changed by the pointwise Gauss-Seidel method and cannot be reduced by the coarse-grid correction.

The combination of $x-$ and $y-$line Gauss-Seidel is a good smoother for the discretisation (2.4) with all $0 \le \varepsilon < \infty$. However, it fails for the diagonally anisotropic problem

$$L_l = h_l^{-2} \begin{bmatrix} -\varepsilon & 0 & -1 \\ 0 & 2+2\varepsilon & 0 \\ -1 & 0 & -\varepsilon \end{bmatrix} . \tag{3.5}$$

In the latter case, the smoother does not reduce the highly oscillating function of Fig. 3.2c, which belongs to region IV.

Similar problems as for the anisotropic problem occur for the convection-diffusion equation

$$-\varepsilon \Delta u + \vec{c} \cdot \operatorname{grad} u = f , \tag{3.6}$$

which reduces to the hyperbolic equation of the first order, $\vec{c} \cdot \operatorname{grad} u = f$, when $\varepsilon = 0$.

Changing $\varepsilon \Delta u$ into $\varepsilon u_{xx} + u_{yy}$ and choosing $\vec{c} = \begin{pmatrix} 1 \\ 0 \end{pmatrix}$, we obtain the more difficult problem

$$-\varepsilon u_{xx} - u_{yy} + u_x = f , \tag{3.7}$$

which reduces for $\varepsilon = 0$ into a parabolic problem. Its discretisation is

$$L_l = h_l^{-2} \begin{bmatrix} & -1 & \\ -\varepsilon - h_l & 2+2\varepsilon + h_l & -\varepsilon \\ & -1 & \end{bmatrix} . \tag{3.8}$$

The ILU methods fail as smoother because of oscillations like in Fig. 3.2a. A possible smoother would be the y-line Gauss-Seidel method, provided that the ordering of the y-lines coincides with the sign of the coefficient of u_x.

Finally, we mention the diagonal five-point scheme

$$L_l = h_l^{-2} \frac{1}{2} \begin{bmatrix} -1 & 0 & -1 \\ 0 & 4 & 0 \\ -1 & 0 & -1 \end{bmatrix} \tag{3.9}$$

of the Poisson equation $-\Delta u = f$. For this problem all usual multi-grid methods fail because of errors of the type as depicted in Fig. 3.2c, unless one uses special prolongations and restrictions.

A multi-grid method would be called "robust" if it works for all (or most of the) foregoing examples. The proposals made in the past do not solve this task completely. The difficulties even increase for three-dimensional problems.

4. A NEW MULTI-GRID METHOD. We explain the new multi-grid iteration for the anisotropic problem

$$-\alpha u_{xx} - \beta u_{yy} + u = f \quad \text{in } \Omega = (0,1) \times (0,1), \ \alpha, \beta \geq 0, \tag{4.1}$$

with periodic boundary conditions. Fig. 4.1 and Fig. 4.2a show the usual grids of size h and $2h$. Instead of only one coarse grid, we consider now four coarse grids Ω_{00}^{2h}, Ω_{10}^{2h}, Ω_{01}^{2h}, Ω_{11}^{2h}, where

$$\Omega_{ij}^{2h} := \{(2\nu + i, 2\mu + j)h \,\varepsilon\, [0,1) \times [0,1) ; \ \nu, \mu \,\varepsilon\, \mathbf{Z}, 0 \leq i, j \leq 1\} \tag{4.2}$$

(cf. Fig. 4.2a-d). Ω_{00}^{2h} is the coarse grid of usual multi-grid methods. For $(i,j) = (0,0)$ Ω_{ij}^{2h} results from Ω_{00}^{2h} by a shift of (ih, jh). Obviously, we have

$$\Omega^h = \bigcup_{i,j=0}^{1} \Omega_{ij}^{2h} . \tag{4.3}$$

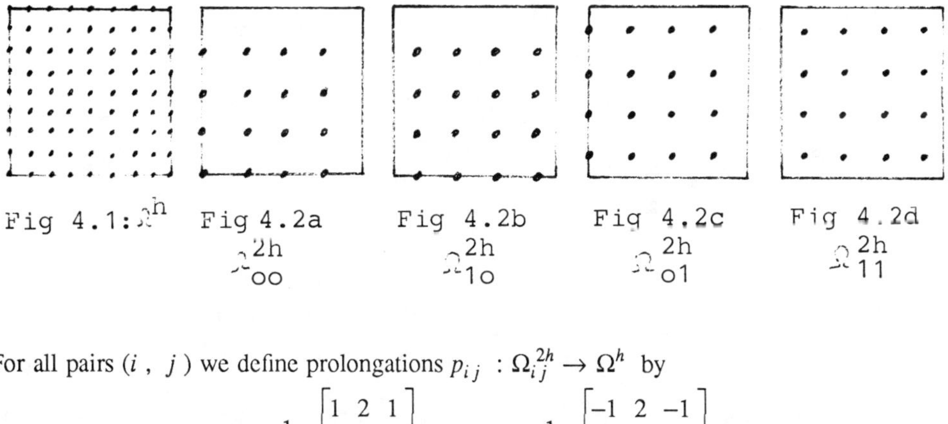

Fig 4.1: Ω^h Fig 4.2a Ω_{00}^{2h} Fig 4.2b Ω_{10}^{2h} Fig 4.2c Ω_{01}^{2h} Fig 4.2d Ω_{11}^{2h}

For all pairs (i, j) we define prolongations $p_{ij} : \Omega_{ij}^{2h} \to \Omega^h$ by

$$p_{00} = \frac{1}{4}\begin{bmatrix} 1 & 2 & 1 \\ 2 & 4 & 2 \\ 1 & 2 & 1 \end{bmatrix}, \quad p_{10} = \frac{1}{4}\begin{bmatrix} -1 & 2 & -1 \\ -2 & 4 & -2 \\ -1 & 2 & -1 \end{bmatrix},$$

$$p_{01} = \frac{1}{4}\begin{bmatrix} -1 & -2 & -1 \\ 2 & 4 & 2 \\ -1 & -2 & -1 \end{bmatrix}, \quad p_{11} = \frac{1}{4}\begin{bmatrix} 1 & -2 & 1 \\ -2 & 4 & -2 \\ 1 & -2 & 1 \end{bmatrix}, \tag{4.4}$$

i.e.

$$(p_{ij}u_{2h})(x,y) = u_{2h}(x,y),$$
$$(p_{ij}u_{2h})(x+h,y) = (-1)^i[u_{2h}(x,y)+u_{2h}(x+2h,y)]/2,$$
$$(p_{ij}u_{2h})(x,y+h) = (-1)^j[u_{2h}(x,y)+u_{2h}(x,y+2h)]/2, \tag{4.4'}$$
$$(p_{ij}u_{2h})(x+h,y+h) = (-1)^{i+j}[u_{2h}(x,y)+$$
$$u_{2h}(x+2h,y)+u_{2h}(x,y+2h)+u_{2h}(x+2h,y+2h)]/4$$

for $(x,y)\varepsilon\,\Omega\downarrow^{2h}_{ijh}$. The adjoint of p_{ij} defines the restriction $r_{ij}:\Omega^h\to\Omega^{2h}_{ij}$ (cf. Eq. (3.1)):

$$r_{ij} := p_{ij}^*, \tag{4.5}$$

i.e.

$$(r_{ij}u_h)(x,y) = \frac{1}{16}\{4u_h(x,y) \tag{4.5'}$$
$$+2(-1)^i[u_h(x+h,y)+u_h(x-h,y)]$$
$$+2(-1)^j[u_h(x,y+h)+u_h(x,y-h)]$$
$$+(-1)^{i+j}[u_h(x+h,y+h)+u_h(x-h,y+h)$$
$$+u_h(x+h,y-h)+u_h(x-h,y-h)]\}.$$

The Galerkin product (3.2) becomes

$$L_{ij}^{2h} := r_{ij}L_h\,p_{ij} \qquad (0\le i,\ j\le 1). \tag{4.6}$$

L_{ij}^{2h} are matrices describing linear mappings in the space of grid functions on Ω_{ij}^{2h}.

The prolongations p_{ij} $(i,\ j=0,1)$ are orthogonal in the sense that

$$r_{ij}\,p_{i'j'} = 0 \qquad \text{for } (i,j)\ne(i',j'). \tag{4.7}$$

REMARK 4.1. The grid function depicted in Fig. 3.2a (3.2b,c) belongs to the range of p_{10} (p_{01}, p_{11}, respectively).

By (4.3) and the positive definiteness of $r_{ij}p_{ij}$ one proves

REMARK 4.2. The linear space of fine-grid functions u_h coincides with the sum of the ranges of p_{ij} $(0\le i,\ j\le 1)$.

For Eq. (4.1) the fine-grid matrix L_h is positive definite. Hence, all coarse-grid matrices L_{ij} from (4.6) are also positive definite; in particular, they are non-singular for all α, $\beta\ge 0$.

The new two-grid method consists of a smoothing part and the new coarse-grid correction

$$\tilde u_h \to u_h^{i+1} := \tilde u_h - \Sigma^*\,p_{ij}\,(L_{ij}^{2h})^{-1}r_{ij}\,(L_h\tilde u_h - f_h), \tag{4.8}$$

where Σ^* denotes the summation over $(i,j)\,\varepsilon\,I^*$, where I^* is an index set with

$$\{(0,0)\}\subset I^*\subset\{0,0),(1,0),(0,1),(1,1)\}. \tag{4.9}$$

REMARK 4.3. (a) If $I^*=\{(0,0)\}$, the coarse-grid correction (4.8) coincides with the standard one (cf. (1.11)). (b) If $(1,0) \varepsilon I^*$, the correction (4.8) reduces error components which are smooth w.r.t. y and highly oscillatory w.r.t. x (cf. Fig. 3.2a). (c) Similarly, $(0,1) \varepsilon I^*$ [$(1,1) \varepsilon I^*$] corresponds to error components smooth w.r.t. x and highly oscillatory w.r.t. y [oscillatory w.r.t. x and y] (cf. Fig. 3.2b [c]).

Proof of (b). Let $\tilde{u}_h = u_h + e_h$ with e_h as in Fig. 3.2a. Then, $d_h := L_h e_h$ has the same structure: d_h is smooth w.r.t. y and oscillatory w.r.t. x. Hence, it is (almost) in the range of p_{10} (cf. Remark 4.1). (4.7) proves

$$u_h^{i+1} \approx \tilde{u}_h - p_{10} (L_{10}^{2h})^{-1} r_{10} L_h e_h .$$

Since e_h is (almost) in the range of p_{10}, Lemma 3.1 shows $u_h^{i+1} \approx u_h$. ∎

The coarse-grid correction (4.8) enables us to eliminate also non-smooth error components, which before had to be treated by the smoothing iteration only. Therefore, the demands on the smoothing iteration combined with correction (4.8) may be lower. In the following we use the damped Jacobi iteration, which is the simplest to perform and, in particular, well-suited for a vector computer.

In the following we study in detail the case of Eq. (4.1). The anisotropic fine-grid matrix is

$$L_h = \alpha h^2 \begin{bmatrix} 0 & 0 & 0 \\ -1 & 2 & -1 \\ 0 & 0 & 0 \end{bmatrix} + \beta h^{-2} \begin{bmatrix} 0 & -1 & 0 \\ 0 & 2 & 0 \\ 0 & -1 & 0 \end{bmatrix} + \begin{bmatrix} 0 & 0 & 0 \\ 0 & 1 & 0 \\ 0 & 0 & 0 \end{bmatrix} . \tag{4.10}$$

The Galerkin product (4.6) yields the coarse-grid matrices

$$L_{00}^{2h} = \frac{\alpha}{(2h)^2} \frac{1}{8} \begin{bmatrix} -1 & 2 & -1 \\ -6 & 12 & -6 \\ -1 & 2 & -1 \end{bmatrix} + \frac{\beta}{(2h)^2} \frac{1}{8} \begin{bmatrix} -1 & -6 & -1 \\ 2 & 12 & 2 \\ -1 & -6 & -1 \end{bmatrix} + E , \tag{4.11a}$$

$$L_{10}^{2h} = \frac{\alpha}{(2h)^2} \frac{1}{8} \begin{bmatrix} 3 & 10 & 3 \\ 18 & 60 & 18 \\ 3 & 10 & 3 \end{bmatrix} + \frac{\beta}{(2h)^2} \begin{bmatrix} -1 & -6 & -1 \\ 2 & 12 & 2 \\ -1 & -6 & -1 \end{bmatrix} + E , \tag{4.11b}$$

$$L_{01}^{2h} = \frac{\alpha}{(2h)^2} \frac{1}{8} \begin{bmatrix} -1 & 2 & -1 \\ -6 & 12 & -6 \\ -1 & 2 & -1 \end{bmatrix} + \frac{\beta}{(2h)^2} \frac{1}{8} \begin{bmatrix} 3 & 18 & 3 \\ 10 & 60 & 10 \\ 3 & 18 & 3 \end{bmatrix} + E , \tag{4.11c}$$

$$L_{11}^{2h} = \frac{\alpha}{(2h)^2} \frac{1}{8} \begin{bmatrix} 3 & 10 & 3 \\ 18 & 60 & 18 \\ 3 & 10 & 3 \end{bmatrix} + \frac{\beta}{(2h)^2} \frac{1}{8} \begin{bmatrix} 3 & 18 & 3 \\ 10 & 60 & 10 \\ 3 & 18 & 3 \end{bmatrix} + E \tag{4.11d}$$

with $E = \frac{1}{64} \begin{bmatrix} 1 & 6 & 1 \\ 6 & 36 & 6 \\ 1 & 6 & 1 \end{bmatrix}$. Specially, we consider the anisotropic case $0 \le \alpha \ll 1$, $\beta \approx 1$.

According to §3, the frequencies in region II of Fig. 3.1 cause the difficulties. Therefore, I^* must contain the index (1,0) besides (0,0). The fact that $(i, j) = (0,0)$ and (1,0) are necessary corrections in (4.8) corresponds to the fact that L_{00}^{2h} and L_{10}^{2h} are both discretisations of $-\beta u_{yy} + u$, when we neglect the term with $\alpha \ll 1$. Vice versa, the fact that $(i, j) = (0,1)$ and (1,1) are unnecessary corrections, corresponds to the fact that L_{01}^{2h} and L_{11}^{2h} have a condition $O(1)$. Even when these corrections are not necessary, they do not deteriorate the coarse-grid correction and, moreover, the coarse-grid equations are easy to solve because of the good condition of the matrices.

The two-grid process consisting of a smoothing step by damped Jacobi iterations and of the coarse-grid correction (4.8) can be investigated by Fourier analysis (cf. [4,§8.1.2]). For the limit case $\alpha \to 0$, $\beta \to \infty$, the (Fourier transformed) iteration matrix takes a very simple form and has the same convergence speed as the usual two-grid method applied to the one-dimensional problem $-u'' = f$ (analysis and results are given in [4,§2.4]).

The fact that for $\alpha = 0$ the new multi-grid method behaves like the standard one for a one-dimensional problem w.r.t. the y-direction can be explained as follows. Let $L_h = \begin{bmatrix} * \\ * \\ * \end{bmatrix}$ be a three point scheme in y-direction as in the limit case $\alpha = 0$ of the anisotropic problem. The damped Jacobi iteration reduces the frequencies being highly oscillatory w.r.t. y (regions III and IV in Fig. 3.1). The coarse-grid correction (4.8) with $(0,0), (1,0) \in I^*$ must eliminate the components which are smooth w.r.t. y, whatever the behaviour w.r.t. x is. Such components are (almost) in the range of p_{00} and p_{10}. Therefore, e_h is an error of the form

$$e_h = p_{00} e_0^{2h} + p_{10} e_1^{2h} . \tag{4.12}$$

The coarse-grid correction (4.8) with $I^* = \{(0,0), (1,0)\}$ corresponds to the matrix

$$C = I - p_{00} (L_{00}^{2h})^{-1} r_{00} L_h - p_{10} (L_{10}^{2h})^{-1} r_{10} L_h . \tag{4.13}$$

Note that p_{00} is a tensor product of the linear interpolations $p_0^x = (1/2\ 1\ 1/2)$ and p_0^y w.r.t. x and y: $p_{00} = p_0^x p_0^y$, while $p_{10} = p_1^x p_0^y$ with $p_1^x = (-1/2\ 1\ -1/2)$. The analogue of Lemma 3.1 is

LEMMA 4.4. $Ce_h = 0$ for all $e_h \in$ range (p_{00}) + range (p_{10}) from (4.12). In particular, C is a projection: $C^2 = C$.

Proof. $Cp_{00} = p_{00} - p_{00}(L_{00}^{2h})^{-1} r_{00} L_h p_{00} - p_{10}(L_{10}^{2h})^{-1} r_{10} L_h p_{00} = 0$, since $L_{00}^{2h} = r_{00} L_h p_{00}$ and $r_{10} L_h p_{00} = r_1^x r_0^y L_h p_0^x p_0^y = (r_1^x p_0^x)(r_1^y L_h p_1^y) = 0$ because of (4.7): $r_1^x p_0^x = 0$. Similarly, $Cp_{10} = 0$ is shown. ∎

Hence, all errors e_h of the form (4.12) are eliminated by the new coarse-grid correction, although e_h may be highly oscillatory w.r.t. x.

	coefficients α	β	index set I^*: 1	1-2	1-3	1-4
T:	1	1	0.562	0.562	0.252	0.252
M:			0.502	0.502	0.418	0.212
T:	$\frac{1}{2}$	2	0.809	0.359	0.228	0.228
M:			0.734	0.373	0.229	0.255
T:	10^{-5}	10^5	0.999	0.250	0.250	0.135
M:			0.816	0.318	0.236	0.165

Table 4.1 Convergence speed of the new two- (T) and multi-grid (M) method applied to Eq. (4.1).

The rows of Table 4.1 denoted by "T" contain the convergence speed of the two-grid iteration with $h_l = 1/16$, $\nu = 2$, S_l = damped Jacobi iteration with damping factor 1/2. The convergence rates are computed by Fourier analysis. The rows marked by "M" show the convergence speed of the multi-grid iteration ($\gamma = 1$, i.e. V-cycle, for $h_l = 1/16, \ldots, h_0 = 1/2$), which is again the recursive application of the new two-grid iteration. The numbers are obtained from a numerical experiment. The index set I^* is abbreviated by " 1", . . . , " 1–4". the meaning is

$$1{:}\ I^* = \{0{,}0)\},\ 1\text{–}4{:}\{I^* = \{0{,}0),\ (1{,}0,\ (0{,}1),\ (1{,}1)\}\ ,$$

$$1\text{–}2{:}\ I^* = \{0{,}0),\ (1{,}0)\},\ 1\text{–}3{:}\ I^* = \{(0{,}0),\ (1{,}0),\ (0{,}1)\}\ .$$

For $\alpha \ll \beta$ it is necessary to have at least (0,0) and (1,0) in I^*. Table 4.1 shows that for such I^* the convergence is uniform for all $\alpha = 1/\beta\ \varepsilon\ [0,1]$. As soon as also (0,1)$\varepsilon\ I^*$, the method works in the same way for all $\alpha = 1/\beta\ \varepsilon\ [1, \infty)$ because of the symmetry in x and y. Hence, for I^* according to " 1–3" and " 1–4" the new multi-grid method is robust for all $\alpha, \beta \geq 0$.

REMARK 4.5. If I^* contains no more than 3 indices, the computational work of the new multi-grid iteration is $O(n_l)$, where n_l = number of unknowns = $O(h_1^{-2})$. If I^* contains all 4 indices, the work is $O(n_l \cdot l) = O(n_l \cdot \log n_l)$.

Similarly, one can analyse the new multi-grid approach for the other problems of §3. For Eqs. (3.5) and (3.9) it is necessary to have (0,0), (1,1)$\varepsilon\ I^*$. By Remark 4.3c error components like in Fig. 3.2c which prevent the usual multi-grid method from fast convergence, are corrected by the coarse-grid correction (4.8).

The convection diffusion equation (3.6) (without artificial viscosity) can be solved for all $\vec{c}$, if I^* contains all 4 indices. For $\varepsilon = 0$ and $\vec{c} = \begin{pmatrix} 0 \\ 1 \end{pmatrix}$ or $\vec{c} = \begin{pmatrix} 1 \\ 0 \end{pmatrix}$ Lemma 4.4 applies again.

For the solution of the parabolic problem (3.7), (3.8), one needs (0,0), (1,0)$\varepsilon\ I^*$. It is worth noting that the coarse-grid matrix L_{10}^{2h} has a condition $O(h_l^{-1})$ although it is not a discretisation of Eq. (3.7). For unsymmetric problems as (3.6) and (3.7/8) it is important to use the matrix-dependent prolongations and restrictions (cf. [4,§10.3]).

If parallel processors are available one may use

REMARK 4.6. The solution of the coarse-grid problems $L_{ij}^{2h} v_{ij} = d_{ij}$ $(0 \leq i\ ,\ j\ \leq 1)$ can be done in parallel.

Conclusion. The new multi-grid method works efficiently for all examples presented without change of its components. It allows to use the simplest smoothing iteration. It is not restricted to elliptic or discrete elliptic problems as could be seen from the examples (3.6-9). The computational work is at most $O(n_l \cdot \log n_l)$.

REFERENCES

[1] R. E. BANK, PLTMG user's guide – Edition 4.0, Technical Report, University of San Diego at La Jolla, 1985.

[2] D. BRAESS, W. HACKBUSCH, U. TROTTENBERG (eds.), Advances in Multi-Grid Methods, Proceedings, Oberwolfach, Dec. 1984, Notes on Numerical Fluid Mechanics, 11, Vieweg, Braunschweig, 1985.

[3] A. BRANDT, Multi-level adaptive solutions to boundary-value problems, Math. Comp., 31 (1977), pp. 333-390.

[4] W. HACKBUSCH, Multi-Grid Methods and Applications, Springer, Berlin, 1985.

[5] W. HACKBUSCH and U. TROTTENBERG (editors), Multi-Grid Methods, Proceedings, Köln-Porz, No. 1981. Lecture Notes in Mathematics, 960, Springer, Berlin, 1982.

[6] W. HACKBUSCH and U. TROTTENBERG (editors), Multi-Grid Methods II, Proceedings, Köln, Oct. 1985. Lecture Notes in Mathematics, 1228, Springer, Berlin, 1986.

[7] G. WITTUM, Distributive Iterationen für indefinite Systeme als Glätter der Stokes-und Navier-Stokes-Gleichungen mit Schwerpunkt auf unvollständigen Zerlegungen. Doctoral thesis, Christian-Albrechts-Universität zu Kiel, 1986.

Model Driven Simulation Systems

JOHN HOPCROFT*

Abstract. Off-line robot programming is necessary to avoid damaging or tying up costly equipment. Model driven simulation will be an essential ingredient of any off-line programming system and will facilitate work cell layout by modeling a design before committing it to hardware. In addition, such a system will be a valuable aid in studying grasping and manipulation strategies. Today, simulation tools are very problem specific. What is needed is a model-driven simulation system to verify robot programs and to investigate such things as gripper design, approach strategies, and understanding physical phenomena. Discussed here will be the development of a model-driven simulation system to support these activities. In particular, areas will be identified where basic research is needed to support progress in this area.

Introduction. A model-driven simulation system simulates the behavior of a physical system from its geometric description. Thus, from a description of the geometry and masses of the rigid components of a robot, the manner in which the components are connected, and the related control algorithms; the system would correctly simulate the behavior of the robot. Such a system would allow one to debug robot programs off-line as well as experiment with the design of robots, grippers, etc. Today there are a number of systems that simulate the dynamic behavior of various robots and more general simulation systems such as ADAMS[4]. What differentiates a model-driven simulation from present systems is that the dynamic behavior is determined from the control programs and the geometrical description. One of the important features of such a system is the ability of the system to automatically modify the dynamics of the collection of objects being simulated whenever a collision between two objects occurs. Thus, the interaction of the robot with its work space is accounted for. Systems such as ADAMS always maintain a contact force between objects where the form of the force is such that the force is negligible except when the separation becomes small.
While this is a powerful technique for many situations, it is not clear that it is viable in situations where there are large numbers of surfaces that potentially could collide but only a few of which actually do collide during the course of a simulation.

*Dept. of Computer Science, Cornell University, Ithaca, NY 14853. This work was supported in part by the NSF under grant DMC-8617355 and the Office of Naval Research under grant N-00014-86-0281.

A model-driven simulation system has widespread applicability. Not only could it be used for off-line robot programming but also for studying such tasks as gripping or rotating an object. Instead of building a gripper such as the Salisbury hand[15] or the Utah hand[6], one would build an electronic version of a gripper and simulate it to study various tasks. The advantages are many. For example, by editing a text file one can change the number of fingers on the hand, alter the length of the fingers, or change their location . This ability to experiment is important in designing effective, dextrous end effectors. Of course, at some point one must experiment with real hardware since a simulation may overlook important details.

Another area of interest is in testing hypotheses considering human behavior. With a simplified model of a human being consisting of rigid limbs hinged at joints, one can experiment with algorithms for walking, balancing, or throwing an object. One could demonstrate that a given theory does indeed predict the way we perform some action or reject the theory as not possible. One could imagine that some day in the future a theory of how a child learns to walk could be programmed and examined. At a minimum, such a system demonstrates the level of detail required for such a theory to be complete. Wide ranging applications in the medical field include determining how forces on the spine are transmitted after a ruptured disk is repaired by fusing it with an adjacent disk and verifying the function of a prosthesis device under some reasonable range of human activities such as stepping off a step or sitting in a chair.

In the manufacturing area one can electronically redesign objects for ease of assembly and then verify various aspects of the design such as the removability of each serviceable component. In addition, certain objects for which it is difficult to build physical prototypes can be modeled. An example of this is an antenna that is to be deployed in space but which will not support its own weight under gravity. If the antenna fails to deploy properly, dejamming procedures can be carried out on the electronic model. An advantage here is that one would be reluctant to try out a dejamming procedure on a physical prototype if there was some danger that the procedure might destroy the prototype. The list of advantages goes on and on.

Science Base. A model-driven simulation system called Newton [5] is under development at Cornell. It is currently being used to simulate the Salisbury hand and to experiment with programming a simple model of a human to stand and sit. However, the lack of a science base to support such activities is inhibiting progress in this area. Work is needed on many fronts. In this talk I will briefly describe the simulation system and then discuss a number of areas where progress is needed. Areas include geometrical modeling, computational geometry, user interfaces, representational and language issues, and control of complex objects. Due to space limitations, I can give only brief examples of how results in these areas are used and directions in which work is needed.

An Overview of Project Newton The Newton project represents objects in multiple domains: abstract, geometric, dynamic, etc. Each domain has the capability to model primitive objects and to combine models of primitive objects to obtain models of composite objects. The abstract domain coordinates the domain specific representations of a given object. The abstract representation consists of a name and a property list providing information on mass, inertia, friction, and color. The geometrical domain maintains 3-dimensional geometric models of objects. It must be capable of representing geometry and updating the position and orientation of objects. In addition, it must be able to answer questions concerning the location of features of objects in 3-space and determining which surfaces touch or intersect.

The dynamic domain maintains state variables, position, and velocity along with the dynamic equations of motion. In the modeler, primitive objects are rigid solids. The primitive objects can be hinged together by specifying that a point in the reference

Figure 1. Line drawing from computer model of Salisbury hand.

frame of one object is constrained to a point, line, or plane in the reference frame of another primitive object. Each primitive object has an equation schema of the form:

$$m\ddot{r} = F, \text{ and}$$
$$J\dot{\omega} + \omega \times J\omega = T.$$

All external forces such as gravity and control forces and all torques are combined into a single resultant force F and torque T, respectively, which are applied at the center of mass. When primitive objects are combined, a kinematic constraint is imposed on the state variables of the objects. The constraint equation relates the accelerations of the two objects; consequently, the dynamics are no longer consistent with the geometry. Thus, the dynamics are modified as follows:

$$m_1\ddot{r}_1 = F_1 - X$$
$$m_2\ddot{r}_2 = F_2 - X$$
$$J_1\dot{\omega}_1 + \omega_1 \times J\omega_1 = T_1 - c_1 \times X$$
$$J_2\dot{\omega}_2 + \omega_2 \times J\omega_2 = T_2 - c_2 \times X$$

The resulting system consists of 12 scalar equations with 15 unknowns. Three unknowns are removed by three geometrical constraint equations. The control domain is not yet completely designed as experience in types of communication required is only now being obtained. Basically, each primitive object has its own local control. The control is able to sense any state variable of the system. What also remains to be worked out are convenient communications between the control algorithms at each joint and the higher level programs that provide global control.

The simulation proceeds by numerical integration of the dynamic equations. At each step, an interference algorithm checks for collisions. An event handler modifies the geometric and dynamic models whenever collisions occur and integrates the impact to obtain the boundary conditions for continuing the simulation.

Although the preceding discussion of Newton is brief, it should allow the reader to understand the structure of the system sufficiently enough to appreciate the following areas where a better science base is needed to support modeling systems such as Newton.

1. Geometrical Modeling

Although much work has been done on geometrical modeling, a number of problems remain to be solved. Among the more important are maintaining topological consistency of the model and handling algebraic curves and surfaces.

A. Consistency of Models

Suppose one has a model of an object in which four planes meet in a point. If one numerically calculates the point of intersection of three of the planes, the intersection point will not lie precisely on the fourth plane because of numerical round off. This poses a serious problem. In making numerical calculations one sets an ε such that if $|x-y|<\varepsilon$, x is deemed equal to y. Thus, it is possible to introduce topological inconsistencies into an object. For example, if three lines l_1, l_2, and l_3 intersect, one may decide that l_1 intersects l_2 at the same point that l_2 intersects l_3 but that the point of intersection of l_1 and l_3 is not coincident with the point of intersection of l_2 and l_3. One can claim that symbolic calculations will resolve difficulties of this nature. In addition to the fact that symbolic computations are too inefficient, they still do not completely solve the problem since there is usually some approximation in the original model. Another approach is never to perform a numerical computation in order to make a decision if the decision can be inferred from earlier computations. Another useful idea is to include topological data along with numerical data as input to a decision. For instance, if one is trying to decide if vertex v of an object lies on plane p, one might ask if all edges incident to v intersect p in the vicinity of v. This prevents difficulties when v is within ε of p but an edge e incident at v is almost parallel to p. One might otherwise claim v is on p and therefore e intersects p at v, only to discover later that e intersects p quite far from v. Of course, if $e=(v,w)$ and v and w are both within ε of p, then the proper decision is that v and w are both on p and the intersection of e with p is not numerically relevant. A major need in this area is a theory of consistency. A calculus that would aid in insuring topological consistency would be extremely useful. Some initial work on modeling consistency for 2-dimensional problems has been done by Milenkovic[10].

B. Algebraic Curves

A considerable amount of nonpolygonal modeling is based on parametric surfaces. Much of the reason for this is that simple algorithms for display of bicubic patches exist. Furthermore, walking along a parameterized curve is a simple task compared to walking a curve in 3-space that is the intersection of two algebraic surfaces. Unfortunately, the intersection of two bicubic patches is not simple. Since a bicubic patch is an algebraic surface of degree 18, the curve of intersection may be as high as 324. For this and other reasons it is desirable to investigate algorithms for surfaces and curves, where surfaces are represented by multivariate polynomial equations such as $p(x,y,z)=0$ and curves are represented by the intersection of two such polynomials. One problem that arises is the difficulty in walking along a curve that has a singularity. This situation can arise even though the curve on a solid might not have any singularity, since one might in the process of projecting the curve onto a plane introduce a singularity.

Consider a planar curve with a singularity at the origin. It is well known that the curve can be desingularized by a quadratic transformation[1,16]. The technique is as follows: Consider a singularity at the origin and assume that no tangent to the curve at the origin is aligned with the x-axis. This assumption can always be achieved by a rotation of the plane. Consider the transformation $x' \leftarrow x$ and $y' \leftarrow \frac{y}{x}$. The transformation maps all points in the xy plane with the exception of the origin to points in the $x'y'$ plane. The points on the line $y=0$ are mapped to infinity. The inverse transformation, $x \leftarrow x'$ and $y \leftarrow x'y'$, maps points in $x'y'$ plane one to one onto the xy plane with the exception that the entire line $x'=0$ is mapped to the origin. A continuous curve in the xy plane not going through the origin is mapped to a continuous curve in the $x'y'$ plane. For a curve through the origin, we map the origin to the limit of the image of points on the curve as we approach the origin. Thus, a straight line through the origin of slope m is mapped to the horizontal line $y'=m$. Points on the line $x'=0$ can be thought of as directions through the origin.

The quadratic transformation, $x' \leftarrow x$ and $y' \leftarrow \frac{y}{x}$, removes the singularity when it is applied to a curve with an ordinary singular point at the origin in which each branch through the origin has a distinct tangent line. Figure 2 illustrates the result of applying the transformation to the curve $x^3 - x^2 + y^2$. If the curve has a singular point with nondistinct tangents, then it can be shown that a finite number of applications of the transformation suffice to remove the singularity.

One way to trace a curve is to walk along the curve $f(x,y)=0$ using a Newton method until the partial derivatives f_x and f_y become smaller than some predetermined $\varepsilon > 0$. This signals a singularity. Locate and transform the singularity to the origin, apply the quadratic transformation, and walk the transformed curve mapping the path back to the original curve until safely past the singularity[3]. This method handles plane curves in an efficient and reliable fashion. Further work needs to be done in order to apply the technique to space curves.

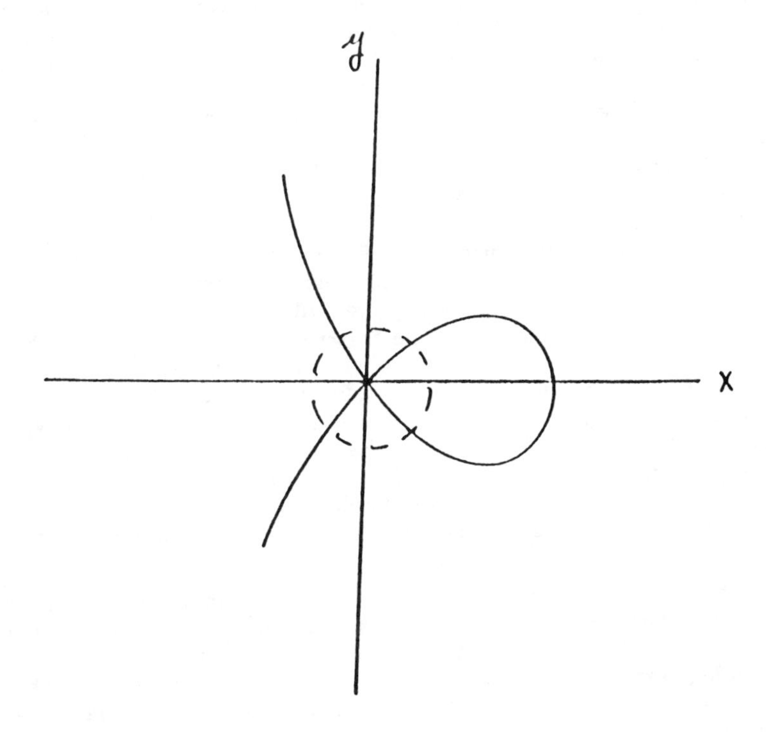

Figure 2. The curve $x^3 - x^2 + y^2$ and its transformation.

2. Computational Geometry

A number of algorithms from computational geometry[12] play an important role in speeding up computations and substantially increasing the size of problems with which one can work. Perhaps one of the most important contributions is the plane sweep algorithm. Suppose we are trying to intersect two objects represented by their boundaries. Normally one would intersect every face with every other face, a process that requires time proportional to the number of faces. If each face is a region of an algebraic surface, then just the process of intersecting two faces is in itself a complicated matter. An obvious way to speed up the algorithm is to box each face, that is find an enclosing rectangle. We can then intersect two such rectangles to see if they intersect. If they do intersect, we continue the process of trying to intersect the contained faces. This will significantly reduce the computation involved if for most pairs of faces we determine that the enclosing boxes do not intersect. However, it does not solve the problem that the complexity grows as the square of the number of faces since we still must intersect n^2 boxes. However, computational geometry provides an asymptotically efficient algorithm for finding all intersections among n^2 boxes.

Consider sweeping a plane in the z-direction. At any instant of time the plane will intersect some set of the boxes in rectangles. A z-coordinate at which a box begins or ends will be called a critical point. In order to determine if any two boxes intersect, it suffices to determine for each of the n critical points whether any two rectangles intersect.

In particular, we only need to consider situations where we are adding a new rectangle to the sweep plane. Thus, we again have an asymptotically worst case of order n^2. However, the performance may be better since we intersect two boxes only if they overlap in the z-direction. The real savings comes by maintaining the rectangles in the sweep plane in a balanced tree structure where we can insert, delete, and test for intersection in time order of $\log^2 n$.

I will only provide a sketch of the ideas, details can be found in texts on computational geometry. Instead of considering a two-dimensional problem of intersecting rectangles, first consider a one-dimensional problem of intersecting line segments. In particular, we will need to maintain a data structure which will store a set of line segments, allow us to efficiently add or delete a segment, and determine if a new segment intersects any segment in the data structure. Let (a_i, b_i) be the ith segment. To determine if a segment (a,b) intersects any segment in the set (a_i, b_i) $1 \le i \le n$, we need to answer two questions:

(1) Does there exist an i such that $a_i \le a \le b_i$, and
(2) does there exist an i such that $a \le a_i \le b$?

Both questions can be answered by maintaining balanced tree structures. To answer the first question, maintain a balanced search tree with the b_i's at the leaves. At each internal node, in addition to information necessary to carry out a search for a b_i, is stored the smallest value of any a_j corresponding to a b_j in the subtree. To determine if there exists an (a_i, b_i) such that $a_i \le a \le b_i$, start at the root of the balanced tree and search for the value a. We will end up at the rightmost $b_i \le a$. As we move down a path from the root to the leaf, whenever we take a left child, we check to see if the a_j value of the right child is less than a. If so, the right subtree has a leaf with $a_j \le a \le b_j$. The procedure is easy to modify to obtain all (a_i, b_i) satisfying (1) in time $O(\log n + k)$, where n is the total number of intervals and k is the number of intervals satisfying (1).

The second condition can be answered by maintaining a second balanced structure. In this structure the a_i's are stored at the leaves. Two searches are performed, one for a and one for b. Both searches terminate at leaves: the first search at the leftmost leaf less than a and the second at the rightmost leaf greater than b. All leaves strictly between these two contain a_i such that $a \le a_i \le b$. Again the search time is $O(\log n + k)$.

The two-dimensional problem is solved in an analogous manner. There exists a rectangle $(a_i,b_i)\times(c_i,d_i)$ that overlaps $(a,b)\times(c,d)$ if and only if there exists an i such that the interval (a,b) overlaps (a_i,b_i) and (c,d) overlaps (c_i, d_i). This two-dimensional problem is solved in a manner analogous to the 1-dimensional problem except that the data structure is more complex; since at each internal node of the balanced tree, we must have a data structure to solve the 1-dimensional problem. The search time is $O(log^2 n+k)$. Details can be found in [9].

The plane sweep algorithm is only one of many algorithms in computational geometry that is useful in geometric modeling. Unfortunely, most of the work in computational geometry is based on the asymptotic worst-case paradigm[2]. It is critical to build up an understanding of techniques for geometric modeling that are efficient and effective for real world problems. Clearly if there is an asymptotically efficient solution that works, then all is well. However, many problems not having good worst-case solutions still can be effectively solved in practice.

3. Dynamic simulation. Dynamic simulation, particularly model-driven simulation, will be important to a wide range of areas. Simulation will be a powerful tool in the understanding of physical phenomena and in the design of mechanical objects such as dextrous manipulators. As mentioned earlier, a simple simulator can be structured as follows: Objects are made up of components where each component is a rigid solid having a geometry, mass, and inertia. Each component has its own coordinate frame. Objects are constructed by constraining one or more points in the coordinate frame of a component to a point, line, or plane in the coordinate frame of a second component. Such constraints are called hinges. Thus, a hinge might be of a ball and socket type if a single point of one component is constrained to a single point of another component. A hinge such as on a door is obtained by constraining two points on one component to two points on a second component.

The dynamic equations for an object can be formulated by taking the equations of motion for each component and modifying them to account for hinges. A simulation is carried out by integrating the dynamic equations iteratively. After each step, a collision detection algorithm is run. If a collision occurs between a vertex of one object and a planar face of another, a hinge is created to constrain the vertex to the plane as long as there is a positive force between the vertex and the plane. The hinge is temporary in that it would automatically be removed should the control force go to zero or become negative. In this way, the correct behavior of the two objects is maintained during simulation.

In order to program a model of a human to walk up stairs or perform some other task, a language for programming the model is needed. We have chosen to control the torque on each joint locally. Consider a joint at an elbow that we wish to move so that the arm moves from vertical to horizontal. To do this, we could place a torque $K_d(\boldsymbol{\theta}-\boldsymbol{\theta}_d)+K_v\dot{\boldsymbol{\theta}}$ at the elbow joint. The K_d term produces a torque proportional to the distance to travel, and the K_v provides the necessary damping. If we were to use this method, we would need to add a term to integrate the error; otherwise the arm would come to rest at an angle at which $\boldsymbol{\theta}-\boldsymbol{\theta}_d$ was just sufficient to offset the gravity load. Furthermore, if one suddenly changed the load by placing an object in the hand, the arm would drop until the $\boldsymbol{\theta}-\boldsymbol{\theta}_d$ term was sufficient to offset the additional load. Then the integral of the error term would gradually restore the arm to the desired horizontal position. To increase the stiffness and have the arm quickly react to the additional load, we chose an adaptive control strategy instead. In this method, the arm is modeled as a rigid object with a given mass and moment of inertia acted upon by some unknown external force. The control algorithm calculates the value of the unknown forces by examining the behavior of the arm and then supplies the appropriate control torque for the model. This simplified adaptive control seems to suffice to control individual joints.

One thing that is different from controlling the one legged robot of Raibert[13,14] is that the number of degrees of freedom is much greater than the number of functions such as balance and posture that we need control. Raibert was able to map each degree of freedom to a function thereby simplifying his control strategy. Another observation is that in attempting to throw a ball, one brings the arm back. This process will unbalance the model and cause it to fall over unless we compensate by monitoring the center of gravity of the person and adjusting other components so as to maintain balance. This suggests building an environment with programs such as balance running in the background. In programming a person to perform a dive, one might have the model bend forward and rotate the arms backward with the balance mode on. Then as the arms swing forward, balance is turned off and the dive progresses. Programming such tasks will need significant work on building up a hierarchical control environment and the background algorithms necessary to relieve the programmer of many details.

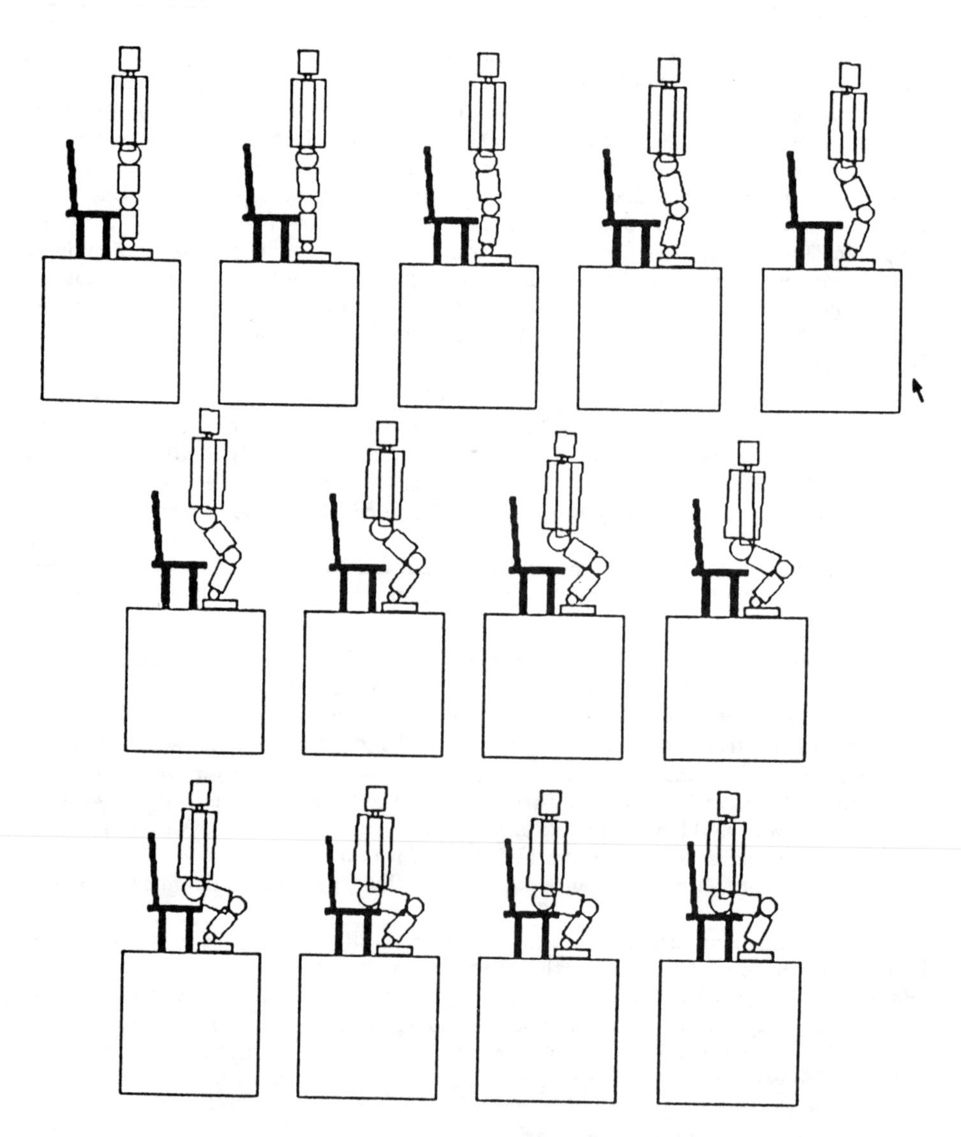

Figure 3. The simulation of a person standing up.

Figure 3 illustrates a simulation of a person standing up from a sitting position in a chair. The basic algorithm involves rotating the torso forward then locking the waist. As the waist is locked, the momentum of the torso is transferred to the body and the seat lifts off the chair. If no further torques were applied, gravity would overcome the momentum and the seat would drop back to the chair. However, by applying torque at the ankle and knee, the center of gravity can be moved directly over the ankle. By applying a reverse torque at the appropriate moment, the forward momentum can be reduced to approximately zero just as the center of gravity reaches the position directly above the ankle. At this point the knee and back can be straightened, and the person stands.

In order to program more complex algorithms for walking, climbing stairs, encountering a wind force, etc., we need some manner of higher level control. For example as mentioned earlier, a balance program that would run in background and constantly rebalance the model would relieve the programmer of many details. There are a large number of parameters that can be used to rebalance a model; selecting a strategy is important. Similarly, programs for other simple tasks such as bending, stopping, etc. could be written.

4. User interfaces. In order for a modeling system to be usable, it must have an interface that is convenient and easy to use. Although this sounds like a simple engineering problem, significant research needs to be done before we will have the understanding necessary to develop such user interfaces. I will mention only two topics here. The first is symbolic specification. Human beings are not good at transforming coordinate systems and manipulating numerical data. Thus in a modeling system, one should have the capability of naming features of one objects and then assembling objects by specifying that certain features should align. To position one cube on top of another, after having named the vertices of the two cubes, one might request that w_1 and v_5, w_2 and v_6, and w_3 and v_7 coincide. The three constraints uniquely determine the position of the second box on top of the first. However, one need not give a set of constraints that completely determines the positioning. This brings us to our second observation about user interfaces.

Suppose we had a model of a human with a 100 degrees of freedom, and we wished to place this model on a diving block and have it execute a dive. At first, it might appear that initializing the diver would require specifying all 100 degrees of freedom. Note, however, that if we were to tell a programmer to initialize the diver, we would simply say "place the diver on the diving block." The reason human communication can be so simple is that humans have a world model with which they interpret statements. Since I understand in general terms how a human is going to interpret what I say, I can make use of this information in communicating. The same principle can be used in human-machine communication to improve efficiency. At present, one cannot build a world model in the computer with which to interpret statements. However, the world model is not the crucial element. What is important is that we understand how the person interprets what is said. Thus, we can capture the essence by building a simple algorithm for solving constraints. One would specify a set of constraints, and the constraint solver would determine a set of values for the parameters that satisfied the specified constraints. What is important is that the algorithm used by the constraint solver be understood by the user. One such algorithm might be the following: If the constraints can be satisfied by a translation of the model, then that is done. If not, use the smallest possible rotation. If no rotation suffices, then modify joint angles according to some simple set of rules. With such a system, a user quickly determines three or four constraints whose solution by the above algorithm will result in the system specifying all 100 constraints in the manner desired, a significant improvement in user productivity.

Summary. In conclusion, we believe that model-driven simulation will play an important role in the future. However, there are a fair number of problem areas in mathematics and computer science whose development will significantly impact future systems.

References

(1) Abhyankar, S.S., "Historical Ramblings in Algebraic Geometry and Related Algebra," *American Mathematical Monthly 83:6*, June-July 1976, 409-448.

(2) Aho, A.V., J.E. Hopcroft and J.D. Ullman, *The Design and Analysis of Computer Algorithms,* Addison-Wesley Publishing Company, Reading Mass., 1974.

(3) Bajaj, Chanderjit, J.E. Hopcroft and C. Hoffmann, in preparation.

(4) Chace, M., "Modeling of Dynamic Mechanical Systems," CAD/CAM Robotics and Automation Institute and Intl. Conf., Tucson, Feb. 1985.

(5) Hoffmann, Christoph M., and John E. Hopcroft, "Simulation of Physical Systems from Geometric Models," *IEEE Journal of Robotics and Automation, Special Issue on Robot Motion Planning,* June 1987, ed. by V. Lumelsky.

(6) Jacobsen, S.C., J.E. Wood, D.F. Knutti, K.B. Biggers, and E.K. Iversen, "The Version I Utah/MIT Dextrous Hand," in *Robotics*, H. Hanafusa and H. Inone, eds. M.I.T., Cambridge, MA. 1985.

(7) McGhee, R.B., "Control of Legged Locomotion Systems," *Proc. 18th Automatic Control Conf.*, San Francisco, June 1977, 205-215.

(8) McMahon, T.A., "Mechanics of Locomotion," *IJRR* 1984, 3 (2-Summer), 4-28.

(9) Mehlhorn, K., *Data Structures and Algorithms 3: Multi-dimensional Searching and Computational Geometry*, Springer-Verlag, New York 1984.

(10) Milenkovic, V.J., "Verifiable Implementations of Geometric Algorithms Using Finite Precision Arithmetic," Dept. of Computer Science, CMU, Pittsburgh, Pa., 15213.

(11) Orin, D.E., "Supervisory Control of a Multilegged Robot," *IJRR* 1982, 1 (1-Spring), 79-91.

(12) Preparata, F.P., and Shamos, M.I., *Computational Geometry*, Springer-Verlag, New York, 1985.

(13) Raibert, M.H., H.B. Brown, and M. Chepponis,, "Experiments in Balance with a 3D One-legged Hopping Machine," *International Journal of Robotics Research*, 3 (2-summer) 1984, pp. 75-92.

(14) Raibert, M.H., H.B. Brown, M. Chepponis, E. Hastings, J. Koechling, K.N. Murthy, S.S. Murthy, A.J. Stentz, *Dynamically Stable Legged Locomotion: Progress Report* October 1982 - October 1983 (CMU-RI-TR-83-20), The Robotics Institute, Carnegie-Mellon University, Pittsburgh, 13 December 1983(b).

(15) Salisbury, J.K., "Kinematics and Force Analysis of Articulated Hands," Report No. STAN-CS-82-921, Stanford University, July 1982.

(16) Walker, R.J., *Algebraic curves*, Springer-Verlag, New York, 1978.

Mathematics and Computing

PETER D. LAX*

Abstract. Mathematics has enriched computing by many original and ingenious methods and concepts. The favor has returned in the form of exploratory, experimental computing that revealed astounding new phenomena — strange attractors, almost periodic behavior of nonlinear systems, solitons, complete integrability, universal laws, chaotic behavior, and many others. At the same time, computations are becoming integral parts of mathematical proofs.

How much and what kind of confidence can one place in a computation? It depends on what its purpose is. Many calculations owe their significance to statistical factors which are poorly understood, or not at all. In the numerical exploration of severely instable processes that is danger that the computation becomes a substitute for a non-existent theory.

1. Introduction. Computational science today rivals in importance and intellectual respectability its older sisters, the theoretical and the experimental. It enables theorists to obtain numerical predictions from their theories which can be tested against measurements of experimentalists and observers; in those fields where theory is known to rest on firm foundations, computational modeling makes it possible to eliminate much experimentation, saving time, money, and exposure to danger. Computing also makes it possible to extract hidden information from massive amounts of data by subtle mathematical manipulation.

In this talk I will give examples of what mathematics has done for computing, and what computing has done for mathematics.

2. Mathematical Methods in Computing. Just as the experimental sciences are creatures of instruments of ever increasing resolving power, so the computational sciences are dependent on computers of ever increasing speed and capacity. This is not all; for the effective utilization of computers, efficient operating systems high level programming languages, and intelligent compilers are needed. But most of all, effective computing needs appropriate choice of variables for representing the system under study, ways of replacing these continuous variables by discrete ones with as little discretization error as possible, and clever algorithms for solving

* Courant Institute of Mathematical Sciences, 251 Mercer Street, New York, NY 10012.

the resulting equations. In a study carried out a few years ago [15], John Rice has estimated that the speedup in solving elliptic boundary value problems gained between the years 1945 and 1975 due to increase in machine speed — we are pitting Cray I against an IBM 650 — has been exceeded by the gain due to improved numerical methods. Here is a partial list of the algorithms that brought such impressive improvements:

Line relaxation

Successive overrelaxation of Young

The conjugate gradient method of Lanczos, Hestenes and Stiefel

The optimal algorithm of Wachspress.

The multigrid method, proposed by Federenko and Bahvalov, and developed by Brandt and others.

The capacitance matrix method of Widlund, which exploits the fast Poisson solvers of Bunemann and Hockney.

Sparse martix partitioning by George.

Fluid Dynamics is another area where difficult problems have been made easy and unthinkable ones brought within hailing distance by mathematical ideas such as these:

Method of alternating directions due to Peaceman, Ratchford and Douglas, and the method of fractional steps due to Godunov, Yanenko, Marchuk and Strang.

Second and higher order difference schemes such as Lax-Wendroff, McCormack, Kreiss and others.

Adaptive grids, such as advocated by Thompson, Brackbill and others.

Mesh refinement, as preached and practiced by Ciment, Berger, Jameson, Colella and others.

Implicit difference schemes, such as those proposed by Hirt, Beam and Warming, Harned and others.

Finite element methods are a significant alternative to finite differences for coping with geometric complications.

Spectral methods, pioneered by Leith and put on the map by Orszag and Gottlieb, thanks to the fast Fourier transform of Cooley and Tukey.

The vortex method of Chorin uses an unusual description and approximation to generate and propagate vorticity. Among its many successes is Peskin's calculation of the flow of blood around valves, real and artificial, in the beating human heart.

The method of complex coordinates, devised by Garabedian, allows a unified treatment of subsonic and supersonic flows, and has been used successfully to design shockless transonic airfoils and compressor blades.

The challenge of calculating flows with shocks has generated a number of mathematical ideas. One line of thought is shock capturing, accomplished by v.Neumann and Richtmyer with the aid of artificial viscosity; to this was added the notion of difference approximation in conservation form and numerical flux. Godunov showed how to thread together, by averaging, Riemann initial value problems. Glimm replaced direct averaging by random sampling, and furthermore proved the a. e. convergence of his method. Liu proved convergence of the method when random sampling is replaced by sampling along well distributed sequences. Chorin showed that Glimm's method is very effective for problems that are sensitive to excessive artificial viscosity.

Van Leer, Colella and Woodward have introduced higher accuracy into Godunov's method, to yield astonishingly detailed and accurate calculations of complicated shock patterns.

Jameson has developed an intricate, rapidly converging iterative technique for the calculation of steady transonic flows, with shocks, around complicated aerodynamic shapes.

There is a tendency for spurious oscillations to develop in the calculation of shock fronts; these can be prevented, as Harten has shown, by using variation diminishing schemes. He and Osher have shown how to achieve even greater resolution by using schemes that avoid oscillations through appropriately located interpolation.

Contact discontinuities offer just as great a computational challenge as shocks; the problem here is to avoid loss of resolution through smearing. This difficulty was successfully tackled early on by Harlow's PIC method; another fix is Boris' flux corrected transport, and yet another is the compression method of Harten.

An entirely different way of tackling shocks and contact discontinuities is front tracking. Pioneer users of this method are Moretti, Richtmyer and Lazarus, and Zhang. Glimm, McBryan and their collaborators have put front tracking on the map by demonstrating convincingly that it is a flexible and versatile procedure, adaptable to many problems. Further recent interesting contributions to this approach are due to Colella and I-Lang Chern.

Thanks to these and other ideas, computational fluid dynamics is an enormously active and successful enterprise. There is a danger though that too strong an advocacy of their chosen methodology will keep the various groups from interacting with each other. One of the roles of Congresses such as the present one is to prevent separation by serving as a forum for all groups.

3. Experimental Computing. In a prophetic lecture delivered in Montreal in 1945, see [13], v. Neumann concluded that "really efficient highspeed computing devices may, in the field of nonlinear partial differential equations as well as in many other fields which are now difficult or entirely denied of access, provide us with those heuristic hints which are needed in all parts of mathematics for genuine progress."

In precisely this fashion have Fermi, Pasta, and Ulam discovered the remarkable, almost periodic, behavior of the vibrations of nonlinear chains, and Kruskal and Zabusky the generation and interaction of solitons. The complete integrability of the Toda Lattice became plausible through very careful numerical calculations by Joe Ford; Mitchell Feigenbaum discovered his remarkable universal laws on iterations by analyzing numerical experiments. Numerical studies led Lorenz to the concept of a strange attractor; the understanding of chaotic behavior of simple dynamic systems coexisting with islands of stability has been much enhanced by numerical studies. Not a bad prediction by v. Neumann, especially since the computers he spoke of in 1945 were then merely figments of his imagination.

Already Legendre and Gauss used tables of primes to guess the asymptotic distributions of prime numbers; with the advent of modern computers the quest for asymptotic laws can be and is pursued with a vengeance. So are the roots of Riemann's zeta function; recently van de Lune and te Riele have shown that the zeta function has exactly 300,000,001 zeros whose imaginary parts lie between 0 and 119,590,809.282; all of them have real part ½.

In [12], Mostow did a lot of computing to construct fundamental domains for certain special discrete groups; the computations do not enter the final proof but were essential in discovering what was to be proved.

In [14], Phillips and Sarnak used numerical computations to estimate the lowest eigenvalue of the Laplace-Beltrami operator; they were able to prove mathematically the phenomena discovered experimentally.

In [2], Brezis and Nirenberg were looking for positive solutions of the nonlinear elliptic equation

$$-\Delta u = u^5 + \lambda u$$

in a domain Ω in $\mathbf{R}^n$. Clearly, a necessary condition is that λ be less than the lowest Dirichlet eigenvalue of Δ over Ω. For $n > 3$ this condition turned out to be sufficient for the existence of a positive solution; but the proof could not be extended to $n = 3$. Numerical calculations by Glowinsky for Ω the unit ball in $\mathbf{R}^3$ revealed that λ cannot be less than a certain quantity λ_o. This hint supplied by calculations enabled the authors to prove the existence of a positive solution u under the condition $\lambda_0 < \lambda < \lambda_1$. Another example where the true state of affairs was suggested by a numerical calculation, occurred in the theory of liquid crystals. On the simplest level these are described in terms of harmonic maps from a domain Ω in $\mathbf{R}^3$ to S^2. When the prescribed boundary correspondence $\partial\Omega \to S^2$ has nonzero degree there can be no continuous solution. Numerical investigation by Luskin, Kinderlehrer, et al. [11] of energy minimizing configurations indicate the appearance of a finite number of defects, i.e. discontinuities at a finite number of points. The calculations disclose the nature of the discontinuity at these points. Armed with these suggestions furnished by the calculation, Brezis, Coron and Lieb, [1] succeeded in proving a series of sharp results about the nature of energy minimizing solutions.

In a series of papers David Hoffman, William Meeks and James Hoffman embarked on a search for minimal surfaces that are both complete and embedded without self-intersection in three dimensional space. To the classical tools of explicit representation formulas and asymptotic analysis the authors added visualization of surfaces with the aid of computer graphics. The struggle, eventually crowned with success, is described dramatically in [4]. Hoffman observes: ''we used computer graphics as a guide to verify certain conjectures about the geometry of the surface. We were able to go back and forth between the equations and the images. The pictures were extremely useful as a guide to the analysis''.

Yet another example where a computer calculation has helped to convince a mathematician that he is on the right track is deBrange's celebrated proof of Bieberbach's conjecture. To clinch his argument deBrange needed to know that each of a certain sequence of polynomials is positive on the unit interval. For low degrees this could be done by hand; to deal with polynomials of degree >6 deBrange asked Walter Gautchi to use a computer to check positive for some cases. This was accomplished up to degree 30, at which point Richard Askey was called in, who was able to report that he and George Gasper have some years before proved the positivity of all these polynomials. For a more detailed account see [5].

Many other examples, in geometry, in combinatorics, can be given of successful computing for gaining insight. Experimental computing has truly become a way of life in most branches of mathematics.

4. How Much Confidence Can One Place in a Computation. There are many kinds of computations and many kinds of confidences.

a) Some computations are part of the logical structure of a proof. If real numbers are used, the computations have to furnish ironclad bounds such as are provided by interval arithmetic; we give some examples:

i. Lanford [7] proves Feigenbaum's conjectures concerning iterates of a map. The conjecture states, roughly that high-order iterates of all unimodal maps of an interval into itself, when rescaled appropriately, have very nearly the same shape. This shape is characterized by a functional equation of the form

$$Tf = f \, .$$

Lanford proves the existence and stability of a fixed point f by first computing, by iteration, an approximate fixed point f_ε and then demonstrating the contractive character of a quasi-Newton type iteration in some neighborhood f_ε.

ii. In [3], Fefferman studies the question of formation of atoms in a mixture of electrons and nuclei. In one part of the proof the positivity of a quadratic form has to be proved; this is reduced to showing the nonvanishing of a function defined in terms of an ordinary differential equation. The ordinary differential equation is solved numerically; the task remains to show that errors introduced into the numerical solution, due to truncation and roundoff, do not affect the conclusion. The latter is accomplished by using interval arithmetic; to prevent the intervals from growing too fast, stable algorithms have to be used, see pp. 98-100 of the above quoted work.

iii. The most famous proof relying on machine computations is Haken and Appel's celebrated proof of the four-color theorem. Here all the needed calculations are discrete and can be carried out with absolute accuracy. This proof has been criticized by some for providing no insight why the result is true; this criticism is valid, but can be leveled with equal force against many other proofs which employ no computer. Others have criticized the proof because it is so difficult for a single reader to verify its correctness; true, but equally true of other, hand-carved, elaborate arguments extending over thousands of pages. There are also some who refuse on principle to accept a computer-assisted proof. This strikes me as headed in the wrong direction; after all, logicians agree that an unassailable mathematical proof is one executable by a Turing machine. There is something faintly ridiculous about holding up as the epitome of exactitude proofs carried out by imaginary computers and then balking at a proof carried out by a real one.

b) Many, perhaps most, calculations are carried out to provide quantitative information about problems that are reasonably well-understood theoretically, such as initial and boundary value problems for ordinary and partial differential equations which have been shown to have unique solutions that depend continuously on the data. When solving such problems approximately we are looking for realistic, rather than ironclad, error estimates.

Realistic error estimates are, alas, not easy to come by for problems of real interest, so one proceeds otherwise. A sequence of discretized problems is set up, depending on one or several parameters Δ which measure the scale of discretization; the solution of the discretized problem is denoted by u_Δ. We require the scheme to be stable and to be compatible with the original problem; stability means that u_Δ remain bounded as Δ tends to zero; compatibility means that if a sequence u_Δ converges in some topology as Δ tends to zero, its limit is a solution of the original problem.

It is easy to show that if all u_Δ are contained in a compact set, then the scheme converges. Compactness, however, does not always hold, and even where it does it is hard to prove. Stability, necessary for compactness is not easy to prove either but at least we possess a number of heuristic criteria of practical value. Experience shows that stability and compatibility are valuable design principles for discretizing well-posed and reasonably well-understood problems.

c) There is a large class of calculations, perhaps larger than we suspect, whose significance is only statistical. The investigations of the behavior of the iterates of volume preserving and other maps belong to this class. On chaotic regions, high-order iterates of such maps depend extremely sensitively on the starting point. Clearly, one cannot compute a sequence of thousands of iterates of such transformations. Nevertheless, for transformations about which theory has something to say, numerically computed strings of pseudoiterates behave very much as genuine iterates are supposed to. An explanation for this similarity of behavior has been given by Yorke and associates, see, e.g. [16], based on the so-called Anosov-Bowen shadowing lemma, applicable to mappings whose Jacobians have no eigenvalues of absolute value 1.

Another kind of possible explanation might be given along the following lines:

Suppose for simplicity that the transformation T in question maps the n-dimensional torus into itself. A computer operating with n digit numbers maps n cubes of side length 10^{-n} into cubes of the same size. Denote this transformation by T_n. Numerical experimentations suggest that iterates of such approximations T_n behave, in a sense to be made precise, like iterates of the transformation T which they approximate. In fact, if this were not so, at least in an overwhelming number of cases, it would serve no purpose at all to carry out numerical experimentations!

A result of the above type would be like the KAM theory, with the additional complication that the perturbation that changes T into T_n changes the domain of the map from a manifold to a finite set. No such results are known; I would, however, like to call attention to [9], where it is shown that if T is volume-preserving, T_n can be chosen to be volume preserving too; this is a result in the right direction, but far from what is wanted.

d) An increasing number of calculations attempt to deal with unstable phenomena, such as the interface instabilities of Helmholtz and Rayleigh-Taylor, turbulent flows at high Reynolds number, turbulent multiphase flows, and turbulent combustion. Calculations of this kind typically contain algorithms that are discrete analogues of physical processes. The success of such modeling is measured by the extent to which the approximate flow patterns resemble actual flows observed in laboratories and in nature. There is nothing wrong with such a criterion; a scheme which fails in comparison with reality has to be rejected resolutely. Yet there is something unsatisfactory when a computational scheme usurps the place of a theory. The discrete calculations ought to be the shades cast by a Platonic continuum theory of unstable phenomena.

As more researchers carry out more and more numerical experiments, looking for guidance it is inevitable that some will be misguided. A spectacular example from the past concerns the distribution of prime number. The number of primes less than x, denoted as $\pi(x)$, is, according to the Prime Number theorem, asymptotic to $x/\log x$. Already Gauss has suggested that the logarithmic integral, denoted as $\mathrm{li}(x)$ and defined by

$$\mathrm{li}\,(x) = PV \int \frac{du}{\log u}$$

is in some sense a better approximation to $\pi(x)$ than $x/\log x$. Numerical experiments suggest that this is indeed so; furthermore, for all values of x for which $\pi(x)$ has been computed, it appears that

$$\pi(x) < \mathrm{li}\,(x)\,.$$

This was generally believed to be true for all x, until Littlewood, in 1914, proved the contrary: $\pi(x) > \mathrm{li}\,(x)$ for infinitely many values of x. Littlewood's proof was indirect and gave no indications where the first violation might occur. A student of Littlewood's came up with a constructive proof that there is a number x less than 10 raised to the power 10 four times, for which $\pi(x) > \mathrm{li}\,(x)$. Sherman Lehman has whittled this number down to $10^{1,165}$, still way out of the range for which $\pi(x)$ is expected to be computed exactly in the foreseeable future.

An example of more recent vintage concerns the existence of breather solutions of semilinear wave equations

$$u_{tt} - u_{xx} + f\,(u) = 0, \quad f\,(0) = 0$$

i.e. solutions that are periodic in time and decrease exponentially as functions of x as $|x| \to \infty$. Such solutions have been constructed explicitly by Skyrme when $f\,(u) = \sin u$. Numerical calculations seem to indicate that breathers also exist for other nonlinearities, e.g. $f\,(u)$ cubic. Yet present theory leads toward the conviction that such breathers don't exist, and that the numerical examples are not truly time periodic but decay. The rate of decay is so excruciatingly slow that it cannot be verified by a numerical calculation, see Kruskal and Segur [6].

REFERENCES

[1] Brezis, H., Coron, M. and Lieb, E., Harmonic Maps with Defects, preprint Institute for Mathematics and Applications, University of Minnesota, #253, July 1986.

[2] Brezis, H. and Nirenberg, L., Positive solutions of nonlinear elliptic equations involving critical Sobolev exponents, Communications on Pure and Applied Mathematics, Vol. 36, pp. 437-477, 1983.

[3] Fefferman, C., The N-Body Problem in quantum Mechanics, Communications on Pure and Applied Mathematics, Vol. 30, Num. S, Supplement, 1986.

[4] Hoffman, D. The Computer-Aided Discovery of New Embedded Minimal Surfaces, The Mathematical Intelligencer, Vol. 9, No. 3, pp. 8-21, 1987.

[5] Korevaar, J., "Ludwig Bieberbach's Conjecture and Its Proof by Louis de Branges," The American Mathematical Monthly, Vol. 93, pp. 505-514, 1986.

[6] Kruskal, M., and Segur, H., Nonexistence of Small-Amplitude Breather Solutions in ϕ^4 Theory, Physical Review Letters, Vol. 58, #8, pp. 747-750, 1987.

[7] Lanford, O., A computer assisted proof of the Feigenbaum Conjecture, Bulletin of the American Mathematical Society, **6**, pp. 427-434, 1982.

[8] Lax, P. D., Mathematics and Computing, Journal of Statistical Physics, Vol. 43, No. 5 and 6, pp. 749-756, 1986.

[9] Lax, P. D., Approximation of Measure Preserving Transformations, Comm. on Pure and Applied Mathematics, Vol. XXIV #2, pp. 133-135, 1971.

[10] Lehman, R. S., On the Difference $\pi(x) - \mathrm{li}(x)$, ACTA Arithmetica XI, pp. 397-410, 1966.

[11] Luskin, M., Kinderlehrer, D., and Ericksen, J. L., Remarks about the Mathematical Theory of Liquid Crystals, and Theory and Application of Liquid Crystals, IMA Preprint Series #276, 1986, and IMA Volumes in Math. and Appl. #5, 1986.

[12] Mostow, G. D., Pacific Journal of Mathematics, Vol. 86, pp. 171-276, 1980.

[13] v.Neumann, J. and Goldstine, H. H., On the Principles of large scale computing machines, Collected Works, Vol. V, (Pergamon Press, The MacMillan Co., N.Y. 1963).

[14] Phillips, R. and Sarnak, P., The Laplacian for domains in hyperbolic space and limit sets of Kleinian groups, ACTA MATH. Vol. 155, 1985.

[15] Rice, John, Numerical Methods, Software and Analysis, McGraw-Hill, 1982.

[16] Yorke, J. A., Hammel, S., and Grebogi, C., Do Numerical Orbits of Chaotic Dynamical Processes Represent True Orbits? Jr. of Complexity 3, pp. 136-145, 1987.

[17] Van de Lune, J. and Te Riele, H. J., Math. Comp. (to appear).

Geometry of Numbers: An Algorithmic View

L. LOVÁSZ*

Abstract. The classical field of the geometry of numbers arose as a tool to handle diophantine approximation and diophantine equations; but its main problem, namely finding lattice points in convex bodies, is also the main theme of integer programming and combinatorial optimization. However, these two fields have been developing rather independently. Recent results in the algorithmic treatment of these questions may bridge this gap: they provide efficient means to compute simultaneous diophantine approximation of several numbers, and also yield applications of these algorithms to combinatorial optimization.

1. Introduction. In many algorithmic problems in mathematics, we are looking for a "solution", a vector satisfying certain constraints. Some of the constraints for the solution are "nice and smooth": linear or algebraic equations or inequalities. Often, the set of vectors satisfying these constraints is a convex set; the feasibility of this set of constraints, as well as most other basic problems (e.g. optimization), can be solved by well-known methods from linear and non-linear programming quite efficiently even for rather large problems.

Another typical set of constraints imposes discreteness on the solution: e.g. the solution must have integral entries. Geometrically, these constraints determine a lattice, and we are looking for a common vector of the lattice and the convex set.

Number theorists, motivated by questions from the theory of diophantine approximation and diophantine equations, have studied this problem for over a century. Diophantine approximation, i.e., the approximation of numbers by rational numbers with small denominators, has held out promises of important applications in numerical mathematics. However, the influence of this deep and beautiful theory on applied mathematics has been minimal.

The same general question arises in integer programming; here the lattice is the standard lattice and we are looking for a lattice point satisfying certain linear inequalities. Very little

* Department of Computer Science, Eötvös Loránd University, Budapest, Hungary, H-1088 and Department of Computer Science, Princeton University, Princeton, NJ 08544.

general theory of integer programming was known, however; the general integer programming problem is very hard (NP-complete). In particular, methods from the "geometry of numbers" have been considered useless for integer programming.

In 1979, H. W. Lenstra, Jr. (12) showed that if we fix the number of variables, then integer programs can be solved in polynomial time. He applied methods from the geometry of numbers, in particular, basis reduction in lattices. His work (and some independent development in combinatorial optimization) raised interest in the algorithmic aspects of diophantine approximation. Lenstra, Lenstra and Lovász (13) showed that a crucial step in Lenstra's algorithm, the basis reduction, can be carried out in polynomial time even for varying n. This yielded polynomial-time algorithms for simultaneous diophantine approximation, polynomial factorization and a number of further problems. There is, however, an error factor in this algorithm which grows exponentially with the dimension, and thereby blocks several other possible applications.

These algorithmic results have lead to renewed interest in lattice geometry, in particular in the effectiveness of classical results and in the polynomiality of certain classical bounds.

In this paper we survey some of these recent developments from an algorithmic point of view. The rest of this section contains some preliminaries from geometry. Section 2 discusses the "homogeneous" problem, where K is centrally symmetric with respect to the origin and we are looking for a non-zero lattice point in K. Section 3 discusses the "inhomogeneous" version.

By a *convex body*, we mean closed, convex, bounded, full-dimensional set in $\mathbf{R}^n$ vol(K) denotes the volume of K.

Given n linearly independent vectors $a_1, \ldots, a_n$, the *lattice* generated by them is the set of all of their linear combinations with integral coefficients:

$$L(a_1, \ldots, a_n) = \{x_1 a_1 + \cdots + x_n a_n \; : x_1, \ldots, x_n \in \mathbf{Z}\} .$$

The set $\{a_1, \ldots, a_n\}$ is called a *basis* of the lattice. A lattice can have many bases, but the following quantity, called the *determinant* of the lattice, depends only on the lattice:

$$\det(L) = |\det(a_1 \cdots a_n)| .$$

Geometrically, the determinant of L is the area of the parallelepiped spanned by the vectors of any basis of L.

The *standard lattice* in $\mathbf{R}^n$ consists of all integral vectors.

The set of all vectors $v \in \mathbf{R}^n$ such that $v \cdot z$ is an integer for every $z \in L$, is again a lattice, called the *dual* (or *polar*) of L, and denoted by L^*. The dual of the standard lattice is itself. It is not difficult to see that

$$(L^*)^* = L \;, \quad \det(L) \cdot \det(L^*) = 1 .$$

Let K be a convex body centrally symmetric with respect to the origin. Define its *polar* by

$$K^* = \{x \in \mathbf{R}^n \; : x \cdot y \le 1 \text{ for all } y \in K\} .$$

This is again a convex body centrally symmetric with respect to the origin. It is true that $(K^*)^* = K$. The relationship between the volumes of K and K^* is more complicated. The product vol(K) $\cdot$ vol(K^*) is not constant, but it is bounded by functions of the dimension. It is quite natural to conjecture that this product is maximized when K is a ball and minimized when K (or K^*) is a cube; in other words, that

$$\frac{4^n}{n!} \le \mathrm{vol}(K) \cdot \mathrm{vol}(K^*) \le \frac{\pi^n}{\Gamma(\frac{n}{2}+1)^2} .$$

The first of these assertions was proved by Santaló (15), but the second is still unsettled. A recent important result of Bourgain and Milman (2) gives a lower bound of the form $(c/n)^n$, which has the same form as the conjectured lower bound (and the upper bound), and will be good enough for our purposes.

Finally, we have to say a few words about how lattices and convex bodies are given when they serve as inputs to our algorithms. It is natural to assume that a lattice is given by a basis (which, in algorithmic considerations, is assumed to have rational entries). Accordingly, the contribution of the lattice to the size of the input is the number of bits needed to write down the numerators and denominators of entries of the basis vectors. With convex sets, the situation is more difficult. We want to allow solution sets of linear inequalities, convex hulls of given sets of vectors, balls, ellipsoids etc. It turns out that what we need is the following:

(a) we have a subroutine to decide whether or not a point belongs to K;

(b) we have a subroutine to find the maximum of any linear objective function over K.

(We remark that (a) and (b) are closely related; under appropriate technical hypotheses, any oracle (subroutine) solving one yields a polynomial-time algorithm to solve the other. We do not go into the details of this, but refer instead to Lovász (14) and to Grötschel, Lovász and Schrijver (5).)

Our framework to treat algorithms for convex bodies is to require that their running time by polynomially bounded in the running times of these subroutines.

2. Homogeneous Problems. In this version, we assume that K is centrally symmetric with respect to the origin 0, and we are looking for a *non-zero* lattice point in K. We may consider K as the unit ball of a norm $\|\cdot\|_K$, where

$$\|v\|_K = \min\{t : v \in t \cdot K\} .$$

Then we could find a non-zero lattice point in K (or conclude that non exists) if we could find the shortest non-zero lattice vector in this norm. We denote the length of the shortest non-zero lattice vector by $\lambda_1(K, L)$. So K contains a non-zero lattice point if and only if $\lambda_1(K, L) \le 1$.

A classical result of Minkowski asserts the following.

2.1 Theorem. Let L be a lattice and K, a 0-symmetric convex body in $\mathbf{R}^n$. Assume that

$$\mathrm{vol}(K) \ge 2^n \cdot \det(L) .$$

Then K contains a non-zero lattice point.

(In other words, $\lambda_1(K, L) \le 2 \cdot (\det(L)/\mathrm{vol}(K))^{1/n}$.)

This theorem is central in the applications of lattice geometry in number theory. However, if we are interested in the algorithmic aspects, then it has a shortcoming: no efficient algorithm is known to *find* the non-zero lattice point it K (the fact that the condition involves the volume of K is another disadvantage, since the volume of K is difficult to compute even if K is as simple as a polytope described by linear inequalities). Results of Grötschel, Lovász and Schrijver (5,6), based on the work of Lenstra, Lenstra and Lovász (13), imply the following:

2.2 Theorem. Let L be a lattice and K, a 0-symmetric convex body in $\mathbf{R}^n$. Then we can find a non-zero lattice vector v such that $\|v\|_K \le 2^n \lambda_1(K, L)$.

In other words, if K contains a non-zero lattice point, then we can find a lattice point in $2^n \cdot K$ in polynomial time.

To exemplify the applications of this theorem, let us formulate one of its consequences in diophantine approximation. Let $\alpha_1, \ldots, \alpha_n$ be real numbers. Suppose we want to approximate them by rationals $p_1/q, \ldots, p_n/q$ with a common denominator q. This is trivial with error $1/(2q)$; we want something better; more exactly, we want an error ε/q, where $0 < \varepsilon < \frac{1}{2}$ is given. A classical theorem of Dirichlet says that such an approximation always exists with common denominator $q < \varepsilon^{-n}$. Unfortunately, it is not known how to find such an approximation (even if we assume that the α's are rational numbers). However, applying Theorem 2.2 to the lattice generated by the columns of the matrix

$$\begin{pmatrix} 1 & 0 & & \alpha_1 \\ & 1 & & \alpha_2 \\ & & \cdot & \cdot \\ & & \cdot & \cdot \\ & 0 & \cdot & \cdot \\ & & & 2^{-n^2}\varepsilon^{n+1} \end{pmatrix}$$

and to the cube with sides 2ε with center 0 as K, we obtain the following result:

2.3 Theorem. Given n rational numbers $\alpha_1, \ldots, \alpha_n$, and a positive rational number ε, we can compute in polynomial time integers $p_1, \ldots, p_n$ and q such that

$$\left| \alpha_i - \frac{p_i}{q} \right| < \frac{\varepsilon}{q} \quad (i = 1, \ldots, n) \text{ and}$$

$$0 < q < 2^{n^2}\varepsilon^{-n}$$

It is perhaps the most important unsolved problem in this field to replace the exponential factors between the existence and algorithmic results by something smaller. Schnorr (16) proved that the factor 2^n can be replaced by $(1+\delta)^n$ for any fixed positive δ. There is no hope to go down to 1, i.e., to find the shortest lattice vector in polynomial time. This is known to be NP-complete for the cube (van Emde-Boas (3)), and probably for most other bodies. But one could hope that it could be replaced by some polynomial of n, maybe by $O(n)$.

This hope is supported by the following argument. Assume that we have indeed found in polynomial time a non-zero lattice vector which is guaranteed to be at most, say, $O(n)$ times as long as the shortest. Then in particular this algorithm implies a lower bound on $\lambda_1(K, L)$, which is within a factor of $O(n)$ to $\lambda_1(K, L)$, and which is verifiable in polynomial time. Now the following result, which follows from Largarias, Lenstra and Schnorr (11) and Bourgain and Milman (2), shows that such a lower bound does indeed exist. Let $\{b_1, \ldots, b_n\}$ be a basis of the lattice. Define

$$(*) \qquad t_k = \min_{x_1, \ldots, x_{k-1}} \| b_k + x_{k-1}b_{k-1} + \ldots + x_1 b_1 \|_K$$

Then it is easy to see that $\lambda_1(K, L) \geq \min\{t_1, \ldots, t_n\}$. Moreover, there always exists a basis $\{b_1, \ldots, b_n\}$ for which $\min\{t_1, 2t_2, \ldots, nt_n\} \geq \text{const} \cdot \lambda_1(K, L)$. For a given basis, the numbers $t_1, \ldots, t_n$ can be computed in polynomial time (at least with arbitrarily small relative error). This follows from the "ellipsoid method" (see Grötschel, Lovász and Schrijver (5)), since the quantity minimized on the right hand side of (*) is a convex function of $x_1, \ldots, x_{k-1}$.

In many applications, we are in fact using the following corollary of Minkowski's Theorem and the result of Bourgain and Milman:

2.4 Theorem. For every convex body K and lattice L in $\mathbf{R}^n$,

$$\lambda_1(K, L) \cdot \lambda_1(K^*, L^*) \leq \text{const} \cdot n .$$

Note that this theorem asserts the existence of two vectors, one in L and one in L^*. Unfortunately, we do not know how to find (in polynomial time) such vectors, or even two vectors such that the product of their lengths is bounded by polynomial in n. In fact, it can be shown that this can be achieved if a polynomial error factor can be achieved in Theorem 2.2. Of course, Theorem 2.2 implies that we can find (in polynomial time) two non-zero vectors $v \in L$ and $v^* \in L^*$ such that $\|v\|_K \cdot \|v^*\|_{K^*} \leq 4^n$.

We conclude this section with a sketch of some applications of the diophantine approximation results. Let $\alpha_1, \ldots, \alpha_n$ be rational numbers with "huge" numerators and denominators. We want to replace them by "nicer" rationals $\beta_1, \ldots, \beta_n$ so that some basic patterns among these numbers are preserved. For example, many basic problems in combinatorial optimization (shortest path, minimum cut, matching, travelling salesman etc.) have the following framework. We are given a finite set, say $\{1, \ldots, n\}$, and (implicitly) a collection of "feasible" subsets of it. We are also given costs α_i ($i = 1, \ldots, n$). Our task is to find a feasible subset with minimum cost. Several techniques known to solve various classes of such problems in polynomial time (ellipsoidal methods, scaling) have a time bound which depends on the number if digits of the costs. Of course, if the costs are very "ugly" then already a simple operation like comparing two of them will take a long time, if we count bit operations. But if we count an arithmetic operation (addition, subtraction, multiplication, comparison) as one step, then it is natural to require (at least from a theoretical point of view) that the running time be bounded by a polynomial in n. (For technical reasons, it is also usually required that the numbers produced during the computation have only a polynomial number of digits.) Such algorithms are called *strongly polynomial.*

Frank and Tardos (4) developed a method to turn polynomial time algorithms into strongly polynomial ones. A first idea (which does not yet work) is find simultaneous diophantine approximation p_i/q to the α_i with error less than $1/(nq)$. We know such an approximation exists with $q < n^n$ and we can even find one if we only require $q < Q$ where

$$Q = 2^{n^2} \cdot n^n .$$

to the α's and then solve the problem with the p_i's in place of the α_i's. It is quite easy to see that if X and Y are two subsets of S and

$$\sum (\alpha_i : i \in X) \leq \sum (\alpha_i : i \in Y),$$

then also

$$\sum (p_i : i \in X) \leq \sum (p_i : i \in Y).$$

This implies that every optimum solution to the original problem will remain an optimum solution for the new, rounded costs. However, new optimal solutions may arise and when we solve the new, rounded problem, we may find one of these, not knowing whether it was also optimal originally.

Frank and Tardos find a very elegant way out of this difficulty by developing a "diophantine series expansion" (which may be well applicable in many other situations as well). They start with normalizing the costs so that the maximum absolute value is 1. Then they find simultaneous diophantine approximation, subtract the approximating rational numbers from the original costs, normalize and approximate the errors and so on. The procedure terminates in n steps, since in each step a new 0 cost is introduced. As a result, they obtain a representation of the vector $a = (\alpha_1, \ldots, \alpha_n)^T$ in the following form:

$$(**) \qquad a = Pw ,$$

where P is a square integral matrix with ''small'' entries (each entry of P is less than Q and so has $O(n^2)$ digits) and $w = (\omega_1, \ldots \omega_n)^T$ is a vector whose entries decrease very fast:

$$| \omega_{i+1} | \le \frac{| \omega_i |}{n q_{i+1}}$$

where q_i is the denominator in the ith approximation. In particular, q_i is at most Q and at least as large as any entry in the ith column of P. Now if we take any rational vector u whose entries drop in a similar rate and compute a vector $b = (\beta_1, \ldots, \beta_n)$ by

$$b = Pu ,$$

then for any two subsets X and Y of S we will have

$$\sum (\alpha_i : i \in X) \le \sum (\beta_i : i \in Y)$$

if and only if

$$\sum (\beta_i : i \in X) \le \sum (\beta_i : i \in Y) .$$

So the costs α and β will lead to the same optimal solutions. Now the number of digits of the β's depend only on n; hence if we can solve the problem with the costs β in polynomial time, this will be in fact a strongly polynomial algorithm.

The Frank-Tardos procedure can also be used to answer a question raised by N. Megiddo (personal communication): Assume that we can do exact arithmetic with reals, and that we are given n real numbers $\alpha_1, \ldots, \alpha_n$. We want to compute a lower bound on the difference of two subset-sums of the α_i, provided it is not 0. This is trivially solved by computing 3^n linear combinations of the α_i with $\{-1, 0, 1\}$-coefficients and then sorting them. If the α_i are rational, the reciprocal of the product of their denominators is an appropriate bound, which can be computed in polynomial time. Can we compute such a bound for general (real) α_i in a polynomial number of arithmetic operations? The answer is in the affirmative: compute the representation (**), and take the last non-zero entry of w.

3. Inhomogeneous Problems. The general question here is: given a lattice L and a convex body K, find a lattice point in K (or conclude that none exists). Integer programming is a typical example, where L is the standard lattice $\mathbf{Z}^n$ and K is the solution set of a system of linear inequalities.

The polynomial algorithm of H. W. Lenstra, Jr. for integer programming in bounded dimension depends on this geometric approach. It seems that to improve the efficiency of such algorithms, we need better answers to the question: how does a lattice-point-free convex body look like? Note that there is no immediate extension of Minkowski's Theorem: a lattice point free body can have arbitrarily large volume (e.g. consider a flat but very long rectangle between two consecutive lattice lines).

However, there is the following basic property of lattice point free convex bodies.

3.1 Theorem. Let L be a lattice and K, a convex body in $\mathbf{R}^n$ and assume that $K \cap L = \phi$. Then there exists a vector $v \in L^*$, $v \neq 0$ such that

$$\max\{v \cdot x : x \in K\} - \min\{v \cdot x : x \in K\} \le f(n),$$

where $f(n)$ is a constant depending only on the dimension.

We call the minimum of $(\max\{v \cdot x : x \in K\} - \min\{v \cdot x : x \in K\})$, taken over all non-zero dual lattice vectors v, the *lattice width* of K. It is easy to see that this number is just $\lambda_1((K - K)^*, L^*)$.

This result goes back to Khinchine (10), who proved it when K is a ball with $f(n) \approx n!$. Lenstra (12) used an algorithmic version of this result in his integer programming algorithm: he showed that for fixed n, and $f(n) \approx 2^{n^2}$, one can either find in polynomial time either a lattice vector in K or a dual lattice vector v with the property in the theorem. Grötschel, Lovász and Schrijver (6) showed that this can in fact be achieved in polynomial time even for variable n. Babai (1) improved this algorithmic result for $f(n) \approx 3^n$.

If we give up the algorithmic requirement then better results are known. Lagarias, Lenstra and Schnorr (11) and Hastad (7) proved the above theorem with $f(n) \approx n^{2.5}$. Currently the strongest result appears to be that of Kannan and Lovász (9) which asserts that the theorem remains valid with $f(n) = O(n^2)$. The best possible value of $f(n)$ is perhaps linear in n. The convex body K defined by the inequalities $x \ge 0$, $\mathbf{1} \cdot x \le n$ contains no integral interior point, and so shrinking it by arbitrarily little, it will be disjoint from the standard lattice. On the other hand, let v be any non-zero integral vector and assume that (say) its first entry $v_1 \ne 0$. Then the linear objective function $v \cdot x$ assumes the value 0 as well as the value nv_1 on K, and so $f(n)$ certainly cannot be replaced by anything less than n.

Unfortunately, these polynomial bounds are not algorithmic. The main source of exponentiality in the algorithmic results is the exponential error factor in Theorem 2.2.

The way Theorem 3.1 is used in Lenstra's algorithm and its subsequent refinements is the following. Suppose that we want to decide whether or not K contains a lattice point. Let us use an algorithmic version of Theorem 3.1 to find either a lattice point in K (in which case we are done), or a dual lattice vector v as in the theorem. In the latter case, the value of $v \cdot x$ for any possible vector $x \in K \cap L$ is an integer between $\min\{v \cdot x : x \in K\}$ and $\max\{v \cdot x : x \in K\}$. Hence we can "branch" and reduce our problem to at most $f(n) + 1$ $(n-1)$-dimensional subproblems: find a lattice point in $K \cap \{x : v \cdot x = t\}$ ($t \in \mathbf{Z}$, $\min\{v \cdot x : x \in K\} \le t \le \max\{v \cdot x : x \in K\}$).

This recurrence leads to a tree-structured algorithm with $f(n) \cdot f(n-1) \cdot \cdots \cdot f(1)$ nodes. If n is fixed then this is polynomial time; but for variable n, it becomes badly exponential. In number-theoretic problems (inhomogeneous diophantine approximation), n is often not large and the difficulty lies in the structure of the lattice. In these cases, such an approach may be practical. But in most combinatorial optimization problems, the number of variables is large (typically the interesting range starts with hundreds), and so the above algorithm is impractical.

This application also underlines the significance of obtaining an algorithmic result with a polynomial $f(n)$. This would still give an exponentially large tree (as it is to be expected, since integer programming is NP-complete), but at least one could hope for some ways of pruning the tree and thereby improving the running time. As it stands, already one branching is forbidding in most cases.

Another way of improving the running time of Lenstra's algorithm was suggested by Kannan (8) and Kannan and Lovász (9). It is based on the idea that if the lattice width of a lattice-point-free body is close to the upper bound then there will be many vectors in the dual lattice realizing a small lattice width. More exactly, the following is true.

3.2 Theorem. Let L be a lattice and K, a convex body in $\mathbf{R}^n$, and assume that $K \cap L = \phi$. Then there exists a k, $1 \leq k \leq n$, and k linearly independent vectors in the dual lattice $v_1, \ldots, v_k$, such that for each $1 \leq i \leq k$,

$$\max\{v_i \cdot x : x \in K\} - \min\{v_i \cdot x : x \in K\} \leq \text{const} \cdot k^3 \cdot \log^2(k+1) .$$

We have no control over the value of k here. Consider the standard lattice and the convex set in $\mathbf{R}^n$ defined by $x \geq 0$, $x_1 + \ldots + x_m \leq m$, $x_{m+1} \leq N, \ldots, x_n \leq N$. Then only m linearly independent dual lattice vectors realize small lattice width; on the other hand, the lattice width of this body is m. So the choice of k in the theorem must be between $m^{1/3}$ and m.

This theorem too has an algorithmic version: if we replace the right hand side by 2^k then we can find, in polynomial time, either a lattice point in K or k linearly independent lattice vectors as in the theorem.

We can use this result in a Lenstra type algorithm as follows: we either find a lattice point in K and stop, or find, for some k, k linearly independent dual lattice vectors in whose direction K has width at most 2^k. This means that we can branch into $(2^k)^k$ subproblems, each in dimension $n-k$. If k is bounded, this is a bounded number of subproblems. If k is large, we gain a lot because the dimension drops substantially.

There is definitely not enough known about the structure of convex bodies containing no lattice points. The following recent result of Kannan and Lovász (to be published) gives a property involving the volume, a bit in the style of Minkowski's Theorem. Define the *covering radius* $\mu(K, L)$ as the positive real number t for which $t \cdot K + L = \mathbf{R}^n$, i.e., the translates of $t \cdot K$ by the lattice vectors cover the whole space. Clearly if K contains no lattice point then $\mu(K, L) > 1$; in fact, $\mu(K, L > 1$ if and only if K has a translate containing no lattice point.

Trivially,

$$\mu(K, L) \geq \left(\frac{\det(L)}{\mathrm{vol}(K)} \right)^{1/n} .$$

This may be a very bad lower bound since, as remarked, $\mathrm{vol}(K)$ may be arbitrarily large while $\mu(K, L) > 1$. To obtain a stronger version, consider any integer k, $1 \leq k \leq n$, and a projection Π of the whole space on a k-dimensional subspace. Then it is also easy to see that

$$\mu(K, L) \geq \left(\frac{\det(\Pi L)}{\mathrm{vol}(\Pi K)} \right)^{1/k} .$$

Let $\phi(K, L)$ denote the maximum of the right hand side for all choices of k and Π. Then the following holds:

3.3 Theorem. If L is a lattice and K, a convex body in $\mathbf{R}^n$, then

$$1 \leq \frac{\mu(K, L)}{\phi(K, L)} \leq n .$$

We do not know an example where the ratio would exceed $O(\log n)$. The standard lattice, together with the body defined by $x \geq 0$, $\sum x_i / i \leq 1$ gives a ratio of $\approx \log n$. So far, we do not have an algorithmic version of this theorem.

REFERENCES

[1] BABAI, L., (1986): On Lovász' lattice reduction and the nearest lattice point problem, *Combinatorica* **6**, 1-13.

[2] BOURGAIN J. AND MILMAN, V. D., Sections euclidiennes et volume des corps symètriques convexes dans R^n, *C. R. Acad. Sci. Paris* **300**, pp. 435-437.

[3] van EMDE-BOAS, P., (1981): Another NP-complete partition problem and the complexity of computing short vectors in a lattice, Report 81-04, Mathematical Institute, Univ. of Amsterdam, Amsterdam.

[4] FRANK A. AND TARDOS E., (1987): An application of simultaneous Diophantine approximation in combinatorial optimization, *Combinatorica,* **7** pp. 49-65.

[5] GRÖTSCHEL, M., LOVÁSZ, L. AND SCHRIJVER, A., (1987): Geometric Algorithms and Combinatorial Optimization, Springer, Heidelberg-New York-Tokyo.

[6] GRÖTSCHEL, M., LOVÁSZ, L. AND SCHRIJVER, A., (1982): Geometric methods in combinatorial optimization, in Progress in Combinatorial Optimization, W. R. Pulleyblank, ed., Academic Press, New York, pp. 167-183.

[7] HASTAD, J., A good dual witness, *Combinatorica* (to appear).

[8] KANNAN, R., (1983): Improved algorithms for integer programming and related lattice problems, in Proc. 15th Annual Symposium on Theory of Computing, pp. 193-206.

[9] KANNAN, R. AND LOVÁSZ, L., (1986): Covering minima and lattice point free convex bodies, in Foundations of Software Technology and Theoretical Computer Science, K. V. Nori, ed., *Lecture Notes in Comp. Sci.* **241**, Springer, Berlin-Heidelberg-New York-London-Paris-Tokyo, pp. 193-213.

[10] KHINCHINE, A., (1948): A quantitative formulation of Kronecker's theory of approximation, Izv. Acad. Nauk SSSR, *Ser. Math.* **12**, pp. 113-122. (In Russian.)

[11] LAGARIAS, J., LENSTRA, H. W. Jr. AND SCHNORR, C. P., Korkine-Zolotarev bases and successive minima of a lattice and its reciprocal lattice, *Combinatorica* (to appear).

[12] LENSTRA, H. W. Jr., (1983): Integer programming with a fixed number of variables, *Oper. Res.,* **8**, pp. 538-548.

[13] LENSTRA, A. K., LENSTRA, H. W. Jr. AND LOVÁSZ, L., (1982): Factoring polynomials with integral coefficients, *Math. Ann.*, **261** pp. 515-534.

[14] LOVÁSZ, L. (1986): An Algorithmic Theory of Numbers, Graphs and Convexity, *CBMS-NSF Regional Conference Series in Applied Math.* **50**, SIAM, Philadelphia.

[15] SANTALÒ, L. A. (1949): Un invariante afin pasa los cuerpos convexos del espacio de n dimensiones, *Portugal Math.*, **8**, pp. 155-161.

[16] SCHNORR, C. P., (1985): A hierarchy of polynomial time basis reduction algorithms, in Theory of Algorithms, L. Lovász and E. Szemerèdi, eds., North-Holland, Amsterdam, pp. 375-386.

Vortex Dynamics: Numerical Analysis, Scientific Computing, and Mathematical Theory

ANDREW MAJDA*

Abstract. This paper presents an overview of some recent and ongoing developments involving the interplay among ideas and results from large-scale scientific computing, numerical analysis, and the modern mathematical theory of fluid dynamics utilizing concepts from vortex dynamics. Applications to inviscid and high Reynolds number incompressible fluid flow in both two and three space dimensions are discussed here.

Introduction With modern applied mathematics as a rapidly expanding and maturing discipline, it is a wonderful development that all the major applied mathematics societies around the world have organized a first international conference, I.C.I.A.M. 87, and the author considers it a great honor to give one of the invited addresses at this meeting.

The topic of this paper is the use of vortex dynamics in the description of incompressible fluid flows. Problems associated with vortex dynamics are certainly of intense scientific and engineering interest with numerous applications ranging for example from the accurate prediction of hurricane paths to control of the hazardous large vortices shed by landing jumbo jets to the design of efficient internal combustion engines. Here we emphasize the interaction of ideas from numerical analysis, large-scale scientific computing, and the modern mathematical theory of fluid dynamics. Thus, we concentrate on the mutual ongoing interactions depicted in the following diagram:

* Department of Mathematics and Program in Applied and Computational Mathematics, Princeton University, NJ. Partially supported by grants N.S.F.#DMS 8702864 A.R.O. #DAAL03-86-K-003, O.N.R.#N00014-85-K-0507, D.A.R.P.A. — O.N.R. #N00014-86-K-0759.

(0.1)

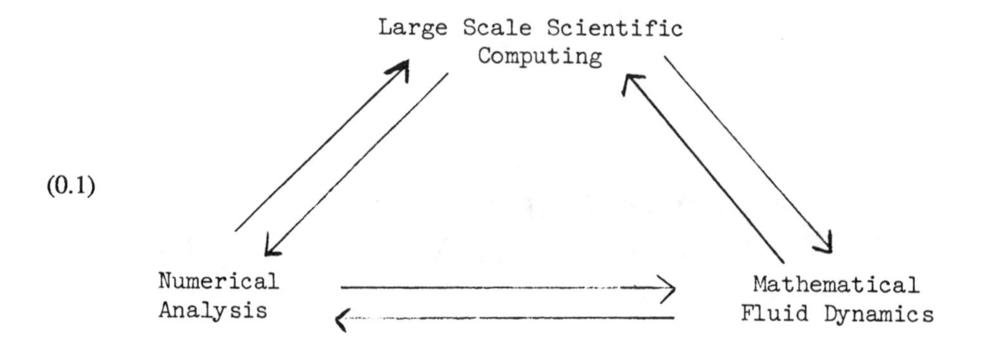

In this paper, we will give several current examples of circumstances where problems in one area have been solved or substantially illuminated through the mutual interaction of these disciplines.

The primitive ancestor of the basic numerical methods which we discuss here is the point-vortex method first used by Rosenhead [63] in 1931 in an attempt to simulate vortex sheets in 2-D inviscid fluid flows. The point-vortex method is an algorithm based on the intuitively appealing idea of approximating a general solution of the 2-D inviscid Euler equations by an interacting finite superposition of elementary exact solutions. However, subsequent attempts with increased resolution to obtain a convergent solution failed ([15]). From the modern viewpoint, the failure of the point-vortex method rests on the choice of the elementary exact solution, a point vortex has the induced velocity field

$$v_\Gamma = \frac{\Gamma}{2\pi|x|^2}\begin{pmatrix}-x_2\\ x_1\end{pmatrix} \tag{0.2}$$

with $|x|^2 = x_1^2 + x_2^2$. Even for extremely singular initial data such as those for vortex sheets, for obvious physical reasons, we are interested in incompressible velocity fields $v = {}^t(v_1, v_2)$ with locally finite kinetic energy, i.e.

$$\int_{|x|\le R} |v|^2 dx \le C(R) \quad \text{for any} \quad R > 0\,. \tag{0.3}$$

On the other hand, point vortices have infinite energy, i.e.,

$$\int_{|x|\le R} |v_\Gamma|^2 dx = +\infty \quad \text{for any} \quad R > 0\,.$$

Thus, one can expect spurious non-physical behavior in approximating solutions with locally finite energy by superpositions of finite numbers of solutions with infinite energy!! This is borne out by careful numerical experiments both for vortex sheets and smooth flows ([18], [6], [49]) — when two point vortices pass nearby, they induce non-physically large velocities which cause instability.

The development of vortex methods into an effective tool for large scale scientific computing began with the seminal paper ([19]) by Chorin in 1973. This paper contains the first large scale simulations of flow past a 2-D cylinder at high Reynolds numbers and these results were achieved through three novel computational ideas which have influenced many of the developments which we discuss in this paper. These three computational ideas are the following:

(0.4)

1) Finite vortex cores are used to cut-off the non-physical singularity in the point-vortex method yielding stable approximations with finite energy.

2) In the interior of the region the effect of diffusion is simulated by a random walk of the finite core vortices.

3) The very attractive idea with strong physical motivation of introducing new vortices at the boundary is utilized in order to satisfy the no-slip boundary conditions — this mimics the way in which a boundary sheds vorticity.

A number of controversial issues (mentioned in [46] for example) regarding Chorin's algorithm initially drove new questions regarding the numerical analysis of vortex methods which began as a field of research in 1978 ([45], [46]). In the 1980's, our knowledge of the numerical properties of vortex methods for inviscid flows in both 2-D and 3-D has expanded at an enormous rate and these developments have led to improved algorithms for large-scale computing (especially in 3-D). In fact, even though the entire field of numerical analysis of vortex methods is less than ten years old — these topics are called the "classical convergence theory" in Section 1 and these developments serve to illustrate the interactions along one leg of the triangle in (0.1). Section 1 also contains a brief discussion of some of the exciting new developments ([41]), [42]) which lead to rapid summation of vortex interactions in $O(N)$ operations — results with potentially great practical importance for large scale simulations. In Section 2, we briefly discuss some of the recent very interesting large scale simulations with vortex methods. We primarily mention only computations which have been done since the appearance of A. Leonard's excellent survey article in 1985 ([51]). We emphasize the details of recent numerical computations with vortex methods for vortex sheets for the 2-D Euler equations. These calculations indicate incredible complexity both as time evolves and as various regularization parameters converge to zero ([50], [49], [66]). Finally, in Section 3 we describe the new mathematical framework which has been developed by R. Dipema and the author ([30], [31], [32]) to elucidate and explain the new phenomena that arise in the evolution of vortex sheets as suggested by the numerical calculations described in Section 2. This ongoing research represents interaction along both the second and third legs of the triangle in (0.1).

We end this introduction by commenting that this paper is not a comprehensive survey article and the author apologizes in advance of any oversights. Although in many respects the article in [52] is obsolete, the author recommends the two excellent review articles by A. Leonard ([52], [51]) as an introduction to large scale computing with vortex methods. Several parts of the four papers ([6], [7], [8], [1]) contain surveys of the numerical analysis of vortex methods for inviscid fluid flow in 2-D and 3-D while the paper [58] provides an introduction to recent developments in the mathematical theory of incompressible flow which utilize vortex dynamics. The remainder of this paper has the following

Table of Contents

0. Preliminaries: The Euler and Navier-Stokes Equations

The Navier-Stokes equations of incompressible fluid flow in three space dimensions are the system of four equations,

$$\frac{Dv}{Dt} = -\nabla p + \nu \Delta v , \quad t > 0 , \quad x \in R^3$$

(0.5)

$$\operatorname{div} v = 0 , \quad v(x, 0) = v_o(x) .$$

Here $v = {}^t(v_1, v_2, v_3)$ is the fluid velocity, $x = (x_1, x_2, x_3)$ denotes the spatial coordinates, p is the scalar pressure, the quantity $\frac{D}{Dt}$ is given by

$$\frac{D}{Dt} = \frac{\partial}{\partial t} + \sum_{j=1}^{3} v_j \frac{\partial}{\partial x_j}$$

and represents convection along the fluid particles, $\nu > 0$ is the reciprocal of the Reynolds number, and Δ is the Laplace operator. When the coefficient ν vanishes in (0.5), we have the incompressible Euler equations. Here we are primarily interested in high Reynolds number and inviscid fluid flows so that $\nu << 1$. For fluid flow in a region Ω with smooth boundary, $\partial\Omega$, the correct boundary conditions for the Navier-Stokes equations involve the no-slip condition,

$$v|_{\partial\Omega} = 0 \tag{0.6}$$

while the boundary conditions for the Euler equations are less stringent and only require that the normal velocity vanishes, i.e.

(0.7) $$v \cdot \vec{n}|_{\partial\Omega} = 0$$

with $\vec{n}$ the outward normal to $\partial\Omega$.

The Vorticity-Stream Formulation

By taking the curl of the first equation in (0.5), we obtain the equation for vortex-stretching,

(0.8) $$\frac{D\omega}{Dt} = (\omega \cdot \nabla)\, v + \nu\, \Delta\, \omega\, .$$

Here $\omega = \operatorname{curl} v$ is the vorticity vector. In all of space, the elliptic system

$$\operatorname{div} v = 0\,, \qquad \operatorname{curl} v = \omega$$

has a solution v given by the familiar Biot-Savart formula,

(0.9) $$v(x, t) = \int_{R^3} K(x - x')\, \omega\, (x', t)\, dx'$$

where $K(x)$ is the singular kernel given by

(0.10) $$K(x) = -\frac{1}{4\pi}\, |x|^{-3} x\, \times\,, \qquad |x| \neq 0\, .$$

The formulae in (0.8) — (0.10) allow us to rewrite the Euler and Navier-Stokes equations in (0.5) as an equation involving only the evolution of vorticity (see [58]).

In two space dimensions, the equations in (0.8) — (0.10) have a well-known simpler form. For 2-D flows, $x = (x_1, x_2)$, $v = {}^t(v_1, v_2)$, the vorticity ω is the scalar quantity,

$$\omega = (v_2)_{x_1} - (v_1)_{x_2}$$

and the equations analogous to (0.8) — (0.10) read as follows.

(0.11) $$\frac{D\omega}{Dt} = \nu\, \Delta\, \omega$$
$$v(x, t) = \int_{R^2} K(x - x')\, \omega\, (x', t)\, dx'$$

with the singular kernel $K(x)$ given by

(0.12) $$K(x) = \frac{1}{2\pi}\, |x|^{-2} \begin{pmatrix} -x_2 \\ x_1 \end{pmatrix} .$$

In fact, with a stream function ψ defined by $v = (\psi_{x_2}, -\psi_{x_1})$, we have the equation

(0.13) $$\Delta\, \psi = -\, \omega\, .$$

For later developments, next we record some simple steady exact solutions for the 2-D Euler equations by utilizing (0.11) — (0.13). We choose ω to be any radial function, $\omega(r)$ with $r = |x|$ for $x \in R^2$ then (0.13) has a radial solution $\psi(r)$ and a corresponding steady exact solution for the 2-D Euler equations given by

$$v = \frac{1}{2\pi} \int_{o}^{|x|} s\omega(s)\, ds\, |x|^{-2} \begin{pmatrix} -x_2 \\ x_1 \end{pmatrix} . \tag{0.14}$$

These are the well-known steady swirling flows. One of the advantages of the vorticity-stream formulation in either 2-D or 3-D (at least for fluid flow in all of space) when compared with the primitive-variable form of the equations in (0.5) is that the incompressibility constraint, div $v = 0$, is automatically satisfied.

Particle Trajectories and Lagrangian Equations for Vortex Stretching

The particle trajectory transformation, $X(\alpha, t)\colon R^3 \to R^3$ marks the location of the fluid particle at time t which began at location $\alpha = {}^t(\alpha_1, \alpha_2, \alpha_3)$ at time $t = 0$. Thus $X(\alpha, t)$ satisfies the nonlinear O.D.E., *the particle trajectory equations*,

$$\begin{aligned} \frac{dX(\alpha,t)}{dt} &= v\,(X(\alpha,t),t) \\ X(\alpha,t)|_{t=0} &= \alpha\,. \end{aligned} \tag{0.15}$$

By differentiating the equation in (0.15), we compute that the Jacobian matrix, $\nabla_\alpha X = \left(\dfrac{\partial X_i}{\partial \alpha_j}\right)$, satisfies

$$\frac{d}{dt} \nabla_\alpha X = \nabla v\, \nabla_\alpha X\,, \quad \nabla_\alpha X|_{t=0} = I \tag{0.16}$$

with ∇v evaluated at $(X(\alpha,t),t)$ in (0.16). One consequence of the incompressibility condition, div $v = 0$, is that det $\nabla_\alpha X \equiv 1$, i.e. the particle trajectory transformations preserve volume. Another consequence of (0.16) is that for inviscid fluid flows, the vorticity equation in (0.8) can be integrated exactly through the particle trajectory transformation. The result is the following formula for the vorticity $\omega(x,t)$:

$$\omega(X(\alpha,t),t) = \nabla_\alpha X \cdot \omega_o(\alpha) \tag{0.17}$$

where ω_o = curl v_o is the vorticity in the initial data. The verification of the identity in (0.17) just involves multiplying the equation in (0.16) on the right by $\omega_o(\alpha)$ and then observing that as a consequence of (0.8), $\omega(X(\alpha,t),t)$ and $\nabla_\alpha X \cdot \omega_o(\alpha)$ satisfy the same O.D.E.; then (0.17) follows immediately from uniqueness of solutions to the O.D.E. Since det $\nabla_\alpha X(\alpha,t) = 1$ but all eigenvalues of $\nabla_\alpha X(\alpha,t)$ do not necessarily have modulus one, the equation in (0.17) provides a direct local interpretation of the "vortex stretching" which plays such a prominent role in the fluid dynamics of inviscid and high Reynolds number flows in 3-D.

Reformulation of the Euler Equations as an Integro-Differential Equation for the Particle Trajectories

Every new formulation of the equations of fluid flow from (0.5) leads to the possible design of new numerical methods. The primitive-variable formulation in (0.5) can be approximated by finite difference, spectral, or finite element methods; the same methods can also be applied to the vorticity-stream formulations in (0.8) — (0.13) in both 2-D and 3-D; for example, Arakawa's scheme for solving 2-D Euler utilizes finite difference approximations to (0.11) and (0.13) with $\nu = 0$. Here we rewrite the inviscid Euler equations in an equivalent form which is particularly suitable for designing inviscid vortex methods in both 2-D and 3-D. First, we express $v(x,t)$ in terms of $\omega(x,t)$ from the Biot-Savart law in (0.9) and insert this expression in the particle trajectory equation to obtain the equation

(0.18) $$\frac{dX(\alpha,t)}{dt} = \int_{R^3} K(X(\alpha,t) - x')\,\omega\,(x',t)\,dx'\,.$$

Next, we change variables in the integral via the transformation, $x' = X(\alpha',t)$ and get

(0.19)A) $$\frac{dX(\alpha,t)}{dt} = \int_{R^3} K(X(\alpha,t) - X(\alpha',t))\,\omega\,(X(\alpha',t),t)\,d\alpha'$$

$$X(\alpha,t)|_{t=0} = \alpha\,.$$

If we recall the formula from (0.17),

(0.19)B) $$\omega(X(\alpha',t),t) = \nabla_{\alpha'}X(\alpha',t) \cdot \omega_o(\alpha')\,,$$

then the equation in (0.19) is a nonlinear integro-differential equation for the particle trajectory transformation in terms of itself. The corresponding equation for 2-D fluid flows is even simpler. With $X = {}^t(X_1, X_2)$ and $\alpha = (\alpha_1,\alpha_2)$, the integro-differential equation for the particle trajectories for 2-D fluid flow is given by

(0.20) $$\frac{dX(\alpha,t)}{dt} = \int_{R^2} K(X(\alpha,t) - X(\alpha',t))\,\omega_o\,(\alpha')\,d\alpha'$$

$$X(\alpha,t)|_{t=0} = \alpha$$

with $K(x)$ defined in (0.12). Although the kernels K in (0.19) and (0.20) are both mildly singular, nevertheless they are bounded operators on suitable function spaces so that the Euler equations as formulated in (0.19) or (0.20) assume the abstract form,

$$\frac{dX}{dt} = F(X),\ X(0) = I$$

where $F(X)$ is a bounded locally Lipschitz operator. It is not difficult to prove that the equations in (0.19) or (0.20) define a well-posed problem through straightforward Picard iteration (see [59] or [35] for the case of bounded domains). This contrasts very strongly with the primitive variable and vorticity-stream formulations from (0.5) and (0.8) which always involve the analysis of unbounded operators like $\frac{D}{Dt}$. In the next section we will show that intelligent discretization of (0.19) and (0.20) leads to the systematic design of inviscid vortex methods. One remarkable consequence of this reformulation (involving only bounded operators) for the numerical analysis of vortex methods is that the time discrete equations for vortex methods do not require time-stepping constraints like C.F.L. conditions to maintain stability — the choice of time step depends on accuracy alone. For the experts in nonlinear P.D.E.'s we comment that the formulations in (0.19) and (0.20) are *not* the same as Arnold's famous reformulation of the fluid equations — that formulation involves $\frac{d^2}{dt^2}X(\alpha,t)$ and has a much more complicated nonlinear structure (see [59]).

I. The Numerical Analysis of Vortex Methods

A) The Design of Inviscid Vortex Algorithms

Here we will use an extremely simple strategy for discretizing the inviscid Euler equations rewritten as an integro-differential equation for the particle trajectories as described in (0.19) and (0.20). This systematic procedure involves the following four steps:

(1.1)

1) Smoothing of the singularity of the kernel K,

2) Approximation of the smoothed integral on the right hand side of (0.19) or (0.20) by a quadrature formula,

3) Approximation of the vortex stretching term, $\omega(X(\alpha',t),t)$

4) Application of the collocation method to obtain a finite system of O.D.E.'s for the approximate particle trajectories.

The above procedure leads to the typical classes of vortex methods used in practice and also yields rather flexible design principles i.e. it will be clear that the same design principles apply in different co-ordinate systems, with different quadrature methods, etc. We begin with

Vortex Methods for Inviscid 2-D Flows

Our objective is to build a discrete approximation to the integro-differential equation in (0.20). To carry out Step 1) from (1.1), we choose a function $\phi(x)$ so that

1) $\phi(x) = \phi(|x|)$, i.e. ϕ is radial

2) $\int_{R^2} \phi(x)\, dx = 1$

3) $\int_{R^2} x^\alpha \phi(x)\, dx = 0\,, \qquad 1 \le |\alpha| \le p-1$ for some fixed p.

We also require that ϕ decays fast enough as $|x| \to \infty$. We define $\phi_\delta(x)$ by $\phi_\delta = \delta^{-2}\, \phi(\frac{x}{\delta})$ and smooth the kernel $K(x)$ by convolution with ϕ_δ, i.e. $K_\delta(x)$ is defined by

$$K_\delta(x) = \int_{R^2} K(x-x')\, \phi_\delta(x')\, dx'\,. \tag{1.3}$$

Before proceeding further, we remark that (1.2) and (0.14) guarantee that $K_\delta(x)$ defines an exact smooth steady swirling flow for the 2-D Euler equations which has the explicit formula

$$K_\delta(x) = K(x)\, f\left(\frac{|x|}{\delta}\right) \tag{1.4}$$

where the *cut-off* function $f(|x|)$ is defined by

$$f(|x|) = \int_0^{|x|} s\, \phi(|s|)\, ds\,. \tag{1.5}$$

For a bounded function $\phi(|s|)$, we have

$$\begin{aligned} f(|x|) &= 0\,(|x|^2) \text{ as } |x| \to 0 \\ f(|x|) &\to 1 \qquad \text{as } |x| \to \infty\,; \end{aligned} \tag{1.6}$$

thus, the function, $f\left(\frac{|x|}{\delta}\right)$, in (1.4) "cuts-off" the local singularity of K and this explains the terminology. The vorticity corresponding to the steady velocity $K_\delta(x)$ is $\phi_\delta(x)$ — since this is a localized smooth patch of vorticity in contrast to the point vortex method with $\delta = 0$, $\phi_\delta(x)$ is called the vortex blob. To carry out Step 1) from (1.1) we approximate the solutions of the particle trajectory equation in (1.20) by the particle trajectory equation for the smoothed version of (1.20) with K replaced by K_δ, i.e. we consider approximate particle trajectories

$X^\delta(\alpha,t)$ which solve the smoothed integro-differential equation,

(1.7)
$$\frac{dX^\delta(\alpha,t)}{dt} = \int_{R^2} K_\delta\,(X^\delta(\alpha,t) - X^\delta(\alpha',t))\;\omega_o(\alpha')d\alpha'$$
$$X^\delta(\alpha,t)|_{t=0} = \alpha$$

Next, we carry out Step 2) from (1.1). Since K_δ is a smooth function, we apply any convenient quadrature formula to the integral on the right hand side of (1.7) involving α'. We illustrate this procedure with the trapezoidal rule. We consider the lattice of integer-valued points in R^N, $N = 2, 3$ defined by $\Lambda_h = \{(i_1, i_2, \ldots, i_N)|\; i_j \in Z\}$ and we introduce the discrete grid $\Lambda = h\Lambda$, $h > 0$. The trapezoidal approximation to the integral $\int g(\alpha')d\alpha'$ is given by

(1.8)
$$\sum_{k \in \Lambda} g(kh)h^N \cong \int g(\alpha')\,d\alpha' \,.$$

By applying the quadrature formula in (1.8) to $g(\alpha') = K_\delta(x - X^\delta(\alpha',t))\;\omega_o(\alpha')$ with x regarded as a parameter, we obtain the approximation

(1.9)
$$\int_{R^2} K_\delta(x - X^\delta(\alpha',t))\;\omega_o(\alpha')\,d\alpha' \cong \sum_{k\in\Lambda} K_\delta(x - X^\varepsilon(kh,t))\omega_o(kh)h^2$$

for the right hand side of (1.9). Since the vorticity does not change along particle trajectories in 2-D inviscid flows, Step 3) in (1.1) is trivial. Finally, we replace the integral on the right hand side of (1.7) by the discrete sum in (1.9) and apply the collocation method to carry out Step 4) of (1.1). Thus, we compute approximate particle trajectories $\{X_j^{\delta,h}(t)\}_{j\in\Lambda}$ which are an approximation to $X^\delta(jh,t)$, i.e. $X_j^{\delta,h}(t) \cong X^\delta(jh,t)$, and satisfy the following system of O.D.E.'s.

The 2-D Vortex Algorithm:

(1.10)
$$\frac{dX_j^{\delta,h}}{dt} = \sum_{k\in\Lambda} K_\delta(X_j^{\delta,h}(t) - X_k^{\delta,h}(t)) - X_k^{\delta,h}(t))\;\omega_o(kh)h^2$$
$$X_j^{\delta,h}\,(0) = jh \;.$$

These are the basic 2-D vortex methods; of course we need to assume that $\omega_o(x)$ has compact support to obtain a finite system of O.D.E.'s. The above system of O.D.E.'s has properties which depend on the cut-off f, and the parameters δ,h. In any case, with the explicit formulae in (1.4) and (1.5), it is extremely easy to implement these methods with a wide range of explicit cut-offs (see [6]). The approximate computed velocity and vorticity are given by

(1.11)
$$v^{\delta,h}(x,t) = \sum_{k\in\Lambda} K_\delta(x - X_k^{\delta,h}(t))\;\omega_o(kh)^{h^2}$$
$$\omega^{\delta,h}(x,t) = \sum_{k\in\Lambda} \phi_\delta(x - X_k^{\delta,h}(t))\;\omega_o(kh)h^2 \,.$$

Vortex Methods for 3-D Fluid Flows

First we quickly repeat Steps 1), 2), and 4) of the design principle outlined in (1.1) for approximating the integro-differential equations in (0.19) — the new ideas occur in constructing the approximation to the vorticity $\omega(X(\alpha',t),t)$ in (0.19). For 3-D, we consider a smooth function $\phi(x)$ satisfying assumptions in R^3 analogous to those in (1.2). We define $\phi_\delta(x) = \delta^{-3}\phi(\frac{x}{\delta})$. The smoothed kernel $K_\delta(x)$ is given by the convolution,

$$K_\delta(x) = \int_{R^3} K(x - x')\, \phi_\delta(x')\, dx' . \tag{1.12}$$

For the moment, we assume that the vorticity $\omega(x,t)$ is a specified function. Then as carried out in detail in (1.7) — (1.10) above by repeating Steps 1), 2), and 4) for the integro-differential equation in (0.19), we arrive at the equations for the approximate particle trajectories, $X_j^{\delta,h}(t) \cong X(jh,t)$, which solve the system of O.D.E.'s described for each j by

$$\begin{aligned} \frac{dX_j^{\delta,h}(t)}{dt} &= \sum_{k \varepsilon \Lambda} K_\delta(X_j^{\delta,h}(t) - X_k^{\delta,h}(t))\, \omega_k(t) h^3 \\ X_j^{\delta,h}(0) &= jh \end{aligned} \tag{1.13}$$

At this stage in the derivation, we have $\omega_k(t)$ defined by the vorticity of the underlying exact solution of the 3-D Euler equations which are trying to compute, i.e. $\omega_k(t) = \omega(X(kh,t),t)$; thus, the equation in (1.13) involves the unknown vorticity $\omega_k(t)$. How do we obtain an approximation to $\omega_k(t)$?

One simple way to achieve this is to use the integrated form of the vortex stretching equation in (0.17) together with a standard finite difference approximation ∇_α^h on the rectangular mesh Λ^h as an approximation to ∇_α. Thus, we have

$$\begin{aligned} \omega(X(\alpha,t),t)|_{\alpha = kh} &= \nabla_\alpha X|_{\alpha = kh} \cdot \omega_o(kh) \\ &\cong \nabla_\alpha^h X_k^{\delta,h}(t) \cdot \omega_o(kh) . \end{aligned} \tag{1.14}$$

By combining (1.13) and (1.14) we obtain *3-D Vortex Algorithms with Lagrangian stretching*:

$$\begin{aligned} \frac{dX_j^{\delta,h}(t)}{dt} &= \Sigma_{k \varepsilon \Lambda} K_\delta(X_j^{\delta,h}(t) - X_k^{\delta,h}(t)\, \nabla_\alpha^h X_k^{\delta h}\, \omega_o\, (kh) h^3 \\ X_j^{\delta,h}(t)|_{t=0} &= jh , \quad \text{for each } j \varepsilon \Lambda . \end{aligned} \tag{1.15}$$

These algorithms depend on the smoothing function ϕ, the finite difference operator ∇_α^h, and the parameters δ,h.

A second way to build an approximation to the vorticity is to utilize an approximation to the Eulerian stretching formula in (0.8). From (0.8) we have

$$\frac{d\omega}{dt}(X(kh,t),t) = \omega \cdot \nabla v|_{(X(kh,t),t)} . \tag{1.16}$$

With a discrete approximation to the 3-D velocity field given by

$$v^{\delta,h}(x,t) = \sum_{k \varepsilon \Lambda} K_\delta(x - X_k^{\delta,h}(t))\, \omega_k^{\delta,h}(t) h^3 , \tag{1.17}$$

we compute $\nabla v^{\delta,h}(x,t)$ directly by differentation to obtain

$$\nabla v^{\delta,h}(x,t) = \sum_{k \varepsilon \Lambda} \nabla K_\delta(x - X_k^{\delta,h}(t))\omega_k^{\delta,h}(t) h^3 . \tag{1.18}$$

We utilize the approximation, $\omega_k^{\delta,h}(t) \cong \omega(X(kh,t),t)$ and (1.16), (1.18) to obtain the *3-D Vortex Algorithms with Eulerian stretching:*

(1.19)
$$\frac{d\omega_j^{\delta,h}}{dt}(t) = \omega_j^{\delta,h} \cdot \sum_{k\varepsilon\Lambda} \nabla K_\delta(X_j^{\delta,h}(t) - X_k^{\delta,h}(t))\omega_k^{\delta,h}(t)h^3$$
$$\frac{dX_j^{\delta,h}}{dt}(t) = \sum_{k\varepsilon\Lambda} K_\delta(X_j^{\delta,h}(t) - X_k^{\delta,h}(t))\, \omega_k^{\delta,h}(t)h^3$$

with

$$X_j^{\delta,h}(0) = jh \ , \quad \omega_j^{\delta,h}(0) = \omega_o\ (jh)\ .$$

For these algorithms, we have six unknowns, $(X_j^{\delta,h}, \omega_j^{\delta,h})$ for each point $jh\, \varepsilon \Lambda^h$.

To construct explicit kernels for the 3-D algorithms in (1.15) and (1.19), it is easier to proceed essentially in the reverse order from the construction in (1.2) — (1.6) which we utilized in 2-D. Thus, we begin with a radial 3-D cut-off function, $f(r)$ and define $K_\delta(x)$ by

(1.20)
$$K_\delta(x) = f\left(\frac{|x|}{\delta}\right) K(x)$$

with $K(x)$ the Biot-Savart kernel for 3-D, i.e $K(x) = -\dfrac{1}{4\pi}|x|^{-3}x\, \times$. If the cut-off function $f(r)$ satisfies

(1.21)
$$f(r) = O(r^3) \qquad \text{as} \qquad r \to 0$$
$$f(r) \to 1 \qquad \text{rapidly as} \qquad r \to \infty$$

then the kernel $K_\delta(x)$ in (1.20) is a smooth approximation to $K(x)$ for $\delta \to 0$ with div $K_\delta = 0$. The corresponding 3-D vortex blob function $\phi(r)$ is given by the formula (see [6]),

(1.22)
$$\phi(r) = \frac{f'(r)}{4\pi r^2}\ .$$

Finally, we note that the numerical approximation to the vorticity for 3-D flows is *not* given simply by the second sum in (1.11) — this term is clearly not an incompressible vector field in 3-D. Instead, the correct approximation for the vorticity for a 3-D vortex algorithm is the curl of $v^{\delta,h}(x,t)$ defined in (1.17). With the form for K_δ in (1.20), this expression is readily computed by explicit formulae.

B) A Summary of the Classical Convergence Theory for Inviscid Vortex Algorithms

The classical convergence theory deals with the following questions: Suppose $v(x,t)$ is a smooth solution of the inviscid Euler equations in (0.5) defined for some interval of time $0 \le t \le T$ with initial data v_o, does the solution of a given vortex algorithm converge to $v(x,t)$ as $\delta,h \to 0$? What is the rate of convergence and the dependence on the blob function ϕ, δ, h, and the difference operator ∇_α^h (for the 3-D methods in (1.15))?

Loosely stated, the main results of the theory are the following

Theorem: With proper choices of the vortex blob function ϕ, the parameters δ, h, (and the difference operator ∇_α^h for the algorithm in (1.15)), the 2-D algorithms in (1.10) and both of the 3-D algorithms in (1.15) and (1.19) have computed velocity fields $v^{\delta,h}(x,t)$ which always converge to the inviscid solution $v(x,t)$ uniformly over the entire time interval [0,T] as $\delta,h \to 0$. The principle restrictions on the convergence require $\delta >> h$ as $\delta,h \to 0$ and that δ satisfies $\delta \le \delta_o(T,v)$. Furthermore, there are appropriate choices of the blob function ϕ, the finite difference operator ∇_α^h, and the integer p in (1.2) so that there is a uniform rate of convergence $O(h^R)$ over the time interval [0,T] for any given $R > 0$ provided that $v(x,t)$ is sufficiently smooth, i.e. vortex methods can be designed with an arbitrarily high order of accuracy for smooth flows. Vortex algorithms are readily discretized in time and standard multi-step and Runge-Kutta finite difference approximations in time converge for smooth solutions on the interval [0,T].

Here is a brief historical summary of the principal developments which lead to this main result. Hald and Del Prete ([45]) gave the first correct convergence proof for vortex methods in 2-D; however, their proof only applied for short times $T_* << T$ with an exponential loss of accuracy. Hald's paper ([46]) on 2-D vortex methods was the first pioneering breakthrough in the numerical analysis of vortex methods; he proved second order convergence for arbitrarily long time intervals [0,T] for a very special class of blob functions from (1.2) with $p = 4$ and $\delta = h^{1/2}$. Beale and the author ([9], [10]) improved Hald's stability argument in an essential way, found that vortex methods could be designed which converge with arbitrarily high order accuracy, and gave the first convergence proof for a class of 3-D vortex methods. These methods are the 3-D vortex algorithms with Lagrangian stretching described in (1.15) although the simplified equivalent formulation in (1.15) was noticed by C. Greengard somewhat later ([1]). Cottet and Raviart ([25]) exploited the analogy between the 2-D Euler equations and the 1-D Vlasov-Poisson equations and gave extremely simple convergence proofs for particle methods for the 1-D Vlasov-Poisson equations. Cottet ([26]) built on these ideas and developed an important simplified approach to the consistency arguments for vortex algorithms. Anderson and C. Greengard ([1]) gave a further simplified version of Cottet's consistency argument for the trapezoidal rule by utilizing the Poisson summation formula; the paper ([1]) also contains an extremely simple proof of convergence of multi-step time discretization for 2-D vortex methods. The 3-D vortex methods with Eulerian stretching described in (1.19) were suggested by Anderson and first presented in [1]. With very different proofs, Beale ([11]) and Cottet ([27]) have independently proved the convergence of the 3-D vortex algorithms in (1.19). Recently, Cottet ([29]) has developed a new approach to the numerical analysis of vortex methods which does not use a direct analysis of the stability and consistency of the O.D.E.'s for the particle trajectories in (1.10) — this approach leads to the convergence of 2-D vortex-in-cell algorithms, etc. Finally, Hald ([27]) has recently given a number of interesting definitive results for the classical convergence theory for 2-D vortex methods building on much of the work already described above — one notable result from [47] is the convergence of Runge-Kutta time-differencing for 2-D vortex methods. The reader interested in a detailed statement and discussion of the convergence proofs can consult the survey papers [6], [8], [1], as well as the references mentioned above. At this stage in the development of the numerical analysis of vortex methods, it is extremely easy for anyone to read the streamlined complete convergence proof for 2-D vortex methods in a few pages. One only needs to combine the argument for consistency on Pages 421-426 of [1] with the general treatment of stability for 2-D vortex methods given in detail on Pages 45-49 of [10]. Careful and detailed numerical studies of 2-D inviscid vortex methods on simple model problems are given in [6], [61].

There has also been an interesting interplay between the numerical analysis and the design of practical vortex methods. The algorithms in (1.15) are new 3-D vortex algorithms initially proposed from theoretical grounds; however, C. Greengard has shown ([43]) that provided one uses a special filament co-ordinate system rather than rectangular co-ordinates in the integro-differential formulation from (0.19), one can follow the design principles in (1.13) – (1.15) to obtain an interesting class of vortex filament algorithms; furthermore, the original proofs from [9] apply with minor modification to yield convergence of these filament algorithms. The filament algorithms obtained in this fashion closely resemble the filament methods for 3-D flows introduced earlier by Chorin ([20]). On the other hand, filament algorithms require the initial vorticity ω_o to have a very special structure i.e. essentially the initial vorticity should break up into a sum of oblique axisymmetric swirling flows. The algorithms in (1.15) can be implemented for arbitrary initial vorticity ω_o. This is not an entirely trivial point since extremely simple looking initial flows such as the celebrated A-B-C flows ([33]) in some parameter regimes can have vortex lines which are everywhere dense. The algorithms in (1.15) and (1.19) are surprisingly close in structure with various advantages and disadvantages for each method – the reader can consult [1] for a clear discussion of these topics. There is a large literature on the numerical evolution of 3-D vortex filaments; however this approach usually does not involve direct discretization of the inviscid equations of 3-D fluid flow. First, local formal asymptotic approximations such as the self-induction approximation are made and then the resulting equations are discretized. The review paper ([51]) contains a discussion of these topics.

We end this subsection by listing some of the computational advantages of vortex algorithms

1) Computational elements are needed only in regions with non-zero vorticity and the algorithms automatically follow the concentration of vorticity as time evolves

2) It is not necessary to compute the pressure during the calculation and the incompressibility constraint is automatically satisfied by the approximation.

3) At least for 2-D fluid flows, vortex methods represent general solutions as superpositions of a finite number of mutually interacting elementary exact solutions of 2-D Euler (see (1.3) and (1.4)

(1.23)

4) Since vortex methods are Lagrangian methods and the fluid velocity is the only characteristic in incompressible fluid flow, these methods have minimal numerical viscosity – this is extremely important since numerical viscosity can swamp and/or falsify the physical phenomena in inviscid or high Reynolds number flows. Other evidence for the lack of numerical viscosity at least for 2-D flows is that despite the smoothing effect of ϕ_δ, 2-D vortex methods conserve a discrete approximation to the energy (see [8]).

The principal disadvantage of vortex methods when implemented in a straightforward fashion is the following:

(1.24) Since each particle interacts with every other particle, there are $O(N^2)$ operations per time step if N is the number of particles.

The recent progress in overcoming the major difficulty in (1.24) without sacrificing the computational advantages in (1.23) is the topic of the next subsection.

C) Rapid Summation Algorithms for Vortex Methods

The traditional approach is overcoming the obstacle of the $O(N^2)$ operation count is to implement a "vortex in cell" or "cloud in cell" algorithm (see [52]); these algorithms involve interpolation of the velocity field onto a rectangular mesh coupled with the use of fast Poisson solvers; such algorithms allow a reduction in operation count to $O(N \log N)$. Unfortunately, some of the attractive features of vortex methods listed in (1.23) can be lost. In particular, such methods often introduce artificial viscosity created by the interpolating mesh. As direct evidence for the presence of artificial viscosity in cloud-in-cell algorithms we mention that Christiansen ([24]) has reported the inviscid merger of constant patches of vorticity using these algorithms; on the other hand, rigorous mathematical theorems guarantee that constant patches of vorticity never merge for inviscid 2-D fluid flows (see [58] Section 3) – thus, the claimed merger is due to artificial viscosity and is a numerical artifact. G. Baker has pointed out other substantial local errors introduced by the interpolating mesh in simulating vortex sheets (see [52]).

With the above defects for cloud-in-cell algorithms, other computational strategies are needed which retain the attractive computational features in (1.23) but lower the operation count. One possible compromise is to use direct summation for nearby interactions and a mesh for the far-field in an attempt to keep the advantages in (1.23) but lower the operation count – this vortex method mimics the P^3M method of Hockney and Eastwood (see [52]) from plasma physics. Anderson ([2]) has developed such a method for 2-D fluid flows. Colella, Baden, and Buttke are currently implementing this method in large scale simulations (private communication). A second possible compromise is to use direct summation for nearby interactions and some other rapid summation algorithm for the distant interactions. In two very interesting recent papers, L. Greengard and Rokhlin ([41], [42]) have shown how to achieve rapid summation in $O(N)$ operations within small error tolerances by using multi-pole expansions in the far-field combined with various nesting procedures.

Following the discussion in [41], we illustrate one of their main ideas in a representative example: We consider two circles C_1, C_2 with radius R so that their centers are distance $3R$ apart. We assume that C_1 is centered at the origin and contains M vortices with strengths $\Gamma_1, \ldots, \Gamma_M$ located at points $z_1, \ldots, z_M$ with $|z_i| < R$. Here we have used the complex notation $z = x_1 + i\, x_2$; since we are interested in distant interactions, we assume for simplicity that the vortices in C_1 are point vortices (see (0.2)). Thus, after a trivial rotation of co-ordinates, the contribution to the 2-D velocity field outside C_1 from the M-points in C_1 is given by the function $w(z)$ defined by

$$w(z) = \sum_{i=1}^{M} \frac{\Gamma_i}{z - z_i} .$$

We note that the same representation would be valid for general vortex blobs provided ϕ in (1.2) has bounded support and $\delta << R$. We assume that there are N-vortices located in C_2 – direct evaluation of $w(z)$ at these N points requires $O(MN)$ operations. A key observation in [41] is the following lemma on multi-pole series with a simple proof which we leave to the reader.

Lemma: For any z with $|z| > R$ we have

$$w(z) = \sum_{k=0}^{\infty} \frac{a_k}{z^{k+1}} , \quad a_k = \sum_{i=1}^{M} \Gamma_i z_i^k .$$

Furthermore, we have the bound,

$$\left| w(z) - \sum_{k=0}^{p} \frac{a_k}{z^{k+1}} \right| < \frac{A}{c-1} \left(\frac{1}{c}\right)^p \frac{1}{|z|}$$

with $A = \sum_{i=1}^{M} |\Gamma_i|$ and $c = \left|\frac{z}{R}\right|$.

The idea to reduce the operation count is to replace the function $w(z)$ by a finite multi-pole series of some fixed order p. The multi-pole series of order p built from the M vortices in C_1 can be constructed in $O(Mp)$ operations while $O(Np)$ operations are required to evaluate this series at the N vortex centers in C_2. Thus, only $O(p(M+N))$ operations are needed to evaluate the velocity field $w(z)$ in C_2 within an error tolerance $O((1/2)^p)$ as guaranteed by the lemma!

D) Simulating Diffusion in High Reynolds Number Flows with Vortex Methods

We would like to mention at the beginning of this section that the goal in designing numerical methods for high Reynolds numbers, i.e. $\nu << 1$, is to design numerical methods with practical error bounds that are independent of ν as $\nu \downarrow 0$. Thus, a perfectly reasonable numerical method for computing viscous fluid flows at a fixed viscosity ν_o might have practical error bounds behaving like $\exp(\frac{1}{\nu})$ as $\nu \to 0$ which force a catastrophic increase in computational labor at large Reynolds numbers or induce numerical artifacts such as artificial viscosity due to inadequate resolution at large Reynolds number. These considerations are essential in designing effective algorithms for high Reynolds number fluid flows.

Simulating Diffusion for Vortex Methods Without Boundaries

As mentioned in (0.4) above, in 1973 Chorin proposed adding an appropriate random walk of the particle trajectories $X_j^{\delta,h}$ in (1.10) to simulate the effects of diffusion away from boundaries in high Reynolds number flows. Thus, a stochastic algorithm is used to solve a deterministic nonlinear P.D.E. This is a very unusual algorithm and its introduction generated a controversy (see the references in [62]) regarding the capability of such algorithms to be useful for high Reynolds number flows in a fashion outlined at the beginning of this subsection. For all of the above reasons, the numerical analysis of the random walk algorithm without boundaries is a very interesting problem. Next, we describe the major progress that has occurred in analyzing these algorithms.

Beale and the author ([12]) established that basic viscous splitting algorithms where the Euler equations and heat equation are solved in alternate steps have convergence properties that improve as $\nu \downarrow 0$. Marchioro and Pulvirenti ([60]) gave the first results yielding very weak convergence of 2-D random vortex methods with unrealistic dependence of δ and h as $\delta, h \to 0$. Through a very complicated argument, Goodman ([40]) succeeded in giving a more realistic convergence rate and strategy for linking δ and h. Recently, D. G. Long ([53]), [54]) in a Ph.D. thesis written under the author's supervision has achieved a major breakthrough in the numerical analysis of random vortex methods without boundaries. For smooth initial data in 2-D, he gives an extremely simple proof that the 2-D random vortex method converges at the rate $O(h\,|\log h|)$ with very high probability and independent of ν as $\nu \downarrow 0$ for $\nu \le \nu_o$. Even without the nonlinear terms, the sharp convergence results for the linear heat equation yield bounds of order $O(h\,|\log h|)$ so Long's estimates for the full nonlinear algorithm are essentially best possible!! Furthermore, the strategy of the proof is extremely simple and parallels the convergence proof for the inviscid case outlined at the end of the second paragraph in Section 1B). First he proves a very simple stochastic version of the consistency argument described on Pages 421-426 of [1] for the inviscid case; then he gives a "grid-free" generalization of the stability theorem on Pages 45-49 of [10] for the inviscid case. Finally, he combines these facts with some simple ideas from the theory of empirical stochastic processes (principally Bennet's inequality) to prove his convergence theorem. The above proof has enormous flexibility and simplicity so Long has generalized his work to include time discretization, and 3-D stochastic vortex algorithms (see [55], [56]). We also mention that Long and the author [57] have studied the random vortex algorithm through explicit analysis for both smooth and discontinuous initial data for simplified fluid solutions describing strained shear layers. Roberts ([62]) has reported on detailed computations with non-smooth initial data for the 2-D random vortex method which directly address the controversy described earlier. Thus, both theoretical results and careful numerical computations confirm that the random vortex method has the desirable features for high Reynold's number flows mentioned in the first paragraph of this section.

Recently, deterministic algorithms for simulating diffusion combined with a vortex algorithm have been proposed ([28]); however at this time, it is unclear whether the requirements needed for high Reynolds number flows are satisfied by these algorithms (the numerical test in [17] does not address this issue in enough detail). Finally, Kuwahara and Takami (see [52]) proposed "core spreading" as a simple way to simulate diffusion in vortex algorithms; C. Greengard ([44]) has given a simple proof that these algorithms are *inconsistent* with the Navier-Stokes equations at *any* Reynolds number – these algorithms approximate the wrong equations!! This is a simple illustration of the use of mathematical theory in displaying the inadequacy of a numerical algorithm that, at first glance, "looks reasonable on physical grounds".

Simulating Diffusion for Vortex Methods at Boundaries

The difference in the boundary conditions from (0.6) and (0.7) leads to the physical generation of vorticity at boundaries. As mentioned in (0.4)3) Chorin's original random vortex algorithm simulates this physical process as a component of the algorithm; Chorin ([21]) introduced the sheet algorithm as a more efficient computational algorithm to simulate vorticity creation at boundaries. Recently, Anderson ([3]) has introduced a finite difference version of these "vorticity-creation" algorithms.

The numerical analysis of algorithms with vorticity creation at boundaries is an extremely important but underdeveloped area of research. In [36] Ghoniem and Sherman introduced a beautiful one-dimensional model problem with shearing and thermal effects which leads to vorticity generation at boundaries. Hald ([48]) has given an analysis of the convergence for these stochastic algorithms which is independent of size of the diffusion coefficients. As regards the Navier-Stokes equation, Benfatto and Pulvirenti ([13]) have given a very interesting analytic formulation of the Navier-Stokes equations in a half-space which can be regarded as the analytic reformulation of the Navier-Stokes equations to display vorticity creation. This work was continued in [14] although the method of proof necessarily gives error bounds with catastrophic dependence on ν like $\exp(\frac{1}{\nu})$ as $\nu \to 0$. Such proofs tend to hide the potentially attractive features of vorticity creation algorithms. Clearly, a great amount of work on this topic remains.

II. Large Scale Scientific Computing with Vortex Methods

Vortex methods are powerful numerical methods even when direct summation has been applied for the interactions; many very interesting physical problems have regions with non-zero vorticity occupying only a small fraction of the total computational domain. Since vortex methods only require computational elements in regions where the vorticity is non-zero, the high computational overhead is often balanced by economical representation of the flow field. Here we discuss some of the recent computational achievements using vortex methods (many other calculations are referenced in [51]).

A) High Reynolds Number Flows in 2-D with Boundaries

When Chorin published his first calculations in 1973 using the random vortex method for flow past a cylinder at high Reynolds numbers, the results were often criticized as looking "qualitatively correct" but lacking any quantitative measure of convergence. All of the calculations we discuss here essentially use the basic random vortex method coupled with Chorin's random sheet algorithm ([21]) to shed vorticity from the boundary. The results we describe here establish the quantitative numerical convergence of random vortex algorithms without any doubt for some rather complex fluid flows. We remark that real turbulence at sufficiently high Reynolds numbers is often three-dimensional in character. Nevertheless, numerical simulations at high Reynolds numbers for 2-D flows provide a very difficult class of computational problems for simulating incompressible flows and the results provide a lot of physical insight into the turbulent structure; the situation is similar to the use of two-dimensional airfoils to simulate the transonic flow past full three dimensional wing-body configurations.

In a series of papers, Ghoniem, Sethian, and their co-workers ([37], [65], [38]) have used the random vortex method to study various channel flows throughout a wide range of Reynolds numbers. The work combines direct comparison with experimental data, systematic numerical parameter checks to ensure practical convergence of the methods, and new predictions regarding the flow structure in various regimes of Reynolds numbers. The reference [65] contains the most detailed convergence studies. There the 2-D random vortex method is used to compute the solution over a backward-facing step for Reynolds numbers 50, 125, 250, 375, 500, and 5,000. The Reynolds numbers 50, 125 yield flows in the laminar regime while for Reynolds number 5,000 the flow exhibits the structure of fully developed turbulence for 2-D i.e. vortex streets, etc. are formed downstream. In the laminar regime Sethian and Ghoniem demonstrate pointwise convergence of the computed velocity fields to the experimental measurements as well as accurate prediction of the size and length of recirculation zones as the Reynolds number varies. At the higher Reynolds numbers, those authors demonstrate numerical convergence of averaged quantities such as average velocity profiles and eddy boundaries. One important consequence of this work is that vortex methods are extremely robust in computing flow fields for a wide range of Reynolds numbers including the laminar regime!! Of course, if one is only interested in flows with Reynolds number of 50, finite difference methods provide a much less expensive solution technique; on the other hand, finite difference schemes are not practical at the high Reynolds numbers even with the largest available supercomputers.

In other interesting work Cheer ([16]) and especially Tiemroth ([67]) have studied flow past bodies at high Reynolds numbers through modern versions of the random vortex method. In a series of careful numerical experiments, these authors have established the quantitative numerical convergence for these problems of important physical quantities such as lift and drag coefficients.

B) Large Scale Simulation in 3-D

The recent review article by Leonard ([51]) contains many references for large scale simulation with 3-D vortex methods. Here we only briefly mention a few results. First Chorin has introduced a 3-D boundary sheet algorithm as well as a 3-D vortex filament method in [20] and applied these algorithms to study boundary layer instability. As we mentioned in Section 1B), Chorin's 3-D algorithm is essentially the 3-D vortex algorithm with Lagrangian stretching (see (1.15)) adapted to a filament coordinate system (see [43]) – thus, minor variants of this algorithm have a theory which guarantees accurate convergent numerical solutions for smooth flow (see 2B) above). In subsequent work, Chorin applied the 3-D filament method to study the breakdown of smooth solutions of 3-D Euler and the rapid accumulation of vorticity ([22], [23]). With the exciting algorithmic developments regarding rapid summation in 3-D (see 1C)), it is likely that numerous highly resolved large scale simulations utilizing 3-D vortex methods on supercomputers will be developed in the near future. Finally, we end this sub-section by mentioning the recent work of Ghoniem, Aly, and Knio (see [39]). These authors use Chorin's 3-D vortex filament method together with the cubic Gaussian cut-offs proposed in [6] to study the modes of inviscid instability for various "vortex tori". They carefully compare the predicted mechanisms of instability from linearized analysis with the results obtained by numerical simulation with vortex methods; also, various numerical convergence studies are presented. This is the first published series of calculations for inviscid 3-D vortex methods which directly utilizes the interplay between the numerical analysis presented in Section 1 with large-scale computing on an interesting physical problem.

C) The Numerical Simulation of Vortex Sheets for Inviscid 2-D Flows

A two-dimensional incompressible velocity field, $v_o(x)$, defines *vortex sheet* initial data provided that there is a piece of a smooth curve C in the plane so that the tangential velocity of $v_o(x)$ jumps across C while the normal velocity for $v_o(x)$ remains continuous. Here, for simplicity, we assume that $v_o(x)$ is a potential flow outside C. Thus, the vorticity $\omega_o = \text{curl } v_o$ has the following structure:

(2.1)A) the vorticity ω_o is a surface Dirac delta measure supported on C_o (this is one prototypical example of a Radon measure used in Section 3).

Another obvious physical requirement for vortex sheet initial data is that v_o has locally finite kinetic energy, i.e.

$$\int_{|x| \le R_o} |v_o|^2 dx \le C_{R_o} \text{ for any } R_o > 0 . \quad \text{(2.1)B)}$$

In the remainder of this paper, when we refer to vortex sheet initial data, we tacitly assume the conditions in (2.1). Vortex sheets occur as excellent approximations for parts of the flow field in the trailing wake behind bodies with sharp edges and there is an enormous engineering and applied mathematics literature on these topics involving both formal asymptotic methods and numerical simulation (see [64]). Numerical computations for vortex sheets are extremely challenging for the following reason: if the equations for the evolution of the vortex sheet itself (the Birkhoff equations) are linearized, these linear equations are ill-posed with catastrophic instability like the initial value problem for the Cauchy-Rieman equations. It is only very recently that vortex methods have given the first successful numerical simulations beyond the critical time when the solutions begin to display enormous complexity. These new computational results suggest several novel problems in the mathematical theory of incompressible flow; the interplay between large-scale computation and this new mathematical framework is the main topic of Section 3.

In [4], Anderson has utilized analogy with 3-D vortex methods and the design principles outlined in Section 1A) to develop a class of vortex algorithms for 2-D fluid flows with variable density in the Bousincsq approximation. Using these vortex methods, he presents inviscid calculations of a piecewise constant rising thermal – the boundary of the thermal is essentially a vortex sheet and this boundary displays enormous intensification of vorticity and roll-up near the edges of the thermal as time evolves. These calculations are developed through a very interesting computational strategy which we outline next for computing vortex sheets for 2-D inviscid fluid flows.

The vorticity ω_o for vortex sheet initial data is too singular for the integro-differential formulation in (0.20) to be directly useful computationally. A better strategy is to replace the kernel K by the smoothed kernel K_δ and utilize the smoothed integro-differential equations in (1.7) – here the smoothness of K_δ for $\delta > 0$ compensates for the singular nature of ω_o for vortex sheet initial data. For a fixed δ, one then applies quadrature specially adapted for the initial data ω_o and obtains a vortex algorithm by discretizing the right hand side of (1.7). In this fashion, approximate particle trajectories $X_j^{\delta,h}$ are obtained for each fixed δ. The numerical strategy is to let $h \to 0$ for a fixed δ thereby obtaining a solution of (1.7) with corresponding vorticity and velocity fields and then to let $\delta \downarrow 0$ through a sequence of calculations of this type. Thus, the computational strategy is a type of "desingularization" algorithm utilizing vortex methods – in the idealized limit $\delta = 0$, the dynamics of the vortex sheet should be recovered.

We describe some of the remarkable computational results achieved by R. Krasny ([49], [50]) recently using the computational strategy described above – we refer the reader to those papers for much more detailed explanation and more computational results. In [50], Krasny considers the roll-up of a vortex sheet which is periodic in one-direction and initially is a small perturbation of a uniform vortex sheet. These calculations are the first which successfully continue beyond the first critical time of singularity formation in the vortex sheet; at this first singularity time a singularity in the curvature of the vortex sheet occurs and this calculation as well as earlier ones and also arguments via asymptotics (referenced in [49], [50]) are in general qualitative agreement (different published calculations use different initial data so a direct comparison is not available). The strong evidence for numerical convergence of the desingularized solutions satisfying (1.7) as $\delta \to 0$ is indicated by the graphs in Figures 5 and 6 from [50]. The computations in [50] are consistent with Pullin's conjecture that the solution beyond the critical time of singularity formation is locally a double branched spiral with an infinite number of turns with vanishing size as time approaches the singularity time from above. We remark that for vortex sheets we expect weak convergence to occur in the limit as $\delta \to 0$ (see Section 3) so we do not expect actual uniform convergence of the vortex sheet shape over the entire region.

In the above calculation, the vorticity had a single sign, i.e. $\omega_o > 0$; in [49], Krasny presents calculations of roll-up of vortex sheets without a distinguished sign and here the subsequent evolution can be incredibly complex. In the first calculations discussed in [49], Krasny studies the numerical solution for vortex sheet initial data corresponding to an elliptically loaded wing. Figures 7-15 of [49] demonstrate the convergence as $\delta \downarrow 0$ of the regularized algorithm in the vicinity of the tips which roll-up. As further evidence for the validity of his approach he also compares the numerical solution with Kaden's self-similar spiral; asymptotic arguments predict this solution controls the behavior of the roll-up. These first calculations are more resolved than earlier ones of Chorin and Bernard (see [18]) on the same problem but confirm the trends from [18]. The second calculations reported by Krasny in [49] have initial data with a vortex sheet strength that changes sign three times. The vortex sheet rolls-up at six different locations and the different pieces of the sheet globally interact like large scale vortices with various signs. Figures 1 and 2 (Figure 19 from [49]) show the incredible small scale complexity in the vortex sheet generated by the large scale coherent structures on the sheet which drive their development. Figures 3, 4, and 5 (Figure 24 from [49]) give closeup views of the incredibly complex portions of the vortex sheet that develop as time evolves. The "fat" portions from the vortex sheet in Figures 4 and 5 are artifacts of the graphics printing; in fact the vortex sheet has folded in several closely packed strips. These last calculations use the crude value of $\delta = .1$ — it is difficult to imagine the structure of the solution as $\delta \downarrow 0$. A mathematical framework designed to address such observed complexity is discussed in the next section.

Finally, we end this section by describing another very interesting numerical computation of perturbed periodic vortex sheet by Shelley and Baker ([66]). They approximate the slightly perturbed initial vortex sheet by a layer with mean finite thickness h and uniform vorticity inside with strength essentially $C_o h^{-1}$ where C_o is a given constant. These authors consider a sequence of calculations with these initial data as $h \to 0$ i.e. as the uniform sheet gets thinner – this is another regularization procedure for the singular vortex sheet. They resolve the behavior of the inviscid fluid flow with fixed h by a sophisticated interface algorithm which tracks the boundaries – these interface algorithms are Lagrangian methods which in some respects are similar to vortex methods (see [5]). The results of their computations with their smallest value of $h = .05$ are presented in Figures 6 and 7. The successive times depicted are $t = 0$, 1.5, 2.0, 2.2, and 2.4. One interesting facet of these calculations is the approximate Kirchhoff ellipse which forms by time 2.4 in the roll-up process. An ellipse with the same ratio of major and minor axes was present in all the resolved runs from [66] and has an area which scales with h as $O(h^{1.6})$ as $h \downarrow 0$. We will discuss the implications of this fact briefly in the next section.

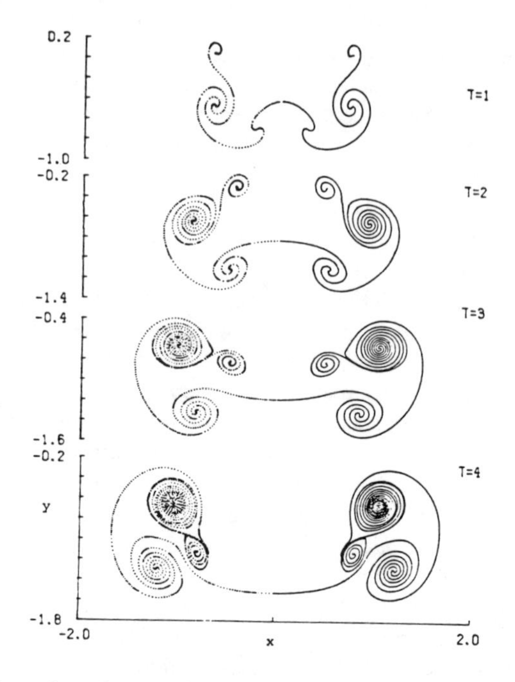

Figure 1: The solution plotted over the time interval $1 \leq t \leq 4$ using $\delta = 0.1$ (R. Krasny [49]) (Journal of Fluid Mechanics)

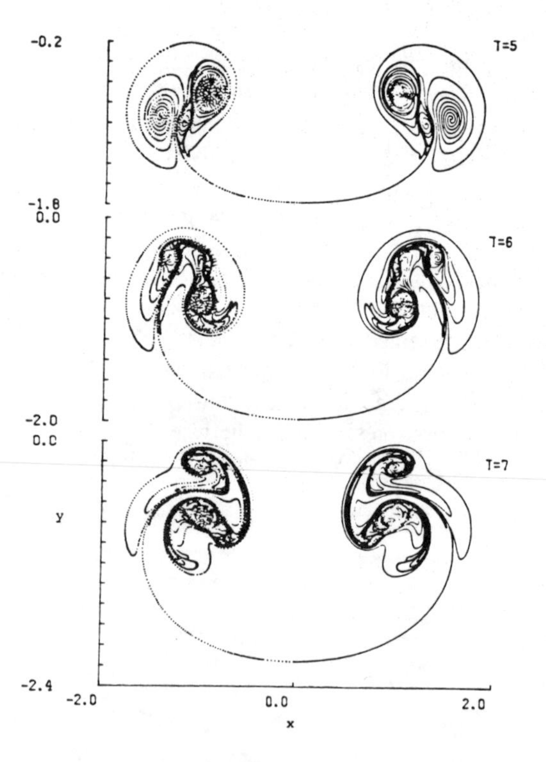

Figure 2: Evolution over the time interval $5 \leq t \leq 7$ for $\delta = 0.1$ (R. Krasny [49]) (Journal of Fluid Mechanics)

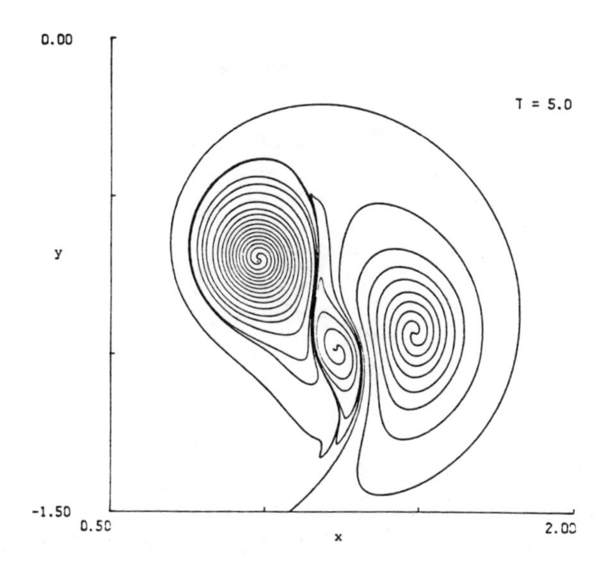

Figure 3: Closeup view of the solution at $t = 5$. (R. Krasny [49]) (Journal of Fluid Mechanics)

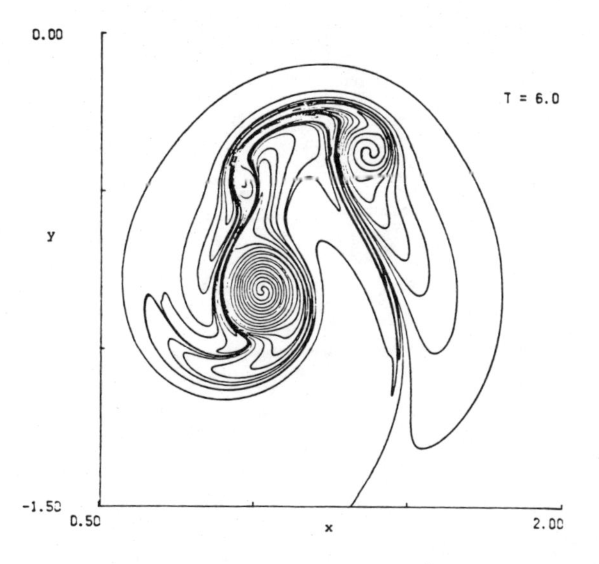

Figure 4: Closeup view of the solution at $t = 6$ (Krasny [49]) (Journal of Fluid Mechanics)

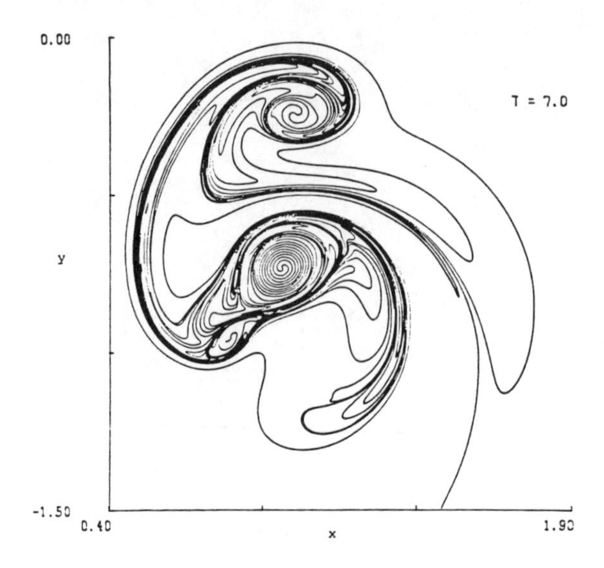

Figure 5: Closeup view of the solution at $t = 7$ (Krasny [49]) (Journal of Fluid Mechanics)

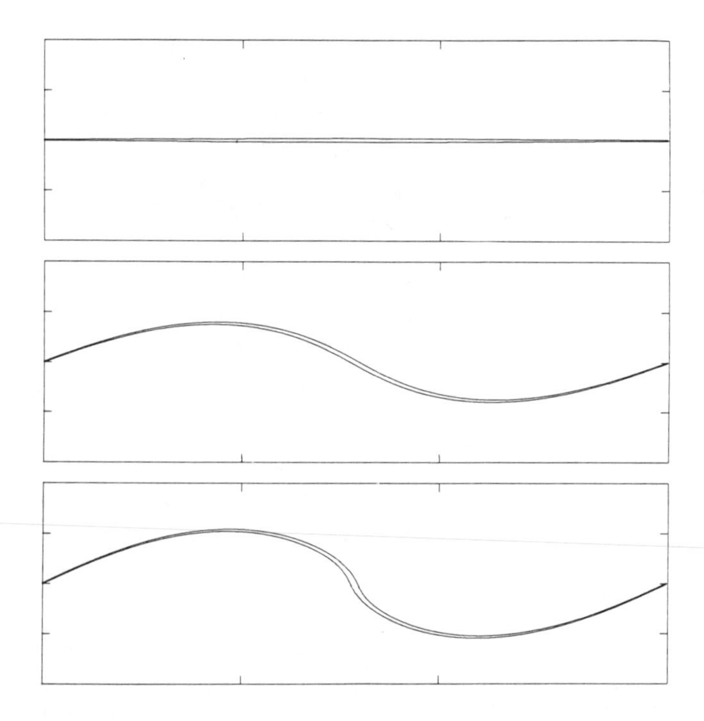

Figure 6: The solution at times $t = 0$, 1.5, and 2.0 with $h = .05$ (Shelley and Baker [66]).

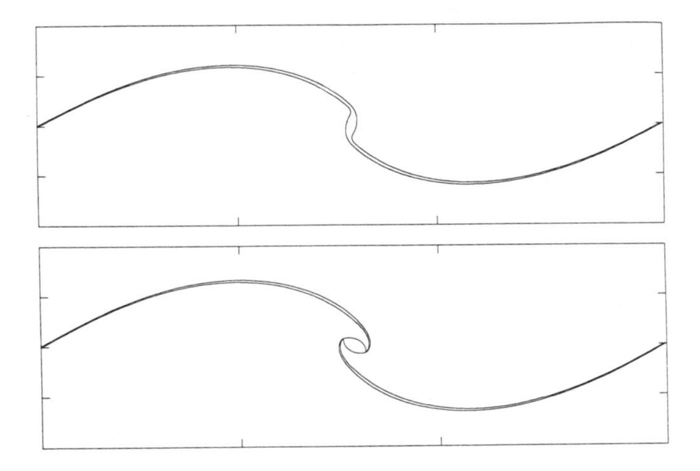

Figure 7: The solution at times $t = 2.2, 2.4$ with $h = .05$ (Shelley and Baker [66]).

III. New Developments in Mathematical Fluid Dynamics Motivated by Results from Large-Scale Computation – Vortex Sheets for 2-D Flow

Here we describe a program developed in joint work by Diperna and the author ([30], [31], [32]) specifically designed to address the new phenomena that arise in taking limits of suitable approximate solution sequences for 2-D and 3-D Euler. We mostly discuss this theory and its implications for approximate solution sequences for 2-D Euler with vortex sheet initial data and indicate how the above theory provides insight into the trends observed in the calculations of Krasny and Shelley and Baker as described in Section 2C) above.

A) **Approximate Solution Sequences for 2-D Euler with Vortex Sheet Initial Data**

We consider an initial incompressible velocity field v_o in two space dimensions for the evolution of a vortex sheet. Thus we assume that v_o has the structure in (2.1), i.e., the vorticity ω_o is a surface Dirac delta measure and the velocity v_o has locally finite kinetic energy. The basic physical problem involved in the evolution of vortex sheets in the high Reynolds number limit is the following:

(3.1) If $v^{\nu}(x,t)$ is the solution of the Navier-Stokes equations in (0.5) with vortex sheet initial data v_o does $v^{\nu}(x,t)$ converge in the high Reynolds number limit as $\nu \to 0$ to a solution of the inviscid 2-D Euler equations? Do new phenomena occur in the limiting process? Do solutions of inviscid 2-D Euler with vortex sheet initial data exist for all time?

In Section 2C), we have described two other ways to generate approximate solution sequences for 2-D Euler with vortex sheet initial data via computational algorithms. In the first method, Krasny uses computational vortex methods to generate approximate solution sequences while in the second method Shelley and Baker smooth the vortex sheet initial data by approximating this data by a finite thickness shear layer and then take the computational limit of the exact solution of inviscid 2-D Euler with patches of constant vorticity as $h \to 0$. The same basic questions described in the first part of (3.1) apply to the other two regularizations for 2-D Euler just described above. Another important question is the following:

(3.2) Do different regularizations for 2-D Euler with vortex sheet initial data converge to the same answer?

In Section 1 of [31], Diperna and the author introduce the concept of approximate solution sequence for 2-D Euler. Loosely speaking, we say that v^ε is an approximate solution sequence for 2-D Euler provided that the following three conditions are satisfied:

(3.3)

(1) v^ε is an incompressible velocity field with *local kinetic energy uniformly bounded independent of* ε, i.e.

$$\max_{0 \le t \le T} \int_{|x| \le R} |v^\varepsilon|^2 dx \le C(R,T)$$

for any $R,T > 0$.

(2) The vorticity $\omega^\varepsilon = \text{curl } v^\varepsilon$ satisfies

$$\max_{0 \le t \le T} \int |\omega^\varepsilon| dx \le C$$

(3) The *sequence of velocity fields* v^ε is *weakly consistent with 2-D Euler*, i.e. for all smooth vector test functions $\Phi(x,t) = (\Phi_1, \Phi_2)$ with bounded support and $\text{div } \Phi = 0$, as $\varepsilon \to 0$

$$\int\int (\Phi_t \cdot v^\varepsilon + \nabla\Phi : v^\varepsilon \otimes v^\varepsilon)\, dxdt \to 0$$

Here $v \otimes v = (v_i v_j)$, $\nabla\Phi = \left(\frac{\partial \Phi_i}{\partial x_j}\right)$, and A:B denotes the matrix product $\sum_{i,j} a_{ij} b_{ij}$. The estimate in 2) of (3.3) is the natural one for 2-D vortex sheet data (see [31]).

We note that for vortex sheet initial data v_o, if 2-D Euler has a solution $v(x,t)$ then this solution is not smooth and only would be a solution in the "weak" distributional sense, i.e. for all smooth Φ with $\text{div } \Phi = 0$

(3.4)

$$\int\int (\Phi_t \cdot v + \nabla\Phi : v \otimes v)\, dxdt = 0$$

and

$$\text{div } v = 0 \text{ in the sense of distributions .}$$

The equation in the first part of (3.4) arises from writing the 2-D Euler equations in conservation form, multiplying by Φ, and integrating by parts in a fashion familiar to the reader perhaps from hyperbolic shock wave theory. With the natural definition in (3.4), we observe that the condition in 3) of (3.3) is the minimum requirement needed for a sequence of approximate solutions to have any chance of converging to a solution of 2-D Euler as $\varepsilon \to 0$. A portion of [31] is devoted to a proof of the following important result:

Theorem: All three of the regularization processes described above with vortex sheet initial data generate approximate solution sequences for 2-D Euler satisfying the conditions in (3.1). For the high Reynolds number limit of the Navier-Stokes equations, $\nu = \varepsilon$; for the regularization of Shelley and Baker, $h = \varepsilon$; while for the class of computational vortex methods described in [31], $\varepsilon = \delta$ with δ and h from Section 1A) suitably linked.

Actually, the approximation of vortex sheets by finite thickness shear layers with constant strength is a minor variant of the approximate solution sequence strategy from [31] called "smoothing the initial data". The approximation used by Shelley and Baker results in approximate solutions generated by patches of constant vorticity – these solutions are not smooth but are mild weak solutions and the theory of Yudovich (see Section 3 of [58]) applies to guarantee global existence and uniqueness of these mild weak solutions.

The complexity observed in the calculations of Krasny reported in Section 2C) indicates that the limits as $\varepsilon \to 0$ of approximate solution sequences can be incredibly complex. Do new phenomena occur in the limiting process for approximate solution sequences? Examples indicate that the answer is yes. The simplest way to generate examples of approximate solution sequences is to take exact solutions for 2-D Euler satisfying the conditions in (3.3) – of course, (3.3)3) is trivially satisfied for these sequences. Two examples discussed in detail in [30], [31] are generated by utilizing the swirling flows in (0.14).

Example #1: Pick a *positive vorticity* distribution $\omega \geq 0$ with bounded support and define v^ε to be the scaled exact swirling flow

$$v^\varepsilon = \left(\log \frac{1}{\varepsilon}\right)^{-1/2} \frac{1}{\varepsilon}\, v\left(\frac{x}{\varepsilon}\right)$$

with v given from ω by (0.14). Then all of the assumptions in (3.3) are satisfied.

Example #2: Pick a *vorticity distribution* with bounded support but *zero total circulation*, i.e. $\int_0^\infty s\, \omega(s) ds = 0$, and define v^ε by

$$v^\varepsilon = \varepsilon^{-1}\, v\left(\frac{x}{\varepsilon}\right)$$

with v given from ω by (0.14), then all of the assumptions in (3.3) are satisfied. These exact solutions are called "phantom" vortices in [31] because these swirling flows vanish identically outside a circle of radius $O(\varepsilon)$ as $\varepsilon \to 0$ (see [30], [31] for further examples).

What happens to the limit as $\varepsilon \to 0$ of these exact swirling flows? First, it is easy to see in both examples that

$$v^\varepsilon \rightharpoonup 0 \tag{3.5}$$

although the convergence is weak and certainly not uniform. On the other hand, if we multiply the nonlinear terms $v_i^\varepsilon\, v_j^\varepsilon$ by a smooth function $\phi(x_1,x_2)$ with bounded support and average,

$$\lim_{\varepsilon \to 0} \int_{R^2} \phi\, v_i^\varepsilon v_j^\varepsilon = C\, \phi(O)\, \delta_{ij}\,, \quad C \neq 0 \tag{3.6}$$

where $\delta_{ij} = \begin{cases} 1, & i=j \\ 0, & i \neq j \end{cases}$. The constant C differs in Example #1 and Example #2 and depends on different averages of the vortex core structure in each of the two different cases. In the language of distributions, (3.6) means that

$$v^\varepsilon \otimes v^\varepsilon \rightharpoonup C \begin{pmatrix} 1 & 0 \\ 0 & 1 \end{pmatrix} \delta(x) \tag{3.7}$$

with $\delta(x)$ the Dirac delta function at the origin. Naively, one might have expected from (3.5) that $v^\varepsilon \otimes v^\varepsilon \rightharpoonup 0$. Instead new phenomena of *concentration* have occurred in the limit. A *finite amount of local kinetic energy* (exactly 2C) *has been lost in the limit* in these examples *and concentrates on a small set of measure zero*, the origin in R^2. Thus, new phenomena of concentration occur in limits of approximate solution sequences. The concept of measure-valued solution for 2-D Euler is introduced in [30], [31] to allow for such potential complexity in the limiting process. One important fact proved in [30], [31] is the following

Theorem: Every approximate solution sequence for 2-D Euler with vortex sheet initial data converges for all time to a measure-valued solution of 2-D Euler. This solution has concentrations but no oscillations.

We remark that the concept of measure-valued solution guarantees only that the 2-D Euler equations are satisfied in a very weak sense involving expected values of certain probability measures. Nevertheless, this is an extremely flexible concept. Thus, not every measure-valued solution for 2-D Euler is a weak solution as defined in (3.4) although the converse is true. The exact solution sequences from Example #1 and Example #2 generate examples of non-trivial measure-valued solutions (see [31]). In fact, the author conjectures that if one takes a limit as $\delta \downarrow 0$ for a sequence of calculations like the one of Krasny depicted in Figures 1-5 with the crude value of $\delta = .1$, the following scenario is possible: there is a critical time t_c so that for $t < t_c$ the limit is a weak solution of 2-D Euler in the standard sense of (3.4) while for $t > t_c$ the weak solution bursts into a much more complex measure-valued solution for 2-D Euler. The guess that a measure-valued solution occurs for $t > t_c$ is based not only on the enormous complexity of the evolving vortex sheet for crude $\delta = .1$ but also because this complexity occurs as a consequence of the fact that *vorticity of different signs concentrates* in a *small* region of *space* and attempts to cancel in an inviscid flow. The simple Example #2 involving phantom vortices necessarily has vorticity with changing signs; as explained in detail in Section 1 of [31], much less singular local behavior occurs in the concentrations from Example #2 than in those from Example #1 with a fixed sign of the vorticity. Since the behavior of concentration is less singular when the vorticity changes sign, the chances of developing concentrations are much greater.

Next, through a combination of theory from [31] and observed trends in the numerical data, we present strong evidence that in the limit as the thickness $h \to 0$ for the calculations of Shelley and Baker depicted in Figures 6 and 7 with $h = .05$, the limit is expected to be an ordinary weak solution of 2-D Euler. The vorticity in this calculation has a distinguished positive sign. Theorem 3.1 of [31] contains a criterion to check whether a given approximate solution sequence converges to a classical weak solution of 2-D Euler; this criterion is especially useful when the vorticity has one sign. For the regularization considered by Shelley and Baker, the far-field condition in (3.5) of [31] is readily verified. Thus, for the specific regularization used by Shelley and Baker, Theorem 3.1 from [31] has the following special form:

Theorem: Assume that the approximate vorticity ω^ε has a distinguished sign, $\omega^\varepsilon(x,t) \geq 0$. Also assume that

$$\max_{\substack{x_o \varepsilon R^2 \\ 0 \leq t \leq T}} \int_{|x - x_o| \leq R} \omega^\varepsilon dx \leq C \log\left(\frac{1}{R}\right)^{-\beta} \text{ for all } R \leq R_o \tag{3.8}$$

for some fixed constant C and some β with $\beta > 1$. Then as $\varepsilon \to 0$ the corresponding velocity fields v^ε of the approximate solution sequence converge strongly to an ordinary weak solution of 2-D Euler satisfying (3.4) for $0 < t < T$.

Next, we check the criterion in the above theorem according to the computational trends mentioned at the end of Section 2C) and depicted in Figures 6 and 7. We recall that the approximate Kirchhoff ellipse that occurs at time $t = 2.4$ was present in a sequence of three resolved runs with roughly the same ratio of major to minor axes in the ellipse as $h \downarrow 0$ while the area of this ellipse was $0(h^{1.6})$. Since vorticity has constant strength $0(h^{-1})$ inside the bounding curves, with the information just presented, the maximum value of the ratio of the left and right hand sides of (3.8) occurs at $t = 2.4$ at the center of the ellipse for $R \approx h^{.8}$. Since

$$h^{.6} << (\log(h^{-.8}))^{-\beta} \text{ as } h \downarrow 0 \,,$$

for any $\beta > 1$, the criterion in (3.8) is satisfied and the limit is expected to be a classical weak solution for 2-D Euler.

We mention here that the criterion in (3.8) is almost sharp; the sequence of swirling flows with vorticity of positive sign from Example #1 satisfies an estimate like (3.8) with the value $\beta = 1/2$ but develops concentrations and does not converge strongly. Thus, a criterion such as $\beta > 1$ is needed and almost sharp. When the vorticity locally has a distinguished sign, formal asymptotic methods ([64]) often predict that the left hand side behaves like $O(h^{\alpha})$ with $\alpha > 0$; thus a classical weak solution is predicted by the Theorem in these instances.

We mention here that it is still possible for the limit of an approximate solution sequence to be an ordinary weak solution of 2-D Euler even though concentrations develop and there is a loss of kinetic energy – this possibility is explored in detail in [32] with several positive results. Also, while the author knows of no explicit rigorous examples where a weak solution for 2-D Euler with vortex sheet initial data bursts at a certain time into a measure-valued solution as expected in the $\delta \downarrow 0$ limit of Krasny's calculations, R. Dziurzynski ([34]) in a Ph.D. thesis written under the author's supervision has found explicit examples where a weak solution of the 1-D Vlasov-Poisson equations bursts at a critical time into a measure-valued solution with concentrations in the charge density. In fact, [34] contains a very careful numerical and analytic study of the evolution of patches of electrons and electron sheets for the 1-D Vlasov-Poisson equations – these are analogous but simpler problems than the evolution of patches of vorticity and vortex sheets for fluid flow so many of the complex numerical and theoretical issues in fluid problems can be readily understood in the context of 1-D Vlasov-Poisson. Finally we mention that Diperna and the author have also introduced the concept of measure-valued solution for 3-D incompressible flow in [30]. With this concept it is proved that every sequence of solutions of the 3-D Navier-Stokes equations converges in the high Reynolds number limit to a measure-valued solution of 3-D Euler. Also, every zero diffusion limit of a sequence of statistical solutions of 3-D Navier-Stokes generates a measure-valued solution of 3-D Euler. This is a remark of C. Foias and J. Keller independently and my current Ph.D. student D. Chae has supplied a rigorous proof. Of course, fluid flows in 3-D are much more complex and approximate solution sequences for 3-D Euler can develop both concentrations and oscillations (explicit examples are constructed in detail in [30]).

Unfortunately, most of the basic problems presented in (3.1) and (3.2) are still unresolved. Nevertheless, we hope that the mathematical ideas presented in [30], [31], [32] will elucidate these issues and contribute to their resolution.

Acknowledgement

The author thanks Robert Krasny, Mike Shelley, and Greg Baker for allowing the author to use figures and data from their as yet unpublished calculations described in Section 2.

BIBLIOGRAPHY

[1] Anderson, C. and Greengard, C., S.I.A.M. J. Num. Anal. 22 (1985), 413-40.

[2] Anderson, C., J. Comput. Phys. 62 (1986), 111-23.

[3] Anderson, C., "Vorticity boundary conditions and boundary vorticity generation for two dimensional viscons incompressible flows" preprint April 1987.

[4] Anderson, C., J. Comput. Phys. 61 (1985), 417-32.

[5] Baker, G. Meiron, D. and Orszag, S., J. Fluid Mech. 123 (1982), 477-501.

[6] Beale, J. T. and Majda, A., J. Comput. Phys. 58 (1985), 188-208.

[7] Beale, J. T. and Majda, A., ''The design and numerical analysis of vortex methods'' in *Transonic, Shock, and Multidimensional Flows* edited by R. Meyer, Academic Press (1982), 329-345.

[8] Beale, J. T. and Majda, A. Contemp. Math. 28 (1984), 221-229.

[9] Beale, J. T. and Majda, A., Math. Comp. 39 (1982), 1-27.

[10] Beale, J. T. and Majda, A. Math. Comp. 39 (1982), 29-52.

[11] Beale, J. T., Math. Comp. 46 (1986), 402-24 and S 15- S20.

[12] Beale, J. T. and Majda, A., Math. Comp. 37 (1981), 243-260.

[13] Benfatto, G. and Pulvirenti, M., Comm. Math. Phys. 96 (1984), 59-95.

[14] Benfatto, G. and Pulvirenti, M., Comm. Math. Phys. 106 (1986), 427-458.

[15] Birkhoff, G. and Fisher, J., Circ. Math. Palermo (1959), 77-90.

[16] Cheer, A., S.I.A.M. J. Sci. Stat. Comp. 4(1983) 685-705.

[17] Choquin, C. and Huberson, S., ''Particle simulation of viscous flow'', preprint 1987.

[18] Chorin, A. J. and Bernard, P., J. Comput. Phys. 13 (1973) 423-428.

[19] Chorin, A. J., J. Fluid Mech. 57 (1973), 785-96.

[20] Chorin, A. J., S.I.A.M. J. Sci. Stat. Comput. I (1980), 1-21.

[21] Chorin, A. J., J. Comput. Phys. 17 (1978), 428-443.

[22] Chorin, A. J., Comm. Pure Appl. Math. 34 (1981), 853-866.

[23] Chorin, A. J., Comm. Math. Phys. 83 (1982), 517-535.

[24] Christiansen, J. P., J. Comput. Phys. 13 (1973), 363-379.

[25] Cottet, G. and Raviart, P., S.I.A.M. J. Num. Anal. 21 (1984) 52-76.

[26] Cottet, G., These de 3e cycle, Universite P. et M. Curie, Paris 1982.

[27] Cottet, G., ''On the convergence of vortex methods in two and three dimensions'' preprint 1985.

[28] Cottet, G. and Mas-Gallic, S., ''A particle method to solve transport diffusion equations, Part I: the linear case'', preprint 1985.

[29] Cottet, G., ''A new approach for the analysis of vortex methods in 2 and 3 dimensions'' print February 1987.

[30] Diperna, R., and Majda, A., Comm. Math. Phys. 108 (1987), 667-689.

[31] Diperna, R. and Majda, A., ''Concentrations in regularizations for 2-D incompressible flow'' Comm. Pure Appl. Math. (in press).

[32] Diperna, R. and Majda, A., ''Reduced Hausdorff dimension and concentration – cancellation for 2-D incompressible flow'' (in press, to appear in first issue of J. Amer. Math. Soc. January 1, 1988).

[33] Dombre, T., Frisch, U., Greene, J., Henon, M., Mehr, A., and Soward, A., J. Fluid Mech. 167 (1986), 353-391.

[34] Dziurzynski, R., ''Patches of electrons and electron sheets for the 1-D Vlasov-Poisson equations'' Ph.D. thesis University of California, Berkeley and Princeton University, May 1987.

[35] Ebin, D., Comm. P.D.E. 9 (1984), 539-559.

[36] Ghoniem, A. and Sherman, F., J. Comput. Phys. 61 (1985), 1-37.

[37] Ghoniem, A. and Sethian J., "Effect of Reynolds number on the structure of recirculating flow" A.I.A.A. J. 1987 (in press).

[38] Ghoniem, A. and Ng, K., "Numerical study of the dynamics of a forced shear layer" preprint May 1986.

[39] Ghoniem, A., Aly, H., and Knio, O., "Three dimensional vortex simulation with application to axisymmetric shear layer" A.I.A.A. 87-0379, Reno January 1987.

[40] Goodman, J., "Convergence of the random vortex method" print 1985.

[41] Greengard, L. and Rokhlin, V., "A fast algorithm for particle simulations" preprint April 1986.

[42] Greengard, L. and Rokhlin, V., "Rapid evaluation of potential fields in three dimensions" preprint January 1987.

[43] Greengard, C., Math. Comp. 47 (1986), 387-98.

[44] Greengard, C., J. Comput. Phys. 61 (1985), 345-348.

[45] Hald, O. and Del Prete, Math. Comp. 32 (1978) 791-809.

[46] Hald, O., S.I.A. J. Numer. Anal. 16 (1979), 726-55.

[47] Hald, O., "Convergence of vortex methods for Euler's equations III" to appear S.I.A.M. J. Numer. Anal.

[48] Hald, O. "Convergence of a random method with creation of vorticity" S.I.A.M. J. Sci. Stat. Comp. 7 (1986), 1081-1094.

[49] Krasny, R., "Computation of vortex sheet roll-up in the Treffitz plane" preprint March 1986 (J.F.M. in press).

[50] Krasny, R., J. Comput. Phys. 65 (1986), 292-313.

[51] Leonard, A., Ann. Rev. Fluid Mech. 17 (1985), 523-59.

[52] Leonard, A., J. Comput. Phys. 37 (1980), 289-335.

[53] Long, D. G., "Convergence of the random vortex method in one and two dimensions" Ph.D. thesis University of California and Princeton University November 1986.

[54] Long, D. G., "Convergence of the random vortex method in two dimensions" preprint December 1986.

[55] Long, D. G., "Convergence of the random vortex method in three dimensions" preprint March 1986.

[56] Long, D. G., "Time discretization of the random vortex method" in preparation.

[57] Long, D. G. and Majda, A., "Explicit analysis of the random vortex method for strained shear layers" in preparation.

[58] Majda, A., Comm. Pure Appl. Math. 39 (1986), S187-220.

[59] Majda, A., *Lectures on Vorticity and Incompressible Fluid Flow*, Princeton University Spring 1985 and Fall 1986 (to appear as research monograph eventually).

[60] Marchioro, C. and Pulvirenti, M., Comm. Math. Phys. 84 (1982), 483-503.

[61] Perlman, M., J. Comput. Phys. 59 (1985), 200-23.

[62] Roberts, S., J. Comput. Phys. (1984) 723-749.

[63] Rosenhead, L., Proc. Roy. Soc. A 134 (1931), 170-192.

[64] Saffman, P. and Baker, G., Ann. Rev. Fluid Mech. 11 (1979), 95-122.

[65] Sethian, J. and Ghoniem, A., "Validation Study of Vortex Methods" J. Comput. Phys. (in press).

[66] Shelley M. and Baker, G., "On the relation between thin vortex layers and vortex sheets: Part 2, numerical study" (in preparation).

Mathematics and Tomography

F. NATTERER*

Abstract. Tomography is a technique for imaging single slices of a 3D object. In its simplest case it requires the computation of a function in $\mathbf{R}^2$ from its line integrals. We give a short survey on the relevant mathematics and on the numerical methods which are being used in practice. We address three specific problems (incomplete data, attenuation correction, resolution) and show how they can be solved by analytical and numerical techniques. Finally we mention some possible future developments.

1. Tomography. "Tomography" is derived from the greek word τομοσ = slice. It stands for several techniques in diagnostic radiology permitting the imaging of a single slice of a 3D object.

Even before the advent of computerized tomography (CT), simple focusing techniques have been used, see Littleton (1976), Barrett and Swindell (1981). In motion tomography, the 3D object lies between two parallel planes. In one of these planes, an X-ray source is moving with a constant speed v_s. In the other plane, a detector (film) is moving
ray source is moving with a constant speed v_s. In the other plane, a detector (film) is moving
in the opposite direction with speed $v_d = \frac{b}{a} v_s$, a, b being the distances of the in-focus plane from the source and detector plane, respectively. Each single point of the in-focus plane is mapped onto one and the same point of the film as the source and the film move. Since a point in any other plane undergoes some blurring, the imaging on the film is a super-position of a sharp image of the in-focus plane and blurred images of the other planes.

The principle of (transmission) CT as introduced by Hounsfield (1973) is completely different. Here, the slice to be imaged is scanned by a thin X-ray beam (see Fig. 1a).

The intensity of the beam is measured by a detector. From all these measurements, an image of the slice is computed and displayed on a screen. The scanning can be done in various ways, see Fig. 1.1b, 1.1c.

A mathematical model for CT is as follows. Let $f(x)$ be the attenuation coefficient of the tissue in the point x of the slice, and let L be the line joining the source with the detector.

* Institut für Numerische und instrumentelle Mathematik, Westf. Wilhelms-Universität Münster, Einsteinstraβe 62, D — 4400 Münster (FRG).

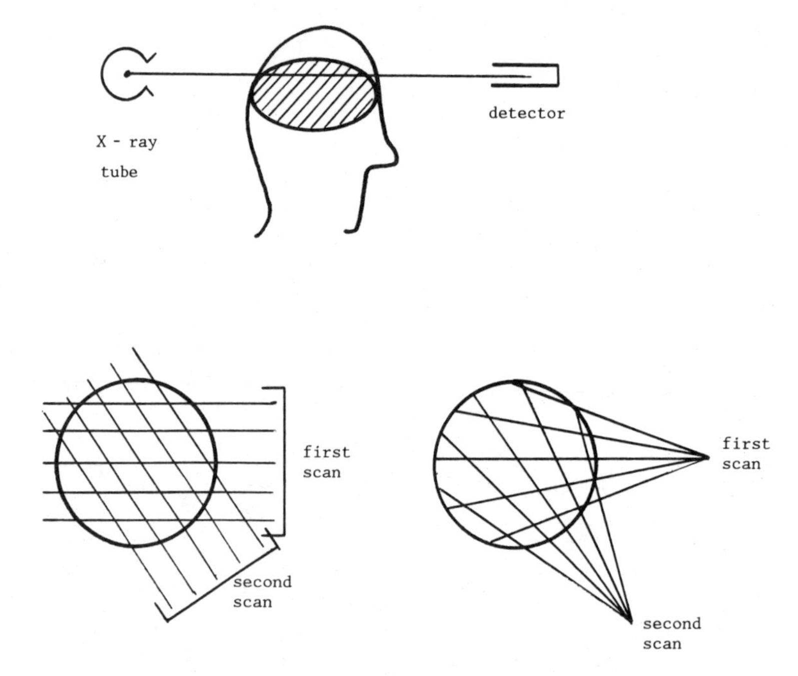

FIG.1.1. a (top) Principle of CT. b (bottom left) Parallel scanning. c (bottom right) Fan-beam scanning.

Let I_o be the incident intensity of the X-ray beam. Then, the intensity past the object which is measured by the detector is

$$I = I_o \exp \left\{ - \int_L f(x) dx \right\}.$$

Thus, the data collected in CT correspond to the line integrals of f along the measured beams. The mathematical problem is to reconstruct a function in $\mathbf{R}^2$ from its line integrals.

There are many other applications of CT. We mention only radio astronomy (Bracewell (1956)), electron microscopy (De Rosier and Klug (1968), Hoppe and Hegerl (1980)), plasma physics and gas dynamics (Kershaw (1962), Kikalov and Preobrazhensky (1987)), satellite remote sensing (Fleming (1982)), geophysics (Dines and Lytle (1979)), radar (Munson, O'Brien and Jenkins (1983) and non-destructive testing (Allan et al. (1985)).

The literature on CT has grown tremendously in the last few years. We mention only the monographs Herman (1980), Deans (1983), Natterer (1986), Pikalov and Preobrazhensky (1987) and the March 1983 issue of the Proceedings of the IEEE.

2. The Radon Transform. The relevant transform in CT is obviously the Radon transform R, which maps a function f in $\mathbf{R}^2$ onto the set of its line integrals. We use the representation $x \cdot \theta = s$ for the straight line, where $\theta \in S^1$ is a unit vector and $s \in \mathbf{R}^1$. Then Rf can be considered as a function on $S^1 \times \mathbf{R}^1$:

$$Rf(\theta, s) = \int_{x \cdot \theta = s} f(x) dx .$$

We need some basic facts on R (for a thorough treatment of the theory of the Randon transform see Helgason (1980)). For ease of exposition we assume f to be sufficiently regular and of compact support. We start with the "projection theorem" (Bracewell (1956)).

$$(Rf)\hat{}(\theta, \sigma) = (2\pi)^{1/2} \hat{f}(\sigma\theta), \quad \sigma \in \mathbf{R}^1 . \tag{2.1}$$

Here, the hat on the left hand side denotes the 1D Fourier transform with respect to the second variable, while the hat on the right hand side stands for the 2D Fourier transform of f. Doing a 2D inverse Fourier transform and introducing polar coordinates yields the inversion formula

$$f = \frac{1}{4\pi} R^* H \frac{\partial}{\partial s} Rf \tag{2.2}$$

of Radon (1917). Here, H is the Hilbert transform acting on the second variable of Rf, and R^* is the backprojection

$$R^* g(x) = \int_{S^1} g(\theta, x \cdot \theta)\, d\theta .$$

R^* happens to be the L_2 - adjoint of R. Considering g as a function on the straight lines, R^* averages over all straight lines going through x. A completely different inversion formula has been given by Cormack (1963) and Kershaw (1962). Let $\theta = (\cos\phi, \sin\phi)^T$ and

$$f(r\theta) = \sum_l e^{il\phi} f_l(r), \quad Rf(\theta, s) = \sum_l e^{il\phi} g_l(s) .$$

Then,

$$f_l(r) = -\frac{1}{\pi} \int_r^\infty (s^2 - r^2)^{-1/2} T_{|l|}\left(\frac{s}{r}\right) g_l'(s)\, ds \tag{2.3}$$

where T_l is the first kind Chebyshev polynomial of degree l. Yet another inversion formula has been given by Vvedenskaya and Gindikin (1984).

Another consequence of (2.1) is the equivalence of the norms

$$\| f \|_{H_0^\alpha(\Omega)}, \quad \| Rf \|_{H^{\alpha+1/2}(S^1 \times \mathbf{R}^1)} \tag{2.4}$$

for compact sets Ω and suitably defined Sobolev spaces, see Smith, Solmon and Wagner (1977), Natterer (1986). It follows that R, as an operator between L_2 - spaces, does not have a bounded inverse, i.e. Randon's integral equation $Rf = g$ is ill-posed. Fortunately, this ill-posedness is not very pronounced. Since the shift in the order of the Sobolev spaces in (2.4) is only ½, the standard problem in CT is only mildly ill-posed. However we shall see that there are also seriously ill-posed problems in CT.

We conclude this section by considering the range of the Randon transform. It is easy to see that

$$\int s^m Rf(\theta, s)\, ds = p_m(\theta), \quad m = 0, 1, \ldots \tag{2.5}$$

is a homogeneous polynomial of degree m in θ. This means that the range of R is highly structured.

3. Reconstruction Algorithms. Many suggestions have been made to solve Radon's integral equation $Rf = g$. We mention only the most important ones. To fix ideas we assume that p parallel projections with $2q + 1$ line integrals each have been taken and that f is supported in the unit disk Ω, i.e. our data is $g(\theta_j, s_l)$, $\theta_j = (\cos\phi_j, \sin\phi_j)$, $\theta_j = \pi j/p$, $j = 0, \ldots, p-1$, $s_l = l/q$, $l = -q, \ldots, q$.

(a) Fourier reconstruction (Bracewell (1956)).

This is a direct implementation of (2.1). For each direction θ_j, we do a 1D FFT of length $2q+1$. This provides us with the values

$$\hat{f}(k\pi\theta_j) = (2\pi)^{-1/2}\hat{g}(\theta_j, k\pi), \quad k=-q,\ldots,q. \tag{3.1}$$

The sampling interval π has been chosen in accordance with Shannons sampling theorem (see e.g. Jerry (1977)) which requires that a function with band-width b be sampled with sampling interval π/b. Thus we get $\hat{f}$ on a polar coordinate grid with radial spacing π. In order to compute f we have to do a 2D inverse Fourier transform for which we need $\hat{f}$ on a cartesian grid, again with spacing π. Thus the second step in Fourier reconstruction consists of an interpolation in the polar coordinate grid, followed by a 2D inverse FFT as final step.

Besides its simplicity, Fourier reconstruction is attractive because of its efficiency. However, the interpolation step has to be implemented carefully. Since the radial spacing is π, quite independently of p, q, simple polynomial or spline interpolation is certainly not good enough, see Bracewell (1979), Rowland (1979). When properly implemented (see Stark, Woods, Paul and Hingorani (1981), Natterer (1985)) Fourier reconstruction is as accurate as other algorithms. The reason why it is not used much in the medical field is a practical one. Since the 2D inverse FFT needs data from all directions, the final step can start only after the scan has been completed. In contrast to this the filtered backprojection algorithm to be described below can be started as soon as the first projection comes in.

(b) The filtered backprojection algorithm (Bracewell (1967), Ramachandran and Lakshminarayanan (1971), Shepp and Logan (1974)). This is essentially an implementation of Radon's inversion formula (2.2). The Hilbert transform and the s-derivative are combined into a 1D convolution yielding a filtered version of g which in turn is backprojected.

A different approach, suggested by K. T. Smith (1983) gives more flexibility and insight. For any pair of functions W on $\mathbf{R}^2$, w on $S^1 \times \mathbf{R}^1$ with $W = R^*w$ we have

$$W * f = R^*(w * g) \tag{3.2}$$

where the convolutions are in $\mathbf{R}^2$ and $\mathbf{R}^1$, respectively. Now choose W as an approximation to the Dirac δ-function. Then, the left hand side of (3.2) provides an approximation to f, while the right hand side can be evaluated in a straightforward manner by numerical integration and interpolation.

(c) ART algebraic reconstruction technique, Gordon, Bender and Herman (1970). This is a completely discrete approach to the reconstruction problem. The reconstruction region is subdivided into little squares ("pixel") numbered $1,\ldots,n$. In pixel i, f is assumed to have constant value f_i. This turns $Rf=g$ into a linear system

$$Af = g \tag{3.3}$$

where, by abuse of notation, $f \in \mathbf{R}^n$, $g \in \mathbf{R}^r$, $r = p(2q+1)$, and A is a sparse $r \times n$ matrix.

ART is an iterative method originally suggested by Kaczmarz (1937) for solving (3.3). Let P_k denote the orthogonal projection onto the k-th hyperplane of (3.3) and put

$$P_k^{\omega} = (1-\omega)I + \omega P_k, \quad P^{\omega} = P_r^{\omega}\ldots P_1^{\omega}$$

where ω is a relaxation parameter. Then, Kaczmarz's method (with relaxation) simply reads $f^t = P^{\omega} f^{t-1}$, $t=1,2,\ldots$ with f^0 arbitrary.

Let $a_k^T f = g_k$ be the k-th equation of (3.3). Then,

$$P_k f = f + \frac{1}{\|a_k\|^2}(g_k - a_k^T f)a_k .$$

The i-th component of a_k is non-zero iff line k meets pixel i. This means that only $\sigma(\sqrt{n})$ components of a_k are non-zero, i.e. one step of ART requires $\sigma(r\sqrt{n})$ operations. This is essentially the operation count for the complete filtered backprojection algorithm. Thus ART is competitive only if a few steps suffice to get an acceptable accuracy.

The convergence properties of ART have been investigated in numerous papers. We mention only Herman and Lent (1976). Björck and Elving (1979) realized that ART is identical with the SOR-method for the system $AA^*h = g$ if the computations are done with $f = A^*h$. So it is not surprising that ART converges for $0 < \omega < 2$ to some vector f_ω. If (3.3) is consistent, f_ω is the minimum norm solution of (3.3) if the initial approximation is in the range of A^*. Otherwise f_ω is a certain least squares solution. See Censor et al. (1983), Natterer (1986) for details. This holds for any linear system (3.3) quite independently of tomography.

Based on an analysis of Hamaker and Solmon (1983), much more can be said in the case of tomography. Let U_m be the subspace of $L_2(\Omega)$ spanned by $U_m(x \cdot \theta_0), \ldots, U_m(x \cdot \theta_{p-1})$, U_m the Chebyshev polynomials of the second kind. Then, the contraction number $\rho_m(\omega)$ of ART on U_m can easily be computed numerically. The results are displayed in Fig. 3.1(a) for the natural ordering of the projections and in Fig. 3.1(b) for a random ordering. In the natural ordering convergence is slow for m small and fast for m large if ω large (i.e. $\omega = 1$). The opposite is the case for ω small (i.e. $\omega = 0.1$). This means, that, for ω large, ART picks up quickly the high frequency components of the picture, while the low frequency parts appear only after a large number of iterations. In view of the (high frequency) noise present in practical applications this behaviour is quite undesirable. Therefore, ART should only be used with strong underrelaxation. Alternatively, one can use orderings of the directions different from the natural one such as random ordering which gives much better convergence, see Fig. 3.1(b) for ART without relaxation.

(d) Direct algebraic methods (Lent (1975)).
As in ART, no analytical tools are used. A regularized solution to Radon's integral equation is computed without any discretization error, the complexity being reduced by FFT. In order to avoid technical difficulties we replace the lines $x \cdot \theta_j = s_l$ by strips L_{jl}. Then we have to solve

$$\int_{L_{jl}} f(x)dx = g_{jl}, \quad j = 0, \ldots, p-1, \quad l = -q, \ldots, q .$$

or $Af = g$, where A is a linear bounded operator from $L_2(\Omega)$ into $\mathbf{R}^r$, $r = p(2q+1)$. Let f_γ, $\gamma > 0$, be the regularized solution to this equation, i.e. f_γ is the minimizer of

$$\| Af - g\|^2_{\mathbf{R}^r} + \gamma \| f \|^2_{L_2(\Omega)} .$$

By standard methods of linear algebra we obtain

$$f_\gamma = A^*h, \quad (AA^* + \gamma I)h = g . \tag{3.4}$$

Here, A^* is the adjoint of A, i.e.

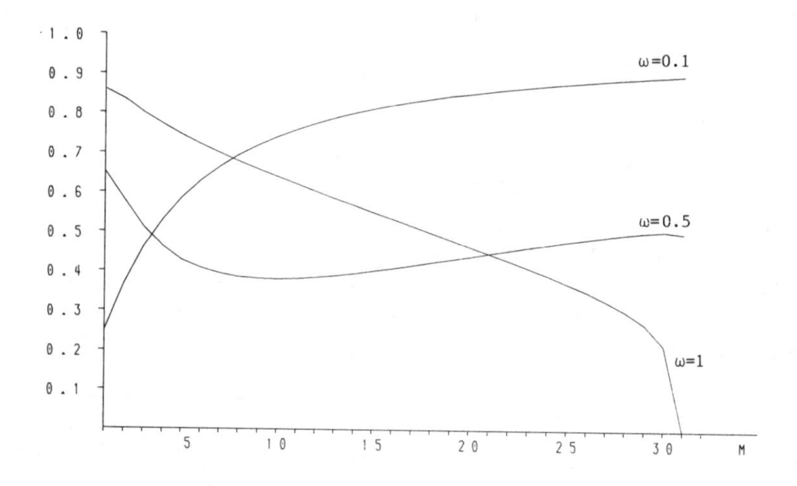

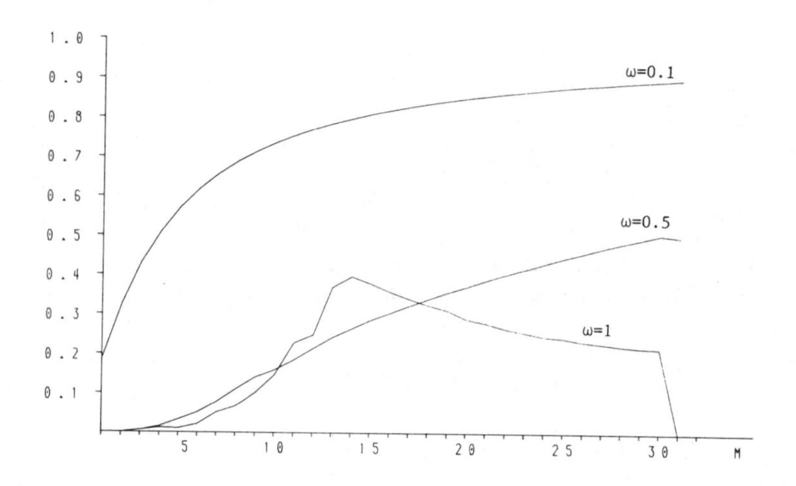

FIG. 3.1. Convergence of ART. a (top): Contraction numbers on U_m for $p = 32$ projections in natural order. b (bottom): Radon ordering of projections.

$$A^*h(x) = \sum_{\substack{j,l \\ x \in L_{jl}}} h_{jl} .$$

This is a discrete backprojection. Thus (3.4) is similar to filtered backprojection, except that the filtering step is replaced by the solution of an $r \times r$ linear system. Of course this system is too large for being solved directly. However, due to the rotational invariance of the scanning geometry, AA^* is a block-cyclic convolution of length p with blocks of size $2q + 1$. Therefore, h can be computed efficiently by FFT.

Direct algebraic methods can be applied whenever the scanning geometry is rotationally invariant. Examples are parallel and fan-beam scanning, but many other geometries in 2D (e.g. in multiplex and pinhole tomography, see Barrett and Swindell 1981) and in 3D (Kowalski (1979), Grangeat (1987)) satisfy this assumption.

4. Incomplete Data. In many cases it is not possible or not desirable to measure a complete data set. We mention only three examples in radiology. In region-of-interest tomography one is interested only in a small part of the body. It is natural to ask if the scanning can be restricted to this region of interest. The opposite case occurs if an opaque implant is present. One then has to reconstruct f outside the implant with the line integrals through the implant missing. Finally, to shorten the scanning process, one sometimes takes only views from a restricted angular range. Each of these examples fits into one of the following classes.

(i) The interior problem. Here, $Rf\,(\theta, s)$ is known only for $|s| \le a$, and one wants to reconstruct $f\,(x)$ in $|x| \le a$.

(ii) The exterior problem. Here, $Rf\,(\theta, s)$ is known only for $|s| \ge a$, and one wants to reconstruct $f\,(x)$ in $|x| \ge a$.

(iii) The limited angle problem. Here, $Rf\,(\theta, s)$ is known only for $\theta = (\cos\phi, \sin\phi)$, $|\phi| \le \phi < \frac{\pi}{2}$.

Problem (i) is not uniquely solvable. Problems (ii), (iii) are uniquely solvable, but there is no substitute for the stability estimate (2.4), not even with the "shift" ½ being replaced by an arbitrary real number. This means that problems (ii), (iii) are severely ill-posed.

In order to show that (2.4) does not hold for problem (ii) it suffices to produce a function f_+ which is not continuous in $|x| \ge a$ but for which Rf_+ is smooth in $|s| \ge a$. With $f \in C_0^\infty$ a radial function such that $f \ne 0$ in a neighbourhood of the disk around 0 with radius a, the function f_+ which coincides with f in the upper half plane and vanishes elsewhere is obviously such a function. This argument is due to Finch (1985). The severe ill-posedness of (iii) follows also from the singular value decomposition, see Davison (1983), Louis (1984).

The structure of Radon's inversion formula (2.2) gives considerable insight into the artefacts caused by incomplete data. It averages over lines through x. The biggest contributions come from the lines in whose neighbourhood Rf varies strongly. This is the case for those lines which are tangent to curves of discontinuity of f. If the integrals over such lines are missing, (2.2) cannot be evaluated accurately on such lines. Thus we come to the following rule of thumb:

Artefacts show up mainly in the vicinity of missing tangents to curves of discontinuity of f.

Fig. 4.1 illustrates this rule of thumb. Fig. 4.1(a) is the original, consisting of 2 disks. Fig. 4.1(d) shows the reconstruction from 256 parallel projections. In Fig. 4.1(b) the reconstruction from exterior data with the filtered backprojection algorithm is displayed. The incompleteness of the data is simply ignored, i.e. the integrals over lines hitting the large central hole (which shows up more clearly in Fig. 4.1(e) below) are put equal to 0. We see that strong artefacts appear along those tangents to the disks which hit this central hole, i.e. which are missing. This is in full agreement with our rule of thumb.

These strong artefacts can be reduced by consistent completion of data, see Lewitt, Bates and Peters (1978). This means we compute from the incomplete data g complete data g^c such that g^c is in the range of R and $g^c = g$ where g is known. In order to make the completion procedure unique and as stable as possible we choose g^c of minimal norm. For the exterior problem, the procedure is as follows. From the incomplete data one can compute the Fourier coefficient

$$\hat{g}_l(s) = \frac{1}{2\pi} \int_{S^1} g\,(\theta, s)\, e^{il\phi}\, d\theta$$

for $|s| > a$. From the consistency condition (2.5) it follows that a complete data set g^c characterized by

$$\int_{-1}^{+1} s^m \hat{g}_l^c(s)ds = 0 , \quad m < |l| .$$

The minimum norm solution to this system subject to $\hat{g}_l^c(s) = \hat{g}_l(s)$ for $|s| > a$ is easily seen to be

$$\hat{g}_l^c(s) = - \sum_{m < |l|}' \frac{2m+1}{a} \int_{|t| \geq a} \hat{g}_l(t) P_m(\frac{t}{a}) dt \; P_m(\frac{s}{a}) . \tag{4.1}$$

Here, P_m are the Legendre polynomials, and the prime indicates that $m + l$ is even. Since the Legendre polynomials increase exponentially outside $[-1, +1]$, (4.1) is highly unstable for m large. Since the exterior problem is severely ill-posed, this instability does not come as a surprise. In practice, (4.1) can only be used for modest values of $|l|$. For larger values of $|l|$, we put $\hat{g}^c(s) = 0$.

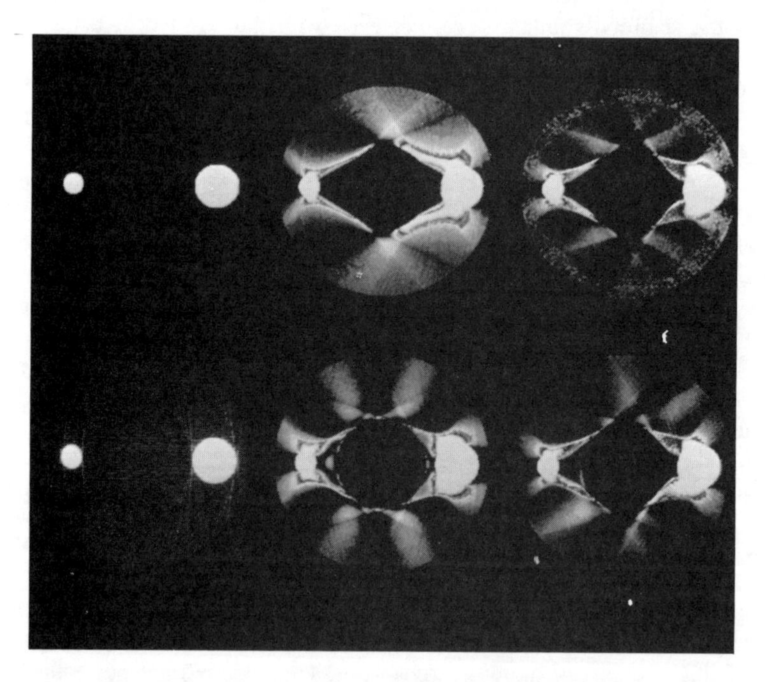

FIG. 4.1. Reconstruction from exterior data. a (top left): Original. b (top middle): Filtered backprojection with incomplete data. c (top right): Direct algebraic method. d (bottom left): Reconstruction from complete data. e (bottom middle): Filtered backprojection from completed data. f (bottom right): ART.

The effect of the completion procedure can be seen in Fig. 4.1(e). It shows the filtered backprojection reconstruction of the completed data set. In Fig. 4.1(c), 4.1(f) the results of the direct algebraic method and of ART are displayed. Both algorithms work directly on the incomplete data, so no preprocessing is necessary. We see that there is not much difference between the various reconstructions. The typical artefacts are still there, but they have been reduced considerably.

Some authors (see e.g. Herman and Lent (1976)) suggest to use a-priori-information on f, such as $f \geq 0$, to further reduce the artefacts. This is easily done in ART by putting

$f = 0$ as soon as it becomes negative.

For the interior problem, K. T. Smith (1983) suggested to backproject the second derivative of Rf, i.e. to compute

$$\Lambda f(x) = \int_{S^1} (Rf)''(\theta, x \cdot \theta) d\theta . \tag{4.2}$$

It turns out that $(\Lambda f)^\wedge(\xi) = |\xi| \hat{f}(\xi)$. Λf can be evaluated locally, i.e. only those lines are needed for $\Lambda f(x)$ which pass through a neighborhood of x. On the other hand, Λ is an elliptic pseudo differential operator which preserves the singular support. This means that Λf is smooth where f is and vice versa. Since Λ amplifies the high spacial frequencies, we expect discontinuities of f to show up even more clearly in Λf. In Fig. 4.2 a reconstruction of Λf for a chest phantom is given. We see that the values of Λf are different from those of f, but the curves along which f has jumps can be made out clearly in Λf. The conclusion is that region-of-interest tomography is possible if only discontinuities of the density are sought for.

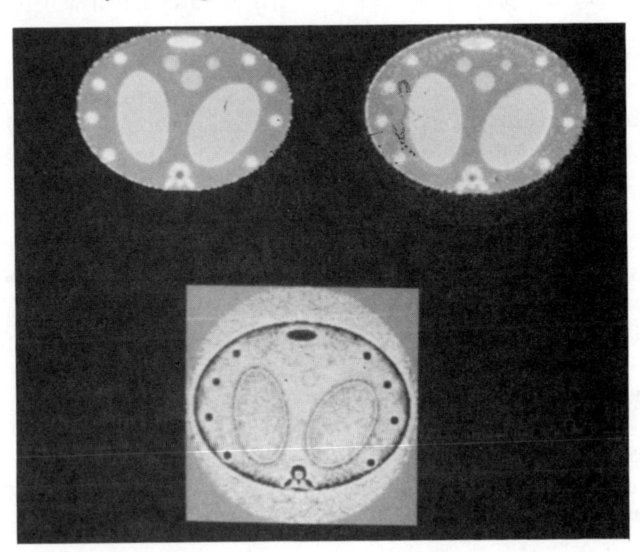

FIG. 4.2. Reconstruction of Λf. a (top left): Original. b (top right): Reconstruction of f. c (bottom): Reconstruction of Λf.

5. Emission Computed Tomography. In nuclear medicine one wants to find the distribution f of a radiopharmaceutical inside the body by measuring the radiation outside the body. If these measurements are done in a tomographic way (i.e. the detectors collimated so as to accept only radiation from a single line in the plane) then we measure again line integrals of f. However, due to the variable attenuation distribution μ of the tissue, these line integrals contain a weight factor depending on μ. In positron emission tomography (PET) the situation is particularly simple. The emission data are simply

$$g = e^{-R\mu} Rf .$$

This equation has to be solved for f. If μ is negligible, then f can be computed by any of the methods of CT. However, in practice μ is not small: $R\mu$ can become as large as 3. Thus the data g has to be corrected for attenuation before the reconstruction. Attenuation correction is one of the three major problems in emission tomography, the other ones being noise (see Shepp and Vardi (1983)) and scatter.

In today's clinical practice, μ is determined by a separate transmission scan prior to the emission scan. This is very time consuming and inconvenient for the patient. Therefore it would be highly desirable to extract all the information on μ one needs to correct for attenuations from the emission data g.

In order to do so we use the consistency conditions in the range of the Radon transform very much in the same way as in Section 4. Since $e^{R\mu}g$ is in the range of R we have from (2.4)

$$\int s^m \, e^{R\mu(\theta, s)} g(\theta, s)\,ds = p_m(\theta)\,, \quad m = 0,1,\ldots\,. \tag{5.1}$$

This is a nonlinear system for μ which can be put up as soon as the emission data g is available. The idea is to solve (5.1) for μ.

Unfortunately (5.1) is seriously ill-posed. Therefore, μ cannot be determined from (5.1) in the same generality as in transmission CT. Fortunately, the situation in PET is much simpler. The particle energy is 511 keV. At this energy, μ is almost constant ($\mu = 0.84\ \mathrm{cm}^{-1}$) in tissue and in bone ($\mu = 0.105\ \mathrm{cm}^{-1}$). So it suffices to distinguish between air, bone and tissue. All we have to do is to find the boundaries between them. For the head, we simply have to determine two curves representing the interior and the exterior boundary of the skull. This reduces drastically the number of degrees of freedom, making (5.1) reasonably well-posed. In Fig. 5.1 we show the results for computer generated phantoms. Fig. 5.1(a) is the activity distribution f inside the skull (not visible) as reconstructed from 72 parallel projections with 88 line integrals each, assuming perfect attenuation correction. This is what we ideally want to see. Fig. 5.1(b) is a display of the reconstruction from the emission data without any attenuation correction. The function values are by far too small, making this reconstruction virtually useless. In Fig. 5.1(c) we see the result of the attenuation correction using (5.1). The improvement over 5.1(b) is obvious, but there are some artefacts at the boundary of the brain which are quite disturbing. They are caused by instabilities in the determination of the interior boundary of the skull. These instabilities disappear if a few artificial sources are added. The reconstruction, shown in Fig. 5.1(d), is fully satisfactory.

So far the experiments with synthetic data. In Fig. 5.2 we show the reconstruction of real patient data of the ECAT-2 scanner of the Kernforschungsanlage Jülich. Fig. 5.2(b) is the reconstruction without correcting for attenuation. Qualitatively it looks similar to Fig. 5.1(b). The reconstruction using attenuation correction is displayed in Fig. 5.2(a).

6. Resolution. We say that a scanning geometry has resolution $2\pi/b$ if a function f containing no details of size $\leq 2\pi/b$ can be recovered reliably from the integrals over the lines of the geometry. By a function f containing no details $\leq 2\pi/b$ we mean — in accordance with image processing (see e.g. Pratt (1978)) — a function for which $\hat{f}(\xi)$ is negligible for $|\xi| \geq b$. Since CT amounts to sampling Rf on a grid on $S^1 \times \mathbf{R}^1$, the question is which grids determine Rf, hence f, essentially uniquely. For parallel scanning it has been known long since that $p \geq b$, $q \geq b/\pi$ suffice to determine f reliably. Thus we need essentially $\frac{2}{\pi} b^2$ data. The question arises if this is best possible.

This is a problem of 2D signal processing. The relevant theorem is the sampling theorem of Petersen and Middleton (1962). In our situation it reads as follows. Let g be a function on $[0, 2\pi] \times \mathbf{R}^1$, and let $\hat{g}$ be its 2D Fourier transform, i.e.

$$\hat{g}(k, \sigma) = \frac{1}{2\pi} \int_0^{2\pi} \int_{\mathbf{R}^1} e^{-ik\phi - i\sigma s}\, g(\phi, s)\,d\sigma\, d\phi, \; k \in \mathbf{Z} \quad \sigma \in \mathbf{R}^1 .$$

Assume that g vanishes on the grid $\{Wl\, , l \in \mathbf{Z}^2\}$ where W is a real 2×2 matrix. If $K \subseteq \mathbf{R}^2$ is an open set such that the sets $K + 2\pi(w^{-1})^T l\, , l \in \mathbf{Z}^2$ are mutually disjoint, then

$$| g | \leq \frac{1}{\pi} \sum_k \int_{(k\, ,\, \sigma) \notin K} | \hat{g}(k\, , \sigma) | \, d\sigma\, .$$

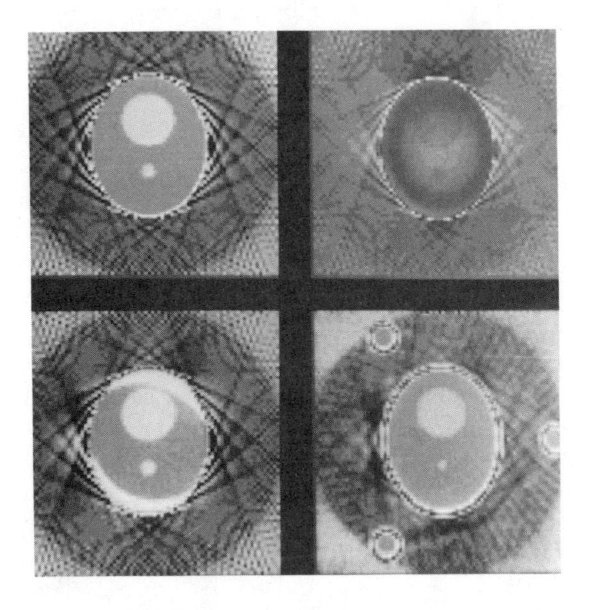

FIG. 5.1. Attenuation correction for synthetic data. a (top left): Perfect correction. b (top right): No correction. c (bottom left): Correction without exterior sources d (bottom right): Correction with exterior sources.

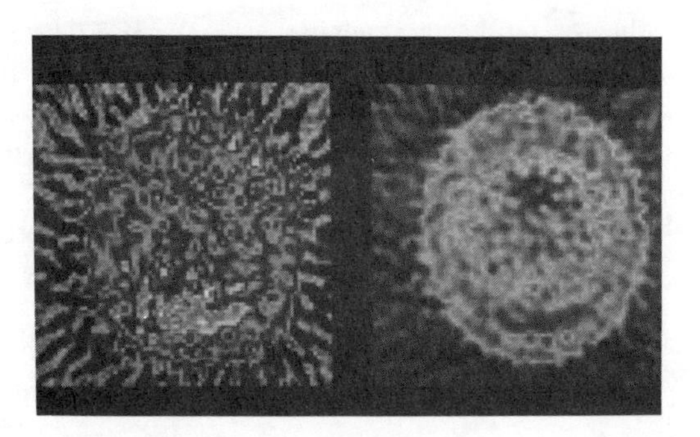

FIG. 5.2. Attenuation correction for real data. a (left): Reconstruction from corrected data. b (right): Reconstruction from raw data.

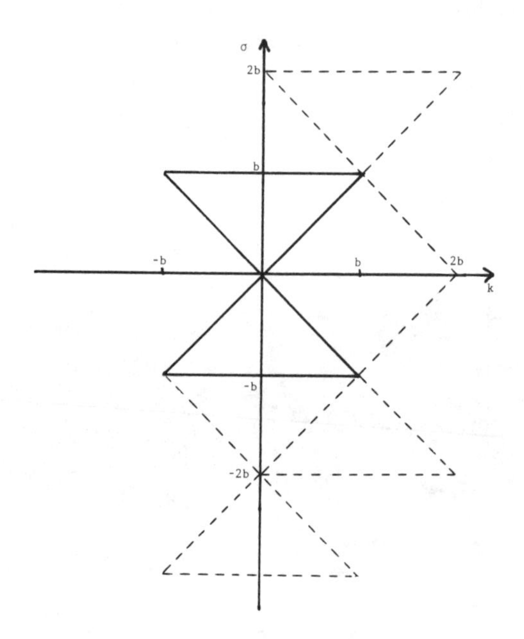

FIG. 6.1. The set K (solid line) and some of its translates (dashed).

We apply this to $g(\phi, s) = Rf(\theta, s)$, $\theta = (\cos\phi, \sin\phi)^T$. It has been shown by Rattey and Lindgren (1981) that $\hat{g}$ is negligible outside a set being slightly larger than the set K in Fig. 6.1. Putting

$$2\pi(W^{-1})^T = p\begin{pmatrix}1 & 0\\ 1 & 2\end{pmatrix}, \quad \text{i.e.} \quad W = \frac{\pi}{p}\begin{pmatrix}2 & -1\\ 0 & 1\end{pmatrix}$$

we see that the sets $K + 2\pi(W^{-1})^T l$ are mutually disjoint if $p \geq b$. The grid generated by W is the interlaced geometry

$$\phi_j = \pi j/p \ , \quad s_l = \pi l/p \ , \quad j + l \text{ even}, \quad j = 0, \ldots, p-1 \ .$$

It needs only $\frac{1}{\pi} b^2$ data, i.e. only ½ of the data of the parallel scanning geometry. A similar analysis can be carried out for fan-beam scanning, see Natterer (1986). The result is that among all the fan-beam geometries, the PET geometry which consists of the lines joining b points evenly distributed on the unit circle is best possible.

A completely different question is if there are algorithms which actually achieve the optimal resolution. This has been answered in the affirmative by Kruse (1986) for the interlaced geometry and for the PET geometry. For the fan-beam geometry the question is open.

7. The Next Steps. Even though there are still problems in CT, the theory has reached a certain state of maturity and it is time to think about what to do next. We mention a few possible extensions.

(a) Generalized Radon transforms.
Let $\Phi(x, \theta)$ be a sufficiently regular function and put

$$R_{\Phi} f(\theta, s) = \int_{x \cdot \theta = s} \Phi(x, \theta) f(x) dx .$$

Such transforms come up in emission tomography (SPECT, see Budinger, Gullberg and Heuesman (1979)) and in range — Doppler radar (see Feig and Grünbaum (1986)). Boman and Quinto (1987) give an account of what is known theoretically about R_{Φ}, see also Beylkin (1984). Inversion formulas are available only in very specific cases (Tretiak and Metz (1980), Feig and Greenleaf (1986)) and the development of numerical methods is still in its infancy. The same is true for generalized Radon transforms integrating over families of curves. Applications come from geophysics (Romanov (1974)) and radar (Hellsten and Andersson (1987)). Inversion formulas are available in cases in which the family of curves enjoys certain invariance properties (Lavrent'ev, Romanov and Shishat — sky (1980), Cormack and Quinto (1980), Cormack (1981), Fawcett (1985)). The theory of this general problem in integral geometry is surveyed in Gelfand, Gindikin and Graev (1980).

(b) The 3D problem.
Typically, one wants to reconstruct a function in $\mathbf{R}^3$ from integrals over lines passing through a curve surrounding the support of the function. The main difficulty is that this problem is severely ill-posed unless the curve satisfies a very restrictive condition which is hard to meet in practice, see Tuy (1983), Finch (1985). Of course all 3D reconstruction algorithms suffer from this severe ill-posedness, see Grangeat (1987) for the state of the art.

(c) Nonlinear problems.
Let $\Gamma_f(x, y)$ be the geodesic with respect to the metric $ds = f(x) \mid dx \mid$ joining x, y. The problem is to recover f from the integrals

$$T(x, y) = \int_{\Gamma_f(x, y)} f(x) ds .$$

See Muhometov (1977) for the theory, Schomberg (1978) for applications and algorithms.

(d) Defraction tomography.
If the acoustical inverse scattering problem (see e.g. Greenleaf (1983)) is solved within the Born or Rytov approximation, then one gets a relation very similar to (2.1), leading to the filtered backpropagation algorithm of Devaney (1982). In the high frequency limit, we end up with CT. Therefore there is some hope that some of the insights and methods of CT are useful for inverse scattering problems.

REFERENCES

[1] ALLAN, C. J., KELLER, LUPTON, L. R., TAYLOR, T. AND TANNER, P.D. (1985): Tomography: an overview of the AECL program. *Applied Optics* **24**, 4067-4074.

[2] BARRETT, H. H. AND SWINDELL, W. (1981): Radiological imaging. Academic Press.

[3] BEYLKIN, G. (1984): The inversion problem and applications of the generalized Radon transform. *Comm. Pure Appl. Math.* **37**, 579-599.

[4] BJÖRCK, A. AND ELVING, T. (1979): Accelerated projection methods for computing pseudoinverse solutions of linear equations, *BIT* **19**, 145-163.

[5] BOMAN, J. AND QUINTO, E. T. (1987): Support theorems for real analytic Radon transforms. *Report No. 6, Department of Mathematics University of Stockholm,*

Sweden.

[6] BRACEWELL, R. N. AND RIDDLE, A. C. (1956): Strip integration in radio astronomy. *Aus. J. Phys.* **9**, 198-217.

[7] BRACEWELL, R. N. (1979): Image reconstruction in radio astronomy, in Herman, G. T.: (ed.), *Image Reconstruction from Projections*, Springer.

[8] BRACEWELL, R. N. AND RIDDLE, A. C. (1967): Inversion of fan-beam scans in radio astronomy. *The Astrophysical Journal* **150**, 427-434.

[9] BUDINGER, T. F., GULLBERG, G. T. AND HUESMAN, R. H. (1979): Emission computed tomography, in Herman, G. T. (ed.), *Image Reconstruction from Projections.* Springer.

[10] CENSOR, Y., EGGERMONT, P. P. B. AND GORDON, D. (1983): Strong underrelaxation in Kaczmarz's method for inconsistent systems. *Num. Math.* **41**, 83-92.

[11] CORMACK, A. M. (1963): Representation of a function by its line integrals, with some radiological applications. *J. Appl. Phys.* **34**, 2722-2727.

[12] CORMACK, A. M. (1980): An exact subdivision of the Radon transform and scanning with a positron ring camera. *Phys. Med. Biol.* **25**, 543-544.

[13] CORMACK, A. M. (1981): The Radon transform on a family of curves in the plane. *Proceedings of the Amer. Math. Soc.* **83**, 325-330.

[14] DAVISON, M. E. (1983): The ill-conditioned nature of the limited angle tomography problem. *SIAM J. Appl. Math.* **43**, 428-448.

[15] DEANS, S. R. (1983): *The Radon Transform and Some of Its Applications.* Wiley.

[16] DeROSIER, D. J. AND KLUG, A. (1968): Reconstruction of three-dimensional structures from electron micrographs. *Nature* **217**, 130-134.

[17] DEVANEY, A. J. (1982): A filtered backpropagation algorithm for diffraction tomography. *Ultrasonic Imaging* **4**, 336-350.

[18] DINES, K. A. AND LYTLE, R. J. (1979): Computerized geophysical tomography. *Proceedings of the IEEE* **67**, 1065-1073.

[19] FAWCETT, J. A. (1985): Inversion of n -dimensional spherical means. *SIAM J. Appl. Math.* **45**, 336-341.

[20] FEIG, E. AND GRÜNBAUM, F. A. (1986): Tomography methods in range — Doppler radar. *Inverse Problems*, **2**, 185-195.

[21] FEIG, E. AND GREENLEAF, F. P. (1986): Inversion of an integral transform associated with tomography in radar detection. *Inverse Problems* **2**, 405-411.

[22] FINCH, D. V. (1985): Cone beam reconstruction with sources on a curve. *SIAM J. Appl. Math.* **45**, 665-673.

[23] FLEMING, H. (1982): Satellite remote sensing by the technique of computed tomography. *J. Appl. Meteor.* **21**, 1538-1549.

[24] GELFAND, I. M., GINDIKIN, S. G. AND GRAEV, M. I. (1980): Integral geometry in affine and projective spaces. Itogi Nauki i Tekniki, *Ser. Sovr. Probl. Mat.* **16**, 53-226 (Russian), English translation in *J. Soviet Math.* **18**, 39-167 (1982).

[25] GORDON, R., BENDER, R. AND HERMAN, G. T. (1970): Algebraic reconstruction techniques (ART) for three-dimensional electron microscopy and X-ray photography. *J.*

Theor. Biol. **29**, 471-481.

[26] GRANGEAT, P. (1987): Analyse d'un système d'imagerie 3D par reconstruction à partir de radiographies X en géométrie conique. Thèse de doctorat, présenteé à l'école nationale superieure des telecommunication.

[27] GREENLEAF, J. F. (1983): Computerized tomography with ultrasound. *Proc. IEEE* **71**, 330-337.

[28] HAMAKER, C. AND SOLMON, D. C. (1978): The angles between the null spaces of X-rays. *J. Math. Anal. Appl.* **62**, 1-23.

[29] HELGASON, S. (1980): *The Radon Transform.* Birkhäuser.

[30] HELLSTEN, H. AND ANDERSSON, L. E. (1987): An inverse method for the processing of synthetic aperture radar data. *Inverse Problems* **3**, 111-124.

[31] HERMAN, G. T. (1980): *Image Reconstruction from Projections. The Fundamentals of Computerized Tomography.* Academic Press.

[32] HERMAN, G. T. AND LENT, A. (1976): Iterative reconstruction algorithms. *Comput. Biol. Med.* **6**, 273-294.

[33] HOPPE, W. AND HEGERL, R. (1980): Three-dimensional structure determination by electron microscopy, in Hawkes, P. W. (ed.), *Computer Processing of Electron Microscope Images.* Springer.

[34] HOUSFIELD, G. N. (1973): Computerized transverse axial scanning tomography: Part I, description of the system, *Br. J. Radiol.* **46**, 1016-1022.

[35] JERRY, A. J. (1977): The Shannon sampling theorem — its various extensions and applications: a tutorial review. *Proc. IEEE* **65**, 1565-1596.

[36] KACZMARZ, S. (1937): Angenäherte Auflösung von Systemen linearer Gleichungen. *Bull. Acad. Polon. Sci. Lett.* **A35**, 335-357.

[37] KERSHAW, D. (1962): The determination of the density distribution of a gas flowing in a pipe from mean density measurements. *Report A.R.L./R1/Maths 4 105, Admirality Research Laboratory, Teddington, Middlesex.*

[38] KOWALSKI, G. (1979): Multislice reconstruction from twin-cone beam scanning. *IEEE Trans. Nucl. Sci. NS-26*, 2895-2903.

[39] KRUSE, H. (1986): Die Auflösung von Algorithmen in der Computer — Tomographie. Dissertation, Münster.

[40] LAVRENT'EV, M. M., ROMANOV, V. G. AND SHIRHAT-SKY (1980): Ill-posed problems of mathematical physics and analysis. Akad. USSR Nauk (Russian), English translation: *Translations of mathematical monographs, Vol. 64,* AMS. Providence, Rhode Island (1986).

[41] LENT, A. (1975): Seminar talk at the Biodynamic Research Unit, Mayo Clinic, Rochester, MN.

[42] LEWITT, R. M., BATES, R. H. T. AND PETERS, T. M. (1978): Image reconstruction from projections: III: Projection completion methods (theory). *Optik* **50**, 180-205.

[43] LITTLETON, J. T. (1976): *Tomography: Physical Principles and Clinical Applications.* The Williams & Wilkins Co., Baltimore.

[44] LOUIS, A. K. (1980): Picture reconstruction from projections in restricted range. *Math. Meth. in the Appl. Sci.* **2**, 209-220.

[45] LOUIS, A. K. (1984): Orthogonal function series expansions and the null space of the Radon transform. *SIAM J. Math. Anal.* **15**, 621-633.

[46] MARR, R. B. (1974): On the reconstruction of a function on a circular domain from a sampling of its line integrals. *J. Math. Anal. Appl.* **45**, 357-374.

[47] MUHOMETOV, R. G. (1977): The problem of recovery of a two-dimensional Riemannian metric and integral geometry. *Dokl. Akad. Nauk USSR* **232**, 32-35 (Russian), English translation in *Soviet Math. Dokl.* **18**, 27-31 (1977).

[48] MUNSON, D. C., O'BRIEN, J. D. AND JENKINS, W. K. (1983): A tomographic formulation of spotlight-mode synthetic aperture radar. *Proc. IEEE* **71**, 917-925.

[49] NATTERER, F. (1986): *The Mathematics of Computerized Tomography.* Wiley — Teubner.

[50] NATTERER, F. (1985): Fourier reconstruction in tomography. *Numer. Math.* **47**, 343-353.

[51] PETERSEN, P. P. AND MIDDLETON, D. (1962): Sampling and reconstruction of wave-number-limited functions in N-dimensional euclidean space. *Inf. Control* **5**, 279-323.

[52] PIKALOV, V. V. AND PREOBRAZHENSKY, N. G. (1987): *Tomographic reconstruction in gas dynamics and plasma physics* (Russian). Novosibirsk: Nauka.

[53] PRATT, W. K. (1978): *Digital Image Processing.* Wiley.

[54] RADON, J. (1917): Über die Bestimmung von Funktionen durch ihre Integralwerte längs gewisser Mannigfaltigkeiten. *Berichte Sächsische Akademie der Wissenschaften, Leipzig, Math.-Phys. Kl.* **69**, 262-267.

[55] RAMACHANDRAN, G. N. AND LAKSHMINARAYANAN, A. V. (1971): Three-dimensional reconstruction from radiographs and electron micrographs: application of convolutions instead of Fourier transforms. *Proc. Nat. Acad. Sci. US* **68**, 2236-2240.

[56] RATTEY, P. A. AND LINDGREN, A. G. (1981): Sampling the 2-D Radon transform with parallel- and fan-beam projections. Department of Electrical Engineering, University of Rhode Island, Kingston, RI02881, Technical Report 5-33285-01, August 1981.

[57] ROMANOV, V. G. (1974): *Integral geometry and inverse problems for hyperbolic equations.* Springer.

[58] ROWLAND, S. W. (1979): Computer implementation of image reconstruction formulas, in Herman, G. T. (ed.), *Image Reconstruction from Projections.* Springer.

[59] SCHOMBERG, H. (1978): An improved approach to reconstructive ultrasound tomography. *J. Appl. Phys.* **11**, L181.

[60] SHEPP, L. A. AND LOGAN, G. F. (1974): The Fourier reconstruction of a head section. *IEEE Trans. Nucl. Sci., NS-21*, 21-43.

[61] SHEPP, L. A. AND VARDI, Y. (1983): Maximum likelihood reconstruction for emission tomography. *IEEE Trans. Med. Imaging* **1**, 113-122.

[62] SMITH, K. T., SOLMON, D. C. AND WAGNER, S. L. (1977): Practical and mathematical aspects of the problem of reconstructing a function from radiographs.

Bull. AMS **83**, 1227-1270.

[63] SMITH, K. T. (1983): Reconstruction formulas in computed tomography. *Proceedings of Symposia in Applied Mathematic (AMS)* **27**, 7-23.

[64] STARK, H., WOODS, J. W., PAUL, I. P. AND HINGORANI, R. (1981): An investigation of computerized tomography by direct Fourier inversion and optimum interpolation. *IEEE Trans. Biomed. Engi., BME-28*, 496-505.

[65] TRETIAK, O. AND METZ, C. (1980): The exponential Radon transform. *SIAM J. Appl. Math.* **39**, 341-354.

[66] TUY, H. K. (1983): An inversion formula for cone-beam reconstruction. *SIAM J. Appl. Math.* **43**, 546-552.

[67] VVENDENSKAYA, N. D. AND GINDIKIN, S. G. (1984): Poisson's formula for the Radon transform. *Dokl. Akad. Nauk USSR 279*, 780-784 (Russian), English translation: *Soviet Math. Dokl.* **30**, 700-704 (1984).

Numerical Flow Simulation in Aerospace Industry

P. PERRIER*

0. INTRODUCTION

The continuous developments of computation in aerospace industry received an impetus of new capacities and of the challenge offered to flow simulations.

It is important to examine first the general improvements of the capacities of computation: a clearer explanation will be found in it for the extraordinary increase in computational fluid dynamics applied to aerospace industry. The encounter of major success and of major challenges has given to aerospace industry the chance of being a prefiguration of what will be the future of CFD and what new requirements it proposes.

So we will review the present status of the impact of computer advances on large computations, then we will give some elements on the main characteristics of present aerospace industry's use of CFD, and we will open our analysis to new advanced projects taking Hermes program as the most typical of european future high technology in that field. In conclusion, a proposal of organizing interactions between research and industry for the most difficult keypoints, will be a review for a realistic european policy of common interest.

1. IMPACT OF COMPUTER REVOLUTION

1.1 — The introduction of computer in aerospace industry occurred a decade after the war, solving simple flows like wing section pressure distribution; complex difficult flows, like blunt-body flow, came soon after as a result of emerging requirements for intercontinental missiles. The simple calculations were a first attempt to improve the speed of routine-minded hand-made computations in repetitive iterations like characteristics integration or 2D flowfield relaxation of Laplace problems.

1.2 — But in fact, the steep increase of computer power has opened progressively the way to large 3D computations on complex configurations not only for simple equations like the incompressible potential problem but for realistic Navier-Stokes equations. We have to be careful about the future capabilities of computer due to the fact that the methods of tomorrow are to be tested today and therefore require a dynamic view of what will be feasible in the next years; the larger the time required for adjustment or finalization of large codes, the bigger will be the computer necessary for tests before their industrial use.

* Avions Marcel Dassault-Breguiet Aviation, 78, Quai Marcel Dassault, 92214 Saint-Cloud, France.

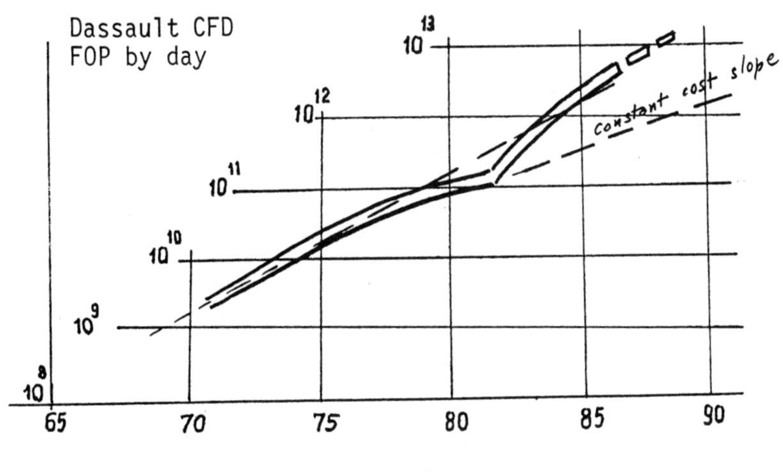

Figure 1

Figure 1 gives a typical idea of the quantity of computations in CFD field in Dassault-Breguet in the last twenty years: it shows a surprisingly regular increase. The lowest figure that we can extract from that general trend is an increase of an order of magnitude (a factor of ten) every 10 years, but the mean realistic value is an increase of an *order of magnitude every 5 years*. Two main reasons can be behind that general trend: one is related to the increase of budget devoted to CFD, due, for example, to the demonstration of its efficiency or its necessity for advanced product definition, the other is the pure increase of the performance of computers. In fact the increase in number or size of computers is currently a small percentage of the increase in productivity of the computer center in the CFD field, and the major part is given by improvement of performance at constant budget. We characterize the production of the computer center as a flux of floating point operations by day. That figure increased in the last twenty years from 10^8 to 10^{12}.

If we take into account the cost, with a final typical cost figure of 5 MF by year of computer rental, we obtain a surprising constant increase of production of *one order of magnitude every 7 years* in the last 30 years. Fig. 2 shows that in fact the selected cost tends to cover one single supercomputer today when it was the value for 10 computers in 1957, and also that the effective power tends to flatten with the new supercomputer based on realistic industrial costs: the highly parallel peak power of computation (theoretically present) is not achieved but great effort is done in that direction. The second dashed curve on that figure is related to the power given by a computer that would require the tenth of the budget of the supercomputer for a day.

Initially, the power extracted was five times smaller, so the size was not cost effective and the optimal number of machine could be less than 5. Around the seventies, the supercomputer was the most efficient in terms of cost for a given computation, but now it is no longer true, except if you can use the maximum power of the computer (typically more than 150 Mflops on a Cray 2). If the total elapsed time for a given computation is not critical, the small computers are again the best way, if they have free access with a small number of users. Such a situation appeared for a short period ten years ago at the beginning of array-processor introduction.

Today a small computer can give the same output as the supercomputer (with a longer time of computation) for a cost 3 times less, and it is the capabilities of large computation that remains the characteristic of large supercomputers. Ten years ago, general purpose computers were almost equal in efficiency; but now the requirement of a highly costly large memory for

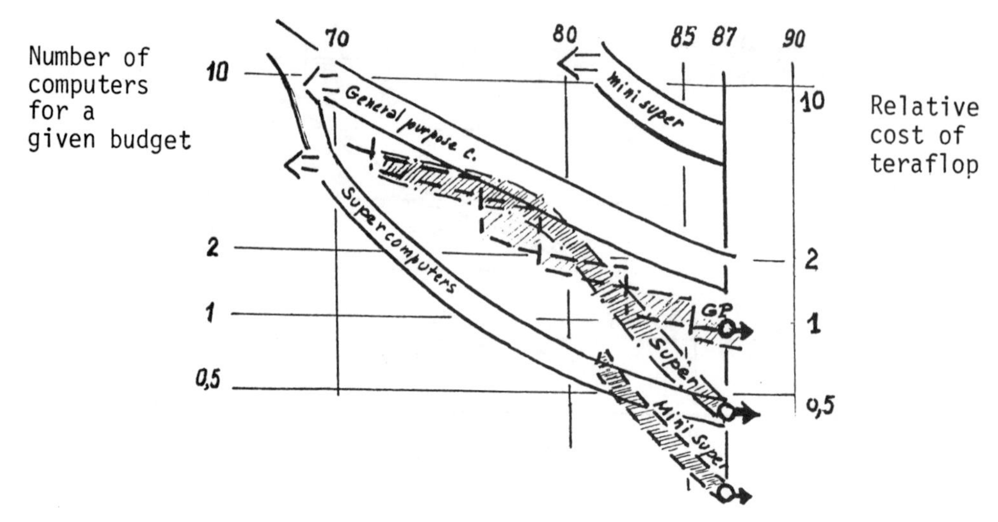

Figure 2

high speed computation gives again the balance of cost in favor of small computers of moderate speed.

Figure 3 gives the general trends of usable memory in millions of words for the supercomputer. The trends are not simple to interpret due to rapid increase in memory required by parallel processing. But this cost of high speed memories gives advantage in cost for smaller computers of lower speed, which no longer remains limited to small 3D problems. If anyway the very large core memory remains too expansive at peak performance required by supercomputers and if they have to rely on 2 or 3 levels of memory hierarchy, intermediate simple chip processors give opportunity to propose cheaper costs than supercomputer.

1.3 — We can define two units that are meaningful for the CFD problem and appear in the precedent figures: the teraflop and the megaword. In fact, we will see that practical Navier-Stokes in 3D corresponds to a number of nodes 10 times greater than 100.000, and that the solution of such a problem tends to make no less than 1.000 floating operations by point on 10 or more variables, with a number of iterations of 100 to 1000. So a typical figure for the total numbers of floating operations in a large CFD simulation is in the order of 10^{12}. We name that quantity *a teraflop*; the necessary memory for such a computation is a multiple of the *megaword*, it is the associated requirement for high speed data storage.

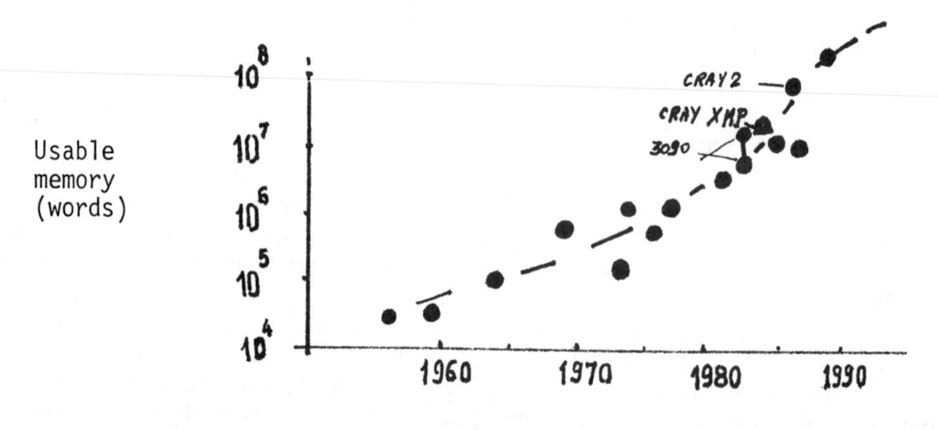

Figure 3

At the end of the seventies, the supercomputer was giving one *megaflops* speed, so it means that such a teraflop total computation was to last 12 days, even if the maximum memory were to be multiplied by 5; the same teraflop computation can now be done repeatedly on a computer giving 10 megaflops. In the future, with a 200 megaflops supercomputer, such a performance could be industrial with a good extraction of parallelism in codes.

From figure 1, it can be seen that such teraflop is now within a typical CFD computer budget. New projects can be designed with the help of such numerical simulation and it can be checked that the percentage of computation in the total aerodynamic simulation cost is increasing and can reach 55% in the last projects. We will come back to that fact in an inspection of CFD requirements for future high technology vehicles like Hermes in Europe or NASP in U.S.

We have to consider also the number of data transferred FDT (Floating Data Transfer) from the disk storage to the main memory: with actual low speed in transfer, the ratio $\frac{\text{FDT}}{\text{FLOP}}$ has to be maintained at values lower than 0.001 to avoid excessive overhead and increase of total elapsed time with poor percentage of CPU time. The overhead can be catastrophic if virtual memory is competing in a multitasking environment.

We have also to consider inside large codes the quantity of logical branches. So we can define the ratio of the number of branch instructions related to floating points operations. Generally speaking that ratio is very small: $\frac{\text{BOF}}{\text{FLOP}} << 1$, except for artificial intelligence codes; if it increases to 0.1 the speed can drop drastically if the architecture is not specially adapted. But, more crucial, the time for debugging can increase drastically. Isn't this time of debugging justifying AI special languages, as well as easy coding of logical expressions? But will not the new extensions of Fortran recover, at the end, that field for large codes?

1.4 — Before extrapolating to the future, we have to consider which computer technology is responsible for enhancement of performance, and which new computer technology will be the origin of future improvements. In addition to some transfer to software of optimization of the use of architecture, we have to notice a large integration of components, the development of efficient pipe-lines and the first tests for solving memory conflicts by software, tests that give limitation to vectorization or optimization. The evolution towards large integration of very high speed and highly integrated components will require a high degree of software and hardware adaptation for the conservation of high speed: optimization of transfers between the different processors, complex hierarchy of memory, good management of highly integrated and specialized scalar processors. All these trends will enforce the mini-super computer efficiency and will give them a probable unequalled cost effectiveness.

So we can imagine a 5-10 years projection in the industry where the mini-super computer will be highly attractive if it reaches the level of the teraflop computation with more than 10 megawords of memory. The advanced computations with 10 to 100 teraflop or repetitive adjustment of a 1 teraflop code will continue to require the use of real supercomputers. In that case, it is clear that the number of users of effective teraflop CFD codes will increase, largely outside aerospace industry and special agencies (atomic energy commissions, meteorology...). A fundamental change will appear when the general trend gives more and more economic and reliable outputs from simulation of industrial flows of interest: will that new generation of tools for evaluation of performance (or checking of limit test-cases) replace a lot of direct experimental qualification by multiple complete numerical-experimental checkings for qualification of mechanical machines? This is quite evident for all the machines where either the risk or the difficulty of simulation actually precludes the experimental qualification. But it can also reduce the number of experimental tests, when their cost is too high, either the models or the ground simulation device too long to build. In that case, the certification laboratories will move towards experiences supporting the basic tests of validation of codes (easier or cheaper). The laboratories will check the true complete physics, doing more fundamental experiments. On the industrial side an approach by modelization (and the process of subsystem qualification) will thus be generalized.

One of the characteristics of validation of codes is the need, *first* for carefully designed test cases for experimental checking of each of the major physical constants of the model, and *second* for careful evaluation of the mathematical solvers for the validation of the discrete equations. So the validation of the codes requires large comparisons inside specialized workshops. It requires also a validation of the effective integration of the program by tests on a complete configuration and on a sufficient number of cases.

One appealing economic approach to the use of numerical codes for simulation of complex apparatus can be the partition of the total problem in subdomains; an elegant way to reduce the cost is to separate a large number of subdomains and to extract from experimental data turbulence modeling when current models are poor. That partition of complex assembly will require additive validation of interface (e.g. in fluid -structure interface) and specially well defined mathematical tools for super element interfaces. The same approach can be valuable with multiprocessor computer for direct physical partition of the problem.

1.5 — The phases of industrial design are also evolving rapidly due to the impact of subteraflop computations. One of the characteristics of code simulation is the accuracy of the answer without limitations due to experimental scattering of data if the convergence and initial value independence is checked. This accuracy allows the step by step improvement of a design, ten steps being an acceptable number of computations in the convergence process towards an experimentally noticeable improvement. If the geometry is a function of a limited number of parameters (typically 10 to 50), the computer simulation can give direct optimum shapes that are completely beyond the reach of any experimental convergence process. That optimization loop requires 1 to 2 orders of magnitude more basic computations and will be opened to complex simulation with teraflop basic time of computation only in the future. But more simple modeling is sufficient for a first step.

A careful analysis of the physical phenomena underlying correct computations is the major problem for any large industrial computation team. This analysis is the best way for understanding success and failure in design and is much better done from complete computer data than from limited experimental measurements. It also allows a reduced number of parameters and so reduces the cost of repetitive computations; but it relies on good physical modeling to be tested by a well chosen workshop during the rebuilding of experimental tests. The proper modeling assumes a mesh generation with adaptation of the size of elements to physical requirements and a clear mathematical modeling not biased by excessive smoothing or damping for fast convergence.

The code heart is the mathematical solver. The quality of results of the discrete problem and of convergence relies on a good numerical analysis. A great effort must be done to improve the mathematical understanding of the work of the solvers; it will give more exact or reliable solutions and also more efficient solvers. We have to keep in mind that the gain obtained in time of computation in the last ten years was typically less than two thirds of the total gain in computer speed, and the gain in algorithmic efficiency more than one third. For Euler codes this is particularly evident with a gain of two orders of magnitude in speed from the rough simple original algorithms when computer speed was increasing more slowly; at the end, the gains have been given by re-formulation of solvers for highly parallel processing.

2. CFD PRESENT STATUS IN AEROSPACE INDUSTRY

2.0 — The CFD is now widely accepted as a basic tool in the aerospace industry. The major steps begin with inviscid transonic design of commercial aircraft and extend now to complete aircraft design in viscous flows without separation. On the mean curve given in Figure 4 one can recognize that the percentage of CFD cost eventually increased to a significant percentage of the total aerodynamic expenditures.

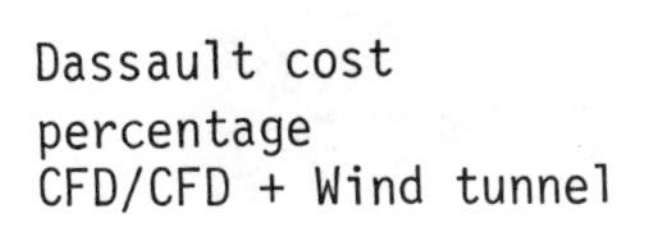

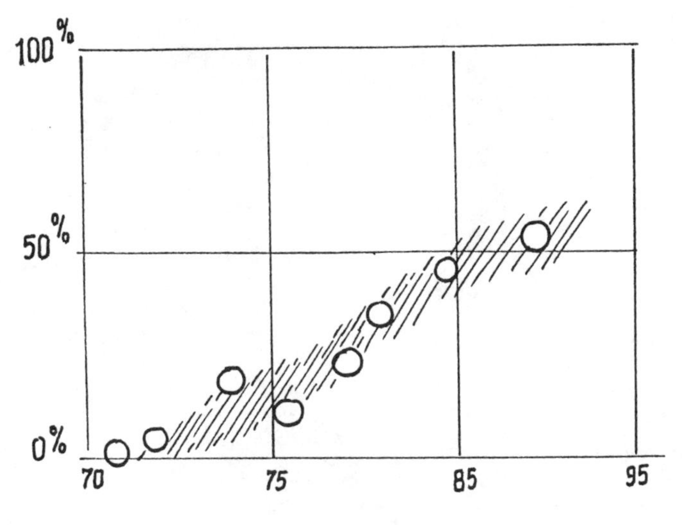

Figure 4

Due to increasing costs of experimental simulations, and decreasing costs of computer simulation, the funding of aerospace industries for computation will probably stabilize — in the future — at around 50% of the one for experimental tests. What is changing, is clearly the effect of the introduction of computers in the design phases, where it will soon cover the largest part of the funds devoted to selection of first realistic configurations. The numerical approach in the preliminary studies is well adapted to large changes of geometry and to the evaluation of derivatives for global selection of vehicle shapes. Another visible trend is the cost increase of validation of codes by comparison with reference computations or experimental results. The cost of development of new codes becomes higher and higher when taking into account more and more realistic geometries, due to the complexity of an industrial chain of computation including mesh generation and results visualization.

2.1 — The codes are evolving from very simple perfect fluid flow computations to real viscous flows, but the evolution is still slow, due to limitations of computers and codes. The integral methods, or panel methods, that give a complete matrix in subsonic by interference of one panel condition to another have been very useful and remain in large use in all the aerospace industry. One of their main advantages is easy meshing on the skin and a good tolerance to large discrete variations of panels in subsonic; this advantage continues to give to panel methods a significative role in the aerodynamic design work, but the increase in precision assumes a new approach for solving the linear operator and for reducing the dependence of time of computation on the number of grid points from $N\,2$ to $N\,*\log N$ and perhaps N. One of the advantages of such methods is related to the possibility of precise discretizations of small parts of the space. If we assume that the complexity of the flow remains in a quasi 2D subset of the space, it means that direct simulation of vortex dynamics can give a much more precise evaluation of what happens in complex unsteady separated flows where diffusion is not the main process. Modeling of viscous flows at high Reynolds numbers with large vortex structures can be the main objective of extension of actual codes if there is new mathematical understanding of vortex dynamics and of their numerical analysis.

Some examples are given in Figures 5 and 6. Corrections for compressibility and bias to linearity can be taken into account by limited source points of intensity $-M^2\frac{\partial V}{\partial s}$ in the space, resulting in exact compressible equation (without enthalpy variation).

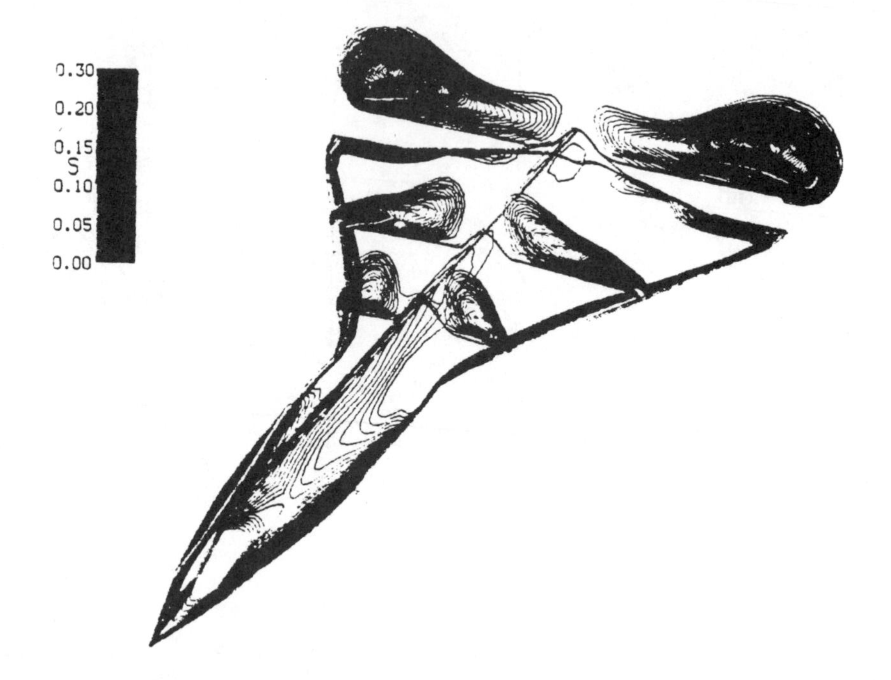

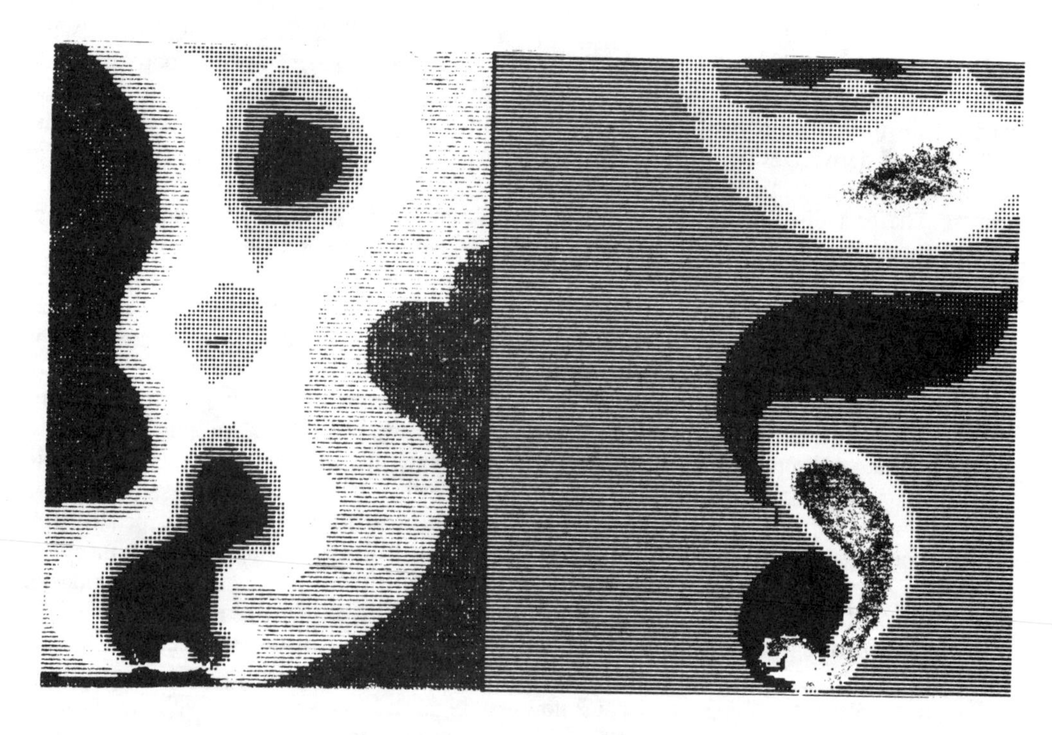

Figure 5

Cut inside flowfield in Hermes wake
(Dassault Navier-Stokes FEM code)

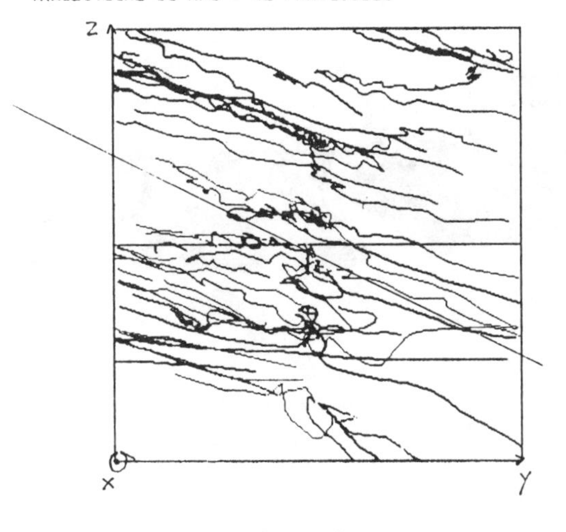

Figure 6

Turbulence simulation for highly distorted flows (Dassault CFD)

On the other extreme of smoothing continuous flows without singularities, the spectral methods have not found their way in the industrial codes and are restricted to very efficient incompressible Navier-Stokes solvers which are very useful for turbulent simulation. An extension as an operator for the smoother part of the Euler solver (with decomposition of flowfield) is of limited use.

2.2 — The largest part of industrial computations has been devoted to finite volume solvers as a more flexible variant of finite difference approximation of the equations. In fact, the complexity of real shapes of airplanes or engines give requirement for large distortions of 3D hexaedra in structured meshes. This means that keeping the i , j , k , complete set of indexes keeps the possibility of vectorization of codes very attractive in cost, but on the other hand the acceptable distortion has to be increased to allow the multiboxes approximation of complex shapes. The realization of such complex codes assumes that the elementary box can be largely distorted with the O , C , H type classification of 3D mesh that we introduced many years ago, and that the connexions between boxes can be effective. The minimum computation in that field (that is wing + body with a C mesh adapted to simulation of the bumps created on the plane of symmetry by the fuselage (Figure 7)) has attained a very high efficiency. Such a code, for some minutes in computation, gives a basis for direct optimization with constraints. On the other side, the maximum size of computation can be well described by a complex assembly of boxes, with transmission of boundary conditions at the interface.

If very easy convergence can be obtained by subdomains relaxation procedures between blocks in subsonic without overlapping of the grids for inviscid flows, the supersonic compatibility seems much more difficult to insure, particularly with potential flows and variable mesh size. The supersonic transfer conditions seem easier to implement with overlapping in Euler codes, where a clear definition of velocity fluxes can be obtained. It is the direction in which the major CFD teams in aerospace industry are working. A typical mesh generation and result is shown in Figure 8. Major difficulties arise from boundary

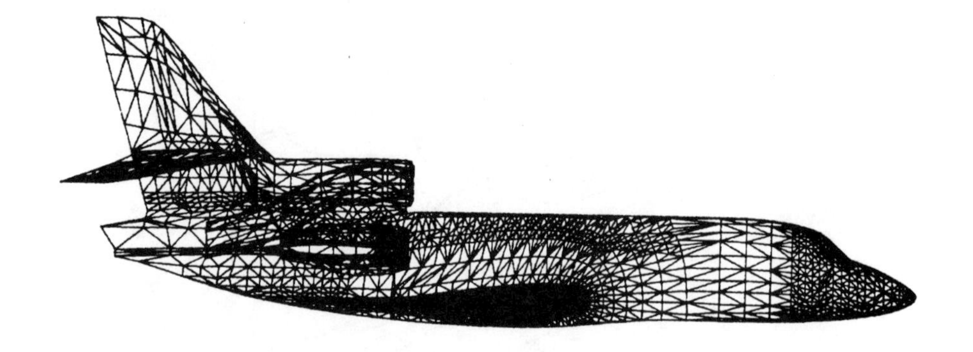

Figure 7

C Mesh on complex configuration
(Dassault Falcon)

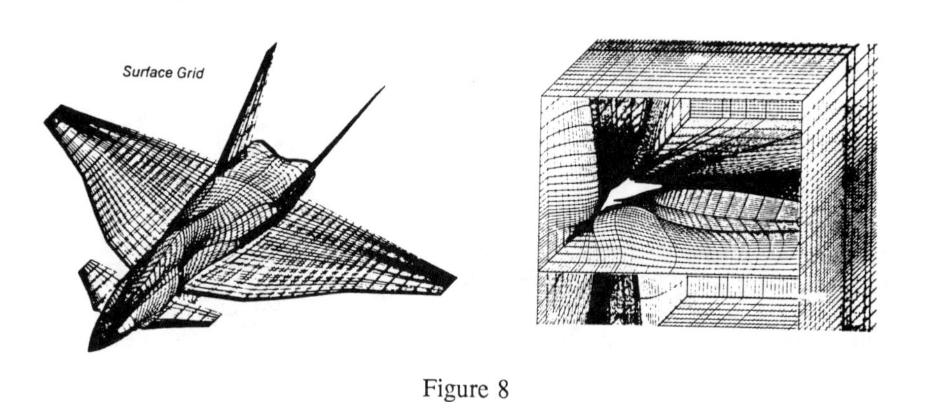

Figure 8

Surface grid complexity for Jameson finite
volume multidomain (Dornier)

conditions requiring conservation of the transport of quantities on the limits. The system of Euler equation can be put on the typical following shape:

$$\frac{\partial}{\partial^t} \iiint_\Omega a \; d\;\Omega + \iint_A \bar{F} \cdot \hat{n} \; dA = 0$$

$$a \equiv \begin{bmatrix} \rho \\ \rho u_i \\ \rho E \end{bmatrix} \quad \bar{F} \cdot \hat{n} = \begin{bmatrix} (\rho u_j)\, n_j \\ (\rho u_i \; u_j + p\, \delta_{ij})\, n_j \\ (\rho u_j \; H) n_j \end{bmatrix}$$

ρ – density	E – total energy	u_i –	Cartesian velocity components
p – pressure	H – total enthalpy	n_j –	unit normal

where a is the vector of density, momentum and energy. The complete set of variables that are to be continuous at the interface requires multi-iteration on multiple block procedures.

An efficient way of damping oscillations modes is to transfer corrections via a multigrid method to subgrids. This approach requires, in 3D, a large number of points multiples of 2, 4 and 8 in each direction. Such a multigrid approach is essential when large readjustment is to be done to the initial values at the beginning of the convergence; for example in inlet or exhaust of engine imposition of airflow, the flowfield can change drastically with boundary conditions requiring an adjustment of all the flowfield in the different boxes. The current sizes of meshes, with the costly way of increasing in three directions the reinforcement of mesh required by geometric singularities (like leading and trailing edges, nose or roots of wings, tips...), require approximately 10^6 mesh points and lead to teraflops Euler computations.

This is the frontier of present supercomputers and also the frontier of mesh size due to complexity of interfaces which can be generated. Adaptation of the mesh to the solution (refinement limited to large shockwave surfaces, or to wake and boundary layers) will allow limited incursion in realistic Navier-Stokes computation. Such computations are not usual and the extension to turbulent flows is not common, even if there are some encouraging tests. The incapacity of turbulence modelization to support complex real flows is, generally speaking, not well known. A lot of people try to convince themselves that the Euler solutions are sufficient to give first reasonable estimates. To give a typical idea of the limitations of this attitude, one can check that no simulation of a complete aircraft with engines has yet been computed in that way. But improvement of speed in the basic box (multigrid, enthalpy damping and high time steps, T V D control,...) is an ingredient of robust solvers: such robust solvers are required for the computation of complex flowfields on complex configurations (Figure 9).

2.3 — From the beginning of finite elements structural stress computation, a general unstructured mesh approach was tested; but it revealed itself ineffective due to many theoretical problems. Dassault-Breguet team was particularly efficient in the first industrial operation of finite elements in fluids, thanks to the INRIA effort with Pr. Lions and Prs. Glowinski, Pironneau and collaborators. The industrial use in transonic complex aircraft design was clearly evident in the Falcon 50 at the end of the seventies (Figure 10). That effort was very efficient for the solution of potential flows in transonic but failed to give very good supersonic solvers. Good success was obtained in subdomain approach for large number of nodes with common elliptic solvers, and also for multiple solutions finding in transonic by mathematical continuation. In the same way, codes able to work on Navier-Stokes with Stokes solvers currently permitted easy Navier-Stokes computation on any complex 3D shapes at low Reynolds number. The extension to compressible and Reynolds averaged flows was more difficult.

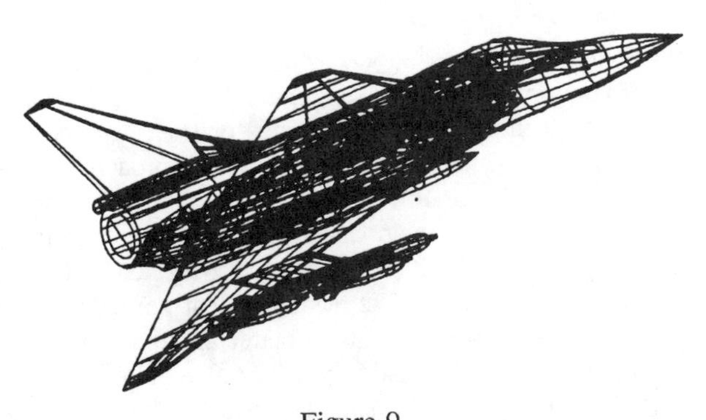

Figure 9

Complex Configuration

Figure 10

Complex complete 3D aircraft
(Dassault potential flow FEM code)

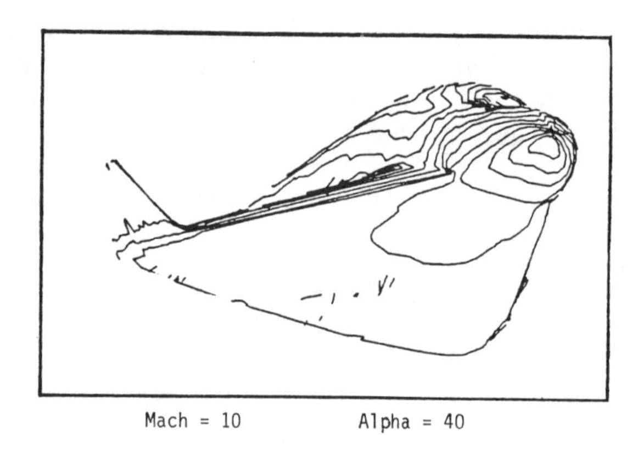

Figure 11

Hermes Hypersonic pressure distribution
(Dassault FEM Euler Code)

New solvers for Euler and Navier-Stokes equations were derived from finite elements approach and are now of common industrial use (Figures 11 – 12). Parallel to that work, and also using tetraedral P1 test functions on finite volumes or true unstructured meshes, Jameson recently published first results of Euler codes in Galerkin approach for complex shapes. T. Hughes does a similar effort with Petrov-Galerkin approximation for stiff Navier-Stokes equation. Two main advantages and a main disadvantage result from the unstructured mesh generation. The disadvantage lies in the increase in cost compared to finite volumes; it is due partly to the consideration of the number of tetraedra against hexaedra, increasing the number of computed fluxes for a given number of points; a better approximation for distorted flows or distorted geometries is a gain, but is to be paid by a loss of regularity in the files of data, and, consequently, a poor parallelism recovery for supercomputer. The main advantage relies in the ability to cope with any complex shape without distorsion of approximation functions related to flowfield in the corners or singular points of the geometry. The second advantage is the ability

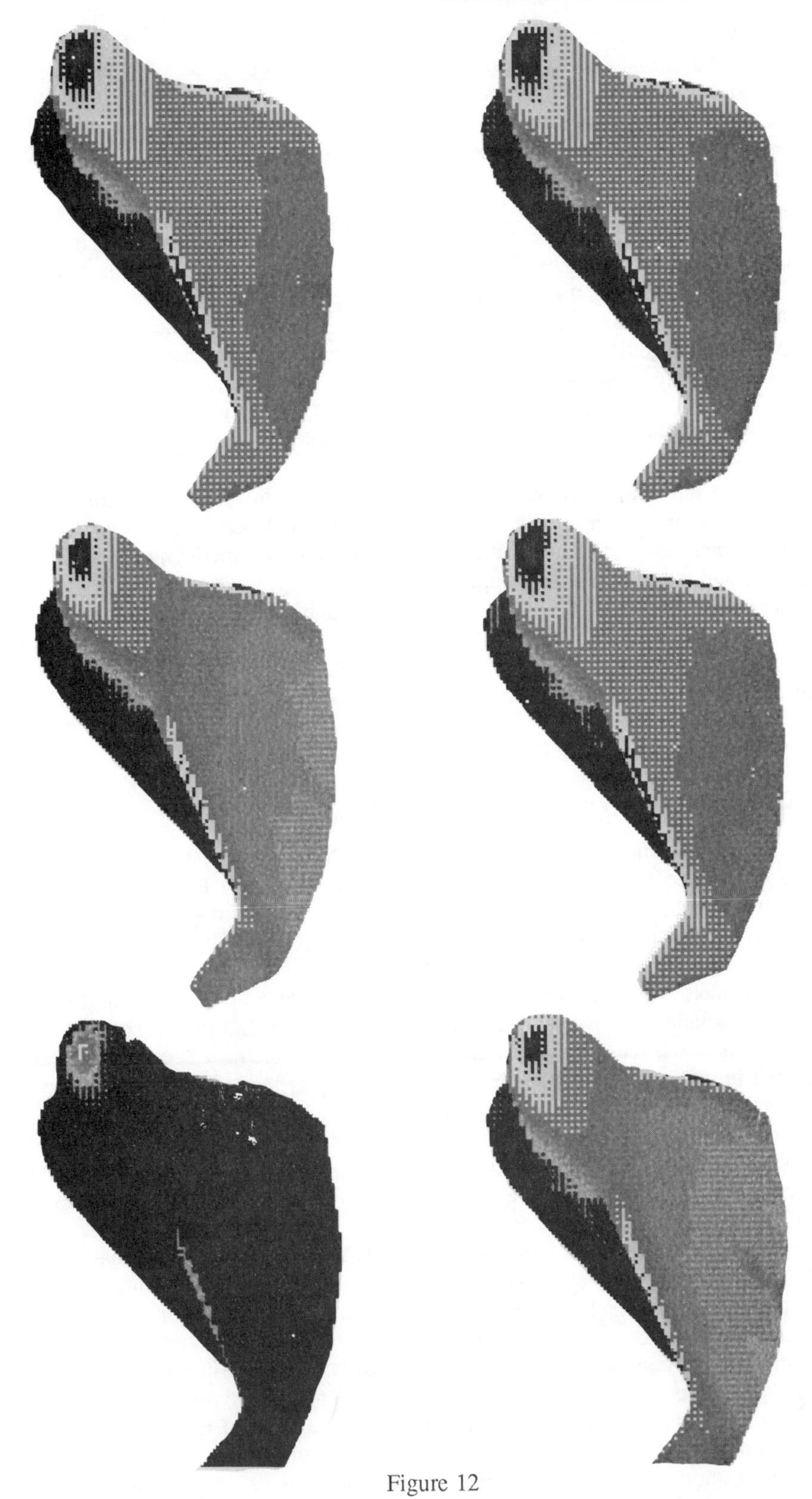

Figure 12

Hermes hypersonic temperature distribution (Dassault N.S. weak interacting code)

to reinforce or reduce *locally* the meshing by addition or suppression of tetraedra where the flowfield requires more precise definition of the solution and of its relation to geometry. Thus any complex geometry of shockwaves, wakes etc... can be precisely located by iterative reinforcement of the mesh without the excessive millions of points required in structured meshes. At the end a gain in cost for a given precision seems probable. Figure 13 gives an example of such refinements, where the effective final flowfield characteristics are very well defined. Unsteady phenomena on complex 3D shapes could be tracked with such an approach via a process of refinement-derefinement. The complexity of mesh generation remains badly unsolved because of the extreme difficulty of solving the topology constraints common in complex flowfield. It remains the main challenge, and it has required the maximum industrial effort in AMD-BA, together with the costly codes to extract visual checking of the mesh and visual analysis of computational data.

It can be seen on the complex geometry given to university teams working on finite elements method that refinement of mesh does not avoid numerical entropy production in corners or poor meshed areas.

2.4 — The future of such computations is in highly non linear flows corresponding to highly viscous separated regimes of flight, for which the turbulence modeling is essential. The effort will be very hard and costly, due to the 3D nature of modeling, and, therefore, of checking. We will return to that point in Section 4.

Figure 14 gives an example of the entropy on the surface of an aircraft and can be a test of the necessary improvements that are to be afforded by better algorithms; this simulation produces spurious entropy in the corner. A reduction is possible with better boundary conditions, or better acceptance of distorted meshes, or with refined meshes. This example shows the way for the second level of research in CFD for aerospace industry to go from 3-D complex computations to high quality computations.

3. CFD FOR ADVANCED PROJECTS LIKE HERMES

3.0 — The coming period in aerospace industry will be characterized by a mix of classical and advanced projects. The classical projects will use the best state of the art procedures, but with a classical approach: an efficient mixing of computational optimization and experimental definition and checking. The selection of each numerical or experimental tool will depend on cost and efficiency. The general trend we have noticed will continue; the part of computation will increase, due to the increasing cost of energy and the increasing quality of computation. However, the major step is done and the revolution is almost achieved, even if it will require a long improvement effort in the next decade. But this change is less crucial in its consequences for each individual project than the complete renewal of the necessary methodology for new advanced projects.

It becomes gradually clear that no sub-system or general design will escape to a new methodology, as far as it is of high level of technology. But it is precisely the characteristic of advanced systems: to use the knowledge of new technologies for new performance in difficult designs, that cannot be done without them.

3.1 — The main idea of this new methodology is to select a rational approach by subdividing systems and problems in the class of the scientific discipline underlined, and by modeling each sub-system and interfaces between systems. The validation of each sub-system will require specific tests at two levels (scientific and technological) and integration tests for the compatibility to the complete system. Let us take an example in CFD. It is easy to see that, whatever will be the system of equations, (Navier-Stokes or Maxwell or Elasticity...) equivalent to real physical or optimality problems, or the system of logic inference equivalent to some control of decisions, the methodology will be the same.

The acquisition of new technology to be confirmed in its application to the advanced project has its own milestones, having its own logical inference with other technologies or with decision of manufacturing such parts. The decision to produce a precise high temperature

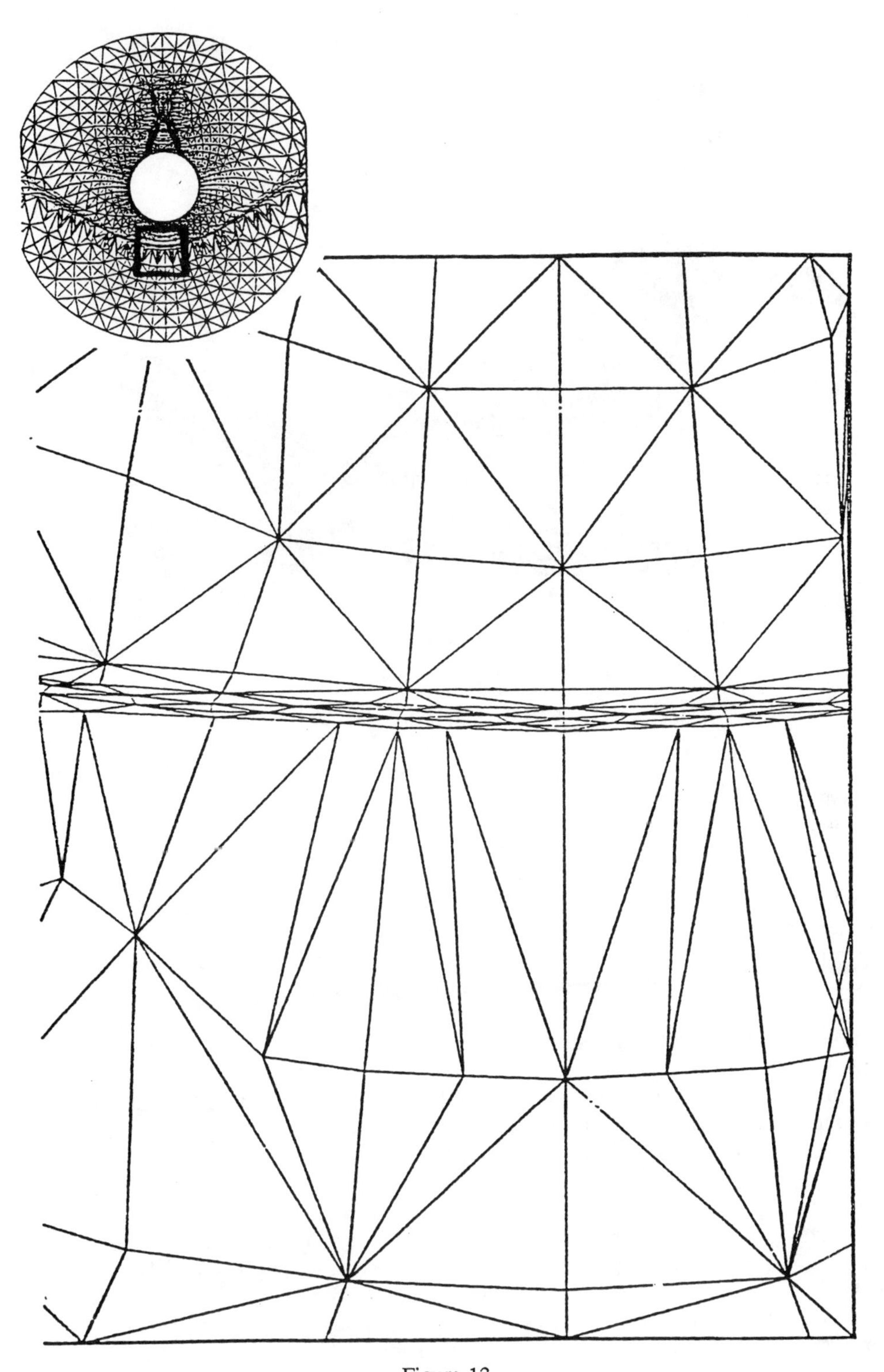

Figure 13

Mesh refinement for hypersonic shock
M = 10 (INRIA)

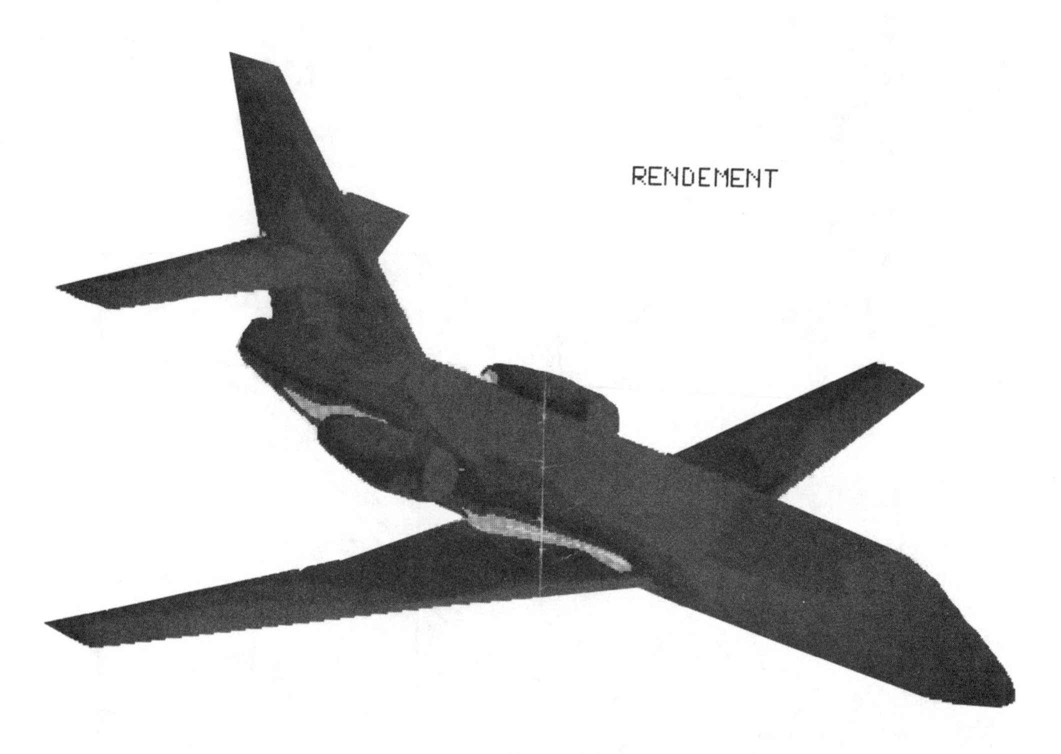

Figure 14

Entropy generation on complex configuration
(Dassault Euler code on Falcon JJ)

material for Hermes cannot be taken without a clear knowledge of loads, temperature, chemical resistance... required. Such a knowledge can be of two types. The first one is connected with two rough estimates of the requirements, only some extra qualities are required and such criterion is hoped to relax when better knowledge (theoretical or experimental) will be available. The second one can be a go-no-go answer for a given technology; in that case, the complete program, except alternative design, will rely on the progress of knowledge or technology in the concerned area. It is the first step for many projects where the feasibility itself of the project or its time schedule rely heavily on technology or on science. All the program has to be delayed if an antioxydation surface layer without abrupt variations with time of its catalycity cannot be realized at the target temperature and with given maximum flux of atomic oxygen, because no confidence can be given on that flux or those catalytic phenomena: it is a critical point.

3.2 — The modelization of the physics has progressed rapidly with the improvement of computational capacities. Beginning with simple solids, modelling has left the linear stress-deformation tensor analysis for non linear dependence and from simple shapes to very complex shapes with mesh refinement where critical behavior appear; the fluid modelling tries in the same way to leave the simple perfect inviscid flows to reach modelization of complex turbulent flows. In the same approach, supported by the quasi 2-D subset of the space occupied by fluid quasi discontinuities (shock waves, shear layers, vortices), the fluid dynamics has tried to easily handle local phenomena in the same approach as smooth evolutions of parameters. The coexistence of different scales of phenomena can give way to partition of approximated equations by subdomains.

One characteristic of the Hermes re-entry trajectory is that it covers the domains where all the parameters are the order of 1 and the domains where they are very far from 1: the viscous layer can be the size of the distance from the body to the shock (viscous interaction parameter), the mean free path can cover that distance and be only 10^{-6} in continuous flight (Knudsen number), in the same way, the characteristic length of relaxation of different degrees of freedom of molecular energy, or the characteristic length of reacting flows (Damkhöler number) may have large variations. So the quality of numerical as the one of physical modelization cannot be too poor and the mesh has to be adapted to the phenomena to compute. All the new trends in computational mathematics would be tested with such a stringent requirement.

3.3 — Two ways have been used in the past for continuation of advanced projects. One was to build a first real size machine and see by direct test if it worked, the second was to build a model and make validation on that model. We have said that physical models tend to be largely replaced by numerical models, and validation, using geometric coherent models, tends to be replaced by physical relevant test cases on references cases. In fact the two ways are complementary because the level of physical modelization and mathematical approximation in discrete codes may be not sufficient, and, moreover, needs to be proved. So the modelization is used with its two layers: one scientific layer (physical understanding + numerical analysis of pertinent set of equations) and one technical layer (geometry discretization, boundary conditions). The scientific layer is to be proved by basic experiments and by workshops, and the technological one by demonstrators of the global value of modelization. The scientific layer is supported by a research program and the technological supported by successive demonstrations. The methodology for Hermes is therefore the following.

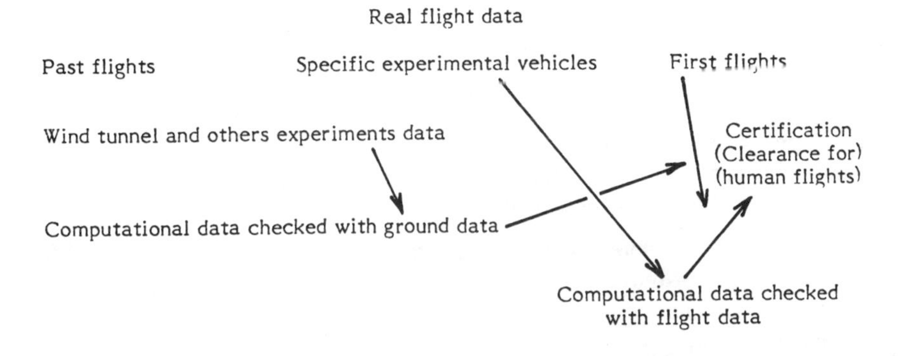

The first line is the technology demonstration line, the second line is wind-tunnel testing, either for basic modelization or for complete real geometry models, the third line is for software development. The different steps are related to inclusion of better physics or mathematical solvers and these steps occur when a research work in a laboratory has opened new understanding or methods for solving equations selected. If the necessary level of useful results doesn't come from research work, a delay can appear in the aircraft milestone. In the same way, the quality of final product, or its safety, or eventually its ability to fulfill the mission, relies on the success of previous research.

If the total number of european wind-tunnels involved in Hermes is 19, the total number of computing industrial teams will be 5, and will cover, in our present estimation, 12,000 teraflop. With an actual figure of 10 hours of computer time for a teraflop, it means 3 supercomputers for 10 years. In fact it will be much less because an improvement of computer speeds is expected, with a critical level in five years if no sufficient improvement has been realized.

3.4 — For Hermes, a large group of contracts has been proposed to the best of european researchers in critical fields of hypersonic aerodynamics and corresponds to 33 contracts in physics and 35 in numerical analysis.

At the end of an initial 3 years period, computation of test cases will check the quality of results. Moreover, complete computations on critical points will be conducted in an alternate way to give other answers for the same conditions with different modelizations and/or algorithms.

The following table gives the repartition of the different actions in numerical analysis, and can be an example of actual topics in that field although especially oriented on hypersonic problems in nine european countries:

Perfect flows	7
Euler and Navier-Stokes flows	5
Boundary layers and viscous flows modelling	7
Thermal effects with chemistry or real gases	14
Transition and turbulence modelling	5

The figure gives an example of computations in the high enthalpy domain that characterizes re-entry. An effort in the reduction of the time of computation can be seen as a direct effort towards discovery of efficient algorithms for vector and parallel computers, and of reduction of unnecessary degree of freedom and/or transfer to crude mesh of necessary diffusion of errors by multigrid or multilevel approach. The coexistence of different equations, solved directly or with modelization of their untractable small scale behavior, will give to Hermes computations the characteristic to be the most complex industrial computations and therefore the most challenging.

It covers:

— Going from rarefied flows with Monte Carlo approach to Boltzmann and Navier-Stokes laminar or modelized turbulent flows P.D.E.

— Mixing simple state equations to chemical stiff coupled systems of O.D.E. in Lagrangian variables (solved with direct explicit multigrid unsteady analysis and implicit highly preconditioned operators with or without artificial viscosity and flux limiters).

— Taking into account complex boundary conditions and using the best refinement or adaptation of the mesh to reduce the cost of computation.

The Hermes computational success requires the top of what can be done in numerical analysis in the next future.

4. INTERFACE RESEARCH — INDUSTRY FOR THE FUTURE

4.1 — If we consider the future of complex fluid flows computations, it is clear that the scientific necessary support for code implementation will be increased and that common effort is of interest for all the applications. So the sharing of the knowledge is not sufficient; a continuous effort has to be given to discover common problems between industrial computational teams as between research laboratories. It is clear on one side that the fluid flow problems are common for all the mechanical industry using fluid flows and not only the aerospace industry, and on the other side that a sufficiently large scientific community is required to support common effort (national effort is too small except in the case of U.S.A.). It becomes evident that common european effort, as for Hermes or other European Space Agency programs, is necessary for success of high technology future programs involving sufficient size, number and quality of industrial and scientific teams.

An analysis of the use of supercomputers shows that at least 40% of total CPU time is devoted to solving problems directly connected with turbulence and combustion in research and industry. Moreover, the gain in performance or in fiability, and therefore in cost, to be obtained by small improvements in that field is much more important, in next years, than the gain that can be hoped for future energy production; energy production economy will probably also rely on intermediate flow optimization.

It is clear that CFD extended to complex turbulent and reacting flows will afford new high quality design capabilities only if a good support for modelling and algorithm selection is given by the scientific community. Today, it becomes a necessity when the cost of teraflop gives access to that 3-D computations for large number of design teams in aerospace, automotive, energy or chemical industry.

Two ingredients seem necessary in that interface between industrial codes and scientific work. First, it is necessary to restrict the exchange to a *pure scientific layer* putting away any problems of concurrence or secrecy. Second, it is necessary to open the exchange to the *best of european researchers* in whatever field and without reducing national local research.

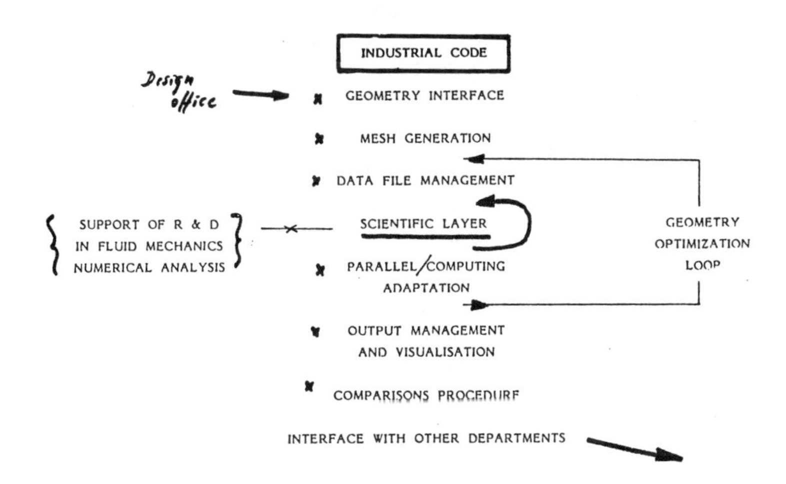

4.2 — Creating a new european center as a large european laboratory will not answer to the necessary future bilateral exchange between local industry and local research for adjustment of common scientific layer to specialized use in a definite industrial use. Asking european researchers to create and operate along common research programs will not be sufficient to adjust the research to future needs and to study common problems arising in the industrial use of first level modelization. A structure of "club" is necessary for the meeting of research and industry men, for the definition of future work and for the checking of work by experimental and numerical workshops. That structure of "club" has to be hierarchical, using a confederation of researchers around pilot centers for local groups of laboratories.

In such a neuronal center, industry is to be represented either by major industrial teams or by cooperative industrial research centers. In such a way, the club will be limited to an efficient nucleus of representative experts. Such a nucleus is beginning to exist in Europe in a certain number of large universities. We can notice for example in France, the Lyon-Grenoble sub-center for turbulence, the Toulouse-Cerfacs sub-center for education, the Marseille-Sophia-Antipolis one for combustion, and in the same direction the Lausanne EPFL center in Switzerland and the Stuttgart and Aachen centers in Germany.

But the club cannot be efficient if a good correlation does not exist between the pilot centers and the different laboratories federated in each pilot center. Such correlation is to be improved by a light structure for meeting and a good network for exchange of data.

Besides the club and giving more efficiency to it, it seems that a center of computation and data retrieval is an important part of future efficiency for such a cooperative european effort. All the workshops and their results (experimental and industrial data) are to be accumulated in a data library at the disposal of every body for checking or comparison. All new computations require high quantity of supercomputer hours and a computational center exclusively devoted to very high level computation. Emphasis on future capabilities is mandatory. Different from any local research center, where there coexists large quantities of small computations, with a free access rule, and some large computations, such a computer center would be devoted to large computations. For example, a 10 to 50 hours computation would be ranked as a first priority because it will be ahead of industrial computations of 1 to 5 years (five years later computations will take place in industry with an order of magnitude cost improvement as we have pointed out in first part). The next generation codes is to be tested within the preceding one is in current use and gives the size of industrial computers!

Such a center with its club of pilot centers and its support by a large data library and supercomputers is actually in the process of constitution in Europe. Its general acronym is actually ERCOFTACS for European Research Community For Flow Turbulence And Combustion Simulation. Its aim is to help european industry to really compute flames and winds. Such a decentralized structure, letting each laboratory or industrial team develop its own excellence, using or, at least, informed of, the best of other european works, seems to be an essential element of future successes in computational physics.

REFERENCES

— CHAPMAN D. R. — Computational Aerodynamics Development and Outlook J. AIAA, Vol. 17, Dec. 1979.

— PETER D. LAX. — Report of the panel on large scale computing in Science and Engineering, Dec. 1982.

— RICHARD G. BRADLEY — Current Capabilities and Future Direction in computational Fluid Dynamics. National Academy Press, 1966.

— JAMESON A. — Successes and challenges in computational Aerodynamics AIAA 87 — 1184.

— JAMESON A. and BAKER T. J. — Multigrid Solution of Euler Equations for Aircraft configurations AIAA Paper 84-0093.

— STOUFFLET, PERIAUX, FEZOUI, DERVIEUX — Numerical Simulation of 3D Hypersonic Euler Flows around Space Vehicles using Adapted Finite Elements AIAA 87-560.

— McCORMACK R. W. — Numerical Solution of the Compressible Viscous Flow Field about a complete Aircraft in flight. Recent advances in Numerical Methods in Fluids, Vol. 3 — Pineridge Press 1984.

— R. GLOWINSKI — Numerical Methods for Non-linear variational Problems, Springer-Verlag, 1984.

— J. F. THOMSON — Numerical Grid Generation — North-Holland, 1984.

— T. R. J. HUGHES — The Finite Element Method, Prentice-Hall, 1987.

— The 1980-81 AFOSR-HTTM — Stanford Conference on Complex Turbulent Flows, Stanford University, 1982.

— P. PERRIER — Euler codes in non-linear aerodynamics — AGARD CP.412.

— GULLO K. — Supercomputing in the real world, Datanation 1.86.

A Review of Algorithms for Nonlinear Equations and Unconstrained Optimization

M.J.D. POWELL*

Abstract. The Newton iteration for the solution of systems of nonlinear equations requires a good starting point in the space of the variables and the calculation of first derivatives, but often these requirements are highly restrictive in practice. Therefore techniques have been developed to force convergence from poor initial estimates of the solution, and to approximate derivatives automatically. They are surveyed in this paper. We find that "trust regions" are more useful than "line searches", and that few extra function values are needed to provide suitable estimates of Jacobians, especially when the Jacobian is sparse. Similar techniques for unconstrained optimization calculations are also reviewed. We consider the relevant software that is available in subroutine libraries, and some recent work is mentioned that may provide substantial improvements to current algorithms.

1. Introduction. The solutions of many calculations satisfy known equations. For example, there may be flow conservation conditions at the junctions of a network of pipes, Kuhn-Tucker conditions may have to hold at the solution of a constrained optimization problem, or a dynamical system may have to be in equilibrium. Moreover, most algorithms for the solution of differential and integral equations employ discrete approximations to derivatives and integrals. Therefore systems of equations of the form

$$\underline{f}(\underline{x}) = \underline{0} \tag{1.1}$$

occur frequently, where $\underline{f}(\cdot)$ is a function from $\mathbf{R}^n$ to $\mathbf{R}^n$.

This paper surveys some general numerical techniques that have been developed to calculate an $\underline{x} \in \mathbf{R}^n$ that solves (1.1). The most usual technique for this calculation is the Newton iteration, which replaces $\underline{x}_{k+1}$ by $\underline{x}_k$, where $\underline{x}_{k+1}$ is the solution of the linear system

$$\underline{f}(\underline{x}_k) + J(\underline{x}_k)(\underline{x} - \underline{x}_k) = \underline{0}, \ \underline{x} \in \mathbf{R}^n . \tag{1.2}$$

Here $J(\underline{x})$ is the $n \times n$ Jacobian matrix, whose elements are the derivatives

* Department of Applied Mathematics and Theoretical Physics, University of Cambridge, Silver Street, Cambridge CB3 9EW, England.

$$J_{ij}(\underline{x}) = df_i(\underline{x})/dx_j, \quad i, j = 1,2, \ldots, n. \tag{1.3}$$

We assume throughout this paper that $\underline{f}(\cdot)$ is continuously differentiable.

In most cases, recursive application of the Newton iteration generates a sequence of points $\{\underline{x}_k : k = 1,2,3, \ldots\}$ that converges rapidly to a solution of the system (1.1), provided that the starting point $\underline{x}_1$ is sufficiently close to the solution. When $\underline{f}(\cdot)$ arises from the discretization of a system of differential equations, for example, one can usually achieve a successful starting point by refining the discretization if necessary. Therefore many algorithms in this field use only the Newton iteration for solving the nonlinear equations that arise, but this approach can be highly inefficient if the smallness of the mesh size is dominated by the need for the Newton iteration to converge.

This is one of several reasons for taking notice of the diligent research of the last 20 years on extensions to Newton's method that force convergence from poor starting points. It is the subject of Section 2. We take the view that each iteration should satisfy the condition

$$\|\underline{f}(\underline{x}_{k+1})\| \leq \|\underline{f}(\underline{x}_k)\| \tag{1.4}$$

for some norm in $\mathbf{R}^n$, and that $\underline{x}^*$ should be a limit point of $\{\underline{x}_k : k = 1,2,3, \ldots\}$ only if the least value of the convex function $\{\|\underline{f}(\underline{x}^*) + J(\underline{x}^*)\underline{d}\| : \underline{d} \in \mathbf{R}^n\}$ occurs at $\underline{d} = 0$. We cannot expect $\underline{f}(\underline{x}^*) = \underline{0}$ in general, because the system (1.1) may be inconsistent. Here and in equation (1.2) we are making use of the "linear model"

$$\underline{f}(\underline{x}) \approx \underline{f}(\underline{x}_k) + J_k(\underline{x} - \underline{x}_k), \tag{1.5}$$

where $J_k = J(\underline{x}_k)$. This model guides the calculation of $\underline{x}_{k+1}$ from $\underline{x}_k$.

Similar considerations apply to algorithms for unconstrained optimization problems, such as the adjustment of parameters in data fitting to minimize the sum of squares of residuals, or finding the configuration of least energy in an atomic system. We let $\{F(\underline{x}) : \underline{x} \in \mathbf{R}^n\}$ be the objective function whose least value is required, so $\underline{x}$ is optimal only if it is a solution of the system

$$\underline{g}(\underline{x}) = \underline{0}, \tag{1.6}$$

where $\underline{g}(\underline{x}) \in \mathbf{R}^n$ is the first derivative vector $\underline{\nabla} F(\underline{x})$. Applying the Newton iteration to this system is equivalent to replacing $\underline{x}_k$ by $\underline{x}_{k+1}$, where $\underline{x}_{k+1}$ is the value of $\underline{x} \in \mathbf{R}^n$ at which the "quadratic model"

$$F(\underline{x}) \approx F(\underline{x}_k) + \underline{g}(\underline{x}_k)^T(\underline{x} - \underline{x}_k) + \tfrac{1}{2}(\underline{x} - \underline{x}_k)^T G_k(\underline{x} - \underline{x}_k) \tag{1.7}$$

is stationary, G_k being the second derivative matrix $\nabla^2 F(\underline{x}_k)$. Now, however, the extensions to Newton's method give changes to the variables that satisfy the inequality

$$F(\underline{x}_{k+1}) \leq F(\underline{x}_k) \tag{1.8}$$

instead of condition (1.4). Another important difference is that G_k (but in general not J_k) is symmetric. However, as pointed out in the excellent book of Dennis and Schnabel [8], algorithms for nonlinear equations and for unconstrained optimization have so much in common that it is advantageous to consider them together.

Usually there is no need to calculate the matrices $\{J_k : k = 1,2,3, \ldots\}$ or $\{G_k : k = 1,2,3, \ldots\}$ explicitly, because suitable methods of approximating them automatically have been developed, which are reviewed in Section 3. We note that finite difference approximations to derivatives can be very useful for estimating sparse derivative matrices, but in general it is inefficient to regenerate all the approximations regularly unless n

is small. Therefore most algorithms for nonlinear equations form J_{k+1} by making a "small" change to J_k to satisfy the "secant equation"

$$J_{k+1} \underline{s}_k = \underline{y}_k , \tag{1.9}$$

where $\underline{s}_k$ and $\underline{y}_k$ are the vectors

$$\left. \begin{aligned} \underline{s}_k &= \underline{x}_{k+1} - \underline{x}_k \\ \underline{y}_k &= \underline{f}(\underline{x}_{k+1}) - \underline{f}(\underline{x}_k) \end{aligned} \right\} . \tag{1.10}$$

Equation (1.9) is suitable because it holds when the system (1.1) is linear and J_{k+1} is the true Jacobian matrix. Moreover, $\underline{f}(\underline{x}_{k+1})$ and $\underline{f}(\underline{x}_k)$ are available because they occur in the test (1.4). Thus some efficient algorithms have been developed for the solution of nonlinear equations without calculating any derivatives.

In unconstrained optimization, the analogy of the secant equation is the relation

$$G_{k+1} \underline{s}_k = \underline{\gamma}_k , \tag{1.11}$$

where G_{k+1} is now an approximation to $G(\underline{x}_{k+1})$, and where $\underline{\gamma}_k$ is the difference

$$\underline{\gamma}_k = \underline{g}(\underline{x}_{k+1}) - \underline{g}(\underline{x}_k) . \tag{1.12}$$

Successful formulae have been discovered for calculating G_{k+1} from G_k, $\underline{s}_k$ and $\underline{\gamma}_k$ so that equation (1.11) and the symmetry condition

$$G_{k+1}^T = G_{k+1} \tag{1.13}$$

are satisfied. Some of them also preserve positive definiteness. They are fundamental to many good algorithms for unconstrained optimization that require only values and first derivatives of the objective function.

Section 4 addresses an assortment of topics, including unconstrained optimization without the calculation of any derivatives. Its main purpose, however, is to mention some recent work that may become important in practice. There is an outline of the tensor methods of Schnabel and Frank [21, 22] for nonlinear equations, that add some second order terms to the model (1.5). Usually the steepest descent method for unconstrained minimization converges very slowly, but we note the discovery by Barzilai and Borwein [1] of two choices of steplength that can provide superlinear convergence. We also mention some numerical results of Conn, Gould and Toint [5], because they suggest that more attention should be given to the "rank one formula" for updating second derivative approximations.

Finally, Section 5 reviews briefly the software for nonlinear equations and unconstrained optimization that is available in the Harwell, IMSL, NAG and MINPACK libraries.

2. Line Searches and Trust Regions. Line searches are the most usual way of extending the Newton iteration to force convergence from poor starting approximations. Here the new vector of variables has the form

$$\underline{x}_{k+1} = \underline{x}_k + \alpha_k \underline{d}_k , \tag{2.1}$$

where α_k is a positive steplength that is calculated by the line search, and where $\underline{d}_k$ is a trial change to $\underline{x}_k$ that is derived from the model (1.5) or (1.7). The search direction may be the vector

$$\underline{d}_k = -J_k^{-1} \underline{f}(\underline{x}_k) \tag{2.2}$$

or

$$\underline{d}_k = -(G_k + \beta_k I)^{-1} \underline{g}(\underline{x}_k) \tag{2.3}$$

in nonlinear equation solving or unconstrained optimization respectively, where β_k is a nonnegative parameter to ensure that the matrix $(G_k + \beta_k I)$ is positive definite. Thus, for nonzero $\underline{g}(\underline{x}_k)$ in unconstrained optimization, the function $\{F(\underline{x}_k + \alpha \underline{d}_k) : \alpha \geq 0\}$ decreases initially, so condition (1.8) can be satisfied as a strict inequality. Moreover, assuming that J_k is nonsingular and $\underline{f}(\underline{x}_k) \neq \underline{0}$, equation (2.2) implies that the function $\{\|\underline{f}(\underline{x}_k) + \alpha J_k \underline{d}_k\| : 0 \leq \alpha \leq 1\}$ decreases strictly monotonically to zero, so condition (1.4) can hold as a strict inequality for any choice of norm when $J_k = J(\underline{x}_k)$. It is important to successful convergence, however, to ensure that the reduction $[\|\underline{f}(\underline{x}_k)\| - \|\underline{f}(\underline{x}_{k+1})\|]$ or $[F(\underline{x}_k) - F(\underline{x}_{k+1})]$ is not much smaller than the greatest reduction that can be achieved by adjusting α_k. Ways of choosing suitable steplengths are discussed in many publications; for example, see Fletcher [11] and Gill, Murray and Wright [13].

Line searches have proved to be very successful in unconstrained optimization calculations. Mainly this is because, when first derivatives are available and $\underline{g}(\underline{x}_k) \neq \underline{0}$, it is easy to choose search directions that provide the reduction $F(\underline{x}_{k+1}) < F(\underline{x}_k)$. For example, many algorithms employ approximations $\{G_k \approx \nabla^2 F(\underline{x}_k) : k = 1,2,3,\ldots\}$ that are always positive definite, and on every iteration their search direction is the vector (2.3) with $\beta_k = 0$, although the second derivative approximations may be highly inaccurate.

For nonlinear equations, however, accurate Jacobian information may be needed to achieve $\|\underline{f}(\underline{x}_{k+1})\| < \|\underline{f}(\underline{x}_k)\|$, especially when the least value of $\{\|\underline{f}(\underline{x})\| : \underline{x} \in \mathbf{R}^n\}$ is positive, and $\|\underline{f}(\underline{x}_k)\|$ is near to this least value. Moreover, if $J_k = J(\underline{x}_k)$ is nearly singular, then usually equation (2.2) makes $\|\underline{d}_k\|$ very large, which causes the linear model

$$\underline{f}(\underline{x}_k + \alpha \underline{d}_k) \approx \underline{f}(\underline{x}_k) + \alpha J_k \underline{d}_k = (1-\alpha)\underline{f}(\underline{x}_k) \tag{2.4}$$

to be unsuitable unless $\alpha << 1$. When α is small, however, then the predicted reduction

$$\|\underline{f}(\underline{x}_k)\| - \|\underline{f}(\underline{x}_k + \alpha \underline{d}_k)\| \approx \alpha \|\underline{f}(\underline{x}_k)\| \tag{2.5}$$

is small too, so the search direction (2.2) tends to be highly inefficient when J_k is ill-conditioned. Unfortunately, singular Jacobians are possible in every calculation having the property that the signs of $\det(J(\underline{x}_1))$ and $\det(J(\underline{x}^*))$ are opposite, where $\underline{x}^*$ is the required solution. Further, the Newton iteration with line searches can cause $\{\underline{x}_k : k = 1,2,3,\ldots\}$ to converge to a point that is not a stationary point of $\{\|\underline{f}(\underline{x})\| : \underline{x} \in \mathbf{R}^n\}$ [19]. On the other hand, it should be noted that the addition of line searches to the Newton iteration increases greatly the range of nonlinear equation calculations that can be solved.

The main purpose of trust regions is to avoid calculating from a model a trial step $\underline{d}_k$ that is much longer than the range of validity of the model. For example, a very useful extension to the choice (2.2) of the search direction is to let $\underline{d}_k$ be the value of $\underline{d}$ that minimizes $\{\|\underline{f}(\underline{x}_k) + J_k \underline{d}\| : \underline{d} \in \mathbf{R}^n\}$ subject to the bound

$$\|\underline{d}\| \leq \Delta_k , \tag{2.6}$$

where Δ_k is a parameter whose value is chosen automatically. Here the first norm should be the one that occurs in condition (1.4), but the norm of $\|\underline{d}\|$ can be different. Duff, Nòcedal and Rèid [9], for instance, choose the norms $\|\cdot\|_\infty$ and $\|\cdot\|_1$ respectively in their algorithm for the solution of large, sparse systems of nonlinear equations.

A major difference between trust regions and line searches is the response to an unsuccessful trial step. If a line search method is trying to satisfy inequality (1.4) and a steplength of one gives the increase

$$\|\underline{f}(\underline{x}_k+\underline{d}_k)\| > \|\underline{f}(\underline{x}_k)\|, \tag{2.7}$$

then the next trial vector of variables is on the straight line segment between $\underline{x}_k$ and $\underline{x}_k+\underline{d}_k$. A trust region method, however, would set $\underline{x}_{k+1}=\underline{x}_k$ and $\Delta_{k+1}<\Delta_k$, so the next vector of variables is calculated by the next iteration and is not confined to the line segment. In this case it is also helpful to let J_{k+1} satisfy the secant equation $J_{k+1}\underline{d}_k=\underline{y}_k$, where $\underline{y}_k$ is now the vector

$$\underline{y}_k=\underline{f}(\underline{x}_k+\underline{d}_k)-\underline{f}(\underline{x}_k), \tag{2.8}$$

because then the new linear model predicts that the trial step $\underline{d}_k$ gives an increase in $\{\|\underline{f}(\underline{x})\| : \underline{x}\ \varepsilon\ \mathbf{R}^n\}$. Thus, even if we set $\Delta_{k+1}=\Delta_k$, the next trial step would be different from $\underline{d}_k$. Similar remarks apply to trust region algorithms for unconstrained optimization.

Methods for choosing the bounds $\{\Delta_k : k=1,2,3,\ldots\}$ automatically are reviewed by Shultz, Schnabel and Byrd [24]. One calculates the ratio, ρ_k say, of $[\|\underline{f}(\underline{x}_k)\| - \|\underline{f}(\underline{x}_k+\underline{d}_k)\|]$ (or $[F(\underline{x}_k)-F(\underline{x}_k+\underline{d}_k)]$ in unconstrained optimization) to the value that is predicted by the model (1.5) or (1.7). Then a typical rule for generating Δ_{k+1} is the formula

$$\Delta_{k+1}=\begin{cases} \tfrac{1}{2}\|\underline{d}_k\|, & \rho_k\le 0.1\\ \Delta_k, & 0.1<\rho_k\le 0.7\\ \max[2\|\underline{d}_k\|,\ \Delta_k], & \rho_k>0.7, \end{cases} \tag{2.9}$$

the main idea being that the trust region radius is made smaller if the model is unsuccessful, and it is increased if a substantial reduction in the objective function is found. Shultz et al [24] recommend setting $\underline{x}_{k+1}=\underline{x}_k$ instead of $\underline{x}_{k+1}=\underline{x}_k+\underline{d}_k$ whenever $\Delta_{k+1}<\Delta_k$, but I prefer $\underline{x}_{k+1}=\underline{x}_k$ if and only if $\rho_k\le 0$, because it is surely sensible to work from the vector of variables that has given the least calculated value of the objective function. It is more difficult, however, to prove convergence theorems in the latter case.

Another important question that has received much attention recently is the calculation of the trial step subject to the bound (2.6). We seek only a suitably small predicted value of $\|\underline{f}(\underline{x}_k+\underline{d})\|$ or $F(\underline{x}_k+\underline{d})$ (according to the model (1.5) or (1.7)), because *minimizing* the predicted value is not cost-effective. A highly useful method comes from the remark that, if β_k has any nonnegative value such that $(G_k+\beta_k I)$ is positive definite, and if $\underline{d}_k$ is the vector (2.3), then the least predicted value of $\{F(\underline{x}_k+\underline{d}) : \|\underline{d}\|_2\le\|\underline{d}_k\|_2\}$ occurs when $\underline{d}=\underline{d}_k$; therefore the method adjusts β_k until $\|\underline{d}_k\|$ is acceptably close to Δ_k (see Moré [15], for instance). Alternatively, Dembo and Steihaug [7] recommend applying the conjugate gradient algorithm to minimize the quadratic model function (1.7) starting at $\underline{x}-\underline{x}_k$, where any steps along directions of negative curvature are allowed to be of infinite length, except that the procedure is stopped as soon as the calculated piecewise linear path intersects the trust region boundary $\|\underline{x}-\underline{x}_k\|=\Delta_k$; the trial step $\underline{d}_k$ is then chosen to be the final value of $\underline{x}-\underline{x}_k$. This approach is also suitable when solving systems of nonlinear equations, because one can replace expression (1.7) by the square of the linear model function (1.5). Further work on inexpensive choices of $\underline{d}_k$ is described by Byrd, Schnabel and Shultz [2].

In the opinion of the author, trust region methods are superior to line searches when solving nonlinear equations, because they are more robust in the presence of nearly singular or inaccurate Jacobians. Further, they have strong advantages in unconstrained optimization calculations when G_k is not necessarily positive definite.

3. Jacobian and Hessian Approximations. Two different views can be taken of the accuracy of the approximations $J_k \approx J(\underline{x}_k)$ and $G_k \approx \nabla^2 F(\underline{x}_k)$ for the models (1.5) and (1.7). An obvious one is to try to satisfy the requirement on most iterations that the models are sufficiently good to generate successful trial steps. It provides some useful algorithms, but in the opinion of the author a quite different view has more merit in practice and is more interesting in theory.

Let a model with a rather poor approximation to a derivative matrix be used to calculate a trial step. Usually a good reduction in $\|\underline{f}(\cdot)\|$ or $F(\cdot)$ is predicted by the model, and if this prediction is realized then the poor accuracy of the approximation does not matter. Alternatively, a large discrepancy between prediction and reality provides exactly the right kind of information to improve the derivative approximation. Thus one can usually ensure that each iteration improves substantially either the approximation or the vector of variables (or both). A sequence of iterations can provide an excellent $\underline{x}$ without obtaining a good estimate of all of $J(\underline{x})$ or $\nabla^2 F(\underline{x})$. One expects only the parts of J_k or G_k that are directly relevant to the adjustments of $\underline{x}$ that have occurred to have good final accuracy.

The efficacy of the second approach is often shown by unconstrained minimization calculations when n is large. It is usual to set $G_1 = I$ (the $n \times n$ unit matrix) for the first iteration, and then to calculate G_{k+1} from G_k and the secant equation (1.11) for each k. Successful convergence of $\{\underline{x}_k : k = 1,2,3,\ldots\}$ is often achieved in far fewer than n iterations. Thus one solves the minimization calculation using fewer gradient evaluations than would be needed to form just one good approximation to $\nabla^2 F(\cdot)$, assuming that one cannot take advantage of any sparsity.

When solving systems of nonlinear equations, it is often suitable to form J_1 by finite differences along the coordinate directions. Then on each iteration the secant equation (1.9) can be satisfied by applying "Broyden's formula"

$$J_{k+1} = J_k + \frac{(\underline{y}_k - J_k\,\underline{s}_k)\,\underline{s}_k^T}{\|\underline{s}_k\|_2^2}. \tag{3.1}$$

This approach seems to work well in practice [19], and it combines the two views of derivative approximation that have just been mentioned.

There is a reason for suggesting the choice $G_1 = I$ in an optimization calculation but the use of finite differences to generate J_1. It is that, due to its first derivative term, the quadratic model (1.7) can provide downhill search directions when there are large errors in G_k, but the model (1.5) is unhelpful without some accuracy in J_k. When J_k is a very poor approximation in a trust region method, the bounds $\{\Delta_k : k = 1,2,3,\ldots\}$ may shrink rapidly to zero, causing a severe increase in the number of iterations that are needed to move from $\underline{x}_1$ to the required solution.

The use of sparsity when forming finite difference approximations to $J(\underline{x}_k)$ or $\nabla^2 F(\underline{x}_k)$ is considered by many authors, including Curtis, Powell and Reid [6] and Coleman and Moré [4]. Here the aim is to use as few differences as possible of the vector $\underline{f}(\underline{x})$ or $\underline{g}(\underline{x})$, because it is assumed that it is more convenient to calculate the complete vector at once rather than a few prescribed components of the vector. To explain the technique of Curtis et al [6], we let S be any subset of the integers $\{1,2,\ldots,n\}$, and we let $J(\underline{x}_k, S)$ be the matrix that is composed of the columns of $J(\underline{x}_k)$ whose indices are in S. Then, if S is chosen so that the sparsity structure of the Jacobian implies that there is at most one nonzero element in each row of $J(\underline{x}_k, S)$, and if $\underline{h}$ is a linear combination of the coordinate vectors $\{\underline{e}_i : i \in S\}$, then the single difference equation

$$J(\underline{x}_k)\,\underline{h} \approx \underline{f}(\underline{x}_k + \underline{h}) - \underline{f}(\underline{x}_k) \tag{3.2}$$

provides approximations to all the nonzero elements of $J(\underline{x}_k, S)$. Thus the number of differences that are needed to form J_k is often not much greater than the maximum number of nonzero elements in a row of the Jacobian matrix. This remark applies also to the approximation of $\nabla^2 F(\underline{x}_k)$, except that further savings can be made by using the additional information that $\nabla^2 F(\underline{x}_k)$ is symmetric. These savings are very great when $\nabla^2 F(\underline{x}_k)$ includes a full row, because symmetry allows it to be treated as a full column instead. Further details are given by Coleman and Moré [4].

There is also an extension of formula (3.1), due to Schubert [23], that is suitable for sparse Jacobians. It is derived from the fact that J_{k+1} in expression (3.1) is the matrix that minimizes $\|J_{k+1} - J_k\|_F$ subject to the secant condition (1.9), where the subscript "F" denotes the Frobenius matrix norm

$$\|A\|_F = [\sum_{i=1}^{m} \sum_{j=1}^{n} A_{ij}^2]^{1/2}, \quad A \in \mathbf{R}^{m \times n}. \tag{3.3}$$

Therefore, when J_{k+1} is to inherit a sparsity structure from J_k, we let it minimize $\|J_{k+1} - J_k\|_F$ subject to equation (1.9) and the sparsity conditions. This variational problem is easy to solve as there is no interaction between the rows of J_{k+1}. For $i = 1,2,\ldots,n$, the i-th row of $[J_{k+1} - J_k]$ is the multiple of $\underline{s}_k(i)^T$ that satisfies the secant equation, where $\underline{s}_k(i)^T$ is formed from $\underline{s}_k^T$ by setting to zero the elements that are known to be zero in the i-th row of the Jacobian.

The analogue of expression (3.1) for generating a symmetric matrix G_{k+1} that satisfies equation (1.11) is obtained by minimizing $\{\|G_{k+1} - G_k\|_F : G_{k+1} \in \mathbf{R}^{n \times n}\}$ subject to symmetry and the secant condition. Thus one derives the "symmetric Broyden formula"

$$G_{k+1} = G_k + \frac{\underline{r}_k \underline{s}_k^T + \underline{s}_k \underline{r}_k^T}{\|\underline{s}_k\|_2^2} - \frac{(\underline{r}_k^T \underline{s}_k)\, \underline{s}_k \underline{s}_k^T}{\|\underline{s}_k\|_2^4} \tag{3.4}$$

where $\underline{r}_k = \underline{\gamma}_k - G_k \underline{s}_k$, which occurs in several algorithms for unconstrained optimization. One can also impose sparsity conditions on the matrices $\{G_k : k = 1,2,3,\ldots\}$, and then the result of the variational calculation is that the nonzero elements of G_{k+1} have the form

$$(G_{k+1})_{ij} = (G_k + \underline{\lambda}\, \underline{s}_k^T + \underline{s}_k \underline{\lambda}^T)_{ij}, \tag{3.5}$$

where the vector $\underline{\lambda}$ is defined by the secant equation (1.11) (see Toint [25], for instance).

In the case when each G_k is a full matrix, however, the most successful updating technique is the BFGS formula

$$G_{k+1} = G_k - \frac{G_k \underline{s}_k \underline{s}_k^T G_k}{\underline{s}_k^T G_k \underline{s}_k} + \frac{\underline{\gamma}_k \underline{\gamma}_k^T}{\underline{\gamma}_k^T \underline{s}_k}. \tag{3.6}$$

If the choice of α_k in equation (2.1) gives $\underline{\gamma}_k^T \underline{s}_k > 0$, which is a usual condition of a line search, then G_{k+1} inherits positive definiteness from G_k. Therefore it is usual to combine the BFGS formula with $G_1 = I$ and with the search direction $\underline{d}_k = -G_k^{-1} \underline{g}_k$ for each k. These ingredients provide the most widely used algorithms for unconstrained optimization when it is reasonable to store all of G_k. There is a variational derivation of equation (3.6) [8].

Another highly useful algorithm for unconstrained optimization when first derivatives can be calculated is the conjugate gradient method (see Fletcher [11], for instance). Because it requires only a few vectors of storage, it is particularly valuable when n is large.

In very large calculations, however, substantial reductions in the amount of computation can often be achieved by making use of structure that goes beyond sparsity. The commonly occurring case when $\underline{f}(\cdot)$ or $F(\cdot)$ has the form

$$\underline{f}(\underline{x}) = \sum_{t=1}^{m} \underline{f}_t(\underline{x}), \quad \underline{x} \in \mathbf{R}^n , \tag{3.7}$$

or

$$F(\underline{x}) = \sum_{t=1}^{m} F_t(\underline{x}), \quad \underline{x} \in \mathbf{R}^n , \tag{3.8}$$

is studied carefully by Toint [26] and Griewank and Toint [14, 27]. Each of the functions $\{\underline{f}_t(\cdot)\}$ or $\{F_t(\cdot)\}$ should be highly structured, for example they each depend on only a few components of $\underline{x}$ in finite element calculations, and it should be possible to store their Jacobians or Hessians in a compact form. Then a feature of the "partitioned updating" technique of Griewank and Toint is that, whenever a value of $\underline{f}(\underline{x})$ or $\underline{\nabla F}(\underline{x})$ is calculated, then the auxiliary subroutine provides all the separate vectors $\{\underline{f}_t(\underline{x}) : t = 1,2, \ldots, m\}$ or $\{\underline{\nabla F}_t(\underline{x}) : t = 1,2, \ldots, m\}$, leaving their assembly to the optimization algorithm. Thus approximations to each of the matrices $\{\underline{\nabla f}_t(\cdot) : t = 1,2, \ldots, m\}$ or $\{\nabla^2 F_t(\cdot) : t = 1,2, \ldots, m\}$ are formed, which often provides good accuracy in a relatively small number of iterations.

A few years ago it seemed clear that the formulae (3.1) and (3.6) were best for satisfying the secant equation in nonlinear equation solving and unconstrained optimization respectively. Sparsity and partitioned updating, however, have transformed the range of matrices that have to be estimated, and the trust region techniques of Section 2 increase greatly the range of steps $\underline{s}_k = \underline{x}_{k+1} - \underline{x}_k$ that may occur. Therefore much more research on the approximation of derivative matrices is needed. Further support for this suggestion is given in the next section.

4. Further Topics. Algorithms for unconstrained minimization that do not require the calculation of any derivatives are certainly needed in practice, partly because it is a waste of effort to differentiate analytically (and to program and check the results) in cases when the running time on the computer of the main calculation is going to be negligible, and because software for automatic differentiation is not easily available to most computer users. Therefore several optimization algorithms that usually require first derivatives, including the BFGS method, have been extended to form difference approximations instead. Details can be found in Gill, Murray and Wright [13], for example. The other main approach to derivative-free optimization is to employ sequences of line searches that tend to provide conjugacy properties of the search directions. Powell [18] proposes a method of this type that is still in use. It seems that there have been no major advances in this field during the last 15 years.

I have never liked first derivative difference approximations very much, because I view each difference approximation as a tight cluster of at least $(n + 1)$ function values. Instead, when these clusters are far from the solution, it seems sensible to spread them out in order to provide better coverage of $\mathbf{R}^n$. Therefore one of my students, Jian-wei Zhou, has begun to investigate the use of radial basis function methods (see Powell [20], for instance) to approximate $F(\cdot)$ by interpolation at scattered points, the idea being that each interpolant is a model of the objective function that provides a new trial vector of variables. Numerical experiments with this technique, however, have not yet suggested any promising new algorithms, but they have emphasised a crucial point. It is that a key ingredient of the success of most unconstrained optimization procedures is that they require information about the objective function only along the path that is followed in the search for the minimum, which is a tiny part of $\mathbf{R}^n$. Perhaps, therefore, there is not much future in trying to spread out the points at which $F(\cdot)$ is calculated. I am still very keen, however, to try to supplant difference approximations to derivatives.

The tensor model of Schnabel and Frank [21, 22] for the solution of systems of nonlinear equations gives good numerical results and is particularly suitable when the Jacobian matrix is singular at the solution. It adds to the model (1.5) some terms that are second order in $(\underline{x} - \underline{x}_k)$, so the new model has the form

$$\underline{f}(\underline{x}) \approx \underline{f}(\underline{x}_k) + J_k(\underline{x} - \underline{x}_k) + \tfrac{1}{2}(\underline{x} - \underline{x}_k)^T \, \underline{T}_k \, (\underline{x} - \underline{x}_k)$$

$$= \underline{\mu}_k(\underline{x}), \quad \underline{x} \in \mathbf{R}^n, \tag{4.1}$$

say, where $\underline{T}_k$ is a third order tensor, which can be regarded as an $n \times n$ matrix for each component of $\underline{\mu}_k(\underline{x})$. Having chosen J_k as before, for example it might be the Jacobian matrix $J(\underline{x}_k)$, the tensor is calculated to satisfy the interpolation conditions

$$\underline{\mu}_k(\underline{x}_i) = \underline{f}(\underline{x}_i), \quad i \in I_k, \tag{4.2}$$

where I_k is a subset of $\{1,2,\ldots,k-1\}$. The simplest case is when $(k-1)$ is the only element of I_k, and typically there are at most 3 elements. Thus there is usually much freedom in $\underline{T}_k$, which Schnabel and Frank take up by minimizing the Frobenius norm of $\underline{T}_k$. Hence they derive the form

$$\underline{\mu}_k(\underline{x}) = \underline{f}(\underline{x}_k) + J_k(\underline{x} - \underline{x}_k)$$

$$+ \tfrac{1}{2} \sum_{i \in I_k} [(\underline{x} - \underline{x}_k)^T (\underline{x}_i - \underline{x}_k)]^2 \, \underline{a}_i, \tag{4.3}$$

where the vectors $\{\underline{a}_i : i \in I_k\}$, which depend on k, are defined by the equations (4.2). This new model can be incorporated in any of the algorithms for nonlinear equations that have been mentioned already.

The details of using tensor models, including the automatic choice of I_k, are considered by Schnabel and Frank [21, 22]. We see that the calculation of $\{\underline{a}_i : i \in I_k\}$ is separable. Specifically, for $l = 1,2,\ldots,n$, the l-th elements of $\{\underline{a}_i : i \in I_k\}$ are defined by the l-th components of the conditions (4.2), so there are n square systems of linear equations to be solved, each of order $|I_k|$, and the systems have a common matrix. This approach is very suitable when Jacobian singularity occurs in the original system (1.1), because then the steps $\{\underline{x}_{k+1} - \underline{x}_k : k = 1,2,3,\ldots\}$ tend to lie in the null space of the Jacobian, so second order information is included in the model (4.3) where it is needed. The efficacy of tensor methods is shown clearly by numerical results [21].

Another interesting new idea is the choices of steplength that are proposed and analysed by Barzilai and Borwein [1] for the steepest descent method for unconstrained optimization. This classical method is characterized by $\underline{d}_k = -\underline{\nabla F}(\underline{x}_k)$ in equation (2.1) for every k, and usually the steplength α_k is calculated to (nearly) minimize $F(\underline{x}_{k+1}) = F(\underline{x}_k + \alpha_k \underline{d}_k)$. Almost always, however, the rate of convergence of this procedure is linear and very slow. Therefore Barzilai and Borwein [1] seek choices of α_k that give a superlinear rate of convergence, and that can be applied in practice without the computation of second derivatives.

They take the view that a steepest descent iteration is equivalent to letting $\underline{x}_{k+1}$ minimize the quadratic model (1.7) when G_k is a multiple of the unit matrix, and they generate suitable multipliers for $k \geq 2$ from the secant equation $G_k \underline{s}_{k-1} \approx \underline{\gamma}_{k-1}$, where we are using the notation of equation (1.11). Specifically, they let $G_k = \lambda_k I$, where $\lambda_k \in \mathbf{R}$ minimizes $\|\underline{s}_{k-1} - G_k^{-1}\underline{\gamma}_{k-1}\|_2$ or $\|G_k \underline{s}_{k-1} - \underline{\gamma}_{k-1}\|_2$. Hence they derive the steplengths

$$\alpha_k = (\underline{s}_{k-1}^T \underline{\gamma}_{k-1}) / \|\underline{\gamma}_{k-1}\|_2^2, \quad k = 2,3,4,\ldots, \tag{4.4}$$

or

$$\alpha_k = \|\underline{s}_{k-1}\|_2^2 / (\underline{s}_{k-1}^T \underline{\gamma}_{k-1}), \quad k = 2,3,4,\ldots, \tag{4.5}$$

respectively, for use in the formula

$$\underline{x}_{k+1} = \underline{x}_k - \alpha_k \, \underline{\nabla} F(\underline{x}_k) . \tag{4.6}$$

They show that R-superlinear convergence is achieved for some simple objective functions. Thus the exciting possibility exists that the usefulness of the steepest descent method can be improved greatly. However, extending the choices (4.4) or (4.5) to force convergence from poor starting points may be a severe difficulty, because the technique that provides R-superlinear convergence does not guarantee that the sequence $\{F(\underline{x}_k) : k = 1,2,3,\ldots\}$ decreases monotonically for sufficiently large k.

Another important topic in unconstrained optimization is consideration of updating methods to replace the much used BFGS formula (3.6). This subject is raised now because Conn, Gould and Toint [5] found recently that in several calculations the rank one formula

$$G_{k+1} = G_k + \frac{(\underline{\gamma}_k - G_k \, \underline{s}_k)(\underline{\gamma}_k - G_k \, \underline{s}_k)^T}{(\underline{\gamma}_k - G_k \, \underline{s}_k)^T \underline{s}_k} \tag{4.7}$$

is much more successful than BFGS for revising second derivative approximations. These calculations include trust regions for controlling the lengths of trial steps and simple bounds on the variables, so they are more relevant to the remarks at the end of Section 3 than to the discussion that follows equation (3.6). There are 50 test problems, each with 5 methods for generating the matrices $\{G_k : k = 1,2,3,\ldots\}$ but now we note only the comparison of expressions (3.4), (3.6) and (4.7), the BFGS formula being replaced by $B_{k+1} = B_k$ if $\underline{s}_k^T \underline{\gamma}_k \leq 0$, which happened on only 6 of the calculations. The symmetric Broyden, BFGS and rank one updating techniques require fewest iterations on 4, 13 and 38 of the 50 test problems respectively (the total count exceeds 50 because of ties). Therefore the numerical evidence in favour of the rank one formula is very strong.

When expression (4.7) was first proposed it caused much excitement because, if $F(\cdot)$ is quadratic, if the formula is applied n times in exact arithmetic, and if the steps $\{\underline{s}_k : k = 1,2,\ldots,n\}$ have *any* values that avoid $(\underline{\gamma}_k - G_k \underline{s}_k)^T \underline{s}_k = 0$, then $G_{n+1} = \nabla^2 F(\cdot)$. Perhaps this is the explanation of the above numerical results, but it is not entirely clear that $G_{n+1} = \nabla^2 F(\cdot)$ is an advantage. Some doubt arises if one considers the attainment of this condition when G_1 is very different from $\nabla^2 F(\cdot)$ and the directions $\{\underline{s}_k : k = 1,2,\ldots,n\}$ almost lie in a proper subspace of $\mathbf{R}^n$. In this case G_{n+1} must be highly sensitive to changes in the directions $\{\underline{s}_k\}$, so its calculation by the rank one formula is very ill-conditioned. Hence this is another example of a field where further research could be very useful.

5. Available Software. This section mentions and discusses briefly the Fortran subroutines that are available in the Harwell, IMSL, NAG and MINPACK libraries for the solution of systems of nonlinear equations and for unconstrained optimization calculations. We consider the range of algorithms that are included in the catalogues of these libraries [16, 28, 29, 30], rather than the quality of the software, because the author has had no practical experience with the NAG and MINPACK routines.

MINPACK is devoted to nonlinear equations and nonlinear least squares problems in many variables, but the other three libraries all include special software for the $n = 1$ case. Specifically, they provide routines for solving $f(x) = 0$ when values of $\{f(x) : x \in \mathbf{R}\}$ can be calculated for any x, and for minimizing $\{F(x) : x \in \mathbf{R}\}$ in the two cases when either $F(x)$ or $F(x)$ and $\nabla F(x)$ are available for any x. Of course there are also procedures for finding roots of polynomials, and IMSL gives some attention to the solution of $f(z) = 0$ when $f(\cdot)$ is a general analytic function.

The procedures for systems of nonlinear equations employ the model (1.5) and trust regions that bound $\|\underline{x}_{k+1} - \underline{x}_k\|_2$. Routines of this type are available in all four libraries in the two cases when first derivatives are known and when J_k has to be estimated automatically. In the latter case finite differences followed by Broyden's formula are applied, as suggested in the paragraph that includes equation (3.1). Here there are few differences between the software of the four libraries, except that the Harwell routines with exact Jacobians are designed to take advantage of sparsity. Harwell also offers a procedure that can exploit sparsity of the Jacobian when first derivatives are estimated automatically.

The usual approach to overdetermined equations, $\underline{f}(\underline{x}) \approx \underline{0}$ where $\underline{f}$ is from $\mathbf{R}^n$ to $\mathbf{R}^m$ with $m > n$, is to minimize the sum of squares $\{\|\underline{f}(\underline{x})\|_2^2 : \underline{x} \in \mathbf{R}^n\}$. We have not had time to consider algorithms for this calculation, but we note in passing that each of the four libraries provides suitable routines. Further, the minimization of $\{\|\underline{f}(\underline{x})\|_\infty : \underline{x} \in \mathbf{R}^n\}$ receives attention in the Harwell library.

As mentioned already, MINPACK does not yet include any software for the minimization of general functions, but the other three libraries offer subroutines in each of the following categories: (1) first and second derivatives available, (2) first derivatives available, and (3) only function values available. In Category 1 all three libraries employ extensions of the Newton iteration, adding a diagonal matrix to G_k if necessary in order to achieve positive definiteness. This diagonal matrix is a multiple of the unit matrix in the Harwell and IMSL procedures, but the technique of Gill and Murray [12] is used in NAG. Both IMSL and NAG provide versions of these routines in Category 2, the second derivatives being estimated by differencing first derivatives.

The BFGS algorithm is prominent in the Category 2 software of the three libraries. Harwell includes too a variable metric procedure that employs the hybrid updating formula of Fletcher [10], a trust region routine that is based on the symmetric Broyden formula (3.4), and software for the partitioned case (3.8). Further, all three libraries offer conjugate gradient subroutines for unconstrained minimization.

The use of finite difference approximations to derivatives provides Category 1 software for all the algorithms of the previous paragraph, except the symmetric Broydena and conjugate gradient subroutines in the Harwell library. However, Harwell does include a Fortran version of the conjugate direction method of Powell [18]. Moreover, both IMSL and NAG contain an implementation of the Nelder and Mead simplex algorithm [17], which is a highly successful direct search method for minimization without derivatives.

Most of the routines that have been mentioned are either more than seven years old or could have been written in 1980. This remark reflects the fact that many people who provide optimization software have been engaged recently on implementations of algorithms for constrained calculations. Sparse Hessians and Jacobians, however, certainly deserve more attention from libraries. For example, it would be valuable to include the software for difference approximations to sparse Hessians that was developed by Coleman, Garbow and Moré [3]. Further, it is now clear that the tensor method of Schnabel and Frank [21, 22] is highly useful. Therefore, even if we exclude the promising observations at the end of Section 4, we can look forward to substantial improvements to the range of nonlinear equations and unconstrained optimization calculations that can be solved by the software of subroutine libraries.

REFERENCES

[1] J. BARZILAI AND J. M. BORWEIN, *Two-point step size gradient methods,* IMA J. Numer. Anal., 8 (1988), pp. 141-148.

[2] R. H. BYRD, R. B. SCHNABEL AND G. A. SHULTZ, *Approximate solution of the trust region problem by minimization over two-dimensional subspaces,* Math. Programming, 40 (1988), pp. 247-263.

[3] T. F. COLEMAN, B. S. GARBOW AND J. J. MORÉ, *Software for estimating sparse Hessian matrices*, Tech. Memo. 43, Mathematics and Computer Science Division, Argonne National Laboratory, Illinois, 1984.

[4] T. F. COLEMAN AND J. J. MORÉ, *Estimation of sparse Hessian matrices and graph coloring problems*, Math. Programming, 28 (1984), pp. 243-270.

[5] A. R. CONN, N. I. M. GOULD AND Ph. L. TOINT, *Testing a class of methods for solving minimization problems with simple bounds on the variables*, Report 86/3, Department of Mathematics, University of Namur, Belgium, 1986.

[6] A. R. CURTIS, M. J. D. POWELL AND J. K. REID, *On the estimation of sparse Jacobian matrices*, J. Inst. Math. Appl., 13 (1974), pp. 117-119.

[7] R. S. DEMBO AND T. STEIHAUG, *Truncated-Newton algorithms for large scale unconstrained optimization*, Math. Programming, 26 (1983), pp. 190-212.

[8] J. E. DENNIS AND R. B. SCHNABEL, *Numerical Methods for Unconstrained Optimization and Nonlinear Equations*, Prentice-Hall, Englewood Cliffs, New Jersey, 1983.

[9] I. S. DUFF, J. NOCEDAL AND J. K. REID, *The use of linear programming for the solution of sparse sets of nonlinear equations*, SIAM J. Sci. Statist. Comput., 8 (1987), pp. 99-108.

[10] R. FLETCHER, *A new approach to variable metric algorithms*, Comput. J., 13 (1970), pp. 317-322.

[11] R. FLETCHER, *Practical Methods of Optimization, Vol. 1: Unconstrained Optimization*, John Wiley & Sons, Chichester, England, 1980.

[12] P. E. GILL AND W. MURRAY, *Newton type methods for unconstrained and linearly constrained optimization*, Math. Programming, 7 (1974), pp. 311-350.

[13] P. E. GILL, W. MURRAY AND M. H. WRIGHT, *Practical Optimization*, Academic Press, London, 1981.

[14] A. GRIEWANK AND Ph. L. TOINT, *On the unconstrained optimization of partially separable functions*, in *Nonlinear Optimization 1981*, M. J. D. Powell, ed., Academic Press, London, 1982, pp. 301-312.

[15] J. J. MORÉ, *Recent developments in algorithms and software for trust region methods*, in *Mathematical Programming: The State of the Art*, A. Bachem, M. Grötschel and B. Korte, eds., Springer-Verlag, Berlin, 1983, pp. 258-287.

[16] J. J. MORÉ, B. S. GARBOW AND K. E. HILLSTROM, *User guide for MINPACK-1*, Report ANL-80-74, Argonne National Laboratory, Illinois, 1980.

[17] J. A. NELDER AND R. MEAD, *A simplex method for function minimization*, Comput. J., 7 (1965), pp. 308-313.

[18] M. J. D. POWELL, *An efficient method for finding the minimum of a function of several variables without calculating derivatives*, Comput. J., 7 (1964), pp. 155-162.

[19] M. J. D. POWELL, *A hybrid method for nonlinear equations*, in *Numerical Methods for Nonlinear Algebraic Equations*, P. Rabinowitz, ed., Gordon and Breach, London, 1970, pp. 87-114.

[20] M. J. D. POWELL, *Radial basis functions for multivariable interpolation: a review*, in *Algorithms for Approximation*, J. C. Mason and M. G. Cox, eds., Clarendon Press,

Oxford, 1987, pp. 143-167.

[21] R. B. SCHNABEL AND P. D. FRANK, *Tensor methods for nonlinear equations,* SIAM J. Numer. Anal., 21 (1984), pp. 815-843.

[22] R. B. SCHNABEL AND P. D. FRANK, *Solving systems of nonlinear equations by tensor methods,* in *The State of the Art in Numerical Analysis,* A. Iserles and M. J. D. Powell, eds., Clarendon Press, Oxford, 1987, pp. 245-271.

[23] L. K. SCHUBERT, *Modification of a quasi-Newton method for nonlinear equations with a sparse Jacobian,* Math. Comp., 24 (1970), pp. 27-30.

[24] G. A. SHULTZ, R. B. SCHNABEL AND R. H. BYRD, *A family of trust-region-based algorithms for unconstrained minimization with strong global convergence properties,* SIAM J. Numer. Anal., 22 (1985), pp. 47-67.

[25] Ph. L. TOINT, *On sparse and symmetric matrix updating subject to a linear equation,* Math. Comp., 31 (1977) pp. 954-961.

[26] Ph. L. TOINT, *Numerical solution of large sets of algebraic nonlinear equations,* Math. Comp., 46 (1986), pp. 175-189.

[27] Ph. L. TOINT AND A. GRIEWANK, *Numerical experiments with partially separable optimization problems,* in *Numerical Analysis Proceedings, Dundee 1983,* Lecture Notes in Mathematics 1066, D. F. Griffiths, ed., Springer-Verlag, Berlin, 1984, pp. 203-220.

[28] *Harwell Subroutine Library: a Catalogue of Subroutines,* Report AERE-R-9185 (Seventh edition), Computer Science and Systems Division, Harwell Laboratory, England, 1987.

[29] *IMSL Math/Library: Users' Manual,* IMSL, 2500 Park West Tower One, Houston, Texas, 1987.

[30] *NAG Fortran Mini Manual Mark II,* Numerical Algorithms Group, 256 Banbury Road, Oxford, England, 1984.

PART III:
MINISYMPOSIA ABSTRACTS

Minisymposia Abstracts*

• COMPUTATIONAL AND GEOMETRIC ASPECTS OF ROBOTICS

V. AKMAN (Centrum voor Wiskunde en Informatica Amsterdam, The Netherlands)

Robotics is crucial in the reindustrialization of the world. This minisymposium will concentrate on the algebraic and geometric aspects of the family of problems of the sort:

"Given a set of polyhedra and two points, calculate the shortest path between these points under a suitable metric, constrained to avoid intersections with the given objects."

To this end, we shall look at the powerful ideas such as "configuration spaces" and "locus techniques." Our aim is to illustrate the use of classical algebra and computational geometry in path planning and optimization issues arising in robotics.

Motion Planning with Algebraic Objects

C. Bajaj (Purdue University)

Using configuration space (C-space) to plan collision free motion for a single rigid object amongst physical obstacles reduces the problem to planning motion for a mathematical point amongst "grown" C-space obstacles (the points in C-space which correspond to the object overlapping one or more obstacles). Efficient algorithms shall be presented to generate these C-space obstacles for algebraic objects translating in two and three dimensions. Crucial too here is the internal representation of algebraic curves and surfaces, that is whether they are implicitly or parametrically defined. Algorithms which allow one to convert between these two representations shall also be discussed.

Shortest Rectilinear Paths Among Obstacles

J.S.B. Mitchell (Cornell University)

An important problem in robotics and in wire routing is to find a shortest rectilinear path between a source and a destination in the presence of disjoint obstacles in the plane. We give an algorithm that runs in subquadratic time, building a planar subdivision such that, by locating a query point in the subdivision, the length of the shortest path from the source to the query point can be reported in time $O(\log n)$, and a shortest path can be reported in time $O(k + \log n)$, where k is the number of "turns" in the path. The algorithm generalizes to the case of fixed orientation metrics and to the case of multiple source points to build a Voronoi diagram in the presence of obstacles.

---

*The following information was taken from the ICIAM '87 Final Program and Abstracts, which was published prior to the conference. Any changes to the program after that time are not reflected in these Proceedings.

The summary abstract for each minisymposium is designated by a bullet, and the participants' abstracts follow the chairperson's abstract. This listing is ordered alphabetically by chairperson's last name.

Recent Developments in Algorithmic Motion Planning

G.T. Wilfong (AT&T Bell Laboratories, USA)

Researchers have been studying the complexity issues concerning the problem of finding collision-avoiding paths for various objects amongst obstacles for several years. A very general version of the problem was shown to be decidable by Schwartz and Sharir. Since that time many algorithms have been developed to solve special cases and the techniques used have allowed for more efficient algorithms than the more general techniques of Schwartz and Sharir.

More recently there has been an interest in developing algorithms for motion planning in the presence of kinematic constraints. I will overview some of the algorithms for velocity and acceleration bounded motion as well as discuss a solution to the problem of finding a motion when the moving object has a bound on the curvature of its path.

Moving a Disc Between Convex Polygons

H. Rohnert (Universitat Saarbrucken)

We consider a 2-dimensional variant of the piano mover's problem: Given are f dsijoint, convex polygons with a total of n vertices figuring as obstacles. Answer on-line questions of the form: Can a disc of diameter r be moved from point a to point b? Therefore we preprocess the obstacle space and answer one query for specified diameter, start and destination point in time O(log n) if a yes/no answer is required and in time O(k = log n) if a feasible path is required. Here k is the length of the description of the returned path. We do not find a shortest path, rather just a path avoiding the obstacles. But due to the construction the returned path has another useful property which may be important for practical applications: The minimum distance between the disc and the obstacles during the motion is maximal among all feasible paths. Space complexity is O(n). The pre-processing time is O(n log n).

• INVARIANCE AND VIABILITY FOR UNCERTAIN CONTROL PROBLEMS

J.P. AUBIN (Universite Paris IX, Dauphine-Ceremade)

Facing uncertainty control or coping with disturbances become more central issues in control theory. Several techniques can be proposed: invariance using geometric techniques, Liapunov functions revisited, viability and differential calculus of set-valued maps. The proposed talks of this Minisymposium use these various techniques to answer these basic questions.

Singular and Global non Interacting Control Problems

C.I. Byrnes (Arizona State University)

In this talk, we give a survey of the geometric approach to nonlinear control problems such as disturbance decoupling which involve the notions of controlled invariant submanifolds, etc.

After presenting the basic local theory which was developed in the late 1970's under the appropriate regularity hypotheses, we turn to problems which involve singularities and/or global considerations. In the context of several explicit examples, we will indicate how methods from the theories of singularities, noncommutative differential geometry and nonsmooth analysis should play a role in the future development of the subject. As an illustration we solve a singular disturbance decoupling problem for a rigid satellite.

Stabilization of Uncertain Systems: The Effects of Neglected Dynamics

G. Leitmann (University of California, Berkeley), E.P. Ryan (University of Bath)

We consider a class of uncertain dynamical systems which can be decomposed into two coupled subsystems by means of a singular perturbation parameter, μ. We determine a range of this parameter, $\mu \in$ $(0,\mu^*)$, and a corresponding feedback control (possibly depending on the states of both subsystems), which is deduced from a stabilizing control of the uncertain reduced order system ($\mu = 0$) and which assures stability of the full order system ($(0,\mu^*)$).

Contingent Hamilton-Jacobi Equation
H. Frankowska (Universite de Paris-Dauphine)

The value function of an optimal control problem is in general not differentiable. To apply the dynamic programming verification technique, the classical Hamilton-Jacobi equation is replaced by contingent inequalities. Solutions to contingent inequalities are monotone along the trajectories of dynamical systems and allow one to construct optimal solutions through the corresponding optimal feedback.

A Viability Approach to Differential Games
J.P. Aubin (Universite de Paris-Dauphine)

We consider a two-person differential game. We are looking for tubes invariant for one player and viable for the other. These results are applied to a control system under uncertainty, regarded as a player choosing disturbances.

Funnel Equations for Viable Solutions to Differential Inclusions
A. Kurzhanski (International Institute for Applied Systems Analysis, Laxenburg, Austria)

One of the problems relevant to guaranteed estimation and control and also to viability theory is to specify the set of all solutions to a differential inclusion that satisfy an additional state constraint generated by a finite measurement equation. The evolution of the cross-sections of this "informational tube" turns to be governed by a "funnel equation" for set-valued functions. The solution to the latter coincides in the linear-convex case with a respective convolution integral. The "funnel equations" then turn to be the basis for set-valued approximations and other numerical approaches to viability problems.

• STOCHASTIC IMAGE ANALYSIS
R. AZENCOTT (Universite Paris Sud)

In the last three years, sophisticated stochastic methods have been actively explored in imagery involving concepts such as Gibbs fields and simulated annealing (Geman), probabilistic formalization of interaction between multilevel descriptions (Gremander), infinite dimensional diffusions and Ornftein-Uhlendeck process (Mitter), etc. Some of these mathematical tools may seem too refined to seasoned image specialists, but emerging applications should soon help to sort out theoretical excesses. In the meanwhile, even classical imagery feels the need for improved handling of noisy low-level primitive detectors, and image analysis by nets of interacting processors seems a promising field for more dazzling concepts.

A Relational Model for Shape Recognition
E. Bienenstock (Universite de Paris-Sud), S. Geman (Brown University)

The "Bayes-Gibbs" framework for image processing, which combines Bayesian decision theory with the use of the well-known Gibbs-Boltzmann distribution of statistical physics, has been previously applied to low-level vision problems such as image restoration and segmentation (S. Geman and D. Geman, 1984). We now address within the same framework, but using a three-layer model, the higher-level problem of shape-recognition. An important feature of our approach is that it focuses on relational information, i.e. information pertaining to relationships between elements of the shape under consideration. Another important feature is the simultaneity of processing at all levels. Early experiments suggest that the model is extremely robust, accommodating various degradations, as well as changes in brightness, contrast, texturing, and position.

Determination of Spatio-Temporal Primitives In Image Sequences Based on Likelihood Techniques
P. Bouthemy (INRISA-INRIA Rennes, France)

Image understanding schemes require one to extract and to structure low-level intrinsic features from images. In the context of dynamic scene analysis, this problem can be stated as a spatio-temporal segmentation issue. To solve it, two cooperative approaches can be followed, contour-based and region-based ones. We describe methods which can handle various practical situations in both cases, but for each through a well-defined single formalism owing to some modeling principles and likelihood-based frameworks. A spatio-temporal or moving edge is locally modeled as a surface patch in the (x,y,t) space, e.g. planar patch. An hypothesis testing approach leads to the definition of a generalized likelihood test implemented according to appropriate mask convolution. The region segmentation algorithm takes into account 2D motion field models in a hierarchical way. It relies on an explicit but partial motion information. The partition criterion is expressed by a likelihood ratio test within a split-and-merge procedure.

Experiments in Texture Analysis and Segmentation
C. Graffigne, S. Geman (Brown University)

We consider the problem of texture segmentation from two points of view. In the first, we segment a picture containing previously learned textures. We use a two-tiered Markov Random Field (MRF) model with pixel grey levels at one level, and texture labels at the other. The learning step consists of estimating the MRF parameters for each texture. In the second case, the goal is segmentation without labelling, and requires no prior training. Again, we use a MRF model, this time on region-indicators and a set of statistical features that differentiate texture types. In this second case, the MRF neighborhood system defines a random graph, allowing us to give the same label to disconnected, but homogeneous, regions.

On Stochastic Quantization
S.K. Mitter (M.I.T.), V. Borkar (T.I.F.R., Bombay, India), R. Chari (Tufts University)

Using the theory of Dirichlet forms a distribution valued symmetric Markov process is constructed so that it has the phi-four-two measure as its invariant probability measure. This is done both in finite and infinite volume frameworks. Several properties (path continuity, ergodicity, etc.) of this process are proved and a finite to infinite volume limit theorem is established.

The relevance of this work to the probabilistic analysis of images is discussed.

• PHASE TRANSITIONS IN CONTINUA: NONLINEAR THEORY AND COMPUTATION
J. BALL (Heriot-Watt University), D. KINDERLEHRER (University of Minnesota)

The development and analysis of nonlinear constitutive theory and the use of numerical methods to reconcile experiment with theory are the themes of this minisymposium. In recent years, principles have been proposed to explain phase transitions and defect structures in the context of continuum theory which attempt to be consistent with statistical theory. Computation of 3-D configurations which admit this singular behavior and their state functions is underway.

This is an opportune time to survey the present situation and to assess the future.

Transitions and Defects in Ordered Materials: Nonlinear Theory and Computation
R. Hardt, D. Kinderlehrer, M. Luskin, J.L. Ericksen, R. James
(University of Minnesota)

Theory and computational results for transitions and defects in crystals and liquid crystals will be presented. Continuum theories and their analyses will be given for the twinning transition in crystals. These theories lead to the minimization of non-convex and non-lower-semi-continuous functionals. Numerical approaches to these theories will be presented.

Recent analytical and computational results by the authors and Brezis, Leib, and Coron which have led to an understanding of stable defects in liquid crystals will be given.

Fundamental Problems in the Statistical Mechanics of Elastic Solids
O. Penrose (Heriot-Watt University)

Mathematical statistical mechanics contains theorems about equilibrium which correspond closely to reality in the case of fluid systems, but less so in the case of solids. For example it is not entirely clear how a finitely strained crystal, or a pair of (unstrained) twinned crystals, can be fitted into the supposedly rigorous derivation of thermodynamics from first principles, based on the Gibbs formula relating the free energy to the partition function; nor do we fully understand the restrictions placed by general principles of statistical mechanics on the class of possible free energy functions for a strained crystal. This talk will summarize the present state of the subject and draw attention to some of the outstanding unsolved problems.

Phase Transitions in Crystals-Mathematical Considerations
I. Fonseca (Ecole Polytechnique)

Under deformation some materials with crystalline structure show several twin-related phases; their arrangement can be altered by application of stresses or changes of temperature. We describe them by a continuum theory, proposed by ERICKSEN, based on nonlinear thermoelasticity and on the invariance of the energy density with respect to change of lattice basis together with frame indifference. This gives rise to highly unstable variational problems: under dead loading and at constant temperature, only residual stresses can provide a global minimum of the energy functional and there are severe restrictions for metastable configurations. The subenergy function of ERICKSEN & FLORY plays an important role in the thermodynamics of these materials.

Numerical Solutions to the Cahn-Hilliard Equation
D.W. Heermann (Johannes-Gutenberg Universitat)

The Cahn-Hilliard equation is pertinent to many areas in physics. It is a highly non-linear partial differential equation describing, for example, the evolution of a system from a nonequilibrium state to a state of equilibrium. In this contribution we present solutions to this equation obtained by a novel Monte Carlo method. The solutions show that the approach to equilibrium is characterized by a power law behavior.

Glimm's Scheme vs Phase Transition
D. Serre (UFR de Sciences et Technique)

The equations given by classical thermodynamics allow the state of a fluid to take unstable values. The Word Unstable has two meanings: 1. The hyperbolicity of the system of PDE is lost, thus the system become ill-posed. 2. These values are not physically relevant as shown by experience.

So, PDE's must be dropped at phase transition. But the Riemann problem continues to agree to experience. It allows us to use the Glimm's scheme, which gives rather interesting computed values. This scheme must be used in its general form, instead of the usual one.

• **INVERSE PROBLEM FOR THE 2D AND 3D WAVE EQUATION**
A. BAMBERGER (Ecole Polytechnique), G. CHAVENT (Universite Paris IX - Dauphine)

In many applications, it is improtant to solve the inverse problem for the 2D and 3D wave equation. It is also a difficult and exciting problem from the numerical and mathematical point of view... The objective of that minisymposium is to present recent progress in this field. Different approaches to solve the direct and the inverse problem shall be presented and discussed. Theoretical aspects shall also be investigated.

Inversion of the 2-D Wave Equation for Plane-Layer Media
P. Kolb (ELF-Aquitaine France)

We want to make a quantitative analysis of horizontally stratified media using seismic surface data and solving the nonlinear 2-D acoustic inverse problem: we recover the complete parameter set which minimizes the energy of the residuals between the corresponding numerical wave equation solution and the actually observed data.

In order to improve the efficiency of the method, great attention must be paid to the optimization strategy. We will discuss different features of this problem and illustrate them by some numerical results.

A Solution of the 3-D Born Inversion Problem for Electromagnetic Waves in Lossless Media

B.C. Levy (Massachusetts Institute of Technology)

The inverse scattering problem for a 3-D smoothly varying lossless electromagnetic (EM) medium whose permittivity $\varepsilon(\bar{r})$ and permeability $\mu(\bar{r})$ vary with all space dimensions is formulated by using the variable background Born approximation. The medium is probed by a wideband electric dipole and the scattered EM field is recorded along a receiver surface located outside the domain where the perturbations $\delta\varepsilon$ and $\delta\mu$ of the permittivity and permeability are nonzero. The solution that is proposed relies on the introduction of a backpropagated EM field. This field is defined by using a vector Kirchhoff formula, but it is computed by first deconvolving and filtering the scattered EM field and then using a finite-difference scheme backwards in time to propagate it back into the medium. The backpropagated field is imaged at the source travel times, giving a vector image which is then used to recover $\delta\varepsilon$ and $\delta\mu$.

Seismic Modeling and its Role in Seismic Inversion

K.J. Marfurt (AMOCO Production Company, USA), C.S. Shin (University of Tulsa)

Parameter inversion, commonly used to interpret low frequency and DC nonseismic data where imaging principles do not apply, holds several advantages over the high frequency inverse imaging techniques. As in geotomography, there is no explicit backward propagation of the wave field, such that the receiver aperture and intergroup spacing limitations that plague inverse imaging techniques are minimized. More importantly, a priori constraints further regularize and drive the solution to a geologically acceptable answer. The sensitivity of the error between the measured and model results at any iteration is determined by the Jacobian matrix. For this reason the frequency domain finite element forward modeling technique is well suited computationally to drive the inverse problem.

Recent Progress in the Mathematics of Reflection Seismology

P. Sacks (Iowa State University), F. Santosa (University of Delaware), W.W. Symes (Rice University)

We review a number of recent results concerning the mathematical structure of the reflection seismology problem. These include (i) the role of medium roughness and data aperture in band-extrapolation; (ii) the determination of layered elastic models from single component surface measurements; (iii) the codetermination of the source time-dependence together with a layered model from a point-source response; (iv) the effect of incident wavefield caustics in the high-frequency asymptotics of the (nonlayered) acoustic perturbational response; (v) the spectral and approximative properties of perturbational seismograms in 2- and 3-dimensions.

Results in the Linearized Inversion Problem in Seismic Prospecting

A. Bourgeois, B.F. Jiang (Institut Francais du Petrole)

For methods of inverting seismograms in seismic prospecting, the problem comes up of knowing to what extent the reference environment must be close to the actual environment for the linearized approach to be justified. This paper describes one- and two-dimensional numerical results for the modeling of the wave-propagation equation that has been linearized in relation to different parameters, i.e. velocity, density, impedance and bulk modulus.

• NUMERICAL METHODS FOR NONLINEAR TRANSPORT EQUATIONS

C. BARDOS (Ecole Normale Superieure)

The aim of this minisymposium is to present numerical methods for solving nonlinear transport equations: Monte Carlo methods, particle methods, finite element methods.

We intend to emphasize the numerical problems related to the radiative transfer equations, the Boltzmann equations and some boundary layer problems. Indeed, these problems are of particular interest if one approximates transport equations by diffusion or fluid equations (which arises classically when the mean free path of the particles goes to zero).

Convergent Simulation Scheme for the Boltzmann Equation

H. K. Babovsky (University of Kaiserslautern)

In many applications the modified Nambu for the simulation of the Boltzmann Equation has proved a reasonable alternative to the usually applied Monte Carlo scheme.

It may be interpreted as a method to generate paths of a certain Markov process associated to the time discretised Boltzmann equation.

Recently we have been able to prove convergence of solutions of the simulation scheme to those of the Boltzmann equation. In the space independent case with fixed time step Δt it has been shown that the simulated particles system approximate weakly the solution of the time discretised Boltzmann Equation as n tends to infinity. For the full Boltzmann Equation a similar result holds provided there exists a solution with certain regularity conditions.

The stochastic part in Nambu algorithm may be interpreted as a method to construct approximation of certain product measure from given n points distribution. This point of view allows one to introduce certain modifications. In particular one may even construct a completely deterministic version which also converges.

Direct Simulation Monte Carlo (DSMC) Based on Kac-Prigogine Master Equation

S.M. Deshpande (Indian Institute of Science)

The DSMC is a statistical practice-in-cell method meant for obtaining numerical solution of the Boltzmann equation. It exploits the connection between the Kac-Prigogine master equation and the Boltzmann equation. The Master equation reduces to the Boltzmann equation under the hypothesis of molecular chaos, and the chaos if initially valid is perpetuated by the Master equation. The collision term of the Boltzmann equation is simulated by calculating the expected number of collisions per cell and then allowing that many collisions to occur among the particles in the cell. The collision simulation has been extended to multicomponent gas mixtures. It is shown that a proper sequence of various types of collisions is necessary to avoid biased results.

Numerical Schemes for Radiative Transfer Equations

R. Sentis (C.E.A. Limeil-Valenton)

We first consider a simplified form of the radiative transfer equations, that is to say, a nonlinear system of diffusion equations. The first aim of this talk is to describe an iterative scheme for solving this system such that the first iteration gives a result which is very close to the true solution (at least in the very opaque domains).

We will also talk about the methods to deal with the actual form of the equations (especially a "particular corrector" to take into account the angular dependence of the radiative intensity).

Round-Table: Half space problems. Applications to boundary Conditions for Macroscopic Equations

H. Neunzert (University of Kaiserslautern), F. Coron, F. Golse, B. Perthame (Centre de Mathematiques Appliquees)

Dans cette table ronde, on se propose de faire le point sur les theoremes d'existence d'unicite et de comportement asymptotique des solutions d'equations cinetiques dans un demi-espace. On donnera des methodes de calcul numerique, en particulier methodes variationnelles; on discutera de l'application de ces resultats au calcul des conditions aux limites pour les equations macroscopiques.

• MATRIX PROBLEMS IN SYSTEMS THEORY

S. BARNETT (University of Bradford)

During the past two decades a large literature has been developed on applications of matrices to linear systems (see S. Barnett, Matrices in Control Theory, Van Nostrand Reinhold, 1971; Polynomials and Linear Control Systems, Dekker, 1983). This minisymposium brings together some recent British developments. The first two contributions relate to systems in state-space form: That by Aasaraai and Fletcher studies state feedback for descriptor systems subject to disturbances; Gover and Barnett present a generalization of the Bezoutian matrix, which plays a crucial role in many systems concepts. The other two papers use the polynomial matrix representation of multivariable systems, and address problems of equivalence, and invariance under state feedback.

Disturbance Decoupling for Descriptor Systems

A. Aasaraai, L.R. Fletcher (University of Salford)

The disturbance decoupling problem concerns a time-invariant linear multivariable control system

$$E\dot{x}(t) = Ax(t) + Bu(t) + Sq(t), \quad x(0) = x_0$$
$$y(t) = Cx(t)$$

The questions to be answered are: "When and how can a state feedback $u = -Fx$ be constructed so that y is independent of q?"

The state space version of this problem - that is, when A is a square matrix and $E = I$ - has had a great influence on the development of control theory. The generalisation to the case of singular E requires a more abstract approach to the solution of the system of differential equations

$$E\dot{x} = Mx + f(t)$$

then the usual application of the Kronecker canonical form.

A Generalized Bezoutian Matrix

M.J.C. Gover, S. Barnett (University of Bradford)

Bezoutian matrices play an important role in linear systems theory, providing links between such topics as controllability and observability, canonical forms, realization, root location for polynomials, and Hankel matrices. The bezoutian matrix is defined in terms of two polynomials, and the proposed generalised bezoutian is associated with two bivariate functions of the form

$$f(z, \bar{z}) = \sum_{k=0}^{n} \alpha_k \prod_{r=1}^{k} c^r(z)$$ where $c(\ )$ is a conjugate

operator. It is shown that some of the properties and applications of the standard case extend to this new matrix, including a generalisation of a related Sylvester matrix and an extension of the idea of the greatest common divisor.

A Complete Frequency Structure Preserving Transformation for Linear Systems

G.E. Hayton, A. Walker (The University of Hull), A.C. Pugh (Loughborough University of Technology)

Interest in the dynamic behaviour of linear time-invariant systems as frequency tends to infinity has led to the investigation of system transformations preserving such behaviour. Transformations of extended state-space systems are well understood but extensions to general polynomial systems have proved difficult. This paper describes an equivalence relation, "full equivalence", for polynomial matrices which simultaneously preserves finite and infinite zeros. From this the appropriate system transformation, "full system equivalence", is derived and shown to preserve both finite and infinite dynamic structure. Particularly it is demonstrated that any polynomial system matrix is fully system equivalent to one in extended state-space form.

A Local Theory of Linear Systems and Applications

A.C. Pugh, A. Kafai (Loughborough University of Technology), G.E. Taylor (University of Hull)

A generalisation to nonproper systems of a well-known result concerning the invariance under state feedback of the infinite zero structure of a strictly proper transfer function matrix is presented. The proof indicates the requirement for a highly localised approach to the study of linear multivariable systems. Specifically an analysis is required of the system structure at a single frequency $s = s$ to the exclusion of all other frequencies $s \in \mathbb{C} \cup \{\infty\}$. The relevant results of this theory are developed among which is an interesting local state-space realisation of a given transfer function matrix.

- **RELIABILITY AND VULNERABILITY IN NETWORKS**

L.W. BEINEKE (Indiana University-Purde University), J.C. BERMOND (Universite Paris-Sud)

In the study of graphs and networks, reliability research deals with probabilistic aspects such as maintaining communication through a network when connecting links have failure probabilities, whereas vulnerability investigations are concerned with structural graph parameters which measure resistance to disruption. These two closely related topics constitute the focus of two minisymposia.

Presentations will include surveys of these areas, talks on recent research into specific aspects, and a discussion of directions for future research.

A Survey of Measures of Vulnerability in Graphs and Networks

K.S. Bagga, L.W. Beineke, M.J. Lipman, R.E. Pippert (Indiana University - Purdue University)

Recent interest in such areas as reliable communication networks and fault-tolerant VLSI circuits has resulted in the use of a number of graph-theoretic parameters to measure the vulnerability of the underlying graph to disruption. The potential disruption has various forms and is due in some cases to the removal of vertices or edges, or both, from the graph. We survey many of these parameters, categorizing them according to a preliminary classification scheme. Relationships among some of the parameters are investigated, with particular attention directed to those parameters which are widely used as measures of vulnerability and to those which appear to be especially promising.

Diameter Invulnerable Graphs

J.L.A. Yebra (E.T.S.I. Telecommunications, Spain)

The design of interconnection networks for multiprocessor systems leads to the search for large graphs whose maximum degree Δ and diameter D are given. Moreover, fault-tolerant computing requires the diameter of these graphs not to increase (or not to increase significantly) when a vertex or edge is deleted from the graph.

In this talk the attainability of Moore-like bounds on the order of such graphs is discussed for small values of D. In the case of a vertex deletion they can only exist when $D=2$ and $\Delta=m^2+1$ for any integer m that is not a multiple of 4. Analogously, in the case of an edge deletion it should be $D=2$ and $\Delta= m(m-1)+2$ with m=1,2,4,5,11, and 32. Some constructions of the optimal graphs are also presented.

Generating the Most Probable States of a Communication System

D.R. Shier (Clemson University)

It is frequently required to evaluate the overall performance of a communication system composed of failure-prone components. The exact calculation of most performance measures (such as reliability, throughput, or delay) is, however, mathematically intractable. One approach for approximating such complex measures is to generate a relatively small subset of the states of the system that nevertheless covers in probability a large portion of the sample space. Rather surprisingly, there is an elegant discrete structure (a lattice) underlying the state space, whose properties can in fact be exploited to produce a reasonably effective algorithm for generating the most probable states of the given system.

A Fault-Tolerant Routing in Networks

M. Imase, T. Soneoka, Y. Manabe (NTT Electrical Communications Laboratories, Japan)

We consider the problem of constructing a highly fault-tolerant network which has a fixed routing. The diameter of a surviving route graph has been proposed as a fault-tolerant measure by Broder, Dolev, Tischer, and Simons. This paper first surveys results on the diameter of a surviving route graph and gives some results on routings which minimize this diameter, in both a general network and a generalized deBruijn network.

Graph-Theoretical Aspects of Network Reliability

C.L. Suffel, F.T. Boesch (Stevens Institute of Technology)

It is well known that graph-theoretical models play a central role in the design of reliable communication networks. Two areas of interest in this regard are the analysis and synthesis of such networks. Analysis problems are usually concerned with the determination of the reliability of a given network when the probability distribution of the elements is specified. Herein we shall review and summarize the graph-theoretical concepts which are pertinent to efficient techniques for the computation of network reliability. The synthesis of a most reliable network is a combinatorial optimization problem in which certain network parameters are specified. In the general case the optimization problem appears intractable. We shall describe in detail some simplifying assumptions which reduce the general problem to that of characterizing invulnerable graphs.

• NUMERICAL METHODS IN SYSTEM THEORY AND CONTROL

A. BUNSE-GERSTNER, V. MEHRMANN (Universitat Bielefeld)

For computational problems arising in system theory and control it has turned out that a skillful adaption of standard numerical techniques can considerably increase the efficiency. This minisymposium will focus on numerical linear algebra methods and present some recent developments. In particular algorithms are introduced exploiting the special symmetry structures which these problems exhibit. Also problems are discussed which arise for descriptor systems.

Numerical Linear Algebra for Control Problems

S. Hammarling (Numerical Algorithms Group Ltd., UK)

We discuss the use of the tools of numerical linear algebra in the solution of selected control problems, and illustrate why such tools are preferable to the methods often found in the control literature. The emphasis of the talk is on illustrating the basic numerical principles involved so that the ideas can be applied to other control problems. The tools include the stable factorizations, such as the QR, Schur, and Stewart factorization; the singular value and generalized singular value decompositions; error and perturbation analysis. We also indicate some control problems that require the successful application of these tools.

Linear Quadratic Control Problems with Algebraic Differential or Difference Equations Constraints

V. Mehrmann (Universitat Bielefeld)

We study the following two control problems:

(C) Minimize $S(x,u)=\frac{1}{2}\left(x^*(T)Mx(T)+\int_{t_0}^{T}(y(t)^*Qy(t)+2x(t)^*Su(t)+u(t)^*Ru(t))\,dt\right)$

subject to the constraints

$E\dot{x}(t)=Ax(t)+Bu(t) \quad Fx(t_0)=x^0 \quad y(t)=Cx(t)$

and

(D) Minimize $S'(x_k,u_k)=\frac{1}{2}\left(x_K^*Mx_K+\sum_{k=k_0}^{K} y_k^*\,Q\,y_k+2x_k^*\,S\,u_k+u_k^*\,R\,u_k\right)$

subject to the constraints

$$Ex_{k+1}=Ax_k+Bu_k,\quad Fx_{k_0}=x^0 \qquad y_k=(x_k)$$

where $E,A,F\in\mathbb{C}^{n,n},\quad Q=Q^*\in\mathbb{C}^{r,r}$

$$C\in\mathbb{C}^{r,n},\quad R\in\mathbb{C}^{m,m},\quad S,B\in\mathbb{C}^{n,m}$$

and typically E,F are singular matrices. (Problems of this type are often called descriptor systems.) As one expects, many difficulties arise compared with case of E nonsingular. We will discuss theoretical results and numerical solutions.

On Hessenberg Form and Control Systems

G.S. Ammar (Northern Illinois University)

The use of Hessenberg matrices plays an imortant role in a variety of algorithms for problems in systems and control theory. It is often the case in these algorithms that an initial matrix must be reduced by orthogonal similarity to Hessenberg form while simultaneously obtaining some specified form for other matrices in the problem. Such restricted reductions to Hessenberg form, which are in contrast with the so-called free Hessenberg form reduction that is used in more traditional areas of numerical linear algebra, can lead to some numerical difficulties. In this talk, some numerical aspects of Hessenberg forms for problems in control theory will be considered. The correspondence between Hessenberg matrices and certain subvarieties of the flag manifold will be used to study the numerical sensitivity of restricted reductions of a matrix to Hessenberg form.

Solving Algebraic and Discrete Riccati Equations by Nonsymmetric Jacobi Iteration

R. Byers (North Carolina State University)

We describe how to adapt the nonsymmetric Jacobi iteration to the special structure of Riccati equations. This algorithm preserves the Hamiltonian and symplectic structure of these problems without using a condensed form. Adaptations include variant rotation strategies to keep the iteration from breaking down. The algorithm is attractive for vector and parallel computers.

A Divide and Conquer Algorithm for the Unitary Eigenproblem

W.B. Gragg, L. Reichel (University of Kentucky)

One of the great successes of modern numerical linear algebra is the solution of the Hermitian eigenproblem, by the QR algorithm with Wilkinson's shift. Recently an elegant and simple divide and conquer algorithm for this problem has been given by Cuppen '81. Moreover, Dongarra and Sorensen '86 have shown that the new algorithm is extremely competitive, especially in certain parallel computing environments. Relatively little attention has been paid to the unitary (orthogonal) eigenproblem. Although an analog of Hermitian tridiagonal QR exists for unitary Hessenberg matrices (Gragg '86) the theory is not in as advanced a state as in the Hermitian case. We shall describe an analog of Cuppen's algorithm for unitary Hessenberg matrices. The constructions again are not difficult, but they are rather more intriguing than in the Hermitian case. In particular, we found it moderately nontrivial to obtain a suitable analog of Sorensen's rootfinder.

• NUMERICAL METHODS IN STOCHASTIC PROCESSES

N. BOULEAU (CERMA-ENPC Noisy le Grande, France)

Our aim is to give an account of recent developments in numerical methods for models including stochastic processes. The lectures will develop the following subjects:

- discretization of stochastic differential equations.
- Asymptotic Stability of Diffusions.
- computations of conditional laws of processes.
- accelerated stochastic algorithms.
- simulation of processes.

Although many applied problems involving Markov processes are still treated by using deterministic methods on partial differential equations, it seems that purely stochastic approaches are specifically interesting in an increasing number of cases; some of them are illustrated here.

Numerical Computation of Conditional Laws of Processes
F. Le Gland (INRIA-Sophia-Antipolis, France)

The conditional law of a diffusion process, given a noisy observation of it, is the unique solution of a bilinear stochastic PDE.

A time-discretization scheme, with time step Δt, is proposed for this equation, with an error of order $\Delta t^{3/2}$. To get this estimation, a generalization of the stochastic Taylor formula of Wagner-Platen to the case of infinite-dimensional bilinear stochastic differential equations is proved. Moreover, at each iteration, the discretization scheme splits into prediction and filtering steps.

A probabilistic interpretation is also provided, involving a particular sampling procedure for the observation process.

Discretization of Diffusions and Lyapunov Exponents
E. Pardoux, D. Talay (INRIA-Sophia-Antipolis, France)

The aim is the study of the stability of bilinear stochastic systems with coefficients periodic in time.

A typical application is the stability of the motion of an helicopter blade.

One has to compute the Lyapunov exponent, λ, equal to the limit of (Log $||X(t)||$)/t when t goes to infinity, where (X(t)) is the diffusion solution of the system.

One presents efficient algorithms of discretization of the system permitting a good approximation of λ.

One also studies the sensitivity of λ with respect to diverse parameters related to the modelization of the noise (coloration, ...).

Random Number Generators for Ultracomputers
M.H. Kalos, O.E. Percus (Courant Institute of Mathematical Sciences)

We discuss and analyze issues related to the design of pseudorandom number generators for parallel processors. We are concerned to ensure reproducibility of runs, to provide very long sequences, and to assure an adequate degree of independence of the parallel streams. We consider linear congruential generators $x_{n+1,i} \equiv ax_{n,i}\ b_i \bmod m$ and analyze the correlation properties of such streams when different values of b_i are used. We derive a spectral test ν for such generators. From this, we prove that using the largest primes less than $\sqrt{(m/2)}$ gives a lower bound of the order of $\sqrt{m}$ to $\nu_2(i,j)$ for pairs of corresponding numbers in different streams i,j. More general results of this kind will also be given.

Robust Discretizations for Driven Stochastic Differential Equations
J.M.C. Clark (Imperial College)

Results will be presented on the efficient discretization of vector stochastic differential equations driven by scalar, continuous martingales with values measured only at intervals. The discretization schemes are of Runge-Kutta type. They are efficient in that they converge at a maximum rate with decreasing interval length; they are robust in that they are largely independent of the local characteristics of the driving martingale.

On the Acceleration of Robbins-Monro Algorithm
MM. Lapeyre, Pages, Sab (CERMA-ENPC Noisy le Grand, France)

The use of some particular equidistributed sequences (Von Der Conput, for instance) instead of equidistributed random sequences improves the convergence rate of Monte-Carlo algorithms.

One can then try to use the same idea on other algorithms based on computation of integral, such as Robbins-Monro algorithm.

In this paper we present theoretical justification and practical evidence of the efficiency of this approach.

On the Simulation of Stable Processes

N. Bouleau (CERMA-ENPC Noisy le Grand, France)

Although there is no known explicit expression for densities of stable laws, except for some particular cases, it is possible to give exact algorithms of simulation for such laws. This makes easy the simulation of stable stationary processes and stable processes with independent increments especially in the real case but also in the multivariate case for those processes which are attainable by subordination in Bochner's sense. Of course these simulations, because of their relative slowness, have to be used only when the consequences of the stability property are studied by themselves.

• LATTICE GAS HYDRODYNAMICS

H. CABANNES (Universite Pierre et Marie Curie)

The method of lattice gases is a new method to study the motion of a fluid, which is different from classical numerical methods. Instead of writing the Navier-Stokes equations and solving them by discretisation process, one discretises directly the medium.

In the classical kinetic theory, the fluid is modelised by a discrete repartition of mass, but positions and velocities vary continuously. In a second step (kinetic theory of gases with discrete velocities), the velocities space is also discretised. In the third step (lattice gas hydrodynamics) one finally discretises the physical space as well.

Discrete Models of the Boltzmann Equation

R. Gatignol (Universite Pierre et Marie Curie)

The discretisation of the space of velocities in the kinetic theory of gases leads to the replacement of the Boltzmann equation by a system of semilinear hyperbolic equations: the kinetic equations.

For a discrete gas with a small Knudsen number (little rarefied) we have adaptod the Chapman-Enskog method and obtained the macroscopic equations (Euler, Navier-Stockes, Burnett, ... following the order considered) associated to the medium.

We will show the mathematical properties of the kinetic equations (global existence, exponential or bounded estimates, ...) and the physical behaviour of the gas.

For some particular physical problems (flow along a wall, structure of shock waves, ...) we will present and compare results obtained from the kinetic equations (microscopic description) and from the Navier-Stockes equations (macroscopic description).

Lattice Gas: From Microdynamics to Macrodynamics

J.P. Rivet (Observatoire de Nice, E.N.S. Paris, France)

Lattice gas simulations of hydrodynamical phenomena require a precise knowledge of the corresponding large-scale (macro) dynamics. A multi-scale method (analogous to homogenization) is implemented for the case of the 2-D Frisch-Hasslacher-Pomeau triangular model. From the microdynamical Boolean rules of the model, via a lattice Boltzmann equation, we obtain the macrodynamical equations, which in suitable limits reduce to the incompressible Navier-Stokes equations. An explicit expression is obtained for the viscocity.

Numerical Experiments with Lattice Gas Mixtures

P. Lallemand (Ecole Normale Superieure), G. Searby (Universite de Provence)

Lattice gases provide a simple and efficient way to simulate viscous flows. To describe free boundaries, Calvin et al. introduced lattice gas mixtures made of Boolean particles that interact with rules comparable to chemical reactions. Various versions of the lattice gas mixtures will be defined. Phase transitions will be discussed. Interfacial properties will be presented.

A variety of 2-D hydrodynamical situations with free boundaries will be discussed together with results obtained by computer simulations: nucleation and dynamics of bubbles, break-up of jets, Kelvin-Helmholtz instability.

Body forces can be added to the dynamics. Results will be shown for rising bubbles, gravity waves, Rayleigh-Taylor instability.

For exothermic reactions, fronts behave like flames. The structure and dynamics of these fronts will be discussed, together with a simulation of the Landau-Darrieux instability.

Numerical Simulations of Hydrodynamics on Lattice Gases

D. d'Humieres (Ecole Normale Superieure)

A lattice gas is a representation of a gas by its restriction on the nodes of a regular lattice for discrete time steps. It was recently shown by Frisch, Hasslacher and Pomeau (FHP) that such very simple models lead to incompressible Navier-Stokes equation provided the lattice has enough symmetry and the local rules for collisions between particles obey the usual conservation laws of classical mechanics.

In order to illustrate the power of this new approach of fluids mechanics, we shall present several results of numerical simulations obtained on a triangular lattice with a variant of the original FHP model. Extensions of this model for three-dimensional hydrodynamics on lattice gas will be presented.

• INDUSTRIAL MATHEMATICS AT IBM

FRANCOISE CHATELIN (IBM Scientific Center, Paris)

We present in this session three examples of mathematical work performed in an industrial research environment with strong involvement in scientific computing: IBM T.J. Watson Research Center and Paris Scientific Center. Two examples deal with providing better tools for scientific computing (precision control and optimized vectorized algorithms). The other one is devoted to describing the complete mathematical modeling process of an industrial problem (VLSI).

Mathematical Research and the Production of High Performance and Accurate Vector Softw

Dr. Fred G. Gustavson (IBM T.J. Watson Research Center)

We describe how basic research with the new IBM 3090 vector facility led to new novel algorithms in the areas of elementary functions, matrix linear algebra, and signal processing. The interplay between machine architecture and algorithm design will be emphasized. Some performance figures will be given. The elementary functions algorithms are always accurate to the last bit and correctly rounded more than 95% of the time. This research led to a new theory for scalar elementary functions. The new matrix linear algorithms gave peak 3090 vector performance. The FFT algorithms are fully vectorized and give uniform performance for transform lengths n=8 to 2 million.

Mathematical Modeling for VLSI

F. Odeh (IBM T.J. Watson Research Center)

A partial survey will be given of some numerical and analytical questions arising in the simulation of VLSI circuits and components. We briefly describe some of the techniques used at IBM in attempting to resolve such questions, from wave-form relaxation methods for large lumped-parameter MOS circuits to specialized numerical schemes for solving the distributed systems governing the behavior of small semiconductor components.

A Stochastic Tool for Precision Control in Scientific Computing

F. Chatelin (IBM Scientific Center, France)

<u>Qualitative</u> a priori analysis of round-off and method errors propagation in algebraic processes is well understood, based on perturbation theory (Wilkinson, 1963). We propose

to complement this a priori analysis by an accurate a posteriori quantitative analysis: the user is provided with an estimate of the number of exact significant digits of the desired result. Special emphasis will be put on ill-conditioned problems.

• LINEAR SINGULAR SYSTEMS: THEORY AND APPLICATIONS

M.A. CHRISTODOULOU (University of Patras), V.G. MERTZIOS (Democritus University of Thrace)

The area of singular or semistate or generalized state space systems attained a growing interest in recent years, because they find numerous applications in robotics, electric networks, neural models, multisector economy, Leslie population models in biology etc. Many results which were already known for regular linear systems have been extended for the case of singular systems as for example: controllability linear quadratic control, modal control, application of geometric theory, distributed control etc. In this session, current research will be presented that reflects the increasingly important role of semistate systems in the study of many physical real models.

Boundary Control Systems and Singular Control Systems

L. Pandolfi (Politecnico di Torino)

It is well known that the action of a linear state feedback in a distributed parameter control system with distributed and boundary condrol may take a homogeneous linear state space singular form. This is an ill posed problem. In this paper the theory of singular systems is used to give answers to the problems of existence of feedback such that the closed loop is a consistent system, coefficient assignability of the closed loop system and effect of this on the boundary control problem, and finally relationship between the zeros and the invariant subspaces. Finally we present similarities of some problems in singular systems to time delay systems.

On Sensitivity in Semistate Described Linear Systems

B. Dziurla, R.W. Newcomb (University of Maryland)

This paper presents a theory of sensitivity to go with semistate described linear time-invariant systems. We begin with the canonical semistate equations of linear time-invariant systems of the form

$$A\dot{x} + Bx = Du$$

$$y = Lx$$

where u and y are the system input and output variables and A, B, D and L are constant matrices (with A generally singular). Considering a parameter a of the system we investigate the behavior of the system with respect to changes in the a. In particular it is shown that, as with state-variable systems, the vector dx/da satisfies an almost identical set of canonical semistate equations. From this a discussion is given on the sensitivity properties of the system and comparisons are made with similar state variable circuits.

A Brief Survey of Geometric Theory for Singular Systems

K. Ozcaldiran (Bosporus University), F.L. Lewis (Georgia Institute of Technology)

In this paper some recent results on the geometric theory of the linear singular system

$$EX = Ax + Bu$$

will be discussed. The notions of (A,E,R(B)) - invariance will be reviewed, and they will be used to compute the reachable subspace. Preliminary results on extending to singular systems the concept of reachability subspace will be presented. The use of both subspace recursions and matrix algorithms to compute subspaces will be covered. Some open problems will be discussed, including the disturbance decoupling problem using constant and proportional feedback and system inversion.

Recent Results in Realization Theory for Semi State Systems

M.A. Christodoulou (University of Patras), B.G. Mertzios (Democritus University of Thrace)

Several algorithms are presented for the construction of minimal, finite dimensional realizations of linear, time-invariant, discrete- time, singular systems, when the external description, such as the impulse response matrix or the frequency response matrix are given. The presented algorithms use the theory of Markov parameters for singular systems or the Markov parameters and moments combined, or Taylor series expansion about a general point "a", or use simplified version of Taylor series expansion about an arbitrary point "a".

Advantages, disadvantages, and comparisons between each other algorithm is discussed. Finally the applicability of the above to state space models is shown via various illustrative examples.

• ECONOMIC QUALITY CONTROL

E. VON COLLANI (Institut fur Angewandte Mathematik Und Statistik)

The improved reliability and competitive success of Japanese products have led to a global interest in the concept of "Total Quality Control." While concerned with all aspects of "Process Control", this concept, evolved by the Japanese, focuses on "the will to produce good quality." Given this goodwill, the ultimate success or failure of any Q.C. program will depend on the existence of appropriate procedures, technical equipment and backup services.

In this regard, the Japanese have laid considerable emphasis on the use of "Statistical Quality Control" procedures in their approach to Quality Control. The QC-Research Group Wurzburg, which is supported by the German Research Foundation (DF(works on the development of statistical procedures for industrial quality control and the objective of this minisymposium is to present and discuss the results so far achieved.

An Economic Concept for Industrial Quality Control

E. Von Collani (Institut fur Angewandte Mathematik Und Statistik)

Much discussion is currently focused on applications of experimental design to industrial quality control. This discussion frequently fails to acknowledge that there are two distinct but complementary (and sometimes parallel) tasks associated with establishing control over a production process. Both these tasks require the proper implementation of appropriate statistical procedures. In the first instance, and it is here that experimental design theory can be particularly applicable, a thorough Process Capability Analysis must be completed, so that process yields can be maximized and reasonable quality standards established for future production. Secondly, the process must be monitored in order to maintain production at these standards. In this paper an economic concept is proposed, which enables the development of appropriate and effective control procedures. Such procedures could replace the heuristic methods currently in general use.

Economic X-Charts

J. Sheil (Institut fur Angewandte Mathematik Und Statistik)

The $\bar{x}$-chart is the most widely used statistical tool in the control of continuous production processes. In general, the design of such charts is based on purely statistical reasoning (e.g. three standard deviation limits) with no reference being made to the fact that costs of achieving statistical control will vary from one process to another. In this paper, key parameters required to characterize given production processes are derived. In addition, nomograms are presented which enable the selection of economically optimal $\bar{x}$-charts in the case of single-assignable-cause systems. As the results achieved are on the one hand quite general and on the other hand very simple to apply, they can be viewed as defining a particular family of "(economic) standard $\bar{x}$-control charts."

Economic Control of a Multiple Assignable Cause System

B.F. Arnold (Institut fur Angewandte)

One of the main objections levelled against economically devised control charts is the fact that most of the proposed procedures to determine the economic design assume a single assignable cause system. In this paper the case of multiple cause systems is concerned. The optimal economic design of an x-control chart is determined on the basis of a Bayesian approach and the relations to single assignable cause systems are established.

Economic Acceptance Sampling

R. Gob (Institut fur Angewandte Mathematik Und Statistik)

It is generally recognized that acceptance sampling by attributes is an important tool for industrial quality control. In this paper a sampling scheme is proposed which takes into consideration the structure of the underlying process as well as all relevant costs. The assumptions made are fulfilled in many real life situations and the proposed algorithm is simple enough to be applied at workshop level. In addition the proposed sampling plan can be generalized in a natural way so as to produce an adaptive sampling scheme.

• COMPUTER-AIDED GEOMETRIC REASONING

H. CRAPO (INRIA-Rocquencourt)

Computer-aided design has reached the stage where simple 3-D forms can be reasonably well combined and smoothed, then projected for 2D representation. But general strategies for interactive creation and manipulation of n-dim'l geometric objects (n 2) have not yet been set forth.

Two essential difficulties must be overcome; the complex nesting of special cases (singularities) which accompany even the simplest geometric descriptions, and the lack of adequate global topological and combinatorial theorems with which to classify possible geometric solutions.

This minisymposium highlights methods currently being developed for computer-aided geometric reasoning, with particular reference to its application in structural chemistry and in the study of polytopial realizability.

Combinatorial Aspects of Molecular Geometry

A. Dress (Universitat Bielefeld), T. Havel (Research Institute of Scripps Clinic)

The description, classification, and study of molecular conformation, as originally conceived by Pasteur and Van't Hoff, continue to pose interesting mathematical problems and have important applications in chemistry.

We will discuss how one can study these problems by using the structural interdependencies among the various geometric invariants of point configurations in 3-space. Combinatorial structures such as oriented matroids permit us to specify these invariants. This combinatorial approach enables us to acquire some insight into the range of feasible geometrical interpretations of available chemical information.

On the Computer-Aided Search for Geometric Realizations

J. Bokowski (Fachbereich Mathematik Technische Hochschule Darmstadt)

There are many unsolved problems in geometry which can be formulated as follows, in euclidean 3-space:

Decide, for a given 2-dimensional combinatorial complex, whether there is a corresponding geometric 2-dimensional cell complex (with flat cells and without self-intersections) which can be considered as its realization.

The method of the author, in which the structure of oriented matroids is used to tackle such problems, will be discussed. Finally, examples in which the method has succeeded will be shown in a computer-aided film.

Mechanical Geometry Theorem Proving

S.C. Chou (University of Texas)

The most successful and practical methods of mechanical geometry theorem proving will be introduced and illustrated by many elegant examples. These methods are: Wu's method, the Groebner basis method, and a combination of Wu's method with the Tarski-Seidenberg-Collins method. Our prover based on these methods proved hundreds of theorems from various geometries, including Bolyai-Lobachevskian geometry. Theorems such as the Morley trisector theorem and V. Thebault's theorem (which was open for more than 40 years) were mechanically proved. We will also talk about methods for discovering new theorems mechanically.

New Approaches to Computerized Proofs of Geometric Theorems

B. Kutzler (Universitat Linz)

The problem of automatically proving geometric theorems has gained a lot of attention in the last two years. Following the general approach of translating the given geometric theorem into an algebraic one, various solutions to different variants of the problem are presented. All of the algorithms described are based on two general purpose methods in computer algebra: Buchberger's Groebner basis method and Collins' cylindrical algebraic decomposition method. More than 100 non-trivial geometric theorems have been proved using implementations of these algorithms in the computer algebra systems SCRATCHPAD-II and SAC-2. Finally, the applicability of these approaches to computer aided design is discussed.

- **THE AUTOMATIC GENERATION OF SURFACES**

B.E.J. DAHLBERG (Chalmers University)

We will in this group of talks discuss the problem of generating convex C^2-surfaces. This problem arises e.g. in the styling of cars. Applications of a proposed procedure will be presented.

Generation of Convex Surfaces Using Linear Programming

B.E.J. Dahlberg (Chalmers University)

We formulate the problem of fitting a convex parametric spline surface to given measure points as a nonlinear optimization problem. We will discuss how a solution can be obtained by iteratively solving a sequence of linear programming problems.

MakeConvex

R. Andersson, T. Elmroth (Volvo Data AFB, Sweden)

We present an implementation to generate convex curvature-continuous parametric spline surfaces from sets of ordered measure points. The patch topology is determined from the measure points. Patch boundary curves are generated by a least square fit to the measure points. These curves are shifted by translations to obtain a network of curves. A surface with the required degree of smoothness is constructed from the network of curves. This surface is turned into a convex one by transversal modifications. The perturbation algorithm is based on linearization of the normal curvature and linear programming (LP) techniques. Examples from the car industry will be presented.

Convexity Preserving Approximations

B. Johansson (Chalmers University)

This presentation will be focused on the possibilities to preserve convexity in connection with mathematical descriptions of designed convex surfaces. The approximating surfaces will here be composed of patches, which are described by polynomials. We will show the impossibility of generalizing approximation procedures with nice analytical features as in the case of convex curves. Such procedures are exemplified by linear interpolation and Bezier spline approximation. To get a convexity preserving approximation algorithm in the case of surfaces, it must take into account the geometry of the surface. Such a result will also be shown.

An Implementation of the Karmarkar Algorithm

E. Andersson (Volvo Data AFB, Sweden)

An implementation of the Karmarkar algorithm will be presented. This implementation can handle highly degenerated problems with a dense condition matrix.

• NUMERICAL LINEAR ALGEBRA TECHNIQUES IN SIGNALS, SYSTEMS AND CONTROL

B. DATTA (Northern Illinois University), R.J. PLEMMONS (North Carolina State University)

Bridging the gap of communications between linear algebraists, especially numerical linear algebraists and theorists and engineers working in application areas such as signal processing and control and systems theory, was a major objective of several recent conferences. Amongst them the most notable ones are the very successful AMS-IMS-SIAM sponsored summer Research Conference on "Linear Algebra and its Role in Systems Theory" held in Bowdoin College, Maine, 1984 and the SIAM Conference on "Linear Algebra in Signals, Systems and Control" held in Boston, 1986. The purpose of this proposed minisymposium is to help strengthen the ties between mathematicians and engineers already developed at the above-mentioned conferences. The minisymposium will be highly interdisciplinary in nature. The speakers will stress the use of recently developed sophisticated numerical linear algebra techniques in solutions of several important problems arising in Signal, Systems and Control.

Numerical Methods for Solving Some Large Scale Linear Algebra Problems in Control

Y. Saad (University of Illinois - Urbana, Champaign), B.N. Datta (Northern Illinois University)

Often in control engineering, one must solve matrix problems such as Sylvester's equation, Lyapunov's equation, Riccati's equation, frequency response calculations, pole assignment, and so on. So far, standard solution methods have been limited to matrices of small size, but there are now increasingly more complex models that lead to equations with very large sparse matrices. Many of these models are issued from partial differential equation models or models dealing with structures. We will propose some methods for solving a few of these problems. Most of the methods considered are based on Arnoldi and block-Arnoldi algorithms. They consist of projecting the initial problem into a suitable subspace of small dimension, often a Krylov subspace, and then solving the resulting small matrix problem by standard techniques.

Stability of Algorithms for Matrices with Small Displacement Rank

J.M. Delosme, I.C.F. Ipsen (Yale University)

A few years ago the factorization algorithms of Schur and Levinson for symmetric positive definite Toeplitz matrices were extended to matrices with small displacement rank. The intermediate quantities computed by these algorithms, called generalized Schur parameters, depend on the original factorization of the low rank displacement. The general algorithms found in the literature assume a simple standard factorization that leads to an elegant algorithm which, unfortunately, is now always stable. However it is known that, for some important linear least squares problems whose displacements are given in factored form, the algorithms employing these specific factorizations enjoy good numerical properties. A detailed study of the boundedness of the generalized Schur parameters associated to the different factorizations of the displacement is performed that provides a key to the stability of the algorithms. We also present general algorithms whose Schur parameters are always bounded and advocate their use to ensure numerical stability.

Some Applications of Hyperbolic Matrix Reduction Schemes

R.J. Plemmons (North Carolina State University)

Introduced by Golub in 1969 in the context of least squares downdating, hyperbolic transformations are known to provide a weakly stable but fast alternative to orthogonal transformations for certain matrix reduction computations. Additionally, they are often more amenable to parallel implementation. In this presentation we describe some of our recent work on applying hyperbolic plane rotations and/or hyperbolic Householder reflections to the following problems:

(1) fast recursive least squares filtering computations (joint with T. Alexander and C. Pan), (2) minimizing a weighted sum of Euclidean norms (joint with S. Wright), and (3) preconditioning schemes for conjugate gradients (joint with D. Pierce).

Orthogonalization and Parallelism

B. Philippe (INRIA-IRISA, Rennes) W. Jalby (University of Illinois - Urbana, Champaign)

By definition, the orthogonalization of a given matrix A, corresponds to finding an orthogonal matrix Q such that A = QT where the matrix T has a given shape. Usually, the considered factorization is the QR one corresponding to an upper triangular matrix T. The polar factorization where T is symmetric is also to be considered for its optimal properties; in this last case, we say that we perform a symmetric orthogonalization. For the QR factorization, we consider the algorithms which are based on Gram-Schmit procedure. The symmetric orthogonalization can be obtained from the computation of the matrix $(AA^t)^{-1/2}$. Hence, different implementations of the Gram-Schmit method (including block methods) are compared for their suitability (efficiency, stability) for a parallel architecture. For the symmetric orthogonalization, performance of Higham's algorithm is studied; this is compared with another algorithm we propose for orthogonalizing a nearly orthogonal set of vectors. As before, the stability of the algorithms are analyzed. The target architectures are vector or tightly coupled vector machines (Cray, Cray X-MP48, Alliant FX 8).

Toeplitz Matrix Algorithms for Some Control Problems

G. Ammar, B.N. Datta, K. Datta (Northern Illinois University)

There have been some fine recent developments in the area of sequential and parallel matrix computations involving Toeplitz matrices. In this talk, we first present a brief overview of these Toeplitz algorithms and then show that these concepts and techniques of Toeplitz matrices can profitability be used in designing efficient algorithms both parallel and sequential for several important linear algebra problems arising in control theory such as controllability and observability problems, matrix equations problems, feedback problems, stability and inertia problems, frequency response problems, etc. We will present some computational results on the algorithms as well.

•NUMERICAL SOLUTION OF DIFFERENTIAL ALGEBRAIC EQUATIONS

P. DEUFLHARD (Konrad-Zuse-Zentrum fur Informationstechnik)

Implicit ordinary differential equations (ODE's) and differential-algebraic equations (DAE's) arise in many fields of scientific computing, such as chemical applications, electronic circuit design, combustion problems or, generally speaking, in method of lines treatment of parabolic partial differential equations. Numerical treatment of such systems has recently attracted renewed interest: first, multistep methods of BDF-type have been extended from stiff integration; more recently, new one-step methods of extrapolation and of Runge-Kutta type have been studied and implemented. The common insight is that index = 1 DAE's are numerically fairly tractable, whereas general index 1 problems need some special treatment such as index reduction.

Index Reduction for DAE'S

C.W. Gear (University of Illinois-Urbana-Champaign)

The general DAE has the form $F(w', w) = 0$. It has index m if, after m differentiations, we can solve the resulting set of equations for $w'(w)$. The special DAE has the form $y' = f(y, z)$; $g(y, z) = 0$. Its index is similarly defined. Any general DAE has a corresponding special DAE (of index one higher) whose solutions for the state variables agree and vice versa. This implies that results for the special DAE map into theorems for general DAEs of index one lower and vice versa. High index problems are difficult to solve numerically, but it is possible to reduce the index by differentiation and yet not discard constraints by adding dummy variables whose solution is zero. How far can the index be reduced without a change of dependent variables? It appears to be possible to reduce the index to one (two) if the result is to have the general (special) form. However, if the numerical elimination of variables is considered, it is possible to lower the index one further. In some cases this may be a numerically viable procedure.

Order Results for Implicit Runge-Kutta Methods Applied to Differential/Algebraic Systems

L.R. Petzold (Lawrence Livermore National Laboratory, USA)

We study the order, stability and convergence properties of implicit Runge-Kutta methods applied to nonlinear index one and semi-explicit nonlinear index two systems of differential/algebraic equations. These methods often do not attain the same order of accuracy for differential/algebraic systems as they do for purely differential systems. We derive a set of order conditions for index one systems and for index two systems which the method coefficients should satisfy in addition to the usual order conditions to ensure a given order of accuracy, and we present results on the stability and convergence properties of these methods.

Linearly Implicit Methods for Differential Algebraic Equations

E. Hairer (University of Geneva)

Implicit multistep methods (BDF) and implicit Runge-Kutta methods are well adapted for the numerical treatment of differential-algebraic equations. Linearly implicit methods (such as extrapolation methods, Rosenbrock methods, ...) suffer from an order reduction, when they are formally applied to such problems. In order to avoid this phenomenon, the coefficients of the method have to satisfy additional order conditions. Their derivation will be presented and some numerical results are given.

Extrapolation Methods for Differential-Algebraic Equations

P. Deuflhard (Konrad-Zuse-Zentrum, FRG)

The application at the semi-implicit Euler discretization which is known from stiff ODE's, to the case of DAE's of index 1 is studied in detail. For those implicit ODE/DAE systems, that can be transformed into a Kronecker canonical form, the existence of a perturbed asymptotic expansion is shown. Therefore, for the semi-implicit Euler discretization, a reasonable method up to order 5 in the basic stepsize can be constructed by use of polynomial extra- polation. These and several further detailed theoretical consider- ations led to the implementation of a code LIMEX, which also includes an index monitoring device. Finally, numerical comparisons with BDF-type codes are given in terms of examples from chemical combustion (including method of lines treatment of diffusion-reaction systems).

- **SPECTRAL METHODS FOR FLUIDS FLOW PROBLEMS**

M. DEVILLE (Louvain Catholic University)

The minisymposium will focus on spectral approaches of fluid dynamics within the framework of pseudo-spectral techniques and spectral elements.

1. A spectral staggered mesh algorithm is presented for the spectral element case. Proofs of convergence are provided.
2. Domain decomposition techniques are considered and applied to general second order elliptic problems.
3. A 3-D collocation version of the Navier-Stokes equations is solved through a multigrid scheme.
4. Finite elements are used as Navier-Stokes preconditioners with a multi-domain approach.

How to Justify the Spectral Approximation of the Stokes Problem

Y. Maday (Universite de Paris Val de Marne)

For the numerical simulation of the Stokes problem in velocity-pressure formulation it is well known in the finite element context that the various spaces of discrete velocity and discrete pressure have to be chosen in a compatible way. We first recall the abstract statement of this compatibility then analyse some known spectral collocation methods. We detect the presence of incompatibility between the velocity and pressure spaces these methods use. Finally we present some new methods, the formulation of which is directly inspired of the abstract condition. Numerical considerations and results are also presented that prove the viability of these new methods.

Two Spectral Collocation Algorithms for the Incompressible Navier-Stokes Equations with Two Non-Periodic Directions

C.L. Streett, M.Y. Hussaini (NASA Langley Research Center), Y. Maday (Universite Pierre et Marie Curie)

Two algorithms for solution of the time-dependent incompressible Navier-Stokes equations are discussed. Both methods employ Chebyshev or Legendre discretizations in two coordinate directions and are extendable to treatment of a third, periodic direction employing Fourier discretization. One algorithm involves a time-split scheme with consistent boundary-pressure treatment and has a very low operation count. The second method utilizes a consistent staggered-mesh discretization, with an implicit multigrid algorithm for solution. The latter algorithm is also suitable for efficiently obtaining steady-state solutions. A fast direct-solver for the two-dimensional discrete Helmholtz equations which appear in these schemes is also described.

Finite Element Preconditioning of Chebyshev. Pseudospectral Solutions of Incompressible Flows

P. Demaret, M. Deville (Universite Catholique de Louvain)

Collocation techniques are widely used in numerical fluid mechanics. Chebyshev pseudospectral approximations are best solved through finite element preconditioning. New finite discretizations are investigated in order to get derivative continuity at the interfaces of a subdomain decomposition. Therefore, complex geometries can be treated. Two dimensional viscous incompressible flows will be calculated: regularized square cavity, thermal convection at moderate Reynolds number.

Domain Decomposition Methods for Elliptic Partial Differential Equations

A. Quarteroni (Universita Cattolica di Brescia)

Domain decomposition methods for second order elliptic problems are considered. The given problem is reduced to a sequence of mixed boundary-value problems on every subdomain. Then an iterative method with relaxation at the interface is applied. An optimal strategy for the automatic selection of the relaxation parameter at each iteration is available.

As it stands, the above method applies also to any kind of finite dimensional approximation to the given differential problem. In particular, spectral collocation methods and finite element methods are considered and analyzed, both theoretically and numerically.

• LARGE-SCALE SUPERCOMPUTING

I.S. DUFF (Harwell Laboratories)

The CRAY-2 has brought a new dimension to large-scale scientific processing with its 256 Megawords of main memory. The forthcoming ETA-10 should have memory of equal size.

We will discuss this new dimension in supercomputing, paying particular attention to both new techniques and new problem areas made possible by extremely large memories coupled with high processing power. The four principal countries involved in the ICIAM 87 conference each have access to CRAY-2 supercomputers and talks will focus on recent experience with this machine. A panel discussion will consider in general the impact of large memory computing and its influence on multitasking.

Large-Scale Computing at Harwell
I.S. Duff (Harwell Laboratories)

The CRAY-2 has brought a new dimension to large-scale scientific processing with its 256 Megawords of main memory. The forthcoming ETA-10 will have memory of equal size.

We will discuss this new dimension in supercomputing, paying particular attention to both new techniques and new problem areas made possible by extremely large memories coupled with high processing power. We will consider the use of these techniques in solving large sparse equations on the CRAY-2 at Harwell stressing the use of a high level of vectorization and efficient use of local memory.

Large-Scale Computing at the French GCCVR
P. Herchuelz (Ecole Polytechnique - C.C.V.R.)

Le CRAY-2, dote de quatre processeurs vectoriels fortement couples par une grande memoire commune (deux gigaoctets), a d'ores et deja un impact sur la nature des problemes scientifiques abordes et sur l'architecture des codes de calcul. Nous analysons aussi le choix d'un systeme d'exploitation derive d'UNIX pour un superordinateur, tant du point de vue de l'utilisateur que de celui de l'exploitant.

Supercomputer for the Solution of Problems in Nuclear Industry
F. Schmidt (IKE Universitat Stuttgart)

Supercomputers allow the solution of new classes of problems in nuclear industry. Some of the most challenging ones are problems which can be formulated and solved only in the frame of integrated systems. One such system is the integrated planning and simulation system IPSS which is under development at the University of Stuttgart. In the paper we describe the system and discuss some of the hardware and software requirements which have to be met if such systems are to be used successfully in conjunction with supercomputers.

Parallel Cyclic Reduction Algorithm for the Direct Solution of Large Block - Tridiagonal Systems
D. Anderson (University of California)

In several physics and engineering applications block triangular matrix equations arise in which the blocks are dense. This can occur, for example, when the discretization of a 3D partial differential equation (PDE) uses a spectral representation in two of the coordinates together with a simple finite element or finite difference treatment in the third. For some applications in plasma physics equilibrium and stability theory we have built the PAMS code (Parallelized Matrix Solver) which exploits the features of the Cray-2 computer including the very large memory, vectorization, functional unit overlap, and multitasking. Tests on this code have shown average computing speeds in excess of one Gigaflop.

- **PARALLEL PROCESSING ON M.I.M.D. SYSTEMS: UTILIZATION AND PERFORMANCES FOR REALISTIC CODES**

M. ENSELME (ONERA, France)

The proposal aims to present results obtained in coding some applications in order to run on M.I.M.D. architectures; the point of view of the users is predominant and the presentations answer to the questions: how to have a good idea of performances before coding an application; what about the real performances obtained in coding some realistic applications?

The L. Brochard talk gives an expression of speed up depending on both architectures and algorithm parameters.

The talks of E. Clementi, L. Mane and F.X. Roux give results obtained on two M.I.M.D. installations concerning different physical and realistic problems.

Communication and Control Costs on Loosely Coupled Multiprocessors
L. Brochard (Ecole Nationale des Ponts et Chaussees)

Communication and synchronization costs is a key problem in parallel computing. Studying domain decomposition, we give a speed-up formulation depending on both architecture and algorithm parameters. Different algorithms are compared using central and distributed ways.

Experimental results are compared to speed-up simulation on a loosely coupled array of 8 processors. As good coincidence appears, speed-up simulation is extended to more massive parallel systems.

Experimenting with a Parallel Supercomputer

E. Clementi (IBM Corporation, USA)

We intend to review some of our new applications and performance data on our parallel system consisting of 30 nodes, 10 of which are IBM CPUs, the remaining subdivided into 10 FPS 164 and 10 FPS 264.

Particular attention will be given to the performance using the new FPS and the SCA bulk memory. The application examples are taken from either biophysics or from fluid dynamics.

A Parallel Computing Method for Numerical Simulation of the 2D Unsteady Navier-Stokes Equations for High Reynolds Numbers

L. Mane (ONERA, France)

A parallel M.I.M.D. algorithm, based on the subdomain decomposition technique, is proposed for the resolution of the unsteady Navier-Stokes equations and applied for numerical simulations of separated viscous flows around bodies at high Reynolds numbers.

The resolution method combines finite difference schemes of high precision $O(h^4)$ and of order $O(h^2)$ and uses alternating directions implicit techniques. The algorithm implied, having a high degree of parallelism, has been implemented on the ONERA multi-processors systems (four AP120B and a sharable memory).

Numerous simulations of impulsive starting of airfoil in incidence at Reynolds numbers up to 10^5 were then possible for reasonable calculation times on grid systems with more than 10^5 points. Numerical and experimental comparisons allowed the validation of the method and the study of transition phenomenon.

Parallel Algorithm for Linear Elasticity Problem

F.X. Roux (ONERA, France)

Experiences with multi-array-processors systems for solving the linear elasticity equations for a composite three-dimensional beam are reported. These experiences were performed to test an algorithm based on domain decomposition with the hybrid formulation of the interface continuity conditions.

The hybrid parallel method was compared with a parallel algorithm based on distributing matrix blocks of the classic conforming finite element method.

The performance of the nodal computations on FPS array-processors and the speed-up observed with multi-processing are reported.

Finally, the efficiency of both methods is analysed according to the domain decomposition topology and the features of the parallel machine.

• INDUSTRIAL PROBLEMS USING APPLIED MATHEMATICS

A.M. ERISMAN (Boeing Computer Services, USA)

Mathematics plays a critical role in the design and delivery of products for business and industry. It would not be possible to survey all such activities across diverse industries and provide any meaningful insight into the role of mathematics. Thus we have chosen - for this symposium - to look at selected problems from four different industries: aerospace, petroleum, consumer electronics, and metals. Senior mathematicians from each area have chosen a particular problem to discuss the real world requirement - practical constraints and the mathematical tools required. One common theme is that the problems are not necessarily "solved" but the present status of solution contributes to savings in product design - Reliability and/or cost.

Mathematical Modeling of Material Deformation Processes
O. Richmond (Aluminium Company of America)

Deformation processes like rolling extrusion and forging generally alter not only the shape of a material but also the microstructure and its associated properties. Such processes can be modeled by a set of nonlinear partial differential equations consisting of state equations relating current dependent variables, kinetic equations describing the history dependence of material variables, and classical physical conservation laws. Typically, these equations, together with boundary conditions appropriate for a particular process, are solved using methods based on convected, or Lagrangian-based, finite elements. Because deformations are large, elements can be severely distorted, and mesh regeneration is required.

Mathematics of Strapdown Inertial Navigation
J.W. Burrows (Boeing Computer Services, USA)

Inertial navigation systems have been placed in a large number of vehicle types. The fundamental principles and components used in these systems are described, together with the ways the information provided is used in aircraft control, guidance, and navigation.

The newer strapdown inertial systems, now the system of choice for aircraft and missile applications, make considerably more computational and mathematical demands, a central problem being the vehicle angular attitude computation. Parameters, the corresponding differential equations, and algorithms for attitude tracking are described.

Finally, to emphasize system reliability, redundancy management techniques, including parity filtering, choice of thresholds, and reconfiguration, are described.

Mathematical Problems in a High Technology Consumer Business
R. Crane (RCA David Sarnoff Research Center, USA)

The research and engineering effort required to support the consumer television business creates more interesting and challenging mathematical problems. This talk describes several applications that resulted in mathematical models involving partial differential equations, function approximation, and optimization. Also described are particular constraints that affect an applied mathematician working in research and engineering related to a consumer business such as relevance and timeliness of results, product costs and manufacturability.

Large Scale Geophysical Computation Modeling and Migration
N.D. Whitmore, K.J. Marfurt (AMOCO Production Company, USA)

Over the last several decades the majority of seismic exploration techniques have been oriented towards the common midpoint (CMP) methods. In this method the multi-offset seismic data is compressed by summation to mimic a geologic cross section. While the CMP process has proved to be robust, many of the extractable earth parameters are not retained and the summation is non-optimal. In an attempt to retain more of the information available in seismic data, methods of processing and modeling are now being employed which more closely adhere to the physics of wave propagation in the earth.

Seismic modeling and pre-stack migration are two techniques which are being used quite prevalently to glean more information from seismic data. This paper will discuss the seismic modeling and migration problems, emphasizing unresolved issues.

•ADVANCED CONCEPTS IN SEMICONDUCTOR DEVICE SIMULATION
W. FICHTNER (Swiss Federal Institute of Technology)

During the last few years, semiconductor device simulation has been extended to several new areas. In this minisymposium, three tutorials will be given on the following subjects:

1. 3D Simulation Techniques (Bell Labs, U.S.A.)
2. Simulation for III - V Semiconductor Devices (University of Illinois, U.S.A.)
3. Device Simulation and Monte Carlo Methods (University of Bologna, Italy).

For all talks presented, the emphasis will be on the underlying physical and numerical concepts as well as important applications.

Numerical Methods for 3-D Device Simulation
R.E. Bank (University of California), R.K. Smith (AT&T Bell Laboratories)

This talk is concerned with the generation of spatial grids and the linear iterative methods for the solution of the semiconductor equations in three dimensions. A grid generation algorithm which includes local refinement will be presented. Block iterative methods can be effective in both decoupling the systems of equations as well as separating the insulator and semiconductor regions of the device. The efficiency of point iterative methods can be improved on vector and parallel machines by the use of color orderings. Numerical results, for convection-diffusion equations using several iterative techniques will be presented.

Simulations for III-V Compound Semiconductor Structures
K. Hess (University of Illinois)

Simulations of electronic transport in III-V compound and heterolayer devices (such as MESFET's and the high electron mobility transistor) are reviewed. Emphasis is placed on important nonlinear transient effects such as velocity overshoot and hot electron thermionic emission. Methods of investigation and visualization of these effects are discussed.

Several approximations of the basic physics, standard and extended moment equations, and Monte Carlo simulations are compared and their advantages and disadvantages are stated.

Finally we will discuss the possibility of tabulating the precise Monte Carlo results in certain parameter ranges and using the standard moment equations (plus tables) to obtain a faithful representation of virtually all transient effects.

Monte-Carlo Analysis of VLSI Devices
E. Sangiorgi, F. Venturi, B. Ricco (University of Bologna)

An efficient Monte Carlo simulator has been developed as a post processor of a two-dimensional "drift-diffusion" device analyzer. A new method for obtaining the free-flight time distribution resulted in at least one order of magnitude of CPU time saving compared with the traditional approach. Other special techniques have been applied to increase the simulation efficiency so that real devices can be analyzed in a minicomputer environment. Extensive applications on hot electrons in Silicon Mosfets will be presented and compared with experiments.

•RECENT ADVANCES IN THE REALIZATION AND FEEDBACK SYNTHESIS OF NONLINEAR SYSTEMS
M. FLIESS (C.N.R.S. - E.S.E. - L.S.S.)

The four contributions of this minisymposium describe quite different but equally interesting advances in the theory of nonlinear control systems. Differential geometric techniques and some facts from noncommunicative functional expansions are used for computing all possible static state-feedbacks which achieve decoupling. Longstanding questions in optimal control theory are explained thanks to methods stemming from real analytic geometry. Structures of constrained implicit systems are explained. Finally, differential algebra is used for solving the exact nonlinear model matching problem.

Sufficient Conditions for Optimality for Some Classes of Real - Analytic Problems
H.J. Sussmann (Rutgers University)

For optimal control problems with real-analytic right-hand sides and Lagrangians, we give checkable sufficient conditions for a family of extremals to be a family of optimal trajectories. The proofs use the theory of subanalytic sets. The conditions needed for the proofs to go through are the existence of a finite-dimensional reduction that implies that optimal trajectories have to be bang-bang with locally uniform bounds on the number of switchings, or similar results involving singular trajectories as well.

The Model Matching Problem Using a Differential Algebraic Approach

C.H. Moog (Ecole Nationale Supericure de Mecanique - C.N.R.S.),
G. Conte, A.M. Perdon (Universita di Genova)

This paper gives the solution to the model matching problem and its dual for continuous time systems. The necessary and sufficient conditions are obtained in the newly introduced differential algebraic framework.

The solutions are global in the sense that they are valid almost everywhere. An elementary example illustrates that this approach is perhaps the most "natural" one.

No aspect of causality is considered.

Structural Properties of Systems Defined by Higher-Order Nonlinear Differential Equations

A.J. Van Der Schaft (Twente University of Technology)

Many (physical) systems are not a priori given in a state space form or as an input-output map, but, on the other hand, are naturally described by a set of higher-order differential equations involving external variables (inputs and outputs) and internal variables (state or auxiliary variables). For linear systems this has resulted in the polynomial matrix approach (Rosenbrock, Wolovich,...). Only recently some attempts have been made to extend this theory to nonlinear systems. This talk will report about current research on this topic, with an emphasis on the geometric aspects of realization theory and canonical forms.

Noninteracting Control of Nonlinear Systems by Regular Static Feedback

D. Claude (CNRS-ESE-LSS)

Decoupling algebras are essential tools for the study of the decoupling of linear analytic systems. They make up the algebraic counterpart of the set of invariant distributions which are used in a geometric approach.

In the case of ordinary noninteracting control (Morgan's problem), for each output, we show the singleness of the smallest decoupling algebra associated to a decoupling feedback. If, as well, the sum of characteristic numbers with the number of outputs of the system is equal to the dimension of the state manifold, then, for each output, the only decoupling algebra is the maximum algebra.

This explains the latest and interesting results of Ha and Gilbert which make possible the calculation of the whole set of regular feedback laws which ensure an ordinary noninteracting control.

•EXPERT AND KNOWLEDGE BASED SYSTEMS FOR SCIENTIFIC COMPUTING

B. FORD (The Numerical Algorithms Group Ltd.)

This mini-symposium carries forward the themes extensively explored at the IFIP WG2.5 Working Conference on "Problem Solving Environments for Scientific Computing", held at Sophia-Antipolis in 1985. The presentation will report on progress towards advanced environments for numerical and statistical computation. The projects described all have a common goal: namely to allow computer users, typically scientists and engineers, to address their computational problems in convenient ways using familiar terms, and to have access to relevant subject knowledge and expertise. Thus the investigation of expert and knowledge base systems is an essential aspect in the development of these advanced environments.

GLIMPSE a Statistical Expert System
J.A. Nelder (Imperial College of Science and Technology)

GLIM is a statistical package, widely used, for fitting generalized linear models. GLIMPSE is an expert system, written in PROLOG, using GLIM as an algorithmic engine. Statistical analysis is split into activities (e.g., data exploration). Each activity has a set of tasks, expressed in a high level command language. GLIMPSE provides 3 levels of syntactic help in constructing tasks for direct use. Semantic help with these statistics has 2 levels, specific help on a basic procedure, or general help for an entire activity. GLIMPSE is designed to be transparent and libertarian, to exploit the users' own skills, and to be self documenting.

ELLIPTIC Expert: An Expert System for Elliptic Partial Differential Equations
W. Dyksen (Purdue University)

In the past 25 years, despite the revolution in computing hardware, the nature of software for scientific computing has been rather constant - a library of Fortran callable routines. Problem oriented, very high level languages like ELLPACK, a system for solving a large class of elliptic PDE problems, represent a first step towards the modernisation of scientific computing. Building on an interactive form of ELLPACK, work on an enhanced version, Elliptic Expert, is underway. Our goal is to make the extensive problem solving capacity of ELLPACK completely accessible to the nonexpert and to study generally the use of KBS techniques in scientific computing.

NEXUS-Towards a Problem Solving Environment (PSE) for Scientific Computing
P.W. Gaffney (IBM Bergen Scientific Center)

Our experience in the use of information systems in mathematical consulting leads us to believe that numerical analysts must look beyond conventional expert systems if they are to be successful in reaching the general scientific user community. To achieve this goal we must develop interactive environments which directly assist in the solution of the user's problems. This presentation describes one such environment, NEXUS, currently under development. The NEXUS project entails two main activities:

- the development of a set of software modules for constructing and accessing information in a knowledge base

- the production of a set of knowledge bases for specific application areas.

Knowledge-Based Interface for Mathematical Software
C.W. Cryer (Westfalische Wilhelms Universitaet)

There exists today much excellent mathematical software, but it is widely felt that this software is not being fully utilised. Among the reasons for this are: casual users are simply unaware of all the software available; users lack the numerical expertise to select the best routine for their problem; many routines require complicated calling sequences which intimidate occasional users. A natural approach to overcome these difficulties is to provide an intelligent interface to the software. After surveying work in this area we discuss the construction of an expert system for the NAG Library, which first guides the user towards the selection of a routine and then helps in the use thereof.

• ELLIPTIC SINGULAR PERTURBATIONS
L.S. FRANK (University of Nijmegen)

The elliptic theory of singular perturbations has been developing in different directions during the last decennium: uniform two-sided a priori estimates for differential singular perturbations with general coercive boundary operators, convergence when the small parameter vanishes, reduction of coercive singular perturbations to regular ones, extension of results to coercive pseudodifferential (Wiener-Hopf) singular perturbations and some classes of quasi-linear singularly perturbed elliptic boundary value problems, numerical analysis of coercive singular perturbations and so on. The purpose of the minisymposium is to present some results obtained in these directions, and to discuss their applications to several problems with small or large parameters which appear in Mathematical Physics.

Coercive Singular Perturbations: Reduction to Regular Perturbations and Applications

L.S. Frank (University of Nijmegen)

The algebraic concept of coerciveness for a one parameter family $\varepsilon \to \mathfrak{A}^\varepsilon$ of linear singular perturbations was introduced in 1976.

It was shown that the coercivity is equivalent with the validity of two-sided a priori estimates in Sobolev type spaces uniformly with respect to $\varepsilon \in (0, \varepsilon_0]$.

It turns out that coerciveness guarantees the existence of a reducing operator S^ε (constructed explicitly) such that

$S^\varepsilon \mathfrak{A}^\varepsilon = \mathfrak{A}^0 + \varepsilon^\gamma \mathcal{Q}^\varepsilon$, where $\mathfrak{A}^0$ is the reduced operator for $\mathfrak{A}^\varepsilon$, $\gamma > 0$ is some positive constant and $\varepsilon \to \mathcal{Q}^\varepsilon$ is a family of (uniformly with respect to ε) bounded operators in corresponding spaces.

Some applications of the reduction procedure for coercive singular perturbations will be considered, especially asymptotic formulae and numerical treatment.

High Order Asymptotics for Elliptic Singularly Perturbed Eigenvalue Problems

J.J. Heijstek (Catholic University)

The Frank-Wendt method ([1]) based on a constructive reduction of coercive singular perturbations to regular ones, is applied for computing high order asymptotic expansions for eigenvalues of such perturbations. The first two terms were indicated in [2].

The same method turns out to be useful for investigating the asymptotic behaviour of solutions of eigenvalue problems for certain classes of quasi-linear coercive singular perturbations.

The results are summarized in [3,4].

[1] L.S. Frank and W.D. Wendt, Comm. P.D.E., 7, 1982, p. 469-535.

[2] L.S. Frank, C.R. Acad.Sc.Paris, 301, I, 3, 1985, p. 69-72.

[3] L.S. Frank and J.J. Heijstek, High Order Asymptotics for Rayleigh's Eigenvalue Problem, Report 8625, 1986, Cath. Univ. Nijmegen.

[4] L.S. Frank and J.J. Heijstek, On the Eigenvalue Problem for Quasi-Linear Coercive Singular Perturbations (to appear).

Stability and Convergence in Singular Perturbation Problems

D. Huet (Universite de Nancy I)

Two-sided a priori estimates (established for differential coercive singular perturbations by L. Frank and for pseudodifferential ones by L. Frank and W.D. Wendt) in Sobolev type spaces with two indices, can be sharpened under certain natural conditions and lead to the inverse stability of elliptic singular perturbations uniformly with respect to the small parameter.

The inverse stability and the concept of the proper approximation of spaces (introduced by J.P. Aubin and F. Stummel) can be used for establishing convergence results for solutions of elliptic singular perturbations when the small parameter vanishes.

Existence and Multiplicity Results for a Semilinear Elliptic Problem with a Large Parameter

G. Sweers (Delft University of Technology)

Consider the eigenvalue problem

$$\begin{cases} -\Delta u = \lambda . f(u) & \text{in } \Omega \subset \mathbb{R}^N \text{ bounded}, \\ u = 0 & \text{on } \partial\Omega \text{ smooth}, \end{cases}$$

where f is such that there exist two numbers $0<\rho_1<\rho_2$ with $f(\rho_1)=f(\rho_2)$ and $f>0$ in (ρ_1,ρ_2).

If $f(0)>0$ Hess gives a sufficient condition for f, such that if λ is large enough a positive solution u exists with max $u \in (\rho_1,\rho_2)$. Without assuming $f(0)>0$, we show that this condition is sufficient and necessary. Secondly, if such a solution exists and $f'(\rho_2)<0$, there is a unique curve $\lambda \to u(\lambda)$ of stable solutions with max $u(\lambda) \to \rho_2$ if $\lambda \to \infty$. Moreover $u(\lambda,x) \to \rho_2$ if $\lambda \to \infty$ for all $x \in \Omega$.

The Reduced Operators of Singularly Perturbed Wiener - Hopf Matrices
W. Wendt (Institut fur Angewandte Mathematik)

The reduced problem is formulated for singularly perturbed pseudodifferential equations with unknown potentials and with general boundary conditions. It is indicated how the stability theory developed by L. Frank and the asymptotic analysis developed by L. Frank and the speaker for singularly perturbed problems without potentials, can be extended.

• FINITE DIFFERENCE APPROXIMATIONS TO HYPERBOLIC INITIAL-BOUNDARY VALUE PROBLEMS
M. GOLDBERG (Technion Institute of Technology), H.O. KREISS (California Institute of Technology)

The purpose of this minisymposium is to discuss various aspects of finite difference approximations to hyperbolic initial-boundary value problems. This includes stability of grid interferences, far field boundary conditions, stability of translatory boundary conditions, intermediate boundary conditions, error analysis for artificial boundary conditions, boundary conditions for different time scales, boundary conditions for multidimensional flow problems, boundary conditions for conversation laws, instability of inflow-dependent boundary conditions, and ill-posedness of absorbing boundary conditions.

Stability and Conservation of Difference Approximations and Grid-Interfaces
M. Berger (Courant Institute of Mathematical Sciences)

Calculations using grid refinement, where finite grids are embedded in coarser grids, or using multiple grids in different coordinate systems, are becoming more common. Both these methods have grid interfaces in the interior of the domain, where special finite difference schemes need to be applied. We derive several such difference schemes for use at grid interfaces, and discuss their stability and conservation properties.

Far Field Boundary Conditions for Wave Computations over Long Time Intervals
B. Engquist (University of California, Los Angeles)

A class of computational far field boundary conditions for the numerical solutions of hyperbolic partial differential equations are presented and analyzed. These boundary conditions are used in order to restrict the computational domain and are designed such that they will limit artificial boundary effects for time dependent problems as well as for steady state calculations.

Convenient Stability Criteria for Difference Approximations of Hyperbolic Initial - Boundary Value Problems
M. Goldberg (Technion Israel Institute of Technology)

In this joint work with E. Tadmor, a new, convenient stability criteria are provided for a large class of finite-difference approximations to initial boundary value problems associated with the hyperbolic system $\partial u/\partial x = A\partial u/\partial x + Bu + f$ in the quarter plane $x \geq 0$, $t \geq 0$. Using the new criteria stability is easily established for numerous combinations of well-known basic schemes and boundary conditions, thus generalizing many special cases studied in recent literature.

Remarks on Intermediate Boundary Conditions

D. Gottlieb (Brown University)

We will analyze the role of intermediate boundary conditions for hyperbolic equations. In these cases one cannot apply directly the Goldberg-Tadmor stability criteria since the boundary conditions are not necessarily in translatory form. We will present many examples that demonstrate that great care should be taken in formulating these boundary conditions.

Error Analysis for Artificial Boundary Conditions

L. Halpern (Ecole Polytechnique)

In this joint work with J. Rauch, we analyze the error committed in the solution of the wave equation when introducing artificial boundary conditions. The error estimate is a sum of two terms. One is proportional to the largest relevant reflection coefficient, and the other is proportional to 1/k, where k is a measure of the average frequency in the wave considered. The result, and the proof, are quantitative versions of the now standard qualitative results describing the reflection of singularities.

Boundary Conditions for Problems with Different Time Scales

H.O. Kreiss (California Institute of Technology)

NO ABSTRACT

Conservative Interface and Boundary Conditions for Multidimensional Flow Problems

R. LeVeque (University of Washington)

The numerical solution of hyperbolic systems of conservation laws in several dimensions typically requires irregularities in the grid; near irregular boundaries, at shocks or other tracked interfaces, or at mesh refinement interfaces. It is desirable to use boundary and interface conditions that are conservative and consistent with the highly stable and accurate interior schemes that have recently been developed. Additional difficulties in this pursuit arise if there are very small irregular mesh cells (we wish to avoid greatly restricting the allowable time step) or if different time steps are used in different regions (as in mesh refinement). An approach toward deriving such boundary conditions will be discussed.

Instability of Discrete Inflow-Dependent Boundary Conditions for Hyperbolic Systems

E. Tadmor (Tel Aviv University)

It is well known that for mixed initial-boundary hyperbolic systems to be well posed, the boundary conditions should be independent of the inflow eigenspace connected with the problem. We discuss the discrete analogue of the above; namely, we show that numerical boundary conditions which are consistent with the inflow part of the interior equations lead to instability.

Ill Posedness of Absorbing Boundary Conditions for Wave Equation Migration

L.N. Trefethen, L.H. Howell (M.I.T.)

Migration, which is common in geophysics and underwater acoustics, refers to the solution of a p.d.e. that approximates an ideal "one-way wave equation". In 1980 Clayton and Engquist derived absorbing boundary conditions for migration, also based on a one-way idea, that have been successful in many applications. We show that the "B2" Clayton-Engquist equation is ill-posed as a boundary condition for the "45°" migration equation. One reason is that there is a Kreiss-type wave mode which radiates energy in from the boundary. But there is also a second, more unusual mechanism of ill-posedness involving waves propagating back and forth between two boundaries at unbounded speeds. In calculations on fine meshes the instability can be quite dramatic.

• MINISYMPOSIUM DEDICATED TO THE MEMORY OF J.H. WILKINSON

G.H. GOLUB (Stanford University)

The work of J.H. Wilkinson has had enormous impact in matrix computations. He developed a model of roundoff which served well to analyze a number of numerical methods. The model was particularly effective in combination with backward error analysis. In this minisymposium, the speakers review and analyze Wilkinson's work in relationship to error analysis, the solution of linear least squares problems, and the symmetric eigenvalues problem.

Jim Wilkinson and Least Squares Computations
A. Bjorck (Linkoping University)
The work of Jim Wilkinson in error analysis and matrix computation has had a profound influence on methods for solving linear least squares problems. This talk reviews some new and old research inspired by him.

Jim, You Left Us Too Soon
W. Kahan (University of California, Berkeley)
Two kinds of problems in error analysis to whose solutions Wilkinson made massive contributions still persist. One concerns excessively pessimistic error bounds, the other concerns practices spuriously motivated by invoking backward error analysis. An example of the latter is unfortunate compromises in the design of arithmetic engines. An example of the former is element growth in Gaussian Elimination.

$W_{21}^{\pm}$ Revisited
B.N. Parlett (University of California, Berkeley)
Wilkinson constructed two simple tridiagonal matrices to illustrate several aspects of the computation of eigenvectors. We review briefly that material and go on to consider three related topics concerning a tridiagonal symmetric matrix T.

1. At what position should the forward and backward recurrences be matched when solving (T- I)x = 0?
2. The use of submatrices in computing orthogonal eigenvectors belongs to numerically equal eigenvalues.
3. What is the minimal gap between eigenvalues when the off diagonal elements are all 1?

Jim Wilkinson's Contributions to Rounding Error Analysis
G.W. Stewart (University of Maryland)
One of Jim Wilkinson's greatest contributions to numerical linear algebra was to place the analysis of rounding error on a sound basis. In this talk, we will discuss the status of rounding error analysis before Wilkinson's work on the subject and show how his approach resolved a number of outstanding questions.

• MULTIGRID: ROBUST METHODS, EFFICIENT ALGORITHMS AND APPLICATIONS
W. HACKBUSCH (University of Kiel), J.F. MAITRE (Ecole Centrale de Lyon)
The use of Multigrid Methods has been growing rapidly these last years, with a diversification of the applications. The optimal cost of O(N) for solving discretized problems is fundamental, but practitioners ask for robust and efficient programs to solve their problems. Recent theoretical and computational studies have led to notable advances in this direction and made accessible new areas of computation. This series of talks will report on recent developments concerning principally adaptive finite element methods, 3-D simulations, and applications to characteristic problems such as Euler and Stokes equations. Time will be saved to allow open discussion.

Multigrid Approaches to the Euler Equations
P.W. Hemker (Centre for Mathematics and Computer Science)
A consistent sequence of Galerkin discretizations for the steady 2-D Euler equations is obtained by the finite volume technique and Osher's approximate Riemann solver. On the

different levels, non-linear Symmetric Gauss-Seidel relaxation is applied to solve the discrete systems. The FAS-multigrid technique accelerates this iterative process.

In subsonic, transonic, and supersonic problems the multigrid iteration is applied to initial estimates obtained by nested iteration. The rate of convergence appears to be independent of the number of cells in the mesh. The total amount of work to obtain a solution, corresponds to a few relaxation sweeps.

Robust and Efficient Multigrid Methods

P. Wesseling (University of Technology)

The results obtained in multigrid convergence theory are briefly reviewed. Important practical cases not sufficiently covered by the theory are identified. Applicability and limitations of existing multigrid codes are discussed.

Multigrid Methods for Stokes and Navier-Stokes Equations

G. Wittum (Universitat Kiel)

In the present talk several multi-grid methods for the Stokes and stationary and incompressible Navier-Stokes equations are discussed. New methods are proposed by using incomplete factorisations as smoother in a distributive manner, applicable for finite-difference discretisations as well as for finite-element ones. Stability and convergence results are given. For the Stokes equations discretized on staggered grids in the unit square numerical tests and comparisons with the distributive Gauß-Seidel method by Brandt-Dinar and the multi grid method with Uzawa-smoothing by Nigon are presented. As inner iteration in Newton and Quasi-Newton methods the process can be used for solving the Navier-Stokes.

Robust Hierarchical Basis Multigrid Methods

H. Yserentant (Fachbereich Mathematik der Universitat Dortmund)

The efficiency of the classical multigrid methods depends on two assumptions: First, that the number of unknowns increases from level to level at least by a given factor greater than one, and secondly on certain regularity assumptions for the continuous problem and the sequence of grids. Both assumptions are hard to verify for adaptively refined meshes, and there remain some doubts whether they are strictly satisfied in practical applications. In a joint work with Randy Bank and Todd Dupont we developed a multigrid method that does not have these problems due to its construction and that we applied with great success to locally refined meshes. The method looks like a V-cycle multigrid method and can be interpreted as a Gauß-Seidel type iteration with respect to a hierarchical basis of the considered finite element space.

- **EFFICIENT ALGORITHMS FOR THE SOLUTION OF COMBINATORIAL OPTIMIZATION PROBLEMS**

H.W. HAMACHER (University of Florida)

Combinatorial optimization (CO) is a powerful mathematical tool for modeling industrial problems. This minisymposium will deal with efficient algorithms for the solution of some CO problems. Topics include randomized heuristics for maximum weight independent sets, approximation algorithms for parametric network problems, and a new version of the traveling salesman problem, the prize collecting TSP.

Maximum Independent Sets via Randomization

P.M. Camerini (Universita di Siena), F. Maffioli (Politecnico di Milano), C. Vercellis (Universita di Milano)

A general frame for combinatorial optimization problems is offered by the search for a maximum weight independent set in a given weighted independence system S = (E, I, w). We propose a randomized heuristic for this problem and study some of its properties from a probabilistic point of view, under various assumptions concerning the nature of S.

Particular cases such as the traveling salesman problem, the (3-dimensional) matching problem and the (k-parity) matroid problem are also considered.

Efficient Approximation Algorithms for Parametric Network Algorithms

R.E. Burkard, G. Rote (Technische Universitat Graz), H.W. Hamacher (University of Florida)

The objective value function of a min cost flow problem with parametric capacities is known to be a piecewise linear function with a, in general, exponential number of breakpoints. We present two algorithms to approximate this function. The first is a global version of the negative cycle algorithm. The second is a sandswitching algorithm, in which the objective function is bounded above and below by functions whose norms will be less than a given ε when the algorithm terminates.

Remarks on Some Industrial Scheduling Problems

R.E. Burkard (Technische Universitat, Graz)

Two industrial scheduling problems and methods for their solution will be described. The first concerns the sequencing of jobs with due dates in order to minimize sequence-dependent set-up times. The other arises from the manufacturing process in a tool factory and is treated by modified Greedy algorithms. Moreover a mathematical framework will be presented which allows one to treat scheduling problems with different objectives in a unified way.

On The Prize Collecting Traveling Salesman's Problem

E. Balas (GSIA Carnegie Mellon University)

The problem of scheduling the daily operation of a rolling mill has been formulated as follows. A salesman who travels between cities i and j at cost c_{ij} and collects a prize p_i if he visits city i, has to visit enough cities to collect a prescribed amount of prize money while minimizing the cost of travel. We describe several classes of facet inducing inequalities for the associated polytope, as well as solution methods for the problem.

- **WAVES DUE TO A MOVING LOAD ON A FLOATING ICE SHEET**

R.J. HOSKING (University of Waikato)

Recent international interest in the response of a floating ice sheet to a moving load (aircraft, hovercraft, etc.) has enhanced our understanding of the importance of so-called critical speeds for the existence and patterns of waves that can be generated.

Mathematical studies have encompassed both steady and time-dependent calculations and other features such as the influence of stress in the ice, or the possibility of internal waves in the underlying water. Detailed observations made only recently may be compared with the developing theory.

Analysis of Floating Ice Covers Subjected to Moving Loads

A.D. Kerr (University of Delaware)

Introduction to subject by reviewing the evolution of analyses for a simple related problem; a beam on a Winkler base that is subjected to a moving load. Discussion of effect of speed, damping, mass, and axial force. Formulation of the title problem assuming that the ice cover is elastic and the liquid base is inviscid and incompressible. Discussion of effects of moving liquid, depth of liquid base, snow layer on the ice cover, and constrained thermal expansions in the cover. Review of available solutions. Formulation of problem assuming that plate is viscoelastic and liquid base is inviscid but compressible. Discussion of related results.

Flexural and Internal Waves

R.J. Hosking (University of Waikato)

A vehicle travelling over floating ice excites flexural (elastic-gravity) waves in the ice that propagate in well-defined patterns, depending on the vehicle speed. If the underlying water is stratified, internal waves may also be generated in the water. A vehicle travelling faster than the well known "critical speed" (for the dynamic response of the ice) excites both flexural and internal waves, but a slower vehicle produces only internal waves and a static response in the ice. The flexural wave amplitude, frequency, and phase depend on the thickness and elastic properties of the ice.

Time-Dependent Ice Plate Deflections

R.M.S.M. Schulkes (Delft University of Technology)

The nature of waves generated by a load moving with constant speed over floating ice depends on the speed of the load and on the time the load has been applied. An exact expression for the deflection is derived by modeling the system via an elastic plate subjected to a concentrated line load. The long time behaviour and spatial development of the deflection is then studied using asymptotic methods. Disturbances generally approach a steady state but there are certain critical speeds at which deflections may grow continuously with time.

Comparison of Theory and Observations for Loads Moving on Sea Ice

V.A. Squire, P.J. Langhorne (University of Cambridge),
W.H. Robinson, A.J. Heine, T. Haskell (Physics and Engineering Laboratory)

During November 1986 a field programme to investigate strains and acoustic emission due to loads such as aircraft and trucks moving across sea ice was carried out in McMurdo Sound, Antarctica. Strain gauges of two distinct types were used to measure the induced strains, and were deployed in rosettes to enable directional facets of the wave fronts to be found at all reasonable speeds. An acoustic emission device simultaneously recorded induced micro-cracking events within the sea ice. The quality of the resulting dataset is excellent, and enables a full comparison of theory and data to be made.

• SIMULATION OF PETROLEUM RESERVOIRS

J. JAFFRE (INRIA-Rocquencourt)

Due to the physical and chemical complexities of oil recovery processes, simulation of petroleum reservoir is a very rich source of mathematical and numerical problems with important economic consequences. The minisymposium is organized as follows: there is one communication which is a survey of problems arising in petroleum reservoirs; the three remaining talks present some recent work addressing specific issues, viscous fingering, fractured reservoirs and thermal recovery.

A Survey of Reservoir Simulation

R.E. Ewing (University of Wyoming)

The variety of petroleum recovery techniques currently in use often involves multiphase or multicomponent flow of reservoir fluids and or chemicals through porous media. The mathematical models describing these processes involve coupled systems of nonlinear, time-dependent partial differential equations, usually requiring a wide variation in length and time scales. A variety of techniques for discretizing and linearizing these model equations and then efficiently solving the large linear systems which result will be
surveyed. The need for utilizing local, adaptive grid refinement techniques to resolve the critical local phenomena will also be described and promising refinement methods will be presented.

Detailed Simulation of Viscous Fingering in Porous Media

M.A. Christie (B.P. Research Centre, England)

This paper will describe detailed simulation of viscous fingering in a porous medium for and miscible flow. The numerical techniques involved in obtaining physically representative solutions will be discussed. Results will be presented which show good agreement with laboratory experiments in fingering in miscible systems. Further results show fingering obtained in a quarter five-spot geometry. Coinjection of water and solvent in a matched velocity flood is shown to result in almost complete suppression of fingering for an example calculation.

Simulation of Flow in Naturally Fractured Petroleum Reservoirs
J. Douglas, Jr. (University of Chicago and IMA, University of Minnesota)
This lecture reports joint research of Todd Arbogast, Paulo Jorge Paes Leme, and the speaker concerning the composite systems of equations describing the flow of fluids in a naturally fractured reservoir. Physically, the reservoir consists of a layer of sedimentary rock that was very slowly deformed in such a way that it fractured in an almost regular geometric pattern into blocks, usually parallelepipeds, of diameter ranging from a few feet to a few yards. The fractures, though of fingernail width, increase strongly the ability of the fluids to flow, and it is necessary to modify the modelling of such reservoirs to take this into account. The resulting models can be based on a system of differential equations in a Darcy-homogenized version of the fractures, considering the blocks as sources, and a system of equations in each block. In the case of a single phase fluid a model has been analyzed completely. For the two-phase case a model has been constructed and a numerical method proposed. This lecture will describe these results.

Numerical Simulation of In-Situ Combustion
P. Le Thiez, P. Lemonnier (Institut Francais du Petrole)
A two-dimensional model was developed for simulating oil recovery by wet or dry forward combustion at field scale. The simulator describes three-phase flow with gravity and capillary effects and heat transfer. The model includes four components (water, oil, oxygen, inert gas) and one reaction of fuel combustion. Emphasis was put on description of the combustion front (chemical reactions with Dirac formulation or Arrhenius kinetic). A new formulation using a heat release is presented, which improves stability and allows the use of large size grid blocks. A sensitivity study to numerical and physical parameters was done.

•MATHEMATICAL ANALYSIS OF CLIMATE MODELS
J.A. JELTSCH (Institut fur Geometrie und Praktische Mathematik)
This minisymposium will not only show what has already been achieved in the analysis of climate models by mathematical and numerical tools but also point to more complex models which wait to be tackled by mathematical analysis in the future. Climate models aim at casting the interaction of atmosphere, hydrosphere,...and the sun into an overall mathematical system that is able to describe climate, i.e. medium- and large-scale phenomena that occur due to internal dynamics of the geophysical system or as a response to the change of external parameters, e.g. anthropogenic sources. The possibility of a dramatic change of earth's climate due to pollution makes this research area a theme of vital importance.

Predicting the Future of Planet Earth
J.A. Dutton (The Pennsylvania State University)
The development of Earth system models to simulate the past and present and provide predictions of future conditions is essential now that human activities have the potential to induce changes in the planetary environment. A model of the Earth System on the scale of decades to centuries, developed by the Earth System Science Committee (NASA) with the strategy of dividing by time scale rather than disciplines, is presented, mathematical aspects are considered, and the requirements for observations to support the implementation of the model are reviewed.

Diffusion Problems Arising from Simple Climate Models
G. Hetzer (Auburn University)

Various sensitivity aspects of the climate system can qualitatively be treated by simple approaches. Typical examples are energy balance models, where a long-term mean of some yearly averaged atmospheric temperature is taken as the only climatic indicator. That mean is determined by demanding a pointwise equilibrium between incoming and outgoing heat and representing the horizontal heat transport by a diffusion operator. The resulting stationary diffusion problem depends on certain parameter and shows a specific sensitivity structure under reasonable hypotheses. Qualitative results are obtained by means of classical methods of nonlinear analysis such as degree theory or implicit function theorem.

Numerical Sensitivity Studies for a 2D Energy Balance Model

H. Jarausch, W. Mackens (Institut fur Geometrie und Praktische Mathematik)

Two dimensional energy balance models lead to a nonlinear diffusion equation on the surface of the globe. A finite element discretization produces a large set of nonlinear algebraic equations. The involved models of incoming and outgoing radiation depend on several parameters. Thus one has to study the solution set of parameter dependent nonlinear systems of equations. This is done by multiparameter bifurcation techniques as applied to a suitable condensation of the large system.

Modeling the Ice Ages with a Low-Order Dynamical System

B. Saltzman (Yale University)

An inductive model of long term (2 million year) changes in planetary climate is formulated in terms of a three-variable, nonlinear, dynamical system of equations governing the variables and feedbacks thought to be relevant from qualitative physical reasoning. The three variables are global ice mass, atmospheric carbon dioxide concentration, and ocean temperature. The effects of external forcing due to variations of the earth-orbital parameters are included in an attempt to account for the details of the observed paleoclimatic record over the last 500 thousand years.

• QUASI-NEWTON METHODS AND APPLICATIONS TO DIFFERENTIAL EQUATIONS

C.T. KELLEY (North Carolina State University), E.W. SACHS (Universitat Trier)

Discrete analogs of nonlinear boundary value problems and optimal control problems present difficulties when traditional Newton or quasi-Newton methods are used for solution. It is often difficult and expensive to compute the analytic Hessian or Jacobian that one would need for Newton's method and standard quasi-Newton methods do not always offer the rapid convergence that one would expect. This minisymposium looks at methods that are designed for specific problems in differential equations and optimal control. By taking the structure of the continuous problem into account performance can be enhanced.

Optimization Techniques for Inverse Problems in Elliptic Partial Differential Equations

K. Ito (Brown University), K. Kunisch (Technische Universitat Graz)

The problem of estimating functional parameters in elliptic PDEs often leads to an ill-posed inverse problem; i.e., the mapping from the parameters to the observations is not continuously invertible. We formulate the parameter estimation problem as a nonlinear optimization in Hilbert spaces, and develop numerical methods that take into consideration the ill-posedness of the inverse mapping. A hybrid method that combines the output least squares and the equation error methods using the augmented Lagrangian method for unconstrained optimization is proposed and analyzed. We also discuss the use of semi-norms in the regularization technique.

Approximate Quasi-Newton Methods

C.T. Kelley (North Carolina State University), E.W. Sachs (Universitat Trier)

We consider the effect of approximation on performance of quasi-Newton methods for infinite dimensional problems. We show how refinement of an approximation at each iterate can be done so as to preserve good local convergence behavior. This is applied to the numerical solution of nonlinear integral equation, boundary value problems and optimal control problems.

High Rank Quasi-Newton Methods for Constrained Optimal Control BVPs
H.G. Bock (Institut fur Angewandte Mathematik)

For BVP parameterizations or discretizations of Optimal Control problems, which satisfy a separability - or: local linear coupling - condition, high rank update formulae can be deduced, the local nature of which makes the rate of convergence essentially mesh independent. The approach applies to multiple shooting, for which it was originally developed (B./Plitt 81/83, MUSCOD), as well as to collocation or finite difference BVPs in a suitable representation satisfying the separability condition.

Successful applications e.g. in time optimal robot control or aerospace engineering - flight path optimization, low thrust interplanetary satellite missions, reentry manoeuvers - demonstrate superior local and global convergence properties compared to previous approaches.

Optimal Control Problems and Quasi-Newton Methods
C.T. Kelley (North Carolina State University), E.W. Sachs (Universitat Trier)

Quasi-Newton methods have been applied to optimal control problems in various contexts. We survey some of the applications and investigate the difference in these approaches. Several recent approaches to using quasi-Newton methods in optimal control are discussed.

• SOME APPLICATIONS OF MATHEMATICS IN CHEMICAL ENGINEERING
J. KING-HELE (Manchester University)

The minisymposium will consist of four lectures on different aspects of Mathematics as applied to Chemical Engineering within I.C.I. Two of the speakers are Industrial Mathematicians from I.C.I. (Runcorn) and two are from the Mathematics Department at Manchester University. Dr. I. Parker (I.C.I.) will open the symposium by outlining the range of applications of Mathematics within I.C.I. The other speakers will then describe some current research problems. Emphasis will be placed on the basic ideas rather than on the technical details.

The Way Mathematicians are used in I.C.I.
I. Parker (I.C.I.)

In this introductory lecture the various uses of Mathematics within I.C.I. will be described.

These include the modelling of new processes, the analysis of existing processes, and work in Management Services and Planning.

Current By-Pass in Electrochemical Cells
A.F. Jones, J.A. King-Hele (Manchester University)

In an ideal cell an applied potential difference leads to chemical reactions at the electrode surfaces with no current loss. In practical applications however some of the current by-passes the electrodes leading to inefficiency. The various routes for current leakage will be described and Mathematical Models discussed.

Total Variation Diminishing Schemes for High Peclet Number Flows
H.J. Ziman (I.C.I.), A.D. Gosman (Imperial College)

The use of computational fluid dynamics for the prediction of multi-dimensional convection dominated flow problems is hampered by the lack of an accurate bounded discretisation technique. Popular bounded schemes such as upwind differencing suffer from excessive numerical diffusion. Higher order schemes (e.g. QUICK) may produce the characteristic undershoots and overshoots of an unbounded scheme. Recently a promising new class of bounded schemes (total variation diminishing schemes) have been proposed for hyperbolic conservation laws like the Euler equations. This paper reports on the extension of these schemes to the convection-diffusion equation and presents the results from some selected test problems.

Software for the Solution of Chemical Equilibrium Problems
T.L. Freeman, I. Gladwell, W. Byers-Brown (Manchester University)
The mathematical model of a chemical equilibrium problem which describes an explosion can be considered as having three phases - explosion, detonation and adiabatic expansion. Using thermodynamic arguments the chemical properties of these phases are given as the solution of a sequence of nonlinear optimisation problems. We discuss the development of an interface between these problems and standard optimisation software and consider how the reformulation of the equations of state affects the numerical method of solution. In particular we describe how the choice of mathematical software has influenced the development of equations of state based on a hard sphere model.

• SEMI-INFINITE PROGRAMMING APPLICATIONS AND SOLUTIONS
K.O. KORTANEK (The University of Iowa)
This collection of papers illustrates the practicality of models having a finite number of variables in an infinite number of inequalities. The problems represented include: moment problems and integration formulas, two-sided uniform approximation, boundary value and eigenvalue problems, parametric programming, data representations, and engineering plasticity design involving material stress and displacements.

The talks will present numerical techniques developed over the last 20 years applicable to problems of this type, with a concentrated focus on the most recent developments including: a three phase algorithm, discretization methods, globally convergent methods, and use of the LP-Scaling algorithm (ala Karmarkar).

Numerical Treatment of an Ill-Posed Problem in Petroleum Technology
S.A. Gustafson, S.M. Skjaeveland (University of Rogaland)
The problem is represented as a Volterra integral equation of the first kind. The kernel is an analytic expression but the right-hand-side is a table of a limited number (say 7 or 8) of measured values. The physical problem requires the solution and the right-hand-side to be monotonic functions. This gives rise to sets of infinitely many linear equations in finitely many variables. Thus we arrive at a semi-infinite program when we want to determine how the solution is effected by errors in input data.

An Implementation of a Method for the Accurate Computation of Eigenvalues and Eigenfunctions of Elliptic Membranes
R. Hettich, E. Haaren, M. Ries, G. Still (Universitat Trier)
In an earlier paper a defect-minimization method was proposed to compute approximate eigenfunctions of membranes by means of a parametric semi-infinite optimization problem. The method produces approximations and error bounds of eigenvalues and eigenfunctions. An algorithm based on this method has been implemented for the special case of elliptic membranes with variable eccentricity.

A computer code with the following features is presented:

- Calculation of all eigenvalues for fixed eccentricity in a given interval
- Representation of an eigenvalue for a fixed mode of oscillation as a function of the eccentricity
- Graphical representation of eigenfunctions and their modal lines for simple and multiple eigenvalues.

Globally Convergent Methods for Chebyshev Approximation Problems
G.A. Watson (University of Dundee)

It is well known that Chebyshev approximation problems defined on infinite sets can be posed as semi-infinite programming problems. Some recent developments in globally convergent methods for such problems, both real and complex, are presented.

Numerical Treatment of Limit Analysis in Plane Strain
E. Christiansen (Mathematics department Campusvej), K.O. Kortanek (The University of Iowa)
Standard finite elements for the continuum plane strain limit analysis problem yield large dual linear programs which are very sparse with markedly non-unique stresses and displacement solutions. We first present robust solutions based upon the scaling LP algorithm, and then follow with a semi-infinite cutting plane algorithm which successively refines the approximating grids in a non-uniform way for more accurate approximations.

• ADVANCES IN THE THEORY OF INTEGER PROGRAMMING
B. KORTE (Universitat Bonn)
This session focuses on recent developments in the theory of integer linear programming and convex polyhedra. A common theme is the problem of obtaining linear inequality systems sufficient to define the convex hull of the feasible solutions to integer and mixed integer linear programs. In addition, issues of dual integrality are discussed, as are several applications of the theory to problems of combinatorial optimization. These problems include directed graphical Steiner problems and odd join polyhedra.

On Cutting Planes for Mixed Integer Programming
W. Cook (Columbia University)
The Chvatal closure of a polyhedron serves as a framework for approaching many combinatorial and integer programming problems. We discuss extensions of this notion to mixed integer programming. This talk is based on joint work with I. Barany, R. Kannan, and A. Schrijver.

Solving Graph Optimization Problems on Graphs with Forbidden Minors
H. Gan, E.L. Johnson (IBM Research Center, USA)
We consider four optimization problems over graphs and their linear programming duals: Chinese postman and packing of odd cuts; odd cut and packing of postman sets; copostman problem and packing of odd circuits; and minimum odd circuit and packing of co-postman sets. The problems are considered on graphs where Seymour's results on matroids with the max-flow min-cut property assure integer duals, i.e. postman and copostman problems on graphs with no K_4 minor, min odd cut on graphs with no $K^*_{2,3}$ minor and min odd circuit on graphs with no $K_{2,3}$
minor. In all cases we show simple reductions allow easy solution procedures for both the primal and dual problems. Also, all four problems can be solved over graphs with no 4-wheel minor but the dual variables might take on fractional values.

Polyhedral Projection and Steiner Problems
W.R. Pulleyblank (University of Waterloo)
A projection of a polyhedron can be obtained by "ignoring" the values contained in a subset of the coordinate positions. It is sometimes the case that combinatorial optimization problems can be formulated as a small linear program by using extra variables, but when the projection onto the set of "intrinsic" variables is obtained, the size of the linear program grows exponentially. We discuss this with respect to the two terminal Steiner problem in a directed graph. We show how a well-known simple solution procedure yields both a polynomially sized linear programming formulation using extra variables and an exponentially large "natural" formulation. This is joint work with E. Balas, M. Ball, Liu W.G.

Packing Odd Cuts

A. Sebo (Computer and Automation Institute, Hungary)

The problem of finding a maximum packing of cuts each of which is the coboundary of a set of vertices of odd cardinality (or, more generally, intersecting a given subset T of the vertices in an odd-cardinality set) is the common formulation of several well-known optimization problems: in particular, it contains the plane multi-commodity flow problem as a special case. K. Matsumoto, T. Nishizeki and N. Saito have recently shown that the plane multi-commodity flow problem can be solved by solving O(n) matching problems sequentially. We show that the general odd cut packing problem can be solved by solving n weighted matching problems independently of each other. We deduce then some applications, including algorithms that find maximum integral packings of some special cases or "nearly maximum" integral packings in general.

• DIFFERENTIAL ALGEBRAIC AND DIFFERENTIAL GEOMETRIC APPROACHES TO NONLINEAR SYSTEMS

A.J. KRENER (University of California)

The speakers in this minisymposium will discuss how differential algebraic and differential geometric tools can be used to attack a wide variety of problems in nonlinear systems theory.

Further Results on Feedback Linearization and Output Decoupling

T.J. Tarn (Washington University), D. Cheng (Academia Sinica), A. Isidori (Universita di Roma)

This paper presents some new results on the problem of feedback linearization and simultaneous output block decoupling. Necessary and sufficient conditions are obtained for the general case. These conditions simplify our previous results under the same assumptions. A new algorithm which is simpler than the previously obtained one is included in this paper.

On the Representation of Distributed Parameters Systems

M. Fliess (C.N.R.S., E.S.E., L.S.S.)

L'etude des systemes a parametres repartis, c'est-a-dire regis par des equations aux derivees partielles, a suscite de nombreuses etudes ayant souvent pour but de generaliser les concepts et les resultats des systemes modelises par equations differentielles ordinaires. Grace a l'algebre differentielle, dont l'auteur a recemment montre la centralite en automatique, il est montre comment obtenir une representation naturelle par espace d'etat. Des applications a divers problemes importants, comme les observateurs, seront discutees.

The Attitude Control and Stabilization of Rigid Spacecraft

C.I. Byrnes (Arizona State University), A. Isidori (Universita di Roma)

In this talk we present general results on the stabilizability of a class of nonlinear systems, which includes Euler's equations governing the motion of a rigid spacecraft and the rigid body model for an n-link robot arm operating in an environment where gravitational forces can be neglected or cancelled. Somewhat surprisingly, for this class of systems, we find that stabilization by nonlinear feedback is equivalent to linearizability by the nonlinear feedback group. In particular, a rigid spacecraft can be stabilized using m independent actuators if, and only if, $m \geq 3$ thereby answering in the negative the longstanding open problem of stabilizing the attitude of a rigid satellite with two momentum exchange devices. This particular system is known, however, to be locally null controllable. Using our nonlinear frequency domain design methods, we can, however, design a feedback law stabilizing this system about a principal axis.

Asymptotic Inversion of Linear and Nonlinear Systems

A.J. Krener (University of California)

The concept of asymptotic left and right inverse of a system are defined and calculations for the existence of such inverses are given. The close relationship of these inverses with the concept of a zero of a system is discussed.

• MATHEMATICAL COMBUSTION THEORY AND COMBUSTION MODELLING

S.B. MARGOLIS (Sandia National Laboratories)

This minisymposium will describe recent analytical and numerical studies in combustion. The talks will focus on transient processes associated with the deflagration to detonation transition, on stability and bifurcation phenomena in gaseous and condensed-phase combustion, on nonadiabatic flame propagation, and on the numerical computation of unsteady combustion waves.

Thermal Explosion in a Reactive Gas

A.K. Kapila (Rensselaer Polytechnic Institute)

Localized thermal explosions play a crucial role in the evolution of dramatic combustion phenomena such as detonations. This study examines the spatial structure and temporal evolution of a thermal explosion, with special attention to the gaseous nature of the medium. The analysis adopts Arrhenius kinetics, employs a combination of asymptotics and numericals, and terminates when the pressure within the exposion has peaked.

On the Stability of Stretched Flames

M. Matalon (Northwestern University)

When the diffusion rates of heat away from, and mass towards a flame do not differ greatly, the flame front is expected to remain stable. However, thermal expansion, always present in combustion systems, has a destabilizing influence. The combined effects show that long waves remain unstable and that thermo-diffusive effects stabilize the shorter waves. In the present study, we examine the effect of flame stretch which suppresses the long wave hydrodynamic instability.

Dynamics of Nearly Extinguished Non-Adiabatic Premixed Flames

H.G. Karper, G.K. Leaf (Argonne National Laboratory), B.J. Matkowsky, W.E. Olmstead (Northwestern University)

We present a nonlinear stability analysis of a steadily propagating plane flame subject to gravitational forces and small volumetric heat loss. We derive a nonlinear partial differential equation describing the evolution and structure of the flame near the "extinction point," i.e., the point beyond which a steadily propagating plane flame cannot sustain itself. We show that the equation admits solutions which correspond to non-steady non-planar flame fronts beyond the extinction point.

Adaptive Numerical Computation of Steady and Unsteady Combustion Problems

A. Bayliss, B.J. Matkowsky (Northwestern University), M. Minkoff (Argonne National Laboratory)

We employ an adaptive pseudo-spectral method in one and two dimensions to compute bot steady and unsteady solutions of combustion problems. The numerical method dynamically adapts the coordinate system in order to minimize a functional which is a measure of the spectral interpolation error and also adds and subtracts collocation points as required. We describe cellular and oscillatory flames, which exhibit spatial and temporal patterns, including patterns with steep gradients. These flames arise as bifurcations, with each successive bifurcation giving rise to more and more structure. Our computations allow us to describe bifurcation diagrams well into the fully nonlinear regime, far beyond the local regime where analytical solutions can be derived. In particular we are able to describe secondary and higher order bifurcations.

New Modes of Quasi-Periodic Combustion Near a Degenerate Hopf Bifurcation Point

S.B. Margolis (Sandia National Laboratories, USA), B.J. Matkowsky (Northwestern University)

Steady, planar propagation of a condensed phase reaction front is unstable to disturbances corresponding to pulsating and spinning waves for sufficiently large values of a parameter related to the activation energy. We consider the nonlinear evolution equations for the amplitudes of the pulsating and spinning waves in a neighborhood of a double eigenvalue of the problem linearized about the steady, planar solution. In particular, near a degenerate Hopf bifurcation point, we describe closed branches of solutions which represent new quasi-periodic modes of combustion.

• INVERSE PROBLEMS - ON THE NONLINEAR DEPENDENCE OF MATERIAL PARAMETERS ON SPECTRAL DATA

JOYCE R. McLAUGHLIN (Rensselaer Polytechnic Institute)

The understanding of the nonlinear dependence of material parameters on spectral data has an important role in the design of structures and in parameter identification. Applications include, e.g., identification of material parameters in the earth, identification of damping mechanisms in flexible structures, and in nondestructive testing. The covered topics in this mini-symposium will include, e.g., identification of discontinuous material parameters, and the use of new kinds of spectral data in the identification problems for the earth as well as in nondestructive testing. Both numerical and well-posedness results will be presented.

Inverse Problems Using Positions of Extreme Points or Nodal Positions as Data

J.R. McLaughlin (Rensselaer Polytechnic Institute)

A new approach to the inverse spectral problem is to use nodal positions or extreme point positions as spectral data. This data can be determined, e.g., by measurements made when materials are excited at their natural frequencies. Well-posedness results as well as constructive techniques will be discussed.

Inverse Eigenvalue Problems for the Earth

O.H. Hald (University of California, Berkeley)

We will present some uniqueness results for the inverse eigenvalue problem for the Earth and discuss the difficulties in solving the problem numerically.

The Inverse Problems for the Elastic Beam with Transfer Function Data

D.L. Russell (University of Wisconsin)

With a variety of electronic devices such as Fourier analyzers it is possible to obtain the trace on the imaginary axis of the transfer function for a physical system, such as an elastic beam, corresponding to a particular mode of control input and measurement output. It therefore becomes important to be able to estimate the parameters of the system from such data. We explore, using the tools of harmonic analysis, the feasibility of such a program and investigate some of its computational aspects.

A Global Analysis for a Nonlinear Eigenvalue Problem

T. Suzuki (University of Tokyo)

Our problem is to find $h = T(u,\lambda) \in C^2(\Omega) \cap C^0(\bar{\Omega}) \times \mathbb{R}_1$ such that

(1) $-\Delta u = \lambda e^u$ (in Ω) $u = 0$ (on $\partial\Omega$)

for a simply connected and bounded domain $\Omega \subset \mathbb{R}^2$.

From the implicit function theory is constructed the branch of minimal solutions $\underline{C}$ originating from $(\lambda, u) = (0, 0)$, while from the singular perturbation method follows the existence of a branch of "large solutions" C^* as $\lambda \downarrow 0$. Our concern is the connectedness of $\underline{C}$ and C^*, and some geometrical methods will be used in the analysis.

• NUMERICAL SCHEMES BASED ON RIEMANN SOLVERS

B. MERCIER (Aerospatiale), J. OVADIA (C.E.A. Centre d'etudes de Limeil)

Since the pionneering work of S.K. Godunov in 1959, the use of Riemann solvers as a basis for numerical schemes has been shown to be of fundamental interest for time dependent gas dynamics and high speed problems.

The second order extension to Godunov's method due to Van Leer is a major step which has opened the way to accurate and non oscillatory schemes. Van Leer's idea was to replace the piecewise constant states used by Godunov by some picewise linear states.

However this means replacing the standard Riemann problems by Generalized Riemann problems, which has been the subject of interesting recent work. For our proposal we have chosen some people able to review the results obtained in this area, as well as some applications to reactive flows, curvilinear geometries, and multidimensional problems.

The Generalized Riemann Problem for Reactive Flows

M. Ben-Artzi (Israel Institute of Technology)

The system of equations of reactive flow consists of four equations: the three standard conservation laws of compressible fluid flow and a "reaction equation" for the mass fractions of the reactants. A general equation-of-state is assumed (internal energy as a function of density, pressure and mass fraction) as well as an Arrhenius-type model for the reaction equation. The generalized Riemann problem for the system is the IVP (Cauchy problem) where the variables are initially linearly distributed on both sides of a jump discontinuity. An analytic solution is provided for the time-derivatives of the flow variables at the discontinuity. A high-resolution Godunov scheme is presented.

Application of Van Leer's Scheme to Detonation

A. Bourgeade (C.E.A. Centre d'Etudes de Limeil)

The approximate solution of the generalized Riemann's problem for reacting flows allows the extension of Van Leer's scheme to such flows. The presentation will describe this new scheme and show, with some examples, how it applies to detonation.

The examples are the ZND detonation and the study of shock to detonation transition. The results will show how promising is this new method.

Flux Splittng Versus Riemann Solvers: Comparison for Aerodynamic Computations

M. Borrel (ONERA, France)

A second order shock capturing method for van LEER's type has been developed at ONERA for aerodynamic computations with strong shocks.

This method is based on a finite volume approach for the discretization of the unsteady Euler equations in two and three dimensions.

A comparison of flux vector splitting techniques versus Riemann solvers is made and the extension to implicit methods is presented.

Efficient Implementation of High Order Accurate Essentially Non-Oscillatory Shock Caputuring Algorithms

S. Osher (UCLA)

In recent years there has been much successful activity concerning the construction, analysis, implementation, and applications of high resolution methods for problems with shocks. Methods based on TVD (total-variation-diminishing) schemes have become state-of-the-art tools in many areas of computational physics.

While these methods have been quite successful they can still be improved. They necessarily degenerate to first order accuracy at isolated critical points. Moreover the time discretization techniques should be simplified - especially as concerns steady state, multi-dimensional calculations. We shall describe recent work along these lines.

• GEOSTATISTICS THEORY AND APPLICATIONS

D.E. MEYERS (University of Arizona)

Geostatistics originated with the work of Matheron in the 1960's for estimation problems in mining and hydrology. Subsequent areas of application include soil physics, geotechnics, environmental monitoring and petroleum resource prediction. The mathematical foundation of geostatistics is reviewed, new developments and open problems are presented. Of special interest are the problems related to covariance estimation in the presence of drift. Extensions to multi-variate geostatistics are presented. The special nature of the prediction of oil in place is discussed together with numerical examples.

Presentation of Geostatistics

G. Matheron (Ecole Nationale Superieure des Mines de Paris)

The aim of geostatistics is to present regionalized phenomena with the help of topo-probabilistic models (random functions or random sets), and to use these models to predict the results of this or that technical operation applied to the phenomenon. Its originality does not lie in the mathematics it uses - these are fairly classical - but rather in the specificity of the problems that it seeks to solve. As an example, the problem of estimating the recoverable reserves in a selective mining operation (i.e., the support effect and the information effect) is presented.

Multivariate Geostatistics

D.E. Myers (University of Arizona)

Original applications of geostatistics were primarily univariate wherein the variable of interest was ore grade. Subsequent applications to hydrology and more recently to soil physics, environmental monitoring are commonly multivariate. The first formulation of cokriging emphasized the under-sampled problem. Myers [Math. Geology (1982)] gave the general formulation with the under-sampled problem as a special case (1984), and with Carr [Computers and Geosc. (1985), (1986)] provided programs for implementation. This formulation includes an extension to positive definiteness for matrix valued functions. These results are reviewed together with applications to generalizing interpolation by radial basis functions and splines.

Geostatistics Applied to Oil in Place Prediction

H. Omre (Norwegian Computing Center)

Information about oil in place of a petroleum reservoir is important in decision making concerning further exploration, development plans, and utilization. Both predictions of oil in place and prediction uncertainities have to be taken into account in the evaluation. These properties must be determined from available sources of information like general geologic knowledge, seismic reflection times, and data from wells. Hence the prediction problem contains both multivariate and spatial aspects.

In the presentation, an outline of the problem will be given. Alternative approaches for solving it will briefly be discussed. An approach based on theory of regionalized variables will be presented, and the results from a couple of examples will be given.

• NONLINEAR PARAMETRIZED EQUATIONS

H.D. MITTELMANN (Arizona State University)

This session is devoted to multi-parameter dependent nonlinear systems of differential equations and nonlinear algebraic systems. The solution manifolds of these problems may possess folds and bifurcation points.

The topics covered range from adaptive multi-level finite element methods and their use to continue along solution paths for nonlinear systems as they arise, e.g., in VLSI device simulation to the computation of multidimensional solution manifolds and the approximation of singularities on these manifolds. In addition to theoretical results working algorithms and programs will be presented as well as numerical results for problems from various fields of applications.

Adaptive Mesh Refinement for Parameter-Dependent pde's

R.E. Bank (University of California at San Diego)

The use of adaptive local mesh refinement for parameter dependent elliptic pdes poses several interesting computational challenges. Among them are the changing character of the solution as the parameter varies, the interaction of mesh refinement with continuation procedures, and a posteriori error estimates required for the mesh refinement. This talk will discuss some approaches to these problems.

Continuation Methods for Parameter-Dependent Elliptic Systems

H.D. Mittelmann (Arizona State University)

Most systems modeled by partial differential equations depend on several parameters. They may enter the differential equation, the boundary conditions, or the domain may depend on them. With respect to these parameters multiple solutions, hysteresis phenomena, folds, bifurcations etc. may occur. The desired information, in addition to the solution, is frequently the behavior of functionals of the solution, a norm, a definite integral etc. under changes of one or more of the system parameters. A continuation method is presented to solve efficiently the above problems for a general class of nonlinear elliptic systems.

The method is incorporated into a multigrid finite element package and results will be presented for a realistic problem from VLSI device simulation.

Computational Analysis of Multidimensional Equilibrium Manifolds

W.C. Rheinboldt (University of Pittsburgh)

Under certain conditions, the solution set of parameterized equilibrium problem forms a differentiable manifold in the combined state and parameter space, and its dimension equals that of the parameter space. New methods have been developed that allow for an efficient computational triangulation of segments of such manifolds. An essential part of these methods is an algorithm for the computation of moving frames. After a review of these results, various approaches will be presented which use the resulting triangulations for the computation of foldpoints and other features of the manifolds. Several computational examples will illustrate the techniques.

Approximation of Folds and Bifurcation Points for Operator Equations and their Discretizations

A. Griewank, G.W. Reddien (Southern Methodist University)

At certain critical parameter values the solution sets of nonlinear operator equations may change discontinuously, either with respect to their topology or their dynamic stability. These critical values as well as the underlying transition states can be calculated as solutions of defining systems that are obtained by appending the original equations with suitable singularity conditions. Several such systems for the computation of fold, Hopf and cusp points are discussed with respect to their numerical properties. Also considered are the relations between the singular points of discretized operators and those of the underlying continuous problem.

- **ANALYZING 0/1 ARRAYS BY ORDINAL AND COMBINATORIAL METHODS**

B. MONJARDET (Universite Rene Descartes)

In the recent development of the "Combinatorial data analysis" (P. Arabie), one of the main trends is the analysis of 0/1 arrays by methods using tools of discrete mathematics: (finite) graphs or ordered set theories, latticial or boolean algebra, combinatorial optimization. Such methods have been especially used within three teams: the Concept Analysis's group of R. Wille (Darmstadt); a group of psychologists and mathematicians (Ducamp, Doignon, Bruxelles Falmagne, New-York,...); the group "Ordered sets and Social Sciences" (Flament, Guenoche, Marseille, Barthelemy, Monjardet, Paris,...). The aim of the minisymposium is to present and confront the approaches of these three groups.

Median Graph Representation of 0/1 Table

J.P. Barthelemy (Ecole nationale Superieure des Telecommunications)

To each 0/1 table studied as a family of dichotomous variables a graph G is associated. Edges of G represent the variables and the individuals are assigned to some vertices, called real vertices. This paper is devoted to two topics:

- G is a median graph and the vertex set of G is generated by real vertices using the iterated ternary median operation.
- The table corresponding to some special median graphs (tree, cube, Hesse diagram of a distributive lattice...) may be characterized.

Order-Dimensional Analysis of 0/1 Data

J.P. Doignon (Universite Libre de Bruxelles), J.C. Falmagne (New York University)

Some 0-1 data can be explained by a single underlying ordinal factor. Due to Guttman, this idea was then generalized by Coombs and Kao to disjunctive and conjuctive models. Following Doignon, Ducamp and Falmagne, we give a precise definition of the ensuing dimensions as well as some related results. Improving these results, Chubb and Koppen have independently devised reduction algorithms for computing the dimension. We mention also a similar approach based on purely classificatory factors.

Using Logical Functions to Analyse 0/1 Arrays

A. Guenoche (C.N.R.S./G.R.T.C., France)

Frequently in Data Analysis, we are confronted with 0-1 array. A line represents a document and a column a property. The aim of this communication is to present algorithms to built boolean functions of the properties that take the value "true" on the array.

The first kind of function is called "implication rule" because we seek conjunctions of properties that imply other ones. This problem is very close to the simplifying logical functions one.

The second kind is called "dilemma function" that is an exclusive OR of conjunctions of properties. To such functions correspond partitions of the lines of the array. Our algorithm of enumeration of dilemma function provides a clustering method.

Implications and Dependencies of Attributes

R. Wille (Technische Hochschule)

Implications and Dependencies between attributes are of fundamental interest for clarifying the interplay of objects and attributes. In the frame of formal concept analysis such implications and dependencies can be successfully studied. In particular, functional dependencies between attributes may be turned into implications which can be systematically determined and visualized. For the determination and visualization of implications and dependencies between attributes, the concept lattice of a context of objects and attributes is basic.

• BIFURCATION AND NON LINEAR PHENOMENA IN INTERFACIAL PROCESSES

R. NARAYANAN (University of Florida)

The problems of multiple solutions that arise in interfacial phenomena are related to materials processing in space and convection. In these survey talks we shall show how bifurcation may arise through morphological instabilities in solidification and as surface waves in "Marangoni" driven flows. The problem of appropriate boundary conditions in porous media leads us to reconsider the validity of Darcy's equations. Pattern selection through stability, transition to chaotic structures etc. are shown to be deeply connected to nonlinearities that reside in the boundary conditions. Discussion on cascades and multiple bifurcation points will be included.

Nonplanar Interface Morphologies During Directional Solidification of a Binary Alloy = 3-D Calculations

G.B. McFadden, R.F. Boisvert, S.R. Coriell (National Bureau of Standards, USA)

During directional solidification of a binary alloy at constant velocity, the planar crystal-melt interface may become unstable and develop into a cellular, non-planar interface, exhibiting periodic structure transverse to the growth direction. The nonlinear free boundary problem associated with nonplanar growth is solved numerically for a model in which the temperature field is linear. Steady-state solute concentration fields and interface shapes are computed using a finite difference mesh which permits solutions with hexagonal symmetry. The various interface morphologies which bifurcate from the planar interface are calculated for parameters appropriate to an aluminum-chromium alloy.

Benard-Marangoni Instabilities in Porous Media
G. Lebon (Liege University)

Natural convection in a thin porous horizontal medium saturated with a viscous fluid, heated from below, is studied (Benard problem). In contrast with the common attitude, the law of Darcy is replaced by the Brinkman equation. The consequences on the onset of convection are discussed.

Brinkman's model is also used to study instability in a porous layer whose upper face is subject to a surface tension depending on the temperature (Marangoni problem). The influence of this effect is stressed. Both Benard and Marangoni instabilities are analyzed within the framework of a linear perturbation technique and the Serrin-Joseph energy method.

Patterns, Oscillations and the Nonlinear Evolution in Thermocapillary Marangoni-Benard Convection
M.G. Velarde (U.N.E.D.)

Here is a survey of the recent advances in our understanding of pattern formation, oscillatory convection, and other nonlinear features, including the transition to space-time chaos in thermocapillary Marangoni-Benard convection. Singular perturbation techniques have been used to study the evolution of the open deformable interface of a thin liquid layer heated from below or above. The relevance of the results to crystal growth and fluid dynamics experiments in microgravity conditions is emphasized and a comparison with ground-based experimental work is also presented.

Velocity Selection in Dendritic Growth
B. Caroli (Universite Paris VII)

Solidification of a pure undercooled liquid gives rise to the growth of dendrites (with quasi parabolic tip of radius R advancing at constant velocity V). In the absence of surface tension, the solutions of the corresponding free-boundary problem are a continuous family of paraboloids with a given value of the product RV. Experiments show that the physical system selects only one solution among this family.

We show that this can be explained by properly treating surface tension as a singular perturbation on this non-local non-linear problem.

• APPLICATION OF OPTIMAL CONTROL METHODS IN METEOROLOGY - 4-D DATA ASSIMILATION PROBLEMS
I.M. NAVON (Florida State University), B. NETA (Naval Postgraduate School)

The optimal control approach has been recently applied to the problem of variational data assimilation in Meteorology. Using the adjoint equations of the assimilating model, one can compute the gradient of a given distance function and solve the 4-D data assimilation problem as a minimization problem in the space of the meteorological model's initial conditions.

The speakers will discuss the current status of the research and numerical results obtained in assimilation of real meteorological observations.

Variational 4-D Data Assimilation with Use of the Adjoint Equations of the Assimilating Model

O. Talagrand (Ecole Normale Superieure), P. Courtier (E.C.M.W.F., UK)

Four-dimensional data assimilation is treated as a variational problem: one seeks the model solution which fits best, in the sense of a given distance-function, the available observations. This problem is solved as a minimization problem in the space of the model's initial conditions, the gradient of the distance function with respect to the initial conditions being computed through integration of the model ajoint equations.

Numerical experiments performed with different models show both the numerical efficiency of that approach and the meteorological quality of the results it produces.

The links between the variational approach and present operational assimilation procedures are discussed, in particular in terms of their implications for the choice of the most appropriate distance-function.

Potential Applications of Optimal Control Theory to Mesoscale Data Assimilation

J.C. Derber (Princeton University)

The analysis and assimilation of data into mesoscale forecast models is currently one of the most difficult meteorological problems. The application of optimal control theory potentially can solve many of the problems. Examples of the difficulties and possible solutions using control theory will be presented.

Building Time Continuity into a Sequence of Upper Air Analysis by Using Dynamical Constraints

J.M. Lewis (National Severe Storms Laboratory, USA)

Analyses of upper air circulation are becoming available at shorter intervals with the new observation platforms in existence. In particular, the VISSR Atmospheric Sounder aboard GOES routinely produces analyses of temperature and wind at intervals of 3-6 h. These analyses are often plagued with obvious inconsistencies or discontinuities in space and time. To alleviate these problems, a strategy based on simple vorticity conservation laws has been instituted and tested in the tropical Atlantic and in mid-latitudes over the U.S. The sequence of analyses are variationally adjusted subject to the constraint of vorticity conservation using the adjoint model.

Variational Methods for Data Assimilation in Meteorology

F.X. LeDimet (Universite de Clermont 2)

A general formation for analysis and data assimilation of meteorological field is given. The method is founded on application of optimal control theory. The algorithms used are reviewed. An application is done for retrieval of data collected during the COPT experiment in Africa in 1981. Dynamic of a squall line is analyzed. Further developments of optimal control methods to meteorology will be discussed.

• FREE AND MOVING BOUNDARY PROBLEMS

J. OCKENDON (University of Oxford)

This minisymposium has its origins in the triennial interdisciplinary meetings "Free and Moving Boundary Problems: Theory and Application," the latest of which was organised at Irsee (Augsburg) in June 1987. Promient themes at these meetings have been the Stefan problem and its generalisations and free boundary problems in fluid and solid mechanics, chemical reactions and materials science. These areas have posed many problems concerning existence, regularity, numerical analysis, especially the use of fixed domiain methods, and optimal control. This minisymposium will describe just some of the problems and techniques to emerge from the Irsee meeting.

Systems with Multiple Free Boundaries

G.H. Meyer (Georgia Institute of Technology)

Free boundary problems for multi-dimensional reaction-diffusion equations are solved numerically with a sequentially one-dimensional front tracking method. A time-implicit algorithm results which can find distinct free boundaries for each species equation. Convergence of the algorithm will be discussed and numerical experiments with two Arrhenius reaction problems with different reaction thresholds will be presented.

Surface Water Waves with Vorticity
J. Norbury (University of Oxford)
The free boundary problem of the steady flow of an ideal fluid in a channel is considered. The flow possesses vorticity and a free surface under constant atmospheric pressure, and gravity acts at right angles to the channel bottom. Although local bifurcation theory proving existence of small amplitude waves has been with us for many years, global existence theory has remained open, as has the relationship between variational and topological solution techniques.

We will map the problem onto a fixed domain and relate periodic wave solutions of the nonlinear elliptic problem to (non-trivial) saddle points of an energy function.

Free Boundary Problems in Hydrodynamics
J.M. Aitchison (RMCS Shrivenham)
Numerical methods for the solution of free surface problems in hydrodynamics will be reviewed. Typical problems include the calculation of the shape of jets, the cavity formed behind an obstacle and the flow over weirs. The free surface conditions generally involve an unknown parameter and so the problem cannot readily be solved by fixed domain methods. This talk will compare methods for dealing with the resulting variable domain and the calculation of the unknown parameter.

The efficient use of finite element or boundary integral methods will also be discussed.

- **CHARACTERISTICS OF THE COMPACT DISC DIGITAL AUDIO SYSTEM**

J.B.H. PEEK (Philips Research Laboratories, the Netherlands)
In order to appreciate more the next three presentations in this Minisymposium this tutorial presentation intends to explain the characteristics of the Compact Disc Digital Audio System. The following characteristics will be described: the laser optical scanning of the disc, the use of control, and display data and the insensitivity to dust, scratches and fingerprints compared with the normal grammophone. One of the reasons of less sensitivity to imperfections is the use of two Reed-Solomon error correcting codes and the application of interleaving in order to deal with burst errors. A further important characteristic that will be described is that the audio signals are digitally recorded on the disc and that digital signal processing is used in the CD player.

Compact Disc Boundary Value Problems
J.P.H. Benschop (Philips Research Laboratories, the Netherlands)
The signal generated by scanning an optical disc (e.g. a Compact Disc) with a focussed laser spot can only be obtained once the scattering (diffraction) of light by the surface height profile of the disc is calculated with adequate precision.

The standard approach in optics is the application of the Kirchhoff integral to the diffraction problem to the obtain angular distribution of the scattered light amplitude. The boundary conditions are approximated here. Although this approach is quite valid for diffracting structures which have a size (or period) of several wavelengths, the results become less reliable when we are dealing with the detailed structures present on a Compact Disc. The rigorous boundary value problem should now be solved. A comparison between the complexity and the precision of both approaches will be presented.

Applications and Problems of Error Correction Coding with Respect to Storage Channels
C.R.M.J. Baggen (Philips Research Laboratories, the Netherlands)

The Compact Disc system can be seen as a particular implementation of a digital storage channel. It will be shown that the use of error correction coding is inevitable in almost all possible applications of modern digital mass storage systems.

Traditionally much mathematical effort has been invested into the design of good codes. In the current industrial environment, more emphasis is put on the decoding algorithms and performance evaluations of codes.

The use of Reed Solomon codes belonging to the class of Maximum Distance Separable Codes will be elucidated. Finally it will be shown that product codes offer interesting possibilities although both the optimal decoding strategy and the performance evaluation still are open problems.

Adaptive Restoration of Unknown Samples in Certain Discrete-Time Signals; Mathematical Aspects

R.N.J. Veldhuis, A.J.E.M. Janssen (Philips Research Laboratories, the Netherlands)

A method for the restoration of unknown samples in a discrete-time signal is discussed. The signal is modelled as an autoregressive process. The method performs well on digital audio signals. An application is additional error correction for Compact Disc.

If the autoregressive parameters are known, the samples can be estimated by minimizing a quadratic function. However, they vary with time and must be estimated.

Estimating parameters are unknown samples simultaneously comes down to minimizing a fourth order function. An iterative procedure that calculates estimates for parameters and unknown samples alternatingly gives good results.

• AEROSPACE APPLICATIONS OF COMPUTATIONAL FLUID DYNAMICS

J. PERIAUX (AMD-BA), R. GLOWINSKI (University of Houston)

The main goal of this Minisymposium, consisting of six presentations of 20'(15' + 5' for discussion) each, is to discuss the use of numerical methods for the simulation of 3-D complex flows originating from Aerospace Engineering. The speakers, all experts in supercomputing, will present the results of numerical simulation founded on 3-D compressible Euler and Navier-Stokes codes. The applications will be relevant to aircraft, turbomachinery, rockets and shuttle technologies and clearly the complexity of these problems required the use of super-computer such as CRAY XMP and 2, ETA-10, NEC... The organizers are fully convinced that such topics can generate very stimulating discussion and, also in a large extent, highlight new concepts and trends in very large scale scientific computing.

A support from STPA is expected.

Finite Element Methods for 3-D Compressible Euler and Navier-Stokes Flows: Application to Aerospace Engineering

R. Glowinski (University of Houston), J. Periaux (AMD-BA, France)

In order to predict accurately the integral flight of aircraft and space vehicles, powerful numerical methods are required particularly for supersonic and hypersonic regimes.

The main goal of this lecture is to discuss such methods applied to the numerical simulation of real gas flows described by mathematical models related to the compressible Euler and Navier-Stokes Equations.

3-D numerical results, some of them associated with air intakes and also the recent flight of the space shuttle HERMES including winglets and body flaps geometries at very severe test causes (High Mach Number M = 25, angle-of-attack = 40) will be presented.

Engineering Applications of Some Contemporary CFD Methods

S. Chakravarthy (Rockwell International Science Center, USA)

Recently, a class of numerical methods based on total variation diminishing formulations has been developed to solve the Euler and Navier-Stokes equations. Such methods have been widely used for many external and internal flow computations. Some examples of such calculations are presented including the flow past the space-shuttle orbiter.

Flow Simulation in Missile Aerodynamics
J.J. Chattot, M. Bredif (Matra, France)

The usefulness of numerical simulation in missile aerodynamics is highlighted with a few examples of application of the EULER model to a complete configuration.

Comparisons with experimental data indicate that the accuracy of the model is satisfactory up to incidences of 20 degrees.

3-D Computations of Ducts and Turbomachinery Passages
C. Hirsch (Faculteit Teegepaste Wetenschappen)

Three-dimensional models for stationary internal flows are described based on a reduced Clebsch representation.

The space discretization uses Finite Element formulations and multigrid accelerations.

Examples of computation, in ducts, rotating and stationary turbomachinery, passages will be presented.

Numerical Simulation of 3-D Flows with Euler Codes
P. Morice (ONERA, France)

Several Euler codes are currently developed at ONERA for the numerical simulation of complex tridimensional flows past aerodynamic configurations. Each one is well suited to a specific field of applications, namely low speed, transonic or hypersonic flows. The respective features and merits of these codes will be demonstrated by the way of selected results.

Large Scale Simulation of Vortex Flow Around Delta Wings Using the CYBER 205
A. Rizzi (The Aeronautical Research Institute of Sweden)

Solutions to the Euler equations and the Navier-Stokes equations obtained with meshes as dense as one million cells are presented. The discussion will include issues like structured grids and vector processing, the viability of representing vortices upon a mesh solution, and the comparison of Euler versus Navier-Stokes solutions.

- **NEW METHODS FOR NUMERICAL TURBULENCE MODELING**

O. PIRONNEAU (Universite Paris VI)

The purpose of the Mini Symposium is to give one survey of the current status of direct simulation of turbulence and 3 key presentations on new methods and turbulent models for the numerical simulation of large eddies with special emphasis on methods which use the fact that the small eddies are random and universal except for a small number of macroscopic parameters. The speakers have been selected as specialists of the field but the lectures should be accessible to an audience specialized in numerical methods for fluids. It is believed that this area will develop largely in the coming years as a tool for the numerical simulation of industrially important flows. The main interest of such a topic is in the incorporation of the stochastic techniques in traditional finite elements or finite differences codes. The potential applications are in transient turbulent flows with boundaries (aerodynamics in particular) and transport phenomenons such as plumes of pollutants in atmosphere and meteorological flows. The talks are expected to stimulate interesting discussions at the end of the session if time allows.

Renormalization Group: Analytical Theory and Turbulence Modeling
S. Orszag, V. Yakhot (Princeton University)

We develop the dynamic renormalization group (RNG) method for hydrodynamic turbulence. This procedure which uses dynamic scaling and invariance together with iterated perturbation methods allows us to evaluate transport coefficients and transport equations for the large scale (slow) modes. The RNG theory, which does not include any experimentally adjustable parameters, gives good numerical values for important constants of turbulent flows. For example: Kolmogorov constant for the inertial-range spectrum C_K = 1.617; turbulent Prandtl number for high-Reynolds number heat transfer P_t = 0.7179. A differential transport model, based on differential relations between k, ε and ν is derived which is not divergent where k tends to 0 and ε is finite. This model is particularly useful near walls.

Modeling Turbulent Velocity Fields with Stochastic PDEs and Stochastic Boundary Conditions
J. Hunt (University of Cambridge)

The large scale eddying motions in turbulent flows of practical interest are random but are strongly affected by the boundary conditions and the initial condition of the flow. Some different approaches for analysing these notions are described: (i) near obstacles and interfaces, changes in turbulence structure can be analyzed by new equations with stochastic boundary conditions (i.e. the turbulence far from the boundary) (ii) the dispersion of fluid or solid particles in complex turbulent flow can best be modelled with stochastic differential equations (i.e. generalized random walks) which can be derived from moments of Fokker-Planck equations. But these require specification of some statistical properties of the Lagrangian turbulence field (iii) to define these Lagrangian properties, it is necessary to compute the trajectories of fluid particles through a random velocity field, for which an economical approximation is a sum of random nodes with the convert energy (Kraichnan). We present some interesting new results for Lagrangian time scales and checks on random walk methods.

Direct Numerical Solutions of Fluid Dynamic Equations
N. Satofuka (Kyoto Institute of Technology)

A method of line for the numerical simulation of Euler and Navier-Stokes equations is described. Using a discretization in space the governing PDEs are reduced to a set of Ordinary differential equations in time. The resulting system is integrated by a rational Runge-Kutta scheme. The method is fully explicit, requires no matrix inversion and is stable at much larger time step than other explicit methods. With such a code, direct simulation of turbulent flows are possible with the maximum number of points that super computers can take.

Large Scale Eddy Simulations in Stably Stratified Turbulence
M. Leisieur, J.P. Cholett (Institut de Mecanique de Grenoble)

Spectral eddy viscosity and diffusivity are used in a subgrid scale model for the simulations of large eddies in a stably stratified turbulent fluid. The subgrid model is derived from statistical theory of turbulence (Kraichnan). We consider the collapse problem in 3D turbulence by a stable density gradient and we compare the numerical results for the initial phase with experiments.

Modeling of Turbulence in Supersonic Flows
F. Grasso (Universita di Napoli)

Inherent limitations of algebraic turbulence models when applied to supersonic flows are shown; a simple differential turbulence model that properly takes account of compressibility effect is discussed.

- **SEMICONDUCTOR DEVICE MODELLING**

S.J. POLAK (Nederlandse Philips Bedrijven B.V.)

Several aspects of the numerical solution of the partial differential equations used to describe the internal behaviour of semiconductor devices will be discussed. First, a mathematical analysis of the equations will be presented, leading to existence and uniqueness of solutions. Next, discretisation using exponentially fitted box schemes is considered. Then, several techniques for solving the nonlinear discretised equations are discussed. Among these is the recently developed correction transformation algorithm. Finally, several applications are presented showing the practical importance of the techniques developed. It is hoped that this minisymposium will lead to a larger effort on the many remaining mathematical problems.

Mathematical Analysis of the Fundamental Semiconductor Device Equations
P.A. Markowich (Technische Universitat, Wien)

The system of partial differential equations, which models potential distribution and current flow in semiconductor devices, is presented. It is identified as a singularly perturbed elliptic system in the static case and as a singularly perturbed mixed elliptic - parabolic system in the transient case. The singular perturbation parameter is the normed characteristic Debye length of the device under consideration.

Existence and (non)uniqueness results for the static as well as for the transient case are presented. The solution structure is analysed by means of singular perturbation theory. The occurrence of internal spatial layers and of an initial temporal layer is demonstrated.

The Discretisation of the Semiconductor Device Equations
P.J. Mole (GEC Hirst Research Centre, UK)

Numerical simulation of semiconductor devices is currently of interest to industry. The variables of the problem are the electrostatic potential and the two carrier concentrations (electron and hole), but discretisation of the equations is very severe as the carrier concentrations can change by 20 orders of magnitude across small regions of the device. This wide variation prevents the use of simple descretisation schemes. In this talk the commonly used discretisation scheme derived from physical arguments of fluxes of carriers with only a slow spatial variation is presented and the singularly perturbed nature of the problem discussed. Present developments in the field include extending the discretisation to include the carrier temperatures as additional variables of the problem and taking the carrier inertia into account. An approach to the discretisation of these physical effects is discussed.

Solution of the Discretised Semiconductor Equations
W.H.A. Schilders (Nederlandse Philips)

Discretisation of the fundamental semiconductor equations leads to a nonlinear system of equations. Because of the high degree of nonlinearity, straightforward application of Gauss-Seidel or damped Newton methods may not be adequate. An important issue is the choice of variables, which has an influence both on the solution of the nonlinear systems and the solution of the non-symmetric linear systems. In this talk, we treat several aspects of the problem and present the method of correction transformation which allows a different choice of variables for the nonlinear and linear systems.

Simulation of Complicated Non-Planar Devices
M. Pinto (AT & T Bell Laboratories, USA)

High density integrated circuit technology has led to increasingly non-planar device structures, governed by complex, nonlinear effects. The need to accurately model the electrical behavior of these devices has placed stringent demands on simulation tools. In this talk, strategies for analyzing complex semiconductor device structures are presented, using a number of relevant examples as motivation. A brief discussion of the underlying physical equations is first given, and computational implications are highlighted.

Appropriate numerical procedures are then reviewed for discretization, grid generation, and solution, with an emphasis on providing both generality and optimal performance.

• ITERATIVE SUBSTRUCTURING METHODS FOR ELLIPTIC BOUNDARY VALUE PROBLEMS

W. PROSKUROWSKI (University of Southern California)

Iterative substructuring (or domain decomposition) is an important and rapidly developing area of research in numerical analysis of elliptic boundary value problems. Efficient preconditioners have been studied that result in the rate of convergence independent on the number of unknowns, or dependent on it only weakly (logarithmically). The solution of the original problem is obtained at a cost proportional to the cost of the solvers on the subdomains only. These techniques, especially with many substructures and intersecting separators, are well suited for parallel computing environment.

Iterative Substructuring Methods for Elliptic Problems Partitioned into Many Substructures

O.B. Widlund (Courant Institute of Mathematical Sciences)

Finite element problems can often naturally be partitioned into subproblems corresponding to subregions into which the region has been partitioned or from which it was originally assembled. A variety of iterative methods is being developed in which the interaction across the curves or surfaces which divide the region is handled by a conjugate gradient method. Various algorithms of this kind will be discussed with particular emphasis on the case of many substructures and the relation of these methods to multigrid techniques.

On the Tradeoff Between Global Coupling and Communication Cost in Domain Decomposition Algorithms

T.F. Chan (UCLA)

Elliptic problems are characterized by a global coupling of the variables in the computational domain. Domain decomposition algorithms can be viewed as methods for reducing this global coupling to one involving only the interfaces between the subdomains. For simple problems, such as constant coefficient problems on regular domains, this reduced coupling can be handled exactly and efficiently, which leads to fast direct solvers. For more complicated problems, however, the coupling becomes expensive to handle exactly. Moreover, on parallel computers, global coupling requires expensive global communication Therefore, it may be desirable to approximate the global coupling by preconditioners requiring weaker local couplings. We are interested in studying the tradeoff between the goodness of a preconditioner versus the communication cost it requires.

Domain Decomposition Method for Problems with Non-Conforming Finite Elements.

M. Dryja (University of Warsaw)

The domain decomposition method for second order elliptic partial differential equations is considered. The higher order non-conforming finite discretization is employed. A substructuring with intersecting separators is introduced. New results are established. The proposed method allows for efficient solution of the original problem.

The Construction of Preconditioners for Elliptic Problems by Substructuring

J.H. Bramble, A.H. Schatz (Cornell University), J.E. Pasciak (Brookhaven National Laboratory)

We will consider the problem of solving the algebraic system of equations which result from finite element (or finite difference) discretizations of elliptic boundary value problems defined on three dimension Euclidean space. We develop preconditioners for such systems based on substructuring (also known as domain decomposition). The resulting algorithms are well suited to emerging parallel computing architectures. We describe two techniques for developing these preconditioners both of which are applicable to domains with general geometries. A theory for the analysis of the condition number for the resulting preconditioned system will be given.

• NONLINEAR OPTIMIZATION - PRINCIPLES AND TECHNICAL APPLICATIONS

H.P. PRUFER (Ruhr-Universitat Bochum)

The demand for optimal technical design has continuously grown in the last years. Traditional design methods seem to be unable to cope with this challenge. This is partly due to the fact that the empirical methods mainly used today need too long a time to give improved results. More often, completely new solutions to the design problem are needed, which cannot be found by those traditional methods. This undesirable condition, however, can be overcome by computer-oriented numerical optimization methods, which have proved to be extremely efficient. In the following three contributions it is shown how optimization should be integrated into the process of design, and which results can be achieved by applying optimization techniques of various kinds to structural design and the performance design of cams.

Design Methodology and Parameter Optimization

C. Weber (Ruhr-Universitat Bochum)

Design methodology is generally understood as a strategy for the synthesis of technical products and systems, which follows predefined, object-independent rules. Using these rules the possibility of finding solutions to design problems is increased and the solution process itself is made easier. Design methodology proves especially advantageous for the synthesis of complicated systems with a multitude of alternative solutions. The advantages, however, can fully be utilized in combination with optimization methods only. This paper deals with a general explanation of the relation between design methodology and parameter optimization. Further, some hints are given concerning the systematical integration of optimization methods into the design process.

Combination and Application of an Optimization Strategy with a General Finite Element Code

K. Schweizerhof, G. Muller (CAD-FEM GmbH)

Optimization strategies mostly affect the way a solution has to be performed and require often specific data during a solution process. In the combination with large, general finite element codes this can achieved by convenient data base techniques. Then a wide range of applications is present exceeding by far the limited use of special structural optimization techniques. Optimization strategies which do not use gradients do not interfere with the modular structure, which should be achieved to create a reliable code, and are therefore preferred. The Sequential Unconstrained Minimization Technique which is implemented in the Finite Element code ANSYS allows the use of the unaffected body of a very general finite element code with all its capabilities. This optimization strategy is applied to several problems in industry and demonstrates the capabilities of this combination.

Application of Nonlinear Parameter Optimization to the Design of Cams

D. Hiby (Mannesmann DEMAG Fordertechnik)

With regard to a performance criterion, a cam mechanism was optimized by feasible variations of the cam shape, which characterizes the relation between input and output motion. Appropriately chosen spline functions, whose boundary values were defined as design variables, were used to realize the law of motion - causing the dynamic response within the entire system. By evaluating the corresponding system of differential equations, an objective function could be defined. The application of the optimization software ASDO-2 to this mathematical model finally led to results showing significant advantages over traditional methods of cam design.

• APPLICATIONS OF COMBINATORIAL OPTIMIZATION

W.R. PULLEYBLANK (Institut fur Operations Research Universitat Bonn)

This session focuses on recent applications of methods of combinatorial optimization to several different large scale problems. The talks illustrate how recent theoretical developments can be applied to greatly increase the scale of problems practically solvable. Examples of this include VLSI design, the design of printed circuit boards, and determining the ground state of spinglass, a problem in theoretical physics. They also deal with problems of recognizing when a general problem contains special structure permitting efficient solution methods.

Circuit Design Spin Glasses and the Max Cut Problem

F. Barahona (University of Waterloo), M. Grotschel, M. Junger,
G. Reinelt (Universitat Augsburg, F.R.G.)

We study the problem of minimizing the number of vias subject to pin preassignments and layer preferences and the problem of finding a ground state of spin glasses with exterior magnetic field. The former problem arises in printed circuit board design and the latter in solid state physics. Both problems can be reduced to the max cut problem in graphs. Based on a partial characterization of the cut polytope we design a cutting plane algorithm and report on computational experience. Our method has been used to solve max cut problems on graphs with up to 1600 nodes.

Recognizing Generalized Networks

C.R. Coullard, J.G. delGreco, D.K. Wagner (Purdue University)

A generalized network matrix is a real-valued matrix with at most 2 non-zeros per column. A generalized network matrix has an associated gain network; i.e., a directed graph with a real number, called a gain, on each arc. Bicircular matroids are those matroids representable by a generalized network matrix whose associated gain network has no unit-gain cycle, where the gain of a cycle is the product of the gains on forward arcs divided by the product of the gains on reverse arcs. We present a polynomial algorithm for generalized network recognition that succeeds provided the input matrix represents a bicircular matroid.

Combinatorial Optimization and VLSI - Design

B. Korte (Universitat Bonn)

This paper reports on a large VLSI-layout problem which was successfully solved by methods of combinatorial optimization. The problem consists of the placement of cells on a grid graph with about 6 million nodes and a routing of about 100000 input- and output pins in this grid graph. This can be considered as a multicommodity hypergraph flow problem whose realisation in the grid graph should be edge disjoint. Combinatorial optimization methods were used to solve the min-cut, bin packing, knapsack, shortest path, spanning tree, and Steiner problems.

Discrepancy of Hypergraphs: A Geometric Approach

L. Lovasz, K. Vesztergombi (Eotvos Lorand University)

Given a set S and a system of its subsets , we want to split S into two sets S_1 and S_2 such that $||A \cap S_1| - |A \cap S_2||$ is small for each $A \in \mathcal{H}$. This problem has many interesting special cases related to diophantine approximation, numerical integration, randomness testing etc. We present an approach motivated by polyhedral combinatorics. In particular, results from geometry can be applied to obtain bounds on the number of edges of a hypergraph with small hereditary discrepancy and on various generalizations of this problem.

• THE PROPOSED FORTRAN 8X PROGRAMMING LANGUAGE

J. REID (Computer Science and Systems Division Harwell Laboratory)

FORTRAN is the most commonly used language for Mathematical and Scientific Programming. The last major revision resulted in the current version, FORTRAN 77. It is expected that a draft of a new version (dubbed FORTRAN 8X) will be circulated for formal comment around the time of the ICIAM meeting. This language contains a number of changes of great importance to scientific programs, including a large set of facilities for manipulating arrays and new ways of encapsulating program modules and expressing precision requirements.

The language will be described by three members of the FORTRAN Committee and a consultant to it. Discussion will be encouraged throughout.

FORTRAN 8X - The Emerging Standard

M. Metcalf (Data Handling Division CERN, Switzerland)

Since 1979, the Fortran standardization committee, X3J3, has been labouring over a draft for the next version of the standard. Its initial intention of publishing this draft in 1982 was hopelessly optimistic, and at best it may be ready this year. However, a number of fundamental issues have been thrown up over the past two years, such that it is clear the unanimity cannot be achieved: efficiency versus functionality; safety versus obsolescence; small and simple or big and powerful?

This paper reviews the current state and content of Fortran 8x.

The FORTRAN 8X Array Features

J. Reid (Harwell Laboratory, England)

Fortran 8x contains extensive features for handling arrays. These will lead to shorter and more readable source programs that compile to more efficient object code, particularly on vector or parallel target machines. The array features will be summarized in this talk.

Generalised Precision Features of FORTRAN 8X

J.L. Schonfelder (University of Liverpool)

Fortran 8x contains features to provide portable selection of floating point approximation quality, and provision for user definition of procedures that will be generic over machine "precisions".

These facilities are described in some detail along with their possible impact on performance.

FORTRAN 8X: A User's Perspective

S.I. Feldman (Bell Communications Research, USA)

Fortran 8x is a radical change to the Fortran language. Many of the changes add significant functionality and are likely to improve portability, readability, and maintainability of programs.

Other aspects of the language may make it harder to write and maintain programs, due to the complexity of the complete language, the need for compatibility with an older language, or unfortunate design decisions. This talk will give the views of one programming language designer and user on these properites of the new language.

- **GRAPHS AS COMMUNICATION MODELS**

R. RINGEISEN (Clemson University)

This mini-symposium will demonstrate some of the myriad of ways in which the theory of graphs interacts with and influences the analysis of communications networks. The subject matter will vary greatly, encompassing such widely differing studies as theoretical discrete mathematics and actual network analysis.

Cohesion Stability Under Edge Destruction

R.D. Ringeisen (Clemson University)

Cutpoints in a graph are a primary consideration when examining the vulnerability of networks. Here we take this concern a bit further in that we study how near a given vertex is to being a cutpoint. The cohesion of a vertex x is the minimum number of edges whose removal makes x a cutpoint. In earlier work with M. Lipman, the changes in cohesion under edge addition were examined. We extend the concepts developed there by considering the stability of a vertex under edge deletion and then discuss the global stability of the graph relative to cohesion.

A New Computational Approach to the Graph Partitioning Problem and Applications
M. Minoux (University Paris IX Dauphine)

An application of the graph partitioning problem concerns the search for a best hierarchical organization of a large network into subnetworks so as to minimize the interactions between the subnetworks. We propose a new approach to this problem based on (i) reformulating the problems as a large scale set partitioning problem (ii) solving the linear relaxation of this set partitioning problem by means of generalized linear programming techniques (iii) exploiting the lower bounds obtained in (ii) within Branch-and Bound Schemes in order to get exact optional solutions to the Graph Partitioning Problem. We establish a theoretical result showing that the bounds in (ii) are computable in polynomial time. Finally, the extensions to other difficult combinatorial problems will be suggested.

Minimum Time Broadcasting as a Function of Transmission Parameters
A. Proskurowski (University of Oregon)

We are concerned with a model of the broadcasting process, whereby a message initially known only to a set of originators (often assumed to consist of one site) is to be disseminated among all the other sites of the network. We will determine the minimum time necessary to complete this task, depending on the values of transmission parameters: the switching time and the transmission time.

Interconnection Network Extension
J.C. Bermond, J. Bond, J.C. Konig (C.N.R.S. - L.R.I. Universite de Paris Sud)

A well known problem in Interconnection Networks is the (Δ,D) graph problem, which consists in constructing dense graphs having given maximum degree Δ and diameter D. Here we consider an additional constraint: the extensibility of the networks. One wants to construct the networks successively by starting from a small one and adding vertices one by one. Two main cases appear according we allow or not during the extension deletion of some links. With such a possibility we construct good families of networks with given degree and connectivity. We consider extensions with a constant diameter and with no edge deleted.

Networks, Graphs and Security
C. Delorme (C.N.R.S.-L.R.I.Universite de Pris Sud), J.J. Quisquarter (Philips Research Laboratory)

Some problems in data security are closely related to the topology of networks. For instance, if one wants the property of anonymity (only the emitter and receiver of a message know its source and destination), one has to use a network with every two nodes linked by a path of length 3. We propose some constructions of such large networks. Another security problem, where users want to exchange information simultaneously without any previous way of having them encrypted, gives rise to constructions of large graphs with at least two paths of length exactly 2 between any two vertices.

• KNOWLEDGE EXPLOITING SYSTEMS IN NONLINEAR PROGRAMMING
K. SCHITTKOWSKI (Universitat Bayreuth), G. VAN DER HOEK (Erasmus University)

This minisymposium focuses on two approaches to cope with nonlinear problem formulations.

The first approach is to add certain "learning options" to algorithms which use gained information on the iteration process to anticipate expected iterative behaviour in remaining iterations. The other approach is to develop Modeling Systems which do not require detailed knowledge of nonlinear models to be available in the application environment. The system generates model formulations, gives a feedback to the user who can add interactively information (e.g., bounds on variables) which is used to generate a more appropriate model.

Practical Expert Systems in Nonlinear Programming
K. Schittkowski (Universitat Bayreuth)

Expert systems for nonlinear programming can be used for modelling a problem, for selecting a suitable algorithm, for generating a code, and for interpreting results or failures. Two systems will be introduced, the general purpose system EMP and the mechanical structural optimization system LAGRANGE. Both systems are self-learning and use rules, e.g., to select a suitable method or to interpret failures. They run interactively and use data bases for processing problem data and results.

Anticipation on Expected Iterative Behaviour of an NLP Code
W.J. Kribbe (Technical University Delft), G. Van der Hoek (Erasmus University)

During the execution of a conjugate gradient algorithm for constrained NLP, the algorithm collects information on the behaviour of the algorithm on the problem to be optimized. This information can be used to anticipate expected behaviour of the algorithm.

A Modeling System With Knowledge About Economics
A. Drud (World Bank, USA), J.J. Bisschop (Technical University Twente)

The modeling system to be discussed is an extension of the GAMS modeling system. Based on global, user-supplied information the System generates on equilibrium model and reports inconsistencies to the user. In an interactive way the user again supplies additional, improved information to enable the system to generate more appropriate functional specifications.

Rule-Based Active Set Strategies for Design Optimization
P.Y. Papalambros (University of Michigan)

Design optimization procedures for mechanical or structural problems may benefit substantially from implementing active set strategies in the NLP solution method. Such a strategy may utilize not only local (iterative) information but also global information, derived from rigorous model analysis (such as monotonicity and boundedness) and from heuristic engineering experience about the problem. Both types of information are codified in a rule-based system that is interfaced with NLP methods for equality constrained problems. This enhances efficiency and reliability of results. Example applications are included.

• FINITE ROTATIONS IN NONLINEAR PLATES AND SHELLS
R. SCHMIDT (University of Wuppertal)

A current trend in plate and shell theory is the formulation of geometrically nonlinear theories on the basis of well defined and physically meaningful assumptions concerning the magnitude of strains and rotations. Based on exact polar decomposition of shell deformation a hierarchical set of nonlinear theories for isotropic thin elastic shells has been developed by Pietraszkiewicz (see e.g. Int. J. Non-Linear Mech. 19 (1984)) and Schmidt (see e.g. Comp. & Struct. 20 (1985)) for problems in which the shell material elements undergo small strains and either moderate, large, or unrestricted rotations. Based on these results a lot of papers of numerous authors have been devoted recently to various theoretical and numerical investigations in the nonlinear Kirchhoff-Love type shell theory.

An extension of the above novel approach to other advanced plate and shell problems is of current interest in literature, e.g. for anisotropic and laminated composite structures, for elastic-plastic material, and for problems involving large elastic strains. Authors dealing with the appropriate formulation of nonlinear plate and shell theories for these problems are Librescu, Naghdi et al., Reddy, Weichert, Schmidt a.o.

The symposium gives an account of the state of the art and new results in modern geometrically nonlinear plate and shell theory. The problems considered in the above field include new energy-consistent nonlinear theories, variational principles, nonlinear static and dynamic FE-analysis, stability equations, and post-buckling behavior.

On the Geometrically Nonlinear Kirchhoff-Love Type Theory of Thin Elastic Shells

W. Pietraszkiewicz (Polish Academy of Sciences)

In recent years a theory of finite rotations in shells based on exact polar decomposition of shell deformation into pure stretch along the principal axes of strain and rigid-body rotation was given by PIETRASZKIEWICZ. This makes it possible to derive appropriate nonlinear shell equations for problems in which the shell material elements undergo restricted strains and rotations of well defined order of magnitude (e.g. small strains of $0(\eta)$, accompanied by moderate ($0(\eta^{1/2})$), large ($0(\eta^{1/4})$), or unrestricted ($>0(1)$) rotations. Based on this approach in recent literature many authors have treated various aspects of the Kirchhoff-Love type nonlinear theory of thin elastic shells. In the present contribution we give a review of the state-of-the-art in this field and present various extensions of the known results.

The topics under consideration include the formulation of nonlinear shell theories, variational principles etc.

Small Strain and Moderate Rotation Shear Deformation Theories for Elastic Anisotropic Plates and Shells

J.N. Reddy (Virginia Polytechnic Institute and State University)

In geometrically nonlinear analysis of shear deformable plates and shells often only the von Karman-type nonlinearity is assumed, which does not account for true finite rotation effects.

On the other hand consideration of full geometric nonlinearity results in complex equations and is also not necessary for most problems.

The small strain and moderate rotation concept used in refined higher-order theories e.g. by KAUL, WEMPNER and recently by LIBRESCU and SCHMIDT is applied here for the derivation of first- and third-order shear deformation theories for static and dynamic analysis of elastic anisotropic plates and shells.

The effect of taking the geometrically nonlinear contributions correctly into account is demonstrated numerically by finite element results for shear deformable plates undergoing small strains and moderate rotations.

Geometrically Nonlinear Refined Theories for Laminated Composite Plates and Shells. Review and Extension for Small Strains and Moderate Rotations

L. Librescu (Virginia Polytechnic Institute and State University)

The increased use of new composite materials has generated a considerable interest in the development of consistent refined geometrically linear and nonlinear theories for plates and shells, which account for transverse shear and transverse normal effects and for higher-order effects.

Here we present a geometrically nonlinear refined dynamical theory of elastically and thermally anisotropic laminated composite plates and shells based on precise order-of-magnitude considerations for small strains and moderate rotations.

In the light of the present theory a variety of earlier approximate nonlinear variants (e.g. the commonly used von Karman-type shear deformation theory) may be classified from the point of view of the relative errors involved and their range of applicability may be clarified.

Elastic-Plastic Plates and Shells at Moderate Rotations

D. Weichert (Ruhr University, Bochum))

In recent years much progress has been achieved to describe geometrically nonlinear elastic shells by the polar decomposition theorem splitting the deformation into pure rotation and stretches. Whereas for thin elastic shells the derivation of appropriate theories is well advanced, the effect of inelastic material behavior has not yet been conclusively investigated in the light of this decomposition. In this paper we shall present a theory for thin elastic-plastic shells and plates undergoing moderate rotations about tangents and small rotations about normals to the mid-surface basing on an advanced incremental elasto-plastic constructive law.

• ALGORITHMS FOR THE CALCULATION OF BIFURCATIONS

R. SEYDEL (University of Wurzburg)

Algorithms for calculating bifurcations have been rapidly developing. While methods for standard ODE boundary-value problems and algebraic equations (one real parameter) have reached the stage of becoming implemented in sophisticated software, present interest concentrates on algorithms for bifurcation in delay-differential equations, optimal control problems, and multi-parameter problems.

The minisymposium will report on related algorithms. Topics to be handled will include the computation of Hopf bifurcations (onset of periodic orbits), application to partial differential equations, and hints on practical implementations.

Computation of Hopf Bifurcation for Delay-Differential Systems

B.D. Hassard (State University of New York)

A code has been developed which will automatically locate and analyze points of Hopf bifurcation for autonomous delay-differential systems

$$\dot{x}(t) = f(x(t-t_1(\nu)),..,x(t-t_m(\nu));\nu)$$

where x and $f \in \mathbb{R}^n$, $\nu \in \mathbb{R}^1$ is the bifurcation parameter, and the "delays" are non-negative. The code normally determines the direction of bifurcation and stability and an approximation to the family of periodic solutions. A new algorithm for the general delay-differential eigenvalue problem that computes all eigenvalues whose real part exceeds a prescribed value is used to verify the spectral hypothesis.

Applications to control systems in biology and mechanics are described.

Numerical Analysis of Hopf Bifurcation in Two Parameter Problems

D. Roose (Department of Computer Science, K.U. Leuven)

In a nonlinear time-dependent problem a branch of Hopf bifurcation points can emanate from a branch of turning points of the steady state problem at an "origin for Hopf bifurcation" ("B-point"). At such a branching point the Jacobian matrix has a double eigenvalue zero with a one-dimensional nullspace.

Algorithms are discussed for computing origins for Hopf bifurcation during the continuation of a branch of turning points and for the calculation of the emanating branches of Hopf points. Emphasis lies on the efficient implementation of these methods which allows one to exploit the structure of the Jacobian matrix (e.g. in the case of a discretization of a system of partial differential equations).

The performance of the algorithms is illustrated by results for two systems of partial differential equations describing chemical reacting systems.

The Optimal Control of Nonlinear Systems in the Presence of Multiplicity and Bifurcations

E.J. Doedel (California Institute of Technology), J.P. Kernevez (Universite de Technologie de Compiegne)

We consider the optimal control of systems described by ordinary differential equations with boundary and integral constraints. These systems are allowed to exhibit bifurcation phenomena. In particular their solutions need not be unique.

We show how certain problems involving the optimal control of such systems can be treated by existing software for bifurcation problems. Examples include the optimal control of boundary value problems and periodic solutions.

Transformation Methods as Interface Between Bifurcation Problems and Numerical Software

R. Seydel (University of Wurzburg)

Most practical problems of a bifurcation analysis (such as continuation and branch switching) can be reformulated into equations in standard form. Since numerical software needed to solve equations in standard form is widely available, it suffices to transform a bifurcation problem into an appropriate equation. Related transformations lead to methods that are most effective with respect to the relation overhead versus applicability. Algorithmic details are discussed and remarks on bifurcation software are given.

• DYNAMICAL PHENOMENA IN DIFFERENTIAL EQUATIONS WITH TIME-DELAYS

G.R. Sell (University of Minnesota)

The minisymposium will cover several recent advances in the dynamical theory of solutions of differential equations with time-delays. The equations of special interest in this session arise in the mathematical models of a variety of applications including negative feedback systems (control theory), optical bistability, mathematical economics, and mathematical physiology.

Monotonicity and Morse-Smale Differential-Delay Equations

J. Mallet-Paret (Brown University)

The asymptotic behavior of bounded solutions x(t), as $t \to \infty$, of the scalar differential-delay equation

$$(*) \qquad x'(t) = -f(x(t), x(t-1))$$

is studied. Under the monotonicity assumption $\frac{\partial f(\xi,\eta)}{\partial \eta} > 0$, $\quad \xi, \eta \in R$

in the delay term, a Poincare-Bendixson theorem is proved. This in turn implies the absence of period doubling bifurcations, invariant tori, and generally the absence of "chaos". Thus the global dynamics of (*) have a Morse-Smale quality.

Existence, Stability and Secondary Bifurcation of Slowly Oscillating Symmetric Periodic Solutions

S.N. Chow (University of Michigan)

Consider the following scalar equations with time delays:

$$x(t) = \int_{1-\varepsilon}^{1+\varepsilon} f(x(t-\tau))\, d\tau, \qquad 0 < \varepsilon < 1$$

or

$$x'(t) = f(x(t-1)).$$

We will discuss the existence of slowly oscillating symmetric periodic solutions of these equations under some monotonicity conditions on the nonlinear term f(x). We will show how to estimate the characteristic multipliers of these periodic solutions and to obtain orbital exponential stability. Numerical studies of secondary bifurcations will also be shown.

Applications of FDE's to Economic Analysis

A. Rustichini (University of Minnesota)

We present examples of applications of the theory of FDE's to economic analysis. In the classical Solow model (one sector economic growth) we prove existence of cyclical behaviour. We then study the standard model of economic growth when delays in production, investment, or both are present. The value function (with non negativity constraints) is proved to be differentiable; then the Bellman equation and Kurzweil theorems are used to study the dynamic behaviour of the solution. The Euler equation is used to study the solution of prices; we derive a system of mixed equations, and study the bifurcation to periodic solutions with a Liapunov-Schmidt decomposition.

• STOCHASTIC MODELS IN BIOLOGY

P. TAUTU (German Cancer Research Center)

The purpose of the proposed minisymposium is to present some recent applications of stochastic processes in epidemiology, genetics, and cell biology. The papers will deal specifically with population-dependent Markov branching processes as models for the growth of cell populations, martingales for some epidemic processes, and interacting particle systems in genetics and cancer research.

The mathematical results will deliver the description of the qualitative behaviour of the considered biological systems, giving possibly a new insight into the structure and dynamics of these populations.

On the Qualitative Behaviour of Population-Dependent Markov Branching Processes

W. Rittgen (German Cancer Research Center)

With the aid of criteria existing in the theory of Markov chains with discrete and continuous parameter, conditions for regularity, transience, recurrence, ergodicity, and geometric ergodicity of population-dependent Markov branching (PDMB) processes are found. Such conditions were given for both the one-dimensional and the multi-dimensional PDMB processes, with the remark that there is no direct transfer of methods from the former to the latter case.

Invariants and Stochastic Models

P. Picard (Universite de Lyon 1)

When a deterministic process leads to equations too complicated to be fully solved, trying to build invariants for this process may be a good device. For instance partial differential equations are frequently studied following that way.

In a stochastic context the same may be true. In that case martingales are the very tools for getting invariants. Although martingales are frequently used chiefly in connection with the stopping time theorem or to get asymptotic results, structurally they ARE invariants. This will be illustrated with examples from genetics and epidemic theory.

Stochastic Age Dependent Population Dynamics

V. Capasso (Universita di Bari)

Stochastic age-dependent birth and death processes are revisited by using a counting process approach.

Martingale methods are used to describe the dynamics of the process and for the estimation of the parameters of the system.

Central limit theorems for maraingales imply the asymptotic properties of the estimators.

Applicability of the theory to epidemics with age dependence is also presented.

Occupation Times of Some Biological Cell Systems

L. Pilz, P. Tautu (German Cancer Research Center)

In this paper the properties of the occupation times of some biological cell systems with a certain spatial structure are investigated. One possibility is to locate the cells on an integer lattice of dimension d, d=1,2..., or in an area of R^d. Every cell possesses the opportunity to change its type in a set of distinct types. In a simple two-type model of carcinogenesis there are initiated and promoted cells describing a precancerous event as it happens, e.g., in skin or liver cell systems. Under the assumption of a simple interaction process the existence of a limit field for the occupation times of a system for $t \to \infty$ and an appropriate normalization is studied.

• SEMICONDUCTOR FABRICATION

A.B. TAYLER (University of Oxford)

The development of semiconductor devices is of great importance to high technology industry and the modelling of the fabrication steps involved in producing an integrated circuit is a rapidly growing subject. Extensive computation is being applied by engineers to crude models but relatively little effort so far has been made to extract asymptotic information from better models. The four speakers will attempt to show how mathematical modelling and analysis can be applied to such a complicated problem area and how some extremely interesting mathematical questions about nonlinear diffusion arise. All the speakers are involved with the Mathematical Study Groups with Industry project.

Simulation of Thermal Dopant Redistribution and Oxidation in Silicon

G.J. Rees (Plessey Research Limited, UK)

The principal effects which occur during the thermal processing of Si wafers will be described with a view to formulating useful mathematical models, capable of numerical solution.

Dopant Diffusion During Furnace Annealing

C. Please (University of Reading)

Dopant is put into crystalline silicon by ion implanation and heat treatment used to repair any crystal damage. With furnace annealing there is no phase change, but the dopant diffuses with a diffusion coefficient which depends on its concentration. At high dopant concentrations this may be modelled by the equation

$$\frac{\partial c}{\partial t} = \nabla.[(c+\varepsilon)\nabla c].$$

Its asymptotic behaviour as $\varepsilon \to 0$ will be described.

Laser Annealing with Phase Change

A.B. Crowley (Royal Military College of Science)

The laser annealing process is used to repair crystalline damage caused during the dopant implantation. The laser heating causes melting and resolidification at very high growth rates. As a result the dopant is redistributed, both by enhanced diffusion in the liquid phase and by segregation on resolidification. This leads to both novel phase change problems and a highly nonlinear process. Some numerical results for such problems will be discussed.

Mathematics of Nonlinear Diffusion

J.R. Ockendon (University of Oxford)

The diverse properties of the nonlinear diffusion equation $\frac{\partial c}{\partial t} = \nabla(c^n \nabla c)$ will be reviewed and an asymptotic treatment given for large values of n.

• APPLIED STOCHASTIC PROCESSES

C. TIER (University of Illinois)

There is a growing use of stochastic processes in a broad spectrum of applications. This minisymposium deals with the modeling and analysis of stochastic processes. The talks include applications from biology, chemical physics, queueing theory, and resource management.

A Singular Perturbation Approach to First Passage Times for Markov Jump Processes

C. Knessl, B.J. Matkowsky (Northwestern University), Z. Schuss (Tel Aviv University), C. Tier (University of Illinois)

We introduce singular perturbation methods for constructing asymptotic approximations to the mean first passage time and do not involve the use of diffusion approximations. An absorbing interval condition is used to properly account for the possible jumps of the process over the boundary which leads to a Wiener Hopf problem in the neighborhood of the boundary. Examples are presented to illustrate our methods.

Control of an Imperfectly Understood System: Fisheries Management
D. Ludwig (The University of British Columbia)

Fisheries provide a great variety of interesting scientific, management, and institutional problems. Although the management problem can be formulated as a stochastic control problem with imperfect information, it is more complicated than any others, because of the poor quality of the information available, and because the separation principle (whereby estimation and control can be handled separately) does not apply. In view of the great costs involved in obtaining information about the behavior of fisheries under exploitation, it appears likely that future scientific approaches will have to deal with poor or insufficient information. Although the resulting problems are exceedingly intractable (or perhaps impossible), some optimism is justified: the recent decline in the costs of computation makes it feasible to apply massive computing power to statistical and decision problems. Our objective is to match estimation methods and models to data and biological assumptions, and to make our methods accessible to managers and other researchers.

Approximate Methods for Queueing Models
C. Knessl, B.J. Matkowsky (Northwestern University), Z. Schuss (Tel Aviv University), C. Tier (University of Illinois)

Approximations are constructed to performance measures for state-dependent queueing systems with fast arrival processes and small mean service requests. Some important performance measures are the stationary distribution of the buffer content, the mean time to buffer overflow, and the mean response time. These measures satisfy integro-differential-difference equations. Asymptotic and singular perturbation techniques are developed and applied to the equations to obtain approximations to the performance measures. Several specific examples are given.

Differential Equations with Random Excitation and First Exit Time Problems in Mechanics and Biology
F.Y.M. Wan (National Science Foundation, USA)

The Spatial Correlation Method was originally developed as an effective solution technique for linear distributed systems with time varying system parameters and random excitations. It is still uniquely efficient (by orders of magnitude) today for the numerical solution of second (and higher) order response statistics of time varying distributed systems such as flexible rotor blades in forward flights. This talk briefly reviews the salient features of the Spatial Correlation Method. The method is then shown to be useful also for calculating the depolarization satistics of a cable model neuron subject to random postsynaptic excitations. Moreover, the key idea in the Spatial Correlation Method can be used in a different way to provide an elegant solution technique for first exit time problems with a moving threshold. Exact and asymptotic solutions for first exit time problems in neurobiology in terms of simple functions will be presented.

Penetration of Laser Light into Skin
G. Weiss (National Institutes of Health, USA)

A simple multiple scattering analysis is made of the depth of penetration of light into skin. The model is oversimplified, yet leads to excellent agreement with experiment.

• ANALYTICAL AND NUMERICAL SOLUTIONS OF VORTICAL FLOW
L. TING (Courant Institute of Mathematical Sciences)

In this minisymposium we present current status reports on the analytical and numerical studies of vortical type flows occurring in aircraft and rotorcraft aerodynamics. The physical implications and the limitations of the models and their solutions will be emphasized. Different numerical methods for the solutions of the inviscid and viscous model equations will be assessed respectively. Comparisons of the solutions with the experimental data will be presented. Problem areas will be identified for which appropriate vortex models and their numerical and/or analytical solutions are still wanted.

Numerical Simulation of Vortical Flow about Wings
H.W.M. Hoeijmakers (National Aerospace Laboratory NLR)
Methods are discussed for the numerical simulation of the flow about wings with vortex flow, i.e. leading-edge vortex flow and trailing vortex wakes. The methods are discussed with a view towards 3D steady flow applications. Results of various methods in use at present will be compared mutually and with available experimental data. The strengths, weaknesses, and prospects of the methods will be indicated.

Motion, Decay, and Merging of Vortex Filaments
C.H. Liu (NASA Langley Research Center, USA)
The asymptotic solutions of N-S eqs. (SIAM J. 1978, 148-175) for vortex filaments of finite, $0(1)$, strength with small, $0(\varepsilon)$, effective vortical cores are summarized. Special emphasis is placed on the interactions of the decaying core structure of large, $0(\varepsilon^{-2})$, vorticity with the motion of the filament(s) (AIAA J. 1986, 1290-1297). Recent extensions of the analyses for vortex filaments submerged in a background rotational flow of $0(1)$ vorticity show the coupling of the vortex motion and core structure with the variations of the background flow. Finite-difference solutions of N-S equations for the merging of the filament(s) are described with emphasis on the approximate boundary conditions on the computational domain.

Structure of Vortex Core Computed in Solutions to the Euler Equations
A. Rizzi (FFA Aeronautical Research Institute and Royal Institute of Technology, Sweden)
Inviscid theory models a shear layer shed from the leading edge of a delta wing as an infinitely spiralling vortex sheet. A core does not form because of the absence of viscosity. Numerical dissipation however spreads the vortex sheet over several grid points and causes a core to develop at the center of the spiral. The paper examines the results of a number of very dense mesh solutions obtained with as many as one million cells and shows that although the profile of the velocity gradient across the shear layer is artificial, the resulting spiral core is qualitatively correct. The errors incurred by the process, like the observed loss in total pressure, is explained, and it is argued that the correct level of circulation is given by the inviscid solution.

Vortex Wake Problem for Hovering Rotors
D.B. Bliss (Duke University), C. Tung (Ames Research Center, USA)
Accurate predictions of helicopter rotor loads depend strongly on the exact knowledge of the vortex wake geometries. The flow field around the rotors is so complex that even to predict the wake geometry around a hovering rotor is one of the most challenging tasks. The speaker will present a new approach to this problem. The desired solution is found by a procedure which does not involve time stepping. Thus, it avoids the numerical stability problems encountered by the usual time marching scheme. New insight has been gained regarding the nature of the hover free wake problem.

• COMPUTER ALGEBRA
E. TOURNIER, J. DELLA DORA (Laboratorie TIM3 - Calcul Formel)
After a brief survey on Computer Algebra Systems, their possibilities and future trends, some computer algebra works in progress will be presented. They are the following:
- Computer algebra with algebraic numbers.
- Formal solutions of ordinary differential equations:
  The solver DESIR.
- Numerical and graphical extensions to the solver DESIR.

- Computational geometry

General Presentation of DESIR a Linear Differential Equations Solver
J. Della Dora, E. Tournier (Laboratorie TIM3)

The Solver DESIR enables one to compute a basis of formal solutions of linear homogeneous differential equations in the neighborhood of regular and irregular singular points.

The complete context of this solver will be presented, containing the following parts:
- formal part (including problems raised by algebraic numbers)
- numerical and graphical part.

Calcul Formel avec des Nombres Algebriques
C. Dicrescenzo, P. Ozello (Laboratorie TIM3), D. Duval (Institut Fournier)

De meme que dans la resolution des equations differentielles, les nombres algebriques apparaissent dans la plupart des problemes de calcul formel: integration, valeurs propres de matrices, geometrie, etc.

Nous presentons diverses solutions utilisees en calcul formel pour les manipuler de maniere exacte:

- les methodes classiques, fondees sur la factorisation de polynomes,
- et le systeme D5, qui considere le calcul avec des nombres algebriques comme un cas particulier de discussion de probleme avec parametres.

From Formal to Actual Solutions...A Numerical and Graphical Extension of the L.D.E. Solver DESIR
F. Richard (Institut de Mathematiques), J. Thomann (C.N.R.S.)

This extension contains:
- a numerical part including the transformation of the divergent series given by DESIR into convergent generalised factorial series. Then, using the K-summability theory, it gives "actual" solutions.
- a graphical part: graphics designed to "materialise" the results given by the formal and numerical parts, and to emphasize phenomena which will afterwards be exploited on the mathematical or numerical level.

Computational Geometry
J.H. Davenport (University of Bath)

The past few years have seen an explosion of interest in computational approaches to Euclidean geometry, motivated by the needs of Computer Aided Design and robotics. In addition, algorithmic approaches to theorem-proving in Euclidean geometry have reached the point where theorem-proving in complex Euclidean geometry is almost trivial. This talk will survey these recent developments, and explain the great computational complexity of real Euclidean geometry and the handling of inequalities.

• **MATHEMATICAL SOFTWARE FOR MESSAGE BASED HIGHLY PARALLEL ARCHITECTURES**
U. TROTTENBERG (SUPRENUM GmbH)

For large scale parallel systems, message passing has become an increasingly important design concept. In the SUPRENUM-project 22 groups from universities, research institutions, and industrial partners collaborate to develop message based, highly parallel (MIMD) hardware and software for high speed scientific computation and numerical simulation. Hardware architecture and parallel application software, with emphasis on multigrid methods, are tuned as to yield multiplicative speed-up. Main application fields include fluid dynamics, meteorology, nuclear physics, and technical simulation, and software development is conducted both on the research level and in industrial applications. The minisymposium should present the scope of dedicated mathematical software developed in the project and provide an opportunity to discuss parallel software development for message based MIMD machines.

Software Concepts for Message Passing Based MIMD Systems
K. Solchenbach (SUPRENUM GmbH)

A basic distinction in the class of MIMD multi-processor systems is with respect to their memory organization: systems may either have a shared memory or each node computer has only its own load memory. The talk concentrates on the latter ones which are characterized by the fact that transfer of data and synchronization of algorithms are not performed by the hardware but have to be explicitly programmed by the user. In order to overcome the hardware dependence of the programming and to find acceptance by the non-specialists, powerful software concepts and tools are necessary. The talk gives an overview on the static and dynamic process concepts with message based passing and different tools which have been developed for the SUPRENUM computer.

Fine Grained Parallelism in Multigrid Algorithms
C.A. Thole (SUPRENUM GmbH)

Two interesting approaches which have been pursued in order to improve and accelerate the numerical solution of PDEs are:

- the multigrid approach and
- the parallel approach.

In order to take advantage of both ideas simultaneously, we examine the amount of parallelism that is contained in multigrid algorithms. For the multigrid treatment of 3D anisotropic elliptic operations point-, line-, and plane-relaxation might be needed as smoothing procedures for the multigrid methods. For these algorithms we discuss the maximal amount of parallelism, the possible speedup for a certain number of processors, and the efficiency for a particular choice of the size of problem and architecture.

Domain Splitting as a Principle for Parallel Mathematical Software Development
R. Hempel, K. Stuben (Gesellschaft fur Mathematik und Datenverarbeitung)

Important classes of numerical methods (e.g. for partial differential equations) are defined on grids over geometric domains. At each point the data dependencies are local. Splitting up the domain and assigning each subdomain to a process yields an effective parallelization strategy with a high arithmetic/data exchange ratio.

Based on domain splitting and standardized data structures for PDE-solvers, a subroutine library covering all communication tasks has been defined and implemented on the Intel iPSC and on simulators for the SUPRENUM machine.

First experiences with the multigrid code for the poisson equation using that library are presented.

MIMD Parallelization of the Full Potential Equation
A. Schuller (Gesellschaft fur Mathematik und Datenverarbeitung)

In the SUPRENUM project parallel software is developed with special emphasis on flexibility, robustness, and efficiency. Multigrid methods are highly efficient solvers for many problems. One example for this is the full potential equation which describes the behaviour of non-viscous, irrotational flow. Problems and experiences in producing efficient parallel multigrid software are discussed for the corresponding boundary value problem. The principle of parallelization is domain decomposition. A high level of flexibility is achieved by use of a communication library and adaptive programming.

MIMD Parallelization of Euler and Navier-Stokes Solvers for Industrial Application
W. Fritz, S. Leicher, B. Wagner (Dornier GmbH-Abt. Aerodynamik)

A computational approach for the Euler and Navier-Stokes equations will be presented based on the finite volume space discretization using contour conformal meshes and on an explicit Runge-Kutta time stepping. Techniques for accelerating convergence to steady state are implemented. Complex technical geometries are subdivided into regular mesh blocks, where regularity guarantees efficient vectorization, and the block sizes can be chosen to optimally suit message based MIMD architectures and minimize data transfer between processors. Sample calculations will be shown for airplanes, cars in wind tunnel experiments and on the road, and for internal flows.

MIMD Computer Structures with Local Memory
U. Trottenberg (SUPRENUM GmbH)

In the field of supercomputer architectures many distinctions and classifications have been made. Simplifying the situation in some respect - we want to distinguish the following four classes of architecures here: (1) vector machines with a single CPU, (2) vector machines with several CPUs, (3) multiprocessor machines with global memory, (4) multiprocessor machines with local memory units. Typically, the classes (1) and (2) can be regarded as SIMD machines, whereas (3) and (4) are of MIMD type. Furthermore, (1) and (2) are usually based on "supercomputer technology"; (3) and (4) on microprocessor technology. (As the nodes of the multiprocessor systems in (3) may be equipped with vector units, the distinction (2) and (3) is somewhat delicate for some systems.) Architectures of class (4) will be discussed here. For such architectures, it is both possible and technically necessary to decide about a suitable interconnection topology of the nodes. In this decision the typical applications and algorithms - and the communication requirements induced - can be taken into account.

• MATHEMATICAL CONSULTING FOR INDUSTRY IN LINZ

H. WACKER (Johannes Kepler Universitat)

Mathematical consulting at the Department in Linz started in 1975 with a pelletizer problem for the steel company VOEST-Alpine. Since then an increasing number of interesting problems has been offered to our group, mostly of technical nature.

Lecture (1) concerns the optimal control of hydro energy storage plants. Results for two real world power plant systems in Austria are presented.

Lectures (2) and (3) deal with the production of steel. In (2) continuous casting of steel is controlled by a suitable cooling process, in (3) the aim is to minimize the costs for reheating the slabs. In Lecture (4) the thermodynamic equilibrium is determined for a composed system of chemical columns, reactors, etc.

On the Control of the Solidification Front in Continuous Casting of Steel
H.W. Engl, T. Langthaler (Universitat Klagenfurt)

The mathematical modelling of the continuous casting process leads to a nonlinear boundary value problem for a nonlinear heat equation. If one wants to control the front of solidification by regulating the pressure of the cooling water, this leads to an inverse problem for this boundary value problem.

In this talk we report about the implementation of an algorithm for approximating a solution of this inverse problem in an industrial environment (VOEST-Alpine AG, Linz).

Hydro Energy Optimization
W. Bauer, H. Gfrerer, E. Lindner, H. Wacker (Johannes Kepler Universitat), A. Schwarz (HBLA fur Wirt)

The aim is to maximize the monetary value of the production of a (system of) hydro energy storage plant(s). Mathematically this leads to a nonlinear constrained problem of optimal control with a nonlinear nonconvex objective. Several solution techniques are discussed. The problem of optimal control can be transformed into an equivalent finite dimensional optimization problem under the simplifying assumptions that the efficiency function of the power plant is constant. Models and numeric results are presented for two Austrian storage power plant systems.

Reheating of Slabs in a Pusher Type Furnace

D. Auzinger, H. Gfrerer, H. Wacker, (Johannes Kepler Universitat), B. Lindorfer (VOEST-ALPINE AG)

The reheating of slabs for rolling is rather energy consuming. For minimizing this energy consumption, a complex model (a coupled system of the heat equation, some ordinary differential equations, and some nonlinear equations), which describes the reheating process depending on the fuel input, has to be solved. For reasons of numerical stability the solution and its derivations with respect to the control variables are calculated by a multiple shooting method.

These gradients are needed to minimize the total fuel demand by Quasi Newton method.

Calculation of Steady State Conditions of Separation Plants

D. Auzinger, L. Peer, H. Wacker, W. Zulehner (Johannes Kepler Universitat), F. Kokert (VOEST-ALPINE AG)

The aim is to determine the thermodynamic equilibrium for a system of multistage multicomponent chemical columns including for instance pumping activities, side strippers, flashers, reactors etc. Mathematically this leads to a high dimensional system of nonlinear equations with a sparse Jacobian of special structure. The problem is solved by using a Newton-like iteration technique, and exploiting the special structure of the Jacobian.

• REAL WORLD PROBLEMS AND IMPLICATIONS FOR MATHEMATICAL MODELLING AND RESEARCH

F. WALKDEN, M. WADSWORTH (University of Salford)

Complexity is a distinguishing feature of many real problems which engineers and others increasingly wish to model in order to make informed decisions or to gain insights from large amounts of data gathered by observation and measurement. Different ways in which complexity occurs will be described.

Consequences of complexity together with modelling/research implications, including the need for re-visiting existing areas of knowledge which the applied mathematical community may believe to be complete, will be discussed.

The papers in this Minisymposium will illustrate the general ideas and principles outlined here.

Real World Problems and Implications for Mathematical Modelling and Research

F. Walkden (University of Salford)

Complexity is a distinguishing feature of many real problems which engineers and others increasingly wish to model in order to make informal decisions or to gain insights from large amounts of data gathered by observation and measurement.

Different ways in which complexity occurs will be described.

Consequences of complexity together with modelling/research implications, including the need for rewriting existing areas as well as knowledge which the applied mathematical community may believe to be complete will be discussed.

Features of problems which lead from requirements to new theoretical approaches, new numerical research, new solutions to selection specification techniques, and new computer program languages and systems will be considered.

Air Traffic Management Costs and Benefits

M. Hirst (Cranfield Institute of Technology)

Air traffic management is concerned with the efficient management of airspace and the economic operation of aircraft. The latter are very sensitive to profile variations that air traffic control systems impose in applying separation standards that minimise collision risk.

The synthesis of information in air traffic management systems needs to respect separation rules and assure a safe traffic solution in typical operations, while minimising economic penalties. A diverse collection of data types needs to be reconciled, and suitable monitoring and control information must be provided to control staff and aircrew. Examples of information flow complexities will be presented.

Mathematics of Inspection of Civil Engineering Structures

A.H. Christer (University of Salford)

The task of modelling the test and inspection process of major civil engineering structures is a complex task by almost any measure.

Available test techniques are variable in type, specification, cost of use, and accuracy. The potential exists to produce large volumes of test related data over a five to 100 year time zone, even the objective inspecting is a non-simple mixture of the requirements for safety and the need for maintenance cost control in an area where both the rate and physical development of a defect is frequently obscure.

The paper will describe both the nature of the problem and a way in which mathematical models can impose a form of structure to assist in clarifying the consequence of alternative inspection policies.

Fitting Smooth Surfaces to Seismic Data

L.R. Fletcher, M. Wadsworth (University of Salford)

This paper is concerned with the problem of fitting a smooth surface to a set of data points. In off-shore oil exploration, data is often obtained from seismic records with the data points lying along lines or tracks. The tracked nature of the data points complicates the problem of surface fitting.

The work is relevant in oil exploration where standard surface fitting techniques have failed to provide satisfactory results.

The problem is formulated as a set of coupled partial differential equations of elliptic type. Existence of solutions can be guaranteed and a wide family of solutions can be generated by varying different parameters.

Constructing Observers for Gas Networks

M. Chapman, D. Shields (Coventry Polytechnic), L.R. Fletcher (University of Salford), D. Pearson (British Gas, Hinckley Operational Centre)

Monitoring and control of a high pressure gas network involves the estimation of the state of the system on the basis of the partial information that is available. A finite dimensional linear asymptotic observer can be used to obtain the required estimates using a linearised and discretised model of the underlying gas flow equations. Ensuring that the observer has specified asymptotic properties involves a pole assignment problem in which the system matrices are large and sparse. This is achieved by means of a major reformulation of a conventional algorithm, to be described in this presentation.

Fokker Planck Equations for Atomic Mixing

M. Wadsworth, R. Collins, D. Armour, R. Badheka (University ofa Salford)

This paper is concerned with the theoretical determination of the amount of atomic mixing, sputtering, and surface erosion which is produced by a high-energy particle beam incident on a multi-layer structure.

The work is relevant to sputter depth profiling where the initial depth distributions of various constituents are inferred from the sputtered ion yields.

A mathematical model, which allows for the build-up of incident ions, ballistic relocations, and diffusion is described. The governing nonlinear partial differential equations are solved numerically.

Results for a GaAs-GaA1As multi-layer structure bombarded with "oh" ions are compared with results obtained experimentally.

• RESERVOIR MODELING AND SIMULATION - VIEW 1987

M. F. WHEELER (Rice University)

The "WHAT" and "WHY" of petroleum reservoir simulation will be briefly discussed; namely the formulation of the physical problem and the economic incentives for numerical simulation.

Various numerical discretization algorithms for reservoir modeling which include the ndustry standard, shock-front tracking, characteristics methods, finite element-Godunov, and mixed finite element methods will be presented and discussed. In addition advanced linear algebra such as domain decomposition and computer science implementation (vectorization and parallelization on supercomputers) will be addressed. Future research directions, the view from 1987, will be considered.

The What and Why of Reservoir Simulation: Definition of Physical Problems and the Economic Incentive for Simulation

I. Aavatsmark (Norsk Hydro, Norway)

The How of Reservoir Simulation: Discretizations: Shock Front Tracking

J. Glimm (Courant Institute of Mathematical Sciences)

Riemann problems define the interaction of nonlinear hyperbolic waves. In the context of oil reservoir simulation oil, water and chemical concentration fronts, as well as geological discontinuities (layer boindaries), are examples of such waves. Two dimensional front tracking calculations using this theory will be presented. Unstable interfaces in compressible fluids can be studied by the same method. Results concerning late time Rayleigh-Taylor and Richtmyer-Meshkov instabilities will be presented.

The Industry-Standard Formulation and the Practical Advantages of Characteristic Methods

T. Russell (Marathon Oil Company, University of Colorado, Denver)

Reservoir simulation, as practiced in the petroleum industry, has been based on upwind finite differences for 30 years. For black-oil modeling, which constitutes the bulk of actual studies, this procedure is satisfactory. However, for enhanced oil-recovery (EOR) processes other than steam injection, including miscible, chemical, and thermal (<u>in situ</u> combustion) recovery, upwind differences are inadequate. Grid-dependent numerical diffusion that has negligible practical impact on black-oil results causes serious misrepresentations of physical phenomena in EOR. Better methods, such as the characteristic-based schemes discussed in this talk, could restore the physics. Perhaps more importantly, they could provide tools toward basic understanding of viscous fingering, sensitivity of reservoir flows to unknowns such as heterogenious permeability, and how to represent quantities measured at laboratory scales in field-scale simulators in the face of reservoir heterogeneities. Ideas along these lines will be presented.

Finite Element - Godunov Approach

G. Chavent (INRIA-Rocquencourt Universite Paris IX, Dauphine)

Finite Element, Mixed Hybrid

R. Eymard (ELF Aquitaine)

Linear Algebra

R. Kendall (J.S. Nolen and Associates, U.S.A.)

Future Directions:

M.F. Wheeler (Rice University)

PART IV:
LISTING OF CONTRIBUTED PRESENTATIONS

Listing of Contributed Presentations*

Iterative Solution of Large Finite Difference Systems with Random Coefficients
Rachid Ababou (Massachusetts Institute of Technology)
Co-author: L.W. Gelhar

Kinematic Treatment of the Singular Points of the Mutual Enveloping Surfaces
Valentin Abadjiev (Bulgarian Academy of Sciences)
Co-author: Dochka Petrova

Hilbert's Method for the Problem of Free Jets of Water Under Gravity
Mina B. Abd-El-Malek (University of Alexandria, Egypt)

Minimum Spanning Trees vs Minimum Hamiltonian Paths: An Independence System View
James Abello (University of California)

Lab-Optim: An Interactive Optimization Package
D. Abinal (ENSIEG, Laboratoire d'Automatique Grenoble, Saint Martin d'Heres, France)
Co-author: A.Y. Barraud

An Appraisal of Two Parameters Estimation Schemes for a Class of Bilinear Mathematical Models
M.A.S. Abo Elela (State University of Ghent)
Co-author: G.C. Vansteenkiste

A Numerical Strategy for Treating the Pressure Singularity at a Re-Entrant Corner
Gorardo A. Ache (Universidad Central de Venezuela)

Adaptive Holding Control Strategy in Public Transport
Andrzej Adamski (S. Staszic University)

Adaptive Finite Element Methods Using First and Second-Order Approximations for Parabolic Systems
Slimane Adjerid (Rensselaer Polytechnic Institute)
Co-author: Joseph E. Flaherty

Taylor Dispersion in a Fractal Capillary Network
P.M. Adler (C.N.R.S. Laboratoire d'Aerothermique, Meudon, France)
Co-author: R. Delannay

*The following information was taken from the ICIAM '87 Final Program, which was published prior to the conference. Any changes to the program after that time are not reflected in these Proceedings.

This listing is ordered alphabetically by author's last name and is not divided into sessions. The session chairperson's name does not appear unless he/she presented a paper.

Homogenization of Miscible Flow in a Porous Medium
Ibrahim Aganovic (University of Zagreb)
Co-author: A. Mikelic

A Vector Implementation of the Mixed Radix FFT on IBM 3090 VF
R.C. Agarwal (IBM, T.J. Watson Research Center)

New Super Fast Vector Matrix Multiplication Algorithms for the IBM 3090 Vector Facility
R.C. Agarwal (IBM, T.J. Watson Research Center, Yorktown Heights)
Co-author: F.G. Gustavson

An Application of Tangent Coordinates and Perturbation Theory in Dynamic Analysis of Flexible Multi-Body Systems
Om Prakash Agrawal (Southern Illinois University)

Iterative Refinement for Multiple Eigenelements of Compact Integral Operators
Mario Ahues (University of Chile)
Co-author: F.D. d'Almeida

Steps Into a Geometer's Workbench
Varol Akman (Center for Mathematics & Comp. Science, Amsterdam, The Netherlands)

Fully Discrete Galerkin Methods for the Schrodinger Equation with Applications to Underwater Acoustics and Nonlinear Problems
Georgios Akrivis (University of Crete)
Co-author: V.A. Dougalis

Discrete-Time Identification of Non-Linear Systems by Means of State-Fine Models: An Improvement
S. Aksas (Institute Industriel du Nord, Villeneuve d'Ascq, France)
Co-authors: D. Normand-Cyrot, D. Meizel

Elements of a New Theory of Runge-Kutta Methods
Peter Albrecht (University of Dortmund)

Dynamic Force Analysis of Complex Three Dimensional Mechanisms
Marie-Jose Aldon (U.S.T.L., Laboratoire d'Automatique & de Microelectrique, Montpellier, France)

Comparison of Rank Tests and Their Applications
Saleem I. Al-Ghurabi (University of Baghdad)

On A Constrained Stock Cutting Problem
Stefano Alliney (Universita di Bologna)

The Levin-Courant Method of Regularisation for Numerically Unstable Inverse Problems with Applications to Hydrology
Henri Allison (CSIRO, Institute of Energy & Earth Resources, Wembley Western, Australia)

Non-Stationary Control Policies with Applications to Adaptive Control of Queues
E. Altman (Technion, Israel Institute of Technology)
Co-author: A. Schwartz

Solving Nonlinear Systems of Equations. A New Approach
Tom Altman (University of Kentucky)

Detuned Self-Tuning Regulator For Guidance System
M. Gamal Aly (Ain Shams University)
Co-author: A.A. Swief

Free Boundary Motion of a Perfect Fluid with Surface Tension
Youcef Amirat (INRIA-Rocquencourt, France)
Co-author: Patrick Joly

Numerical Simulation of Two-Phase Flow in a Gas Pipeline
Youcef Amirat (INRIA-Rocquencourt, France)
Co-author: N. Ranaivoson

A Special Matrix in Electromagnetics
Eliahu Gera Amos (ELTA Electronics Ind. Ltd., Ashdod, Israel)

A Generalization of the Erlang Formula of Traffic Engineering
V. Anantharam (University of California at Berkeley)
Co-authors: B. Gopinath, D. Hajela

Vortex Ring Interaction by the Vortex Method
Christopher Anderson (UCLA)
Co-author: Claude Greengard

Inverse Eigenvalue Problems with Discontinuous Coefficients, Asymptotic Properties
Lars-Erik Anderson (Linkoping University)

A Multitasked General Circulation Model of the Ocean
P. Andrich (Laboratoire d'Oceanographie Dynamique et de Climatologie, Paris, France)

Methode Directe de Reanalyse Spectrale
Aziz Filali Aoual (Laboratoire Genie Civil, Talence, France)

Approximation of Parametric Surfaces in R3 by Parametric Finite Elements Methods
Dominique Apprato (University of Pau)
Co-author: Remi Arcangeli

Qualitative Analysis of System Dynamics World Models
Javier Aracil (Universidad Seville)
Co-author: M. Toro

A Heuristic Approach for Finding Peripheral Nodes in Graph
Ilona Arany (Computer & Automation Institute, Budapest, Hungary)

Restricted Rank Modification of the Symmetric Eigenvalue Problem
Peter Arbenz (Brown-Boveri Cie, Baden, Switzerland)
Co-authors: W. Gander, G.H. Golub

Lyapunov Exponents and Bifurcation in Stochastic Systems
Ludwig Arnold (Institut Fur Dynamische Systeme, Bremen, F.R.G.)

Non Linear Diffusion Equations in Heterogeneous Media
Michel Artola (Universite de Bordeaux)

A Note on Solving Nonlinear Equations and the "Natural" Criterion Function
Uri M. Ascher (University of British Columbia)
Co-author: Michael R. Osborne

Dynamical Analysis of Nonlinear Harmonic Oscillations
George C. Atallah (Westinghouse Defense & Electronics Center, Baltimore)
Co-author: James F. Geer

Discrete Galerkin Methods for Nonlinear Integral Equations
Kendall E. Atkinson (University of Iowa)
Co-author: Florian Potra

Viability: An Introduction
Jean-Pierre Aubin (Universite Paris IX)

Compactness Methods in The Theory of Homogenization
Marco Avellaneda (New York University)

Optimal Bounds and Microgeometries for Elastic Two-Phase Composites
Marco Avellaneda (New York University)
Co-author: Fang-Hua Lin

Trajectories of Bifurcation Points of the Boussinesq Equations in Benard Convection
Y.Y. Azmy (University of Virginia)
Co-author: J.J. Dorning

Local Stabilizability Via Linear Feedbacks
Andrea Bacciotti (Politecnico di Torino)

Computation of Turbulent Flow Around a Cube on a Vector Computer
Frank Baetke (Technische Universitat Munchen)
Co-authors: H. Wengle, H. Werner

An Improvement of F.F.T. Calculations for Truncated Unlimited Functions. Application to the Analysis of Sums of Lorentzian Curves
Yves Balcou (Universite de Rennes)

Analysis of Multi-Exponential Decays Through Transform Methods. Applications to Physical and Bio-Medical Measurements
Yves Balcou (Universite de Rennes)

Error Estimation for Numerical Solution of Singular Initial Value Problems
Katalin Balla (The Hungarian Academy of Sciences)

Mathematical Analysis of Guided Elastic Waves
A. Bamberger (INRIA-Rocquencourt, France)
Co-authors: P. Joly, M. Kern

Second Order Absorbing Boundary Conditions for the Wave Equation: A Solution for the Corner Problem
A. Bamberger (INRIA-Rocquencourt, France)
Co-authors: P. Joly, J.E. Roberts

Periodicity of Patterns in the Derivative Field of Binary Sequences
A.M. Barbe (Katolieke Universiteit Leuves)

Singular Perturbation of Elliptic Equations and Stochastic Differential Equations with Small Noise Intensities
Martino Bardi (Universita di Padova)

Parameter Identification of Non-Stationary Processes in Colored Noise
Michel Barlaud (Laboratoires de Signaux & Systemes, Nice, France)
Co-authors: G. Alengrin, J. Menez

The Accurate Solution of Equality Constrained Least Squares by Weighting
Jesse L. Barlow (The Pennsylvania State University)
Co-author: Susan L. Handy

A Large Scale Nonlinear Optimization Problem Arising from Grid Generation II
Pablo Barrera-Sanchez (Universidad Autonoma Mexico)
Co-author: Jose Castillo

Wiener-Hopf Factorization, Riemann-Hilbert Boundary Value Problem and Realization
H. Bart (Erasmus University)

The Median Procedure, New Developments
Jean-Pierre Barthelemy (Dept. Informatique, ENST, Paris, France)
Co-author: Bernard Monjardet

Local Controllability in the Plane
Jose Basto Goncalves (Universidade do Porto)

On the Iterative Enclosure of Solutions of Ordinary Initial Value Problems
Hartmut Bauch (Padagogische Hochschule Dresden, G.D.R.)

A Coupling Procedure for Numerical Calculation of Gas-Liquid Film Flow in an Air Blast Fuel Atomizer Nozzel
W.W. Baumann (DFVLR, Turbulenzforschung, Berlin, F.R.G.)
Co-authors: H. Bendisch, F. Thiele

Application of the Homogenization and Perturbations Methods to Some Hydrodynamic Lubrication Problems
M. Bayada (I.N.S.A., Centre de Mathematiques, Villeurbanne, France)
Co-author: M. Chambat

Comparison Theorems Between Sparse Block and Point Factorization Iterative Methods
Robert Beauwens (Universite Libre de Bruxelles)
Co-author: M. Ben Bouzid

A Model for an Auditory Receptor Cell: Analysis of Periodic and Adaptive Behavior
Jonathan Bell (SUNY)
Co-author: Mark H. Holme

Non-Linear Stochastic Evolution Equations in Applied Sciences
N. Bellomo (Politecnico di Torino)
Co-author: R. Monaco

Parallel Numerical Solution of Variational Inequalities
M. Benassi (IBM Scientific, Roma, Italy)
Co-author: R.E. White

Robustness Correspondences Between State Space Models and Frequency Response Models
David Bensoussan (Universite de Quebec)

Population of Budworm in a Square Region
Guo Bon Yu (Science and Technology, University of Shahghai)
Co-authors: A.R. Mitchell, B.D. Sleeman

Chaotic Complex Dynamics and Newton's Method
Harold E. Benzinger (University of Illinois)
Co-authors: Scott A. Burns, J.I. Palmore

A Mathematical Model for the Stepwise Development of Verified Programs
Lee A. Benzinger (University of Illinois)

2D-3D Numerical Tests on a Fast Finite Element Solver for Generalized Stokes Problems
Ph. Beque (Electricite de France, L.N.H., Chatou, France)
Co-authors: J. Cahouet, J.P. Chabard, P. Hemmerich

General Solution of Non-Integer Order Linear Differential Equations
Benoit Bergeon (Universite de Bordeaux I)
Co-author: A. Oustaloup

3-D Finite Element Calculation of Newtonian Flow Through a Rectangular Die with Free Surface
A. Bern (Ecole des Mines, Valbonne, France)
Co-authors: J.L. Chenot, J.P. Zolesio

Shape Optimization of General Thin Shells
Michel Bernadou (INRIA-Rocquencourt, France)

Distributions Over Chaotic Spaces and Applications to Random Vibrations
Pierre Bernard (Universite de Clermont II)
Co-author: P. Kree

Collocation Spectral Methods for Elliptic Problems
Christine Bernardi (Universite Pierre et Marie Curie)

Stabilization of the Wave Equation with Pointwise Actuator and Static Output Feedback
Larbi Berrahmoune (Rabat-Takaddoum, Morocco)

Interpolation by Hyperbolic Functions
Jean-Paul Berrut (University of California - San Diego)

Multiprocessor Schemes for Solving Block Tridiagonal Systems
Michael Berry (University of Illinois)
Co-author: Ahmed Sameh

Linear Visco Elasticity for the Analysis of Plastic Material
Gerard Bertrand (Informatique Internationale, Agence de Saclay, Gif-sur-Yvette, France)
Co-author: Jean-Pierre Saussais

An Automatic Control Method for Approximately Defined Objectives
M. Bertrand (E.N.S.A.M., Automatic Control Laboratory, Lille, France)

The Auction Algorithm: A Parallel Relaxation Method for the Assignment Problem
Dimitri Bertsekas (Massachusetts Institute of Technology)

The Numerical Computation of Homoclinic Bifurcations
Wolf-Jurgen Beyn (Universitat Konstanz)

Curved Displacement Discontinuity Elements
P.K. Bhattacharyya (The Polytechnic of Central London)
Co-author: T. Willement

Boundary Integral Equation Method of Fatigue Crack Growth Process
A. Bia (Universite de Metz)
Co-authors: G. Pluvinage, M. Afzali

Efficient Algorithms for the Evaluation of the Eigenvalues of (Block) Banded Toeplitz Matrices
D. Bini (Universita di Roma)
Co-author: V. Pan

Theory of Bond Graph Modelling
S.H. Birkett (University of Waterloo)
Co-author: P.H.O'N. Roe

Computing the Singular Value Decomposition on a Range of Array Processors
Christian H. Bischof (Cornell University)

Vortex Breakdown Solutions of the Navier-Stokes Equations
Denis Blackmore (New Jersey Institute of Technology)

Existence and Nonexistence of Smooth Solutions to Some Problems of Nonlinear Classical Electromagnetic Theory
Frederik Bloom (Northern Illinois University)

A Multigrid Algorithm for the Investigation of Complex Fluid Flows
M.I.G. Bloor (University of Leeds)
Co-authors: P.H. Gaskell, N.G. Wright

Fluid Mechanics in Paper Making Systems
M.I.G. Bloor (University of Leeds)

The Influence of Entry Conditions on the Flow in Industrial Cyclones
M.I.G. Bloor (University of Leeds)

A Dynamical Analysis of the Breeding Process in Nuclear Engineering
Vinicio Boffi (University of Bologna)
Co-author: Giampiero Spiga

Polynomial Approximations to Operator Inverses: Applications in Optimisation and Probability
R.A.B. Bond (University of Natal)
Co-authors: I.A. McDonald, J. Mika

On a Non-Stationary Filtration Problem
Din Phu Bong (Hanoi Polytechnical Institute)
Co-author: Nguyen Dinh Tri

Stable Time-Dependent Probability Densities and Phase Transitions in Mean-Field Theory
Luis Bonilla (Universidad de Sevilla)

Perturbations of Domains Defined by a Set of Equalities and Inequalities
J. Frederic Bonnans (INRIA-Rocquencourt, France)
Co-author: G. Launay

Generalization of Algebraic Systems Theory and Applications to Robotics and Control
Alain Bonnemay (CEA-IRDI, Saclay, Gif sur Yvette, France)

Cut-Off Frequencies for Guided Modes of an Optical Fiber: Mathematical and Numerical Study
Anne-Sophie Bonnet (GHN-ENSTA, Centre de l'Yvette, Palaiseau, France)

Limit Theorem for Parabolic Equation
Alexander D. Borisenko (Kiev State University)

Impulse Systems Solutions Stability
S.D. Borisenko (Kiev State University)

Sur une Formulation Recurrente Minimale des Systemes Discrets
B. Bouaziz (Faculte des Sciences, Laboratoire d'Electronique et d'Automatique, Casablanca, Morocco)
Co-author: E. El Moudni

Adiabatic Annular Shear Flows With Temperature-Dependent Viscosity
Moses Boudourides (Democritus University of Thrace)

Is Random Function Really Useful?
M. Bouleau (C.E.R.M.A.-E.N.P.C., France)

Pollux: A Computer-Aided Design and Analysis of Control Systems
Z. Bouredji (ENSIEG, Laboratoire d'Automatique Grenoble, Saint Martin d'Heres, France)
Co-author: A.Y. Barraud

Homogenization of Two-Phase Flow in Inhomogeneous Porous Media
A. Bourgeat (U.E.R. de Sciences, St-Etienne, France)
Co-author: B. Amaziane

Heat Conduction Through an Infinite Circular Cylinder Having Temperature Dependent Thermal Parameters
Y.Z. Boutros (University of Alexandria, Egypt)

A Method for Numerical Analysis of Dynamic Processes in the Electrical Networks
Christo Boutzev (UNESCO SC/TER, Science Sector, Paris, France)
Co-author: Lubomir Boutzev

Multigrid Solutions of the Fokker-Planck Equation
B.J. Braams (Princeton University)
Co-author: C.F.F. Karney

Optimal Control of District Heating Networks
Ion Bratasanu (Ecole Superieure Mines de Paris)
Co-authors: R. Lidin, G. Cohen

A Combinatorial Approach to Hydrodynamics and Incompressible Flows
Yan Brenier (INRIA-Rocquencourt, France)

Nonexistence Result for a Singular Perturbation Problem
Mary E. Brewster (Rensselaer Polytechnic Institute)

Numerical Analysis of Hopf Bifurcation in Two Parameter Problems
Morten Brons (The Technical University of Denmark)

Models of Parallel Chaotic Iteration Methods
Rafael Bru (Universidad Politecnica, Spain)
Co-authors: Ludwig Elsner, Michael Neumann

On Discrete Schrodinger Equations and Their Two Component Wave-Equation Equivalents
A.M. Bruckstein (Technion-Israel Inst. of Technology)
Co-author: T. Kailath

Computation of Vortex Flow Past Thin Wings of Low Aspect Ratio
Charles-Henri Bruneau (Universite Paris-Sud)
Co-authors: J. Laminie, R. Temam, J.J. Chattot

The Two-Dimensional Stokesian Entry Flow
V.T. Buchwald (University of New South Wales)

A Stable Limited Memory Minimization Algorithm
A. Buckley (INRIA-Rocquencourt, France)

Three-Dimensional Minimum Directed Distance Algorithm
Charles E. Buckley (ETH-Zentrum, Institut fur Integrierte Systeme, Zurich, Switzerland)

Combined Methods for Elliptic Boundary Value Problems with Singularities and Interfaces
T.D. Bui (Concordia University)
Co-author: Z.C. Li

Symmetric Singular Value Decomposition
Angelika Bunse-Gerstner (Universitat Bielefeld)

Feedback Invariant Approximations of Nonlinear Control Systems
Gabriel Burstein (Polytechnic Institute Bucharest)

On the Generalization of Trotter's Formula for Nonhomogeneous Noncontinuous Stochastic Case
George Butsan (Ukrainian Academy of Sciences)

Semi-Local Analysis of Quasi-Newton Methods for Nonlinearly Constrained Optimization
Richard H. Byrd (University of Colorado)
Co-author: Jorge E. Nocedal

Optimal State Observation of Linear System in Presence of the Worst Disturbances
Witold Byrski (Academy of Mining and Metallurgy, Krakow, Poland)

Vibrations in the Presence of Convex Obstacles
Henri Cabannes (Universite Pierre et Marie Curie)

Topological Form Features Recognition From Boundary Models
Marina Cabella (Politecnico di Milano)

Statistical Ideas for the Assignment Problem
L. Cairo (C.N.R.S.-P.M.M.S., Orleans, France)
Co-authors: M.R. Feix, E. Jamin

Applications of the Restricted Pseudosolution
Daniel G. Callon (Miami University-Middletown)

Stability for the Conducting Dusty Gas Flow
Colette Calmelet (Vanderbilt University)
Co-author: Philip Crooke

Production Planning and Scheduling of Manufacturing Systems
Fernando Antonio Campos Gomide (CTI Engenharia Electrica, Campinas, Brazil)
Co-author: C.A. de Oliveira Fernandes

Application of Riemann's Boundary-Value Problem in Mathematical Physics
Milos Canak (Faculty of Agriculture, Zenum, Yugoslavia)

Some Topics in Mathematical Theory of Music
Milos Canak (Faculty of Agriculture, Zemun, Yugoslavia)

Numerical Simulation of Gas Motion in the Cylinder of a Two-Stroke Engine
Pascal Candau (E.N.S.A.M., Paris, France)
Co-authors: V. Daru, A. Lerat

Flexibility Method Non-Linear Analysis Approach in Structural Optimization
Luigi Cappellari (Universita di Padova)
Co-author: L. Simoni

Dynamic Characterization of Linear Flexible Structures
Alfred Carasso (National Bureau of Standards, U.S.A.)
Co-author: Emil Simiu

Representation de Formes Tridimensionnelles Partiellement Fermees
Claude Carasso (Universite de Saint-Etienne)

The Role of Euler Operators in Integrated Modelling Systems
Monica Carimati (Politechnico di Milano)

Optimization of the Wavelength in Single-Mode Optical Fibers for Nonlinear Propagation
Xavier Carlotti (Ecole Polytechnique, Palaiseau)

Bimond 3: A New Bivariate Monotone Interpolation Algorithm
Ralph E. Carlson (Lawrence Livermore National Lab, U.S.A.)

Two Industrial Applications of Quadratic Optimization
Georges Carraci (Departement de Mecanique, Toulouse, France)
Co-authors: M. Astre, M. Boyor, M. Lafitte

A Globally Convergent Framework for Applying an Expert Systems Approach to Optimization
Richard G. Carter (Rice University)

The Effect of Rotation and an Axial Magnetic Field on Czochralski Crystal Growth
Rosemary Ann Cartwright (Bristol University, U.K.)

Thresholding Grey Level Histograms by Minimum Information Loss
Rossana Caselli (S.A.S.I.A.M., Balenzano-Bari, Italy)
Co-author: B. Forte

Symbolic Computation of the Direct Differential Models and Singularities of Robot Manipulators
J.M. Castelain (Laboratoire de Conception de Systems Mecaniques, Valenciennes, France)
Co-authors: T. Bonduelle, D. Bernier

Non-Negative Solutions for a Class of Non-Positone Problems
Alfonso Castro (North Texas State University)
Co-author: R. Shivaji

Nonlinear Model Identification and Asymptotic State Estimation of a Distillation Process
Rafael Castro-Linares (Centro de Investigacion y de Estudios Avanzados del IPN, Mexico)
Co-authors: H. Sossa, B. Castillo

Countable Infinite Easy Handling
Michel Cayrol (Universite Paul Sabatier, Toulouse)

Existence for Nonlinear Boundary Value Problems in Kinetic Theory
Carlo Cercignani (Politecnico di Milano)
Co-authors: R. Illner, M. Shinbrot

Numerical Simulation of Rarefied Gas Flows Past a Flat Plate: Comparison of Various Gas-Surface Interaction Models
C. Cercignani (Politecnico di Milano)
Co-authors: A. Frezzotti, M. Lampis

Discrete Designs and Codes From Hermitian Varieties and Quadrics in Projective Geometries
Indra M. Chakravarti (University of North Carolina)

Seismic Processing on a Hypercube Multiprocessor
R.M. Chamberlain (Chr. Michelsen Institute, Fantoft, Norway)
Co-author: L.R. Hagen

The Toeplitz-Circulant Eigenvalue Problem
Raymond H. Chan (University of Hong Kong)

Exact Controllability of Age-Dependent Population Equations
W.L. Chan (The Chinese University of Hong Kong)

Exaction Solutions of Soliton Equations
W.L. Chan (The Chinese University of Hong Kong)

On the Application of a Generalisation of Toeplitz Matrices to the Numerical Solution of Weakly-Singular Integral Equations
S.N. Chandler-Wilde (University of Bradford)
Co-author: M.J.C. Gover

A Cryptosystem for Secure Broadcasting
C.C. Chang (National Chung Hsing University)
Co-author: C.H. Lin

The Interface Between Fresh and Salt Groundwater
Jean-Riccel Chan-Hong (Delft University of Technology)
Co-author: D. Hilhorst

Identification and Control Using a Fast-to-Compute Volterra Type Model Applied to a Nonlinear Industrial Process
Pierre Chantre (Conservatoire Nat. Arts et Metiers, Paris, France)

Backlund Transformations and Solutions with Variable Speeds
Gu Chaohao (Chinese University of Shanghai)

Adaptive Management of an Uncertain Natural Resource: An Optimization Model with Bayesian Updating
Anthony T. Charles (Saint Mary's University, Halifax)

A Systolic Algorithm for Riccati and Lyapunov Equations
J.P. Charlier (Philips Research Laboratory, Bruxelles, Belgium)
Co-author: P. Van Dooren

About Parallelism Induced by a Nested Dissection Method
P. Charrier (Universite de Bordeaux 1)
Co-author: J. Roman

Computational Analysis of Single-Server Bulk-Service Queues M-Gy-1
Mohan L. Chaudhry (Royal Military College, Kingston, Canada)
Co-author: G. Briere

Dynamic Optimization of Combined Harvesting of a Two-Species Fishery
Kripasindhu Chaudhuri (Javadpur University, Calcutta, India)

Minimum Overlapping Decomposition
Chiuyuan Chen (National Chiao Tung University)
Co-author: Ruei-Chuan Chang

Simulation of Communication System for a Computerised Hospital
T.S. Chen (National Cheng Kung University)
Co-authors: T.G. Gough, S.M. Wu

Symbolic Computation for Matrix Equations in AC Electrical Circuit
Tse-Sheng Chen (National Cheng Kung University)
Co-author: Gwo-Jeng Yu

On Graph Classification, Decomposition, and Lattice Theory
Wai-Kai Chen (University of Illinois)

The Exponential Stability of Nonautonomous Difference Equations for Adaptive Control
Zong Ji Chen (Beijing Institute of Aeronautics & Astronautics)

Existence of Periodic Solutions for Some Forced Hamiltonian System of Differential Equations
A.A. Cherif (University of Mississippi)
Co-author: O. Arino

Quasi-Real Time Processing By Queue (FIFO) Automata
Alessandra Cherubini (Politecnico di Milano)
Co-authors: C. Citrini, S. Crespi Reghizzi, D. Mandrioli

Differential Geometry and Structure of Dynamics Equations
D. Chevallier (CERMA-ENPC, Noisy le Grand, France)

Perturbed Multiple Bifurcation Techniques
C.S. Chien (National Chung-Hsing University)

Algorithms for Finding the Shortest Addition Chain
Y.H. Chin (National Tsing-Hua University, Hsinchu)
Co-author: Yueh-Hsia Tasi

The Reliability of a Software System
Rouh Janc Chou (National Tsing Hua University)
Co-author: Shu-Chuan Chen

Two Dimensional Functions of Matrices
Manolis A. Christodoulou (University of Patras)
Co-authors: B.G. Merzios, B.L. Smyrmos

On the Numerical Stability of the Schur Method for Solving the Matrix Algebraic Riccati Equations
N.D. Christov (Higher Inst. of Mech. & Electrical Engineering, Sofia, Bulgaria)
Co-authors: P.Hr. Petkov, M.M. Konstantinov

Identification of Discontinuous Parameters in Second-Order Parabolic Systems
Chang-Bock Chung (The University of Michigan at Ann Arbor)
Co-author: Costas Kravaris

Tricubic Spline Functions: An Efficient Tool for 3D Images Modelling
Philippe Cinquin (TIM3 - IMAG, Al Tronche, France)

On the Motion of a String Vibrating Against a Glueing Wall
Caludio Citrini (Politecnico di Milano)
Co-author: Clelia Marchionna

A Convexity Preserving Piecewise Rational Cubic Interpolant
John C. Clements (Dalhousie University)

Algorithms for the Eigensolution of Symmetric Matrices on a Mesh-Connected Computer
Maurice Clint (The Queen's University of Belfast)
Co-author: J.S. Weston

The Stability of Weakly Nonlinear Difference Schemes
A.H.J. Cloot (University of Orange Free State)
Co-author: B.M. Herbst

Controlled Release Pharmaceuticals: Nonlinear, Non-Fickian Diffusion in Polymers
Donald S. Cohen (Caltech)

Fourth Order Schemes for Seismic Prospection
Gary Cohen (INRIA-Rocquencourt, France)

The Action of Stress at the Surface of Water Waves
Raymond Cointe (Institut Francais du Petrole, Malmaison, France)

Transient Waves in Nonlinear Electroelastic Solids
Bernard Collet (Universite Pierre and Marie Curie)

Optimal Decision Support System for the Corsican Power System Through Process Identification and Stochastic Dynamic Programming
Patrick Colleter (EDF - DPT, S.E.T.E., Paris, France)
Co-author: Ch. Trzpit

Equilibrium Fluid Interfaces in the Absence of Gravity
Paul Concus (University of California - Berkeley)

A New Algebraic Approach of Integral Transforms on Finite-Order Algebras
Daniel Condurache (Polytechnic Institute of Jassy)

Dissolution-Growth Processes Modellized by Quasi-Variational Inequalities
Francis Conrad (Ecole des Mines, Saint-Etienne, France)
Co-author: Naji Yebari

A Stable Parallel Algorithm for the Solution of Narrow Banded Systems
John M. Conroy (Supercomputing Research Center, U.S.A.)

Singular Perturbation Techniques for US/110 Abstract Volterra Equations
Constantin Corduneanu (University of Texas)

Influence of Destabilizing Effects on the Cyclic Response of Elastic-Plastic Beams
L. Corradi (Politecnico, Milano, Italy)
Co-authors: C. Poggi, A. Cazzani

On a System Identification Problem Motivated by Robust Control
G.O. Correa (National Computing Laboratory, Rio de Janeiro, Brazil)

Linear Systems Solvers on Processors Networks
M. Cosnard (C.N.R.S., Laboratoire TIM3, Grenoble, France)
Co-authors: M. Daoudi, B. Tourancheau

Analytic Solutions to the Cubic Schrodinger Equation
Peter J. Costa (North East Research Associates, Inc., U.S.A.)

Convergence of Finite Difference Schemes for Nonlinear Parabolic Equations
Peter J. Costa (North East Research Associates, Inc., Woburn, U.S.A.)

Linear Conjugacy and Nonlinear Parabolic Equations
Peter J. Costa (North East Research Associates, Inc., Woburn, U.S.A.)

The Effects of Fluid Properties Inclination and Tube Geometry on Rising Bubbles
Benoit Couet (Schlumberger - Doll Research, Ridgefield, U.S.A.)
Co-authors: Gary S. Strumolo, Warren Ziehl

A-Posteriori Error Estimation and Adaptive Analysis for Piecewise Linear Finite Element Spaces
Alan Craig (University of Durham)

Homological Methods in Scene Analysis and Structural Mechanics
Henry Crapo (INRIA-Rocquencourt, France)

A Model for the Effect of the Wake on the Loading of a Vertical Cylinder Due to Surface Waves
A.J. Croft (Coventry Lanchester Polytechnic)

Embedding Expert Knowledge in CFD Software
M. Cross (Thames Polytechnic)
Co-authors: R. Legett, B. Knight

On the Connection Between a Finite Dimensional Inverse Problem and a Discrete-Time Dynamical System
Giovanni Crosta (Universita Degli Studi, Milano)

Application of Min-Max Differentiability to an Inverse Problem Related to the Wave Equation
Michel Cuer (Universite des Sciences et Techniques du Languedoc)
Co-author: J.P. Zolesio

A Hybrid Lanczos Algorithm for Solving a Non-Symmetric Generalized Eigenvalue Problem
J. Cullum (IBM T.J. Watson Research Center, U.S.A.)
Co-authors: R.A. Willoughby, W. Kerner

3-D Solutions for Source-Drain Regions in Field-Effect Transitors
Ellis Cumbertach (Claremont Graduate School)

Sound Scattering From an Underwater Finite Flexible Strip
J.A. Cuminato (Oxford University)
Co-authors: D. Butler, S. McKee

Wave Digital Filters with Retrieval of Reflected Pseudopower - Analysis and Optimization
Adam Dabrowski (Technical University of Poznan)

Conversation Laws and Jump Conditions for Micropolar Fluids Mechanics
Tian-Min Dai (Liaoning University)

Muse: Multivariate Statistical Expertise
Eric Dambroise (IBM-France, Montpelier)
Co-author: P. Massotte

Rime: Resources Improvement Modelling
Eric Dambroise (IBM-France, Montpelier)
Co-authors: P. Massotte, R. Sabatier

Network Formulation in Structural Analysis
Ramendra Das (Tuv Rheinland, Koln, F.R.G.)

New Exact Nonlinear Filters and Separation of Variables
Frederick E. Daum (Raytheon Company, U.S.A.)

Reduced Models of Singularly Perturbed Systems
Genevieve Dauphin-Tanguy (I.D.N., Labo. d'Automatique et Info. Indus., Villeneuve d'Ascq, France)
Co-author: F. Rotella

Fluid Flow and Extinction of Diffusion Flames
Timothy David (Leeds University)

Stabilizing Spatially Uniform Steady States
Paul Davis (Worcester Polytechnic Institute)
Co-author: Jagdish Chandra

Constitutive Equations for Granular Materials
Reint de Boer (Universitat Essen)
Co-author: W. Gollub

A Nonconvex Minimization Problem in Crystallography
Andree Decarreau (Universite Paris-Sud)
Co-authors: D. Hilhorst, Cl. Lemarechal, J. Navaza

Some Experiences with Structural Identification of Multivariable Linear System
Joao de Farias Neto (Universite Catholique, Brazil)
Co-authors: R. Gouvea, A.B.F. Neves

A Stabilized Galerkin Method for Convection-Diffusion Problems
Pieter De Groen (Vrije Universiteit Brussel)
Co-author: M. Van Veldhuizen

Technology Imports and Politico-Economic Fluctuations in Soviet-Type Economies
Christophe Deissenberg (University of Illinois at Urbana-Champaign)

On Kuhn's Root Finding Method
M.C. Delfour (Universite de Montreal)
Co-author: G. Payre

Modal-Flow Analysis of Damped Structures
Kresimir Delinic (Kraftwerk Union, Erlangen, F.R.G.)

A Variational Formulation of Diatomic Elastic Dielectrics
Hilmi Demiray (Research Institute for Basic Sciences, Gebze-Locaeli, Turkey)
Co-author: Sadik Dost

The Geometry of Ill-Conditioning
James Demmel (Courant Institute)

All Nonnegative Solutions of Sets of Linear Equalities and Inequalities
Bart de Moor (Catholic University of Leuven)
Co-author: J. Vandewalle

A New Variational Approach to Plastic Shakedown for Continua Based on the Concept of Average Dissipation
Gery de Saxce (University of Liege)

A Two-Dimensional Problem of Electromagnetic Induction
Jean Descloux (Ecole Polytechnique Federale de Lausanne)
Co-author: Michel Crouzeix

TVD Schemes for Lagrangian Hydrodynamics
Jean-Claude Desgraz (Centre d'Etudes de Limeil-Valenton, Saint Georges, France)

Coloring Large Graphs
D. de Werra (Ecole Polytechnique de Lausanne)
Co-author: A. Hertz

Multi-Vector Parallel Algorithms for the Preconditioned Conjugate Gradient Method
J.C. Diaz (University of Oklahoma)

The Sem Algorithm, A Probabilistic Teacher Version of the Em Algorithm for Maximum Likelihood
J. Diebolt (Universite Paris VI)
Co-author: G. Celeux

Adaptive Control with Constraints on the Inputs
Jean-Michel Dion (E.N.S.I.E.G.-I.N.P.G., Lab. Automatique de Grenoble, Saint-Marint d'Heres, France)
Co-authors: T. Nguyen-Minh, L. Dugard

Conjugate Bases for Constrained Optimization
L.C.W. Dixon (The Hatfield Polytechnic, U.K.)

Inefficient and Efficient Stabilization Policies in a Simple Macroeconomic Model: An Application of Linear-Quadratic Differential Game Theory
Engelbert Dockner (University of Saskatchewan)
Co-author: Reinhard Neck

Biorders and Preference Modelling
J.P. Doignon (Universite Libre de Bruxelles)
Co-author: J.Cl. Falmagne

The Themis Model: A 2D Integrated Numerical Computation of Fluids and Heat Transfers in a Sedimentary Basin
Brigitte Doligez (I.F.P., Geologie et Geochimie, France)
Co-authors: P. Ungerer, J. Burrus, P.Y. Chenet, P. Lascaux

Tools for Developing and Analyzing Parallel FORTRAN Programs
J.J. Dongarra (Argonne National Laboratory, U.S.A.)
Co-author: D.C. Sorensen

On Bennett's Model on Predicting Chromosome Order
Dietmar Dorninger (Technische Universitat, Wien)

The Parallelized Cell Discretization Algorithm
Milo R. Dorr (Lawrence Livermore National Lab., U.S.A.)

Conservative High-Order Schemes for the Generalized Korteweg - de Vries Equation
V. Dougalis (University of Crete)
Co-authors: J.S. Bona, O.A. Karakashian

A Different Approach to Parallelizing Multigrid Algorithms
Craig C. Douglas (IBM Thomas J. Watson Research Center)

Analysis, Design and Applications of "Neural" Networks
Gerard Dreyfus (E.S.P.C. Laboratoire d'Electronique, Paris, France)
Co-authors: I. Guyon, L Personnaz

Boundary Conditions for Systems of Conservation Laws
F. Dubois (Ecole Polytechnique, Centre de Mathmatiques Appliquees, Palaiseau, France)
Co-author: P. Le Floch

Acceleration of Explicit Multistepping Algorithms
G.S. Dulikravich (Pennsylvania State University)

Blunt Body Sampling
S.J. Dunnett (University of Leeds)
Co-author: D.B. Ingham

The Error Estimation of a Computational Method in Geophysics
Tosic Dusan (University of Belgrade)

Theorems of Existence for Differential Systems with Two Point Boundary Conditions
J. Duvallet (Faculte des Sciences, Pau, France)

Models for Branching Networks
Leah Edelstein-Keshet (Duke University)

Parallel Hermite Interpolation
Omer Egecioglu (University of California, Santa Barbara)
Co-authors: E. Gallopoulos, Cetin K. Koc

What Do Multistep Methods Approximate?
Timo Eirola (Helsinki University of Technology)
Co-author: O. Nevanlinna

Knowledge Based Expert Systems for Pursuit-Evasion Games
M. El-Arabaty (Military Technical College, Cairo, Egypt)

Mathematical Modelling of a Generalized Assignment Problem for Tracking Resources
M. El-Arabaty (Military Technical College, Cairo, Egypt)

The Probability of Success in the Searching Game
M. El-Arabaty (Military Technical College, Cairo, Egypt)

Towards the Optimum Solution of Maintenance Policy
M. El-Arabaty (Military Technical College, Cairo, Egypt)

Conception of Real-Time Expert Supervision Based on Quantitative Model Simulation of the Process
S. Abou El Ata (Adersa, Verrieres Le Buisson, France)
Co-author: J. Brunet

Predictive Control for Invertible Systems
S. Abu El Ata-Doss (ADERSA, Verrieres Le Buisson, France)
Co-author: J.L. Estival

An Algorithm for Constrained Approximation
Tommy Elfving (Linkoping University)

A Multi-Grid Method for 3D Linear Elasticity Computational by the Finite Element Method
M. El Hadj (Ecole Nationale Superieure des Mines de Paris, Valbonne, France)
Co-authors: J.L. Chenot, Y. Demay

Time Distortion Correction for Gated Cardiac Images
S.W. Ellacott (Brighton Polytechnic, U.K.)
Co-author: B. Rahimi

Fourier Analysis of Iterative Methods for Elliptic Problems
Howard Elman (University of Maryland)
Co-author: Tony Chan

Metrization of Group Technology
Ahmed K. El-Sakkary (University of Petroleum & Minerals, Saudi Arabia)

Coupled Strong Shocks in Elasticity
Marek Elzanowski (University of Calgary)
Co-author: Marcelo Epstein

Controllability Analysis of Nonlinear Systems by Foliation Theory and Approximative Group Methods
Stanislav V. Emel'Yanov (International Research Institute for Management SC, Moscow)
Co-authors: S.K. Korovin, S.V. Nikitin

An Efficient Method for the Calculation of the Limits of Sensitivity Reduction
Sebastian Engell (Fraunhofer Institut, Karlsruhe)

Simulation and Optimization of Macro-Econometric Models: A New Computational Method
J. Engels (Facultes Universitaires de Namur)

A Sequential Multiple Shooting Strategy for Stiff Boundary Value Problems
Roland England (South Bank Polytechnic, London)
Co-author: R.M. Mattheij

On Partial Integro-Differential Equations with Singular Kernels
Hans Engler (Georgetown University)

On Incremental Path-Independent Integrals Characterizing Stable Crack-Growth
Heinz-Hermann Erbe (Technische Universitat Berlin)

The Modified Rayleigh Quotient Iteration
Jiang Erxiong (Fudan University)

Multicriteria Optimization Procedure and Its Application in Structural Mechanisms
Hans A. Eschenauer (University of Siegen)
Co-authors: P. Post, M. Bremicker

Convergence Analysis of the Least-Squares Identification Algorithm with Variable Forgetting Factor for Time Varying Linear Systems
Martin D. Espana (Ecole Nationale Superieure Mines, France)
Co-author: R.M. Canetti

On Optimal Recursive Multistep Prediction and Generalized Predictive Control: A State Space Formulation
M. Espana (Instituto de Ingenieria, Mexico)
Co-author: R. Ortega

The Limitation of Navier-Stokes Theory in Long-Term Prediction
C. Essex (Universitat Frankfurt)

The Alternating Group Explicit (AGE) Systolic Array for the Solution of Large Linear System
D.J. Evans (University of Technology, Loughborough, U.K.)
Co-author: G.M. Megson

Upstream Weighting for Pressure Gradient (D.P.G. Scheme) in Reservoir Simulation
R. Eymard (ELF Acquitaine, S.N.E.A.P., DREA Paris)
Co-authors: A. Pfertzel, T. Gallouet

Graph Invariants: Derivation, Computation, Relatedness and Chemical Applications
Irena Fabic (Institut Jozef Stefan, Ljubljana)
Co-authors: B. Jerman-Blazic, Vladimir Batagelj

Asymptotic Behavior and Numerical Approximation for a Convection Diffusion Problem
Pierre Fabrie (Universite de Bordeaux I)

Electrostatic Problem of Several Arbitrarily Charged Unequal Coaxial Disks
V.I. Fabrikant (Concordia University)

Approximate Feedback Optimal Controls for Deterministic Control Problems
Maurizio Falcone (Universita di Roma)

A Variable Order Adaptive Mesh Algorithm for Time Dependent Fluid Problems
S.A.E.G. Falle (University of Leeds)
Co-author: J.R. Giddings

Similarity Solutions Calculated Using Multigrids
S.A.E.G. Falle (University of Leeds)
Co-author: J.R. Giddings

Optimal Matching of Deformed Patterns with Positional Influence
Tzu I. Fan (National Central University, Chung-Li, Taiwan)

An Unified Approach to L Optimization, Hankel Approximation and Balanced Realization Problems
Yeh Fang-Bo (Tunghai University)

Optimum Filter for DC.AC Converter in Electronics
Albert F. Fassler (Ingenieurschule Biel, Switzerland)

Non-Linear Phenomena: On the Links Between Parameters and Pulsations for Periodical Solution of Some Types of Differential Equations of Electronics and Mechanics
Robert Faure (Universite des Sciences et Techniques de Lille)

Theory of Impulses Forces (Percussions) in Nonlinear Phenomena
Robert Faure (Universite des Sciences et Technique de Lille)

The Initial-Boundary Value Problem for the Motion of an Elastic String
Joseph D. Fehribach (University of Alabama)
Co-author: Michael Shearer

Optimal Pulsing in Marketing
Gustav Feichtinger (University of Technology, Vienna, Austria)

Optimization of Thermal Systems and Process Applications to Heat Pumps
Michel Louis Feidt (University Nancy I)

Utilisation de la CAO pour la Verification des VLSI: Resultats Recents
H. Felix (Bull S.A., Paris, France)

Bifurcation and Linearized Stability of Nonlinear Equations
De-Xing Feng (Institute of Systems Science, China)
Co-author: Jia-Quan Lin

Multigrid Methods for Solving General Flow Problems in the Hydrocyclone
J.W. Ferguson (University of Leeds)

A New Method of Inequality Selection for Chernikova's Algorithm
Felipe Fernandez (Universidad Politecnica de Madrid)
Co-author: J. Gutierrez

Numerical Simulation of Turbulence for Viscous Flows: The Finite Element Approach to the M.P.P. Model
Enrique Fernandez-Cara (University of Sevilla)

Linear Convolution by Multi-dimensional Techniques Based on Mersenne Number Transforms
Ricardo Ferre (Lund University)

Modelling FMS: A Birth-Death Approach
Antonio Ficola (Universita di Roma)

A Variational Principle for Free Surface Flow with Applications
W.D.L. Finn (University of British Columbia)

Some New Results in Computational Geometry
Jean-Charles Fiorot (Universite de Valenciennes)
Co-author: P. Jeannin

Surfaces of Class C^m Only Defined by Values at Nodes
Jean-Charles Fiorot (Universite de Valenciennes)

Representation and Computation of the One-Way Scalar Helmholtz Wave Field Propagator
Louis Fishman (Catholic University of America)

On the Eigenvalues of Indefinite Elliptic Problems
Jacqueline Fleckinger (E.N.S.E.E.I.H.T., Toulouse, France)

Properties of Toeplitz Approximation Method (TAM) for Direction Finding Problems
R. Foka (Thomson-Sintra, Activites Sous-Marines, Arcueil, France)

Simulated Annealing Applied To Controlled Hydraulic Fracture Growth
Amaury Fonseca Jr. (M.I.T.)

Shape Preserving Surface Interpolation
Ferruccio Fontanella (Universita di Firenze)
Co-authors: P. Costantini, R. Morandi

Polynomial Matrix Approach to the Analysis and Stabilization of Two Dimensional Filters
Ettore Fornasini (Universita Di Padova)
Co-authors: M. Bisiacco, G. Marchesini

A New Formalization of a Geodimetric Network Coordinate Estimation
Luigi Fortuna (Universita di Catania)
Co-authors: G. Nunnari, C. Guglielmino

On the Stability of Closed-Loop Systems with Reduced Order Compensator
L. Fortuna (Universita di Catania)
Co-authors: C. Guglielmino, G. Nunnari

Effective Condition Numbers for Linear Systems
David Foulser (Saxpy Computer Corporation, Sunnyvale, U.S.A.)
Co-author: Tony F. Chan

A Mathematical Model of Frost Heave
A.C. Fowler (University of Oxford)

A Theory of Glacier Surges
A.C. Fowler (University of Oxford)

Freckling in Binary Alloys
A.C. Fowler (University of Oxford)

The Generation and Transport of Magma in the Asthenosphere
A.C. Fowler (University of Oxford)

System Identification Methodology for Model Building
Larissa Fradkin (University of Cambridge)

Control of Systems with Feedbacks
Halina Frankowska (Ceremac, Universite Paris-Dauphine)

Bifurcations and Symmetry Breaking in Simpler Limited Growth Systems with Diffusive Coupling
Emilio Freire (Escuela Superior de Ingenieros, Sevilla)
Co-author: E. Ponce

Pseudo Ritz Values for Indefinite Hermitian Matrices
Roland Freund (Universitat Wurzburg)

Optimization of Non-Linear Large Scale System with Linear Dynamics - An Application to Load Scheduling
Ana Friedlander (IMECC-UNICAMP, Campinas, Brazil)
Co-author: Ch. Lyra

Hydromagnetic Waves in the Earth's Fluid Core
Susan Friedlander (University of Illinois)

Comparison of Brown's and Newton's Method in the Monotonic Case
Andreas Frommer (Universitat Karlsruhe)

A Splitting Method Applied to Problems in Fluid-Dynamics
Laszlo Fuchs (The Royal Institute of Technology, Stockholm)

Digital Simulation of Mechanical Systems Represented by Differential Algebraic Equations
C. Fuhrer (Technische Universitat, Munchen)
Co-author: R. Schwertassek

A Fluid Dynamical Model of Mucociliary Transport in the Lung
Glenn Fulford (La Trobe University)

Indirect Solution of Nonlinear Prediction Equation
Andrzej Gabor (Technical University of Wroclaw)

Approximations for the Scaled White Noise Linear Quadratic Gaussian Estimation and Control Problems of Singularly Perturbed Systems
Zoran Gajic (Rutgers University)

The Recursive Methods for Singularly Perturbed and Weakly Coupled Control Problems
Zoran Gajic (Rutgers University)
Co-author: N. Rayavarupu

Singular Perturbation of Hyperbolic Type Including Analysis of Large Time Scale
Marc Garbey (Universite de Valenciennes)
Co-author: W. Eckhaus

Persistence in Stochastic Population Models
Thomas C. Gard (University of Georgia)

Turbulent Flow Simulation Using A Nonlinear Discretization Scheme
P.H. Gaskell (University of Leeds)
Co-author: A.K.C. Lau

On the Asymptotic Solution to a Class of Linear Integral Equations
Arthur K. Gautesen (Iowa State University)

Directional Sensitivity for the Optimal Solutions in Nonlinear Mathematical Programming
Jacques Gauvin (Ecole Polytechnique, Montreal)
Co-author: Robert Janin

Buffers, Mailboxes and Data Bases
Donald P. Gaver (Department of Operations Research, Naval Postgraduate School, U.S.A.)

Various Applications of Carleman's Integral Equation
Amos Eliahu Gera (ELTA Electronics Industry Ltd., Ashdod, Israel)

Parallel Iterative Methods for the Solution of Linear Systems and Integral Equations
Apostolos Gerasoulis (Rutgers University)
Co-authors: Nikolaos Missirlis, Israel Nelken

Continuum Damage Mechanics: Some Associated Numerical Problems
Giuseppe Geymonat (E.N.S. Cachan, France)
Co-authors: A. Benallal, R. Billardon, J. Florez

A Hybrid Parameterization of Linear Single Input Single Output Systems
Bijoy Kumar Ghosh (Washington University at Saint Louis)
Co-author: W.P. Dayawansa

Air Pollution Measuring Devices and Related Equation
Narayan Ch. Ghosh (S.B. College of B.U. Mogra)
Co-author: K.M. Ghosh

The Conjugate Gradient Method with Vectorized Preconditioning on the Siemens VP200 Vectorprocessor
Horst Gietl (Siemens AG, Communication & Information, Munich, F.R.G.)

Numerical Schemes and Conservation Laws of Mixed Type
Herve Gilquin (UER de Sciences, St Etienne, France)

Adaptive Continuation Algorithms for Laminar Flames Extinction Problems
Vincent Giovangigli (C.N.R.S., Ecole Central, France)
Co-author: M.D. Smooke

Zonotopes and Mixtures Management
Didier Girard (CNRS, TIM3, USTMG, Institut IMAG, Saint-Martin d'Heres, France)
Co-author: P. Valentin

Stability of Nonlinear Systems with Observers
Torkel Glad (Linkoping University)

A Discrete Gel'Fand-Levitan Method for Band-Matrix Inverse Eigenvalue Problems
G.M.L. Gladwell (University of Waterloo)
Co-author: N.B. Willms

Generalizations of the Flux-Difference Splitting Technique for the Euler Equations
Paul Glaister (University of Reading)

On the Impact of Projective and Affine Methods on Nondifferentiable Optimization
Jean-Louis Goffin (McGill University)

Numerical Computation for the Transient Maxwell Equations in Polarized Media
Yanay Goldmann (INRIA-Rocquencourt, France)

An Improved Algorithm for the Problem of Production Control under Variable Speeds and Inspection Points
Dimitri Golenko-Ginzburg (Ben-Gurion University of the Negev)
Co-authors: Zilla Sinuany-Stern, Moshe Gez, Ruth Fimstein

Symmetry and Bifurcation in the Couette-Taylor Experiment
Martin Golubitsky (University of Houston)

On the Solution of an Optimal Domain Problem
Raul B. Gonzalez de Paz (Universidad del Valle)

Reduction Method in Ordinary Differential Equations Applied to The Stability Study of Control Systems
M.J. Gonzalez-Gomez (Universidad del Pais Vasco)
Co-author: Manuel de la Sen

Direct Control Design Using Nonparametric System Information
Walter H. Gotzmann (Universitat Karlsruhe)

The Generalized Partial Correspondence Principle in Linear Viscoelasticity
G.A.C. Graham (Simon Fraser University)
Co-author: J.M.M. Golden

Low Order Spectral Models of the Atmospheric Circulation
Johan Grasman (University of Utrecht)
Co-author: H.E. de Swart

An Optimization Problem in Reliability Theory
W.B. Grasman (GMI Engineering & Management Inst., U.S.A.)
Co-author: S. Chakravarthy

A New Method for Finding Equilibrium Solutions of Markov Chains
Winfried K. Grassmann (University of Saskatchewan)

Behavior of Slightly Perturbed Lanczos and Conjugated Gradient Recurrences
Anne Greenbaum (New York University)

Threshold Effects and Stability in Simple Age-Structured Epidemic Models
David Greenhalgh (University of Strathclyde)

The Calculation of Complex Chemical Equilibria by Generalized Linear Programming
H. Greiner (Philips GmbH, Aachen, F.R.G.)

Stability of Periodic Solutions of Equations of Thermostat Control
Gustaf Gripenberg (Helsinki University of Technology)

On The Exactness of Penalty Functions for Nonlinear Programming Problems
L. Grippo (Consiglio Nazionale Delle Ricerche, Roma, Italy)
Co-author: G. DiPillo

Domain Decomposition on a Loosely Coupled Array of Processors
William D. Gropp (Yale University)
Co-author: Raphaele Herbin

Monotone Convergent Methods for Weakly Nonlinear RVP's
Christian Grossmann (University of Kuwait)

Numerical Shakedown Analysis of Elastic-Plastic Shells Allowing for Geometrical Non Linearities
J. Gross-Weege (Institut fur Mechanik-Bochum)

Some Combinatorial Features of a Robot Working Space
Sava Grozdev (Bulgarian Academy of Sciences)

Solving Complex Problems With Verified High Accuracy
K. Gruner (Universitat Karlsruhe)

Mathematical Model of Turbulent Flow in Turbo-Machinery
Ryszard Grybos (Silesian Technical University)
Co-author: G. Pakula

Some Existence Results for Flows of Viscoelastic Fluids with Differential Constitutive Equations
Colette Guillope (Universite Paris Sud & CNRS)
Co-author: J.C. Saut

"Natural Interpolation" and the Numerical Solution of Functional Differential Equations
Brigitte Gunterberg (Universitat Bonn)
Co-authors: H.G. Bock, J.P. Schloder

Higher Order Conditionally Stable Explicit Schemes for Equations of the Schrodinger Type
Suchitra Gupta (Pennsylvania State University)

Solution to a Non-Linear Integro-Differential Equation
S.A. Gustafson (Rogaland University)
Co-author: Paul Papatzacos

New Scalar and Vector Elementary Functions for the IBM System 1370
F.G. Gustavson (IBM, T.J. Watson Research Center, U.S.A.)
Co-authors: R.C. Agarwal, J.W. Cooley, J.B. Shearer, G. Slishman, B. Tuckerman

New Vector Linear Algebra Algorithms for the IBM 3090 Vector Facility
F.G. Gustavson (IBM-T.J. Watson Research Center, U.S.A.)
Co-authors: R.C. Agarwal, G. Slishman, B. Tuckerman

The Computation of the Attenuation Factors of Periodic Box Splines
Martin Gutknecht (ETH Zurich, Switzerland)

Models for Parasitic Diseases
K.P. Hadeler (Universitat Tubingen)

Continuation Technique for the Seismic Inverse Problem
Younousse Hadjee (INRIA-Rocquencourt, France)

Upperbounds on the Probability of Ruin for Risk Processes in a Macro-Economic Environment
Jean Haezendonck (University of Antwerp)
Co-author: F. Delbaen

A Globally Convergent Algorithm for Constrained Optimization
William W. Hager (Pennsylvania State University)

Evaluation of Arithmetic Expressions by Solving Nonlinear Equations
R. Haggenmuller (SIEMENS, Muenchen, F.R.G.)

Degenerate Bifurcations in a Two-Dimensional Predator-Prey System
Josef Hainzl (Universitat Kassel)

Counting Points in Hypercubes and Convolution Measure Algebras
D.J. Hajela (Bell Communications Research, U.S.A)
Co-author: P. Seymour

On a Conjecture of Komlos About Signed Sums of Vectors Inside the Sphere
D.J. Hajela (Bell Communications Research, Morristown, U.S.A.)

On Faster Than Nyquist Signaling: Computing the Minimum Distance
D. Hajela (Bell Communications Research, Morristown, U.S.A.)

Artificial Boundary Conditions for Parabolic Equations
Laurence Halpern (Ecole Polytechnique, Palaiseau)
Co-author: J. Rauch

Field Computations in Industrial Applications
Eberhard Halter (Institute Fuer Datenverarbeitung und der Technik, Karlsruhe)

Convergence of the Generalized Overrelaxation Method for Linear Equations
Shamsul Haque (NED University of Eng. & Technology, Pakistan)

On the Global and Quadratic Convergence of Cyclic Jacobi Methods for the Generalized Eigenvalue
Vjeran Hari (University of Zagreb)

A Parallel Algorithm for the Generalized Eigenvalue Problem
Bill Harrod (University of Illinois)
Co-author: Ahmed Sameh

Total Variation Diminishing (TVD) Schemes of Uniform Accuracy
Peter-M. Hartwich (NASA Langley Research Center, U.S.A.)
Co-authors: Chung-Hao Hsu, C.H. Liu

Discontinuous Mobilities in Mechanisms via the Lie's Group of Displacement
J. Harve (Ecole Centrale des Arts et Manufactures, Chatenay, Malabry, France)

An Inviscid Approach to a Phase Transition Problem
H. Hattori (West Virginia University)

Existence of Optimal Controls
U.G. Haussmann (University of British Columbia)
Co-author: J.P. Lepeltier

Use of Systolic Arrays for Finite Element Computations
Linda J. Hayes (University of Texas at Austin)

Nilpotent Approximation in Nonlinear Filtering
Michiel Hazewinkel (C.W.I. Mathematical Center, Amsterdam, The Netherlands)

On the Boundary Element Spectral Method in 3-D Viscous Hydrodynamics
F.K. Hebeker (Universitat - GHS - Paderborn)

Parabolic Ray Theory with Boundaries
G.W. Hedstrom (Lawrence Livermore National Laboratory, U.S.A.)
Co-author: Raymond C. Chin

On the Accuracy of the Ritz Method Applied to Heh and Eckart's and Weinberger's Inequalities
Klaus Helfrich (Technical University of Berlin)

Distributed Expert System for Space Shuttle Flight Control
J. Helly (Aerospace Corp. Systems, U.S.A.)

Indirect Calculation of the Inverse Jacobian Matrix for a Robot Manipulator
Ahmad Hemami (Concordia University)

An Algorithm for Obtaining All Roots of an Algebraic System
Michael E. Henderson (IBM-Thomas J. Watson Research Center, U.S.A.)

A Constructive Method for Deriving Finite Elements of Nodal Type
J.P. Hennart (IIMAS-UNAM, Mexico)
Co-authors: J. Jaffre, J.E. Roberts

The Solution of Partial Differential Equations on Composite Overlapping Grids
William D. Henshaw (IBM Thomas J. Watson Research Center, U.S.A.)
Co-authors: David L. Brown, Ceoffrey Chesshire

Monodromy Eigenvalue Assignment in Discrete-Time Linear Periodic Systems
Vicente Hernandez (Polytechnical University, Valencia)
Co-author: A. Urbano

Discontinuous Mobilities in Mechanisms via the Lie's Group of Displacement
J. Herve (Ecole Centrale des arts et Manufactures, Chatenay)

Robustness Analysis & Optimization in State Space
Diederich Hinrichsen (Universitat Bremen)

The Role of Variational Principles in Applied Mathematics
Jean-Baptiste Hiriart-Urruty (Universite Paul-Sabatier)

A Test of ADA for Finite Element Analyses
Marc Hittinger (Informatique Internationale, Agence de Saclay, Gif-sur-Yvette, France)
Co-author: Trinh Chanh Tri

Soliton Interactions of the Korteweg de Vries Equation
P.F. Hodnett (National Institute for Higher Education, Ireland)
Co-author: Th. P. Moloney

Optimal Control of Structural Phase Transitions in Shape Memory Alloys
Karl-Heinz Hoffmann (University of Augsburg)

An Implicit Navier-Stokes Solver for 2D and 3D Compressible Flows
H. Hollanders (O.N.E.R.A., Chatillon-sous-Bagneux, France)
Co-author: O. Labbe

A Multispecies Gas as a Nonlinear Dynamical System
James Paul Holloway (University of Virginia)
Co-author: J.J. Dorning

Application of the Method of Integral Relations Retarded Boundary Layer Flows
Maurice Holt (University of California - Berkeley)
Co-author: Y.J. Moon

Simulation of Hysteresis Loops
Ulrich Hornung (SCHI, Neubiberg, F.R.G.)

Convergence of Vortex Methods for Highly Oscillatory Vorticity Distribution
Thomas Y. Hou (University of California - Los Angeles)

Isotope Separation by Oscillatory Flow
Garry W. Howell (Florida Institute of Technology)

Heat and Current Flow in Thermistors
Samuel Dexter Howison (Mathematical Institute, Oxford, U.K.)

Effect of Randomness of Parameters on Wave Propagation in Viscoelastic Medium
Z. Hryniewicz (Technical University, Koszalin)

A Multiparameter Study of a Boundary Value Problem from Chemical Reactor Theory
Ling Hsiao (Academia Sinica)
Co-authors: P. Fife, Tong Zhang

Forecasting with a Nonlinear System
Hsih-Chia Hsieh (Chung-Hua Institute for Economic Research, Taipei)

Overview of Solutions for a System of Nonlinear Simultaneous Equations
Hsih-Chia Hsieh (Chung-Hua Institute for Economic Research, Taipei)

Evaluation of Algorithms for Solving Nonlinear Simultaneous Equations
Li-Pi-I. Hsieh (Hsih-Chieng College, Tam-Kang University)
Co-author: H.C. Hsieh

On The Phenomenon of Unstable Oscillation in Population Models
Ying-Hen Hsieh (National Chung-Hsing University)

Regular Decomposition of PG(n,p)
Huang Hua-Min (National Central University, Taiwan)

Multigrid-Finite Volume Method for Computing Two-Dimensional Euler Equations
Guizhen Huangfu (Chinese Aeronautical Establishment, Xian, China)
Co-author: Ping Cheng

CADAC: A Computer Arithmetic with Precision Control and Exception Handling
Thomas E. Hull (University of Toronto)

Subspace Methods for Resolving Chromatographic Data
Joseph M. Humel (Denning Mobile Robotics, Inc., U.S.A.)
Co-authors: George C. Verghese, Ryan R. Kim

Compound Critical States in Autonomous Systems
K. Huseyin (University of Waterloo)

Solution Strategies in Problems of Moving Boundary Creeping Flows Coupled with an Additional Evolution Equation
Kolumban Hutter (ETH Laboratory of Hydraulics, Zurich, Switzerland)
Co-author: R.C. Hindmarsh

DB-LIB: A Tool Box of Data Description and Management
J. Ph. Iafrate (ENSIEG, Laboratoire Electrotechnique Grenoble, Saint Martin d'Heres, France)
Co-authors: O. Santana, J.L. Coulomb

Combined Convection Flows in Vertical Ducts
D.B. Ingham (University of Leeds)
Co-authors: D.J. Keen, P.J. Heggs

Flow Through A Cascade
D.B. Ingham (University of Leeds)
Co-author: T. Tao

Natural Convection from a Heated Body in a Porous Media
D.B. Ingham (University of Leeds)

The Boundary Element Method in Lubrication Analysis
D.B. Ingham (University of Leeds)
Co-authors: J. Ritchie, C.M. Taylor

Viscous Flow Through a Fibrous Filter
D.B. Ingham (University of Leeds)
Co-authors: M. Hildyard, P.J. Heggs

Visual Multidimensional Geometry with Applications
A. Inselberg (University of California, Los Angeles)
Co-author: B. Dimsdale

Isotone Projection Cones in Hilbert Spaces and the Complementarity Problem
George Isac (College Militaire Royal, Quebec)

On the Nonlinear Successive Overrelaxation with Projection Applied to Finite Element Solutions for Radiation Boundary Conditions
Kazuo Ishihara (Kyushu Institute of Technology)

On The Convergence Rate of Hildreth's Algorithm
Alfredo Noel Iusem (Instituto de Mathematica Pura e Aplicada, Rio de Janeiro)

Detecting Stiffness in Explicit Runge-Kutta Codes
Ken Jackson (University of Toronto)
Co-author: Brian Robertson

Discontinuous Finite Elements for Nonlinear Scalar Conservation Laws
J. Jaffre (INRIA-Rocquencourt, France)
Co-author: Veerappa Gowda

A Mathematical Model for Heterogeneous Catalysis
W. Jager (University of Heidelberg)
Co-author: U. Hornung

An Interactive Reference Point Method in Multi-Objective Linear Programming
Johannes Jahn (University of Erlangen - Nurnberg)

A Fully Discrete Finite Element Method for an Unsteady and Unsaturated Flow in Porous Media
Tadeusz J. Janik (Technical University of Warsaw)

Statistic Estimation Error and Its Results at Model Application
Krystof B. Janiszowski (Warsaw Technical University)

Necessary Conditions in a Problem of Calculus of Variations
Vladimir Jankovic (University of Belgrade)

Fault Detection and Localisation by Expert System Applied to an Industrial Pilot Process Pise
D. Jaume (CNAM, Laboratoire d'Automatique, Paris, France)
Co-author: M. Verge

Towards An Order Courant-Friedrischs-Levy Condition for Multistep Difference Schemes for Hyperbolic Problems
Rolf Jeltsch (Rwth-Aachen, Aachen, F.R.G.)
Co-author: K. Raczek

Numerical Solution for Nonlinear Boundary Value Problems by Dimensional Reduction
Soren S. Jensen (University of Maryland)
Co-author: Ivo Babuska

Some Recent Progress in Polynomial Preconditionings on Vector Supercomputers
Hong Jiang (University of Alberta)
Co-author: Yau Shu Wong

An Algorithm for Solving Riccati Systems Arising in Cheap Control
Lucas Jodar (Universidad Polytechnica Valencia)

Fast PDE Solvers on Parallel Architectures
S.L. Johnsson (Yale University)

Pseudo Transparent Boundary Conditions for the Diffusion Equation
Patrick Joly (INRIA-Rocquencourt, France)

Analysis and Optimization of Non-Linear Heat Transfer
Ghislaine Joly-Blanchard (U.T.C., Compiegne, France)
Co-authors: J.P. Kernevez, M. Lory

The Bifurcation of Wilton Ripples and Other Capilla-Gravity Waves
Mark C.W. Jones (Queen's University of Belfast)

Optimal Investment, Financing and Dividends: A Stackelberg Differential Game Approach
Steffen Jorgensen (Copenhagen School of Business Administration)
Co-authors: Geert-Jan C.Th. van Schijndel, Peter M. Kort

Convergence of Finite-Difference Schemes for Parabolic Equations
Bosko S. Jovanovic (Institute of Mathematics, Belgrade, Yugoslavia)

A New Approach to Statistical Decision via the Concept of Observed Entropy
Guy R. Jumarie (Universite du Quebec)

Some Mathematical Questions in Cell Population Dynamics Models in Chemotherapy
M.L. Juncosa (The Rand Corporation, U.S.A.)
Co-author: R. Danchick

Two Efficient Methods of Computing Optimal and Stable Decision in Multicriteria Problems
E.M. Jurkiewicz (Polish Academy of Sciences)

Dealing With A Non-Linear System With 4 Parameters
Christian Kaas-Petersen (University of Leeds)

Generalized Schur Methods with Condition Estimators for Solving the Generalized Sylvester Equation
Bo Kagstrom (University of Umea)
Co-author: L. Westin

Nonunique No-Memory Feedback Equilibria in a Resource Management Game
Veijo Kaitala (Helsinki University of Technology)

Spiral Flows of a Viscoelastic-Fluid
P.N. Kaloni (University of Windsor)
Co-author: A.M. Siddiqui

Two-Dimensional Wind Flow Over Buildings
Peter Kaps (Universitat Innsbruck)

A New Predicting Tool for the Flow Field Inside the Cylinder of an Internal Combustion Engine
Fotios Karagiannis (ENSAE, Toulouse, France)
Co-author: A. Giovannini

Parallel Time Stepping Methods for Evolution Equations
Ohannes Karakashian (University of Tennessee)

An Optimization Algorithm Simulates the Effect of Alternative Economic Policies
Mirek Karasek (Presidency of Civil Aviation IAP, Jeddah, Saudi Arabia)

Runge-Kutta Methods for Nonlinear Parabolic Equations with Adaptive Mesh Selection in the Space Variable
Bulent Karasozen (Middle East Technical University, Ankara)

A Mixed Spectral Finite - Difference Model for Flow Over Complex Terrain
Stephen R. Karpik (Atmospheric Environment Service, Canada)
Co-authors: John L. Walmsley, Anton C.M. Beljaars

Elements of the Theoretical-Applied Aspect Appendix of Topological Model
Ataulla Faradj Ogli Kasimov (Azerbaijan Inst. of Oil & Chemistry, Baku, U.S.S.R.)
Co-author: Elshad Ismiev

Unsteady Potential Flows Around Axisymmetric Bodies and Aircraft Engines
E. Katzer (Institut of Aeroelasticity, DFVLR-AVA, Goettingen, F.R.G.)

Reliable Error Verification for the Solution of PDEs and IEs
Edgar Kaucher (Universitat Karlsruhe)

Robust Pole Assignment - Implementation
J. Kautsky (Flinders University)

Sparse and Banded Cycle Adjacency Matrices
A. Kaveh (Iran University of Science & Technology)

Mathematical Model of Bending Steel Pipes with Large Diameter and Wall Thickness Under Inductive Heating
Werner Kazmirek (Mannesmann-Forschungsinstitut, Duisburg, F.R.G.)

Verified Computation of Parameter Influenced Problems
Rainer Kelch (Universitat Karlsruhe)

Automatic Selection Procedures of ARIMA Models
Andre Keller (Universite Paris II)
Co-author: M. Mourad

Approximate Quasi Newton Methods
C.T. Kelley (North Carolina State University)

Continuous-Time Adaptive Control of a Linear Periodical System
Rafael Kelly (National University of Mexico)

A Coupled Boundary Integral/Finite Element Solution for Transient Electromagnetic Geophysical Prospection
Michel Kern (I.N.R.I.A.-Rocquencourt, France)
Co-author: P. Joly

Shallow Water Waves With Slowly Varying Froude Number Passing Through Critical
J. Kevorkian (University of Washington)
Co-author: J. Yu

Conservation Laws Which Change Type: A Model in Three Phase Porous Medium Flow
Barbara Lee Keyfitz (University of Houston)

Flowfield Modelling in Aluminium Reduction Cells
Essam Eldin Khalil (Cairo University)

Numerical Calculations of Heat & Mass Transfer in Industrial Applications
Essam Eldin Khalil (Cairo University)

Methode Multigrille Pour Modeles de Gisement Partie I: Lisseurs
Mohamed Khalil (ELF Aquitaine (Production), Pau, France)
C-author: P. Puiseux

Predictor Corrector Scheme for Fourth Order Parabolic Partial Differential Equations
A.Q.M. Khaliq (University of Bahrain)

A Mixed Finite Element Method for an Electromagnetic Eigenvalue Problem
Fumio Kikuchi (University of Tokyo)

A Class of Bilinear Estimation Problems in Chemometrics
Ryan R. Kim (ORI, Inc., U.S.A.)
Co-authors: George C. Verghese, Richard M. King

Smooth Curve Fit Using Cubic-Quartic Spline
Ha-Jine Kimn (Ajou University)

Experiments with Concurrent Implementations of Iterative Methods on Multiprocessor Computers
David R. Kincaid (University of Texas - Austin)
Co-authors: Thomas C. Oppe, Gildardo D. Zarza

Computation of the Turbulent Diffusion of Gaussian Puffs
Peter Kirkegaard (Riso National Laboratory, Roskilde, Denmark)
Co-authors: T. Mikkelsen, L. Kristensen

A New Solution for the Strongest Fixed-Fixed Column Problem
Ph. G. Kirmser (Kansas State University)
Co-author: K.K. Hu

Transition to Chaos in the Parametrically Excited Vibrations of a Cylindrical Shell
Sava D. Kisliakov (Higher Institute of Civil Engineering, Sofia, Bulgaria)

Effective Computation of an Ill Conditioned Nonlinear Eigenvalue Problem
Wolfram Klein (Universitat Karlsruhe)

Lyapunov Exponents and Large Deviations of Stochastic Systems
Wolfgang Kliemann (Iowa State University)

Hyperbolic Conservation Laws in Two Space Dimensions
Christian Klingenberg (University of Heidelberg)

The Numerical Simulation of Tunneling Effects in Semiconductors
N. Kluksdahl (Arizona State University)
Co-author: C. Ringhofer

Software Design Strategies for CFD
B. Knight (Thames Polytechnic)
Co-authors: D. Edwards, C.W. Richards

An Algorithm for Computing B-Splines in Tension
Per Erik Koch (University of Trondheim)
Co-author: T. Lyche

Macroelement Methods for the Analysis of the Gravity Wave-Structure Interaction
K. Kokkinowrachos (Technical University of Aachen)

Application of Boundary Collocation Method for the Solution of Boundary-Value Problems in Plane Regions Possessing Symmetry of Shape
Zenon Konczak (Technical University of Poznan)
Co-author: J.A. Kolodziej

Optimal Dynamic Investment Policy Within an Uncertain Environment
Peter M. Kort (Tilburg University)

Optimal Model-Structure Selection for Nonlinear Systems
M. Kortmann (Ruhr University Bochum)
Co-author: H. Unbehauen

Blind Deconvolution of Seismic Data
Clement Kostov (Stanford University)

Diffusion and Convection in Pipes with Slowly Varying Cross-Section
Water P. Kotorynski (University of Victoria)

Degeneracy and Asymmetry in Biology
Alessandro Kovacs (Universita "La Sapienza", Roma)

Smoothing Functionals and Polynomial Approximations of Min-Max Problems
Joseph Kreimer (Ben-Gurion University of the Negev)

Generalized Recurrences and Algorithms for Toeplitz Matrices
Bal Krishna (University of Bahrain)
Co-authors: Hari Krishna, Salvatore D. Morgera

Window Selection for Binary Images
K. Krishna (University of Alabama)
Co-author: H.S. Ranganath

Poisson Structures and Multibody Dynamics
P.S. Krishnaprasad (University of Maryland)

On Some Class of Problems in Electro and Magnetostatics with Constraints for Sources and Field Vectors
S.K. Krzeminski (Warsaw Technical University)

A $C_{(0)}$ Pseudospectral Element Method for the Solution of the Navier-Stokes Equations in Primitive Variables
Hwar C. Ku (Johns Hopkins University)
Co-authors: Richard S. Hirsh, Thomas D. Taylor

Arithmetic for Vector Processors
Ulrich Kulisch (Universitat Karlsruhe)

Numerical Analysis of Uniform Flow Around Two Spheres
Teruo Kumagai (Science University of Tokyo)
Co-author: Masahiro Muraoka

Ellipsoidal Approximations in Problems of Guaranteed Adaptive Control
Alexander Kurzhanski (International Institute for Applied Systems Analysis, Laxenburg, Austria)
Co-author: I. Valyi

Ideal Plastic Flow of Metals
Maija Kuusela (University of Technology, Espoo)

Global Qualitative and Quantitative Analysis of Solutions of Nonlinear Elastostatic Problems
Franz K. Labisch (Ruhr Universitat Bochum)

A Finite Element Model of a Molten Polymer Solidification Problem Using an Adaptive Mesh Strategy
Rene Lacroix (Universite de Laval)
Co-author: Philippe A. Tanguy

Monotone Interpolation by Quadratic Splines
Aatos Lahtinen (University of Helsinki)

Response of Shell's Structure Subjected to Arbitrary Random Pressure Field
Aouni A. Lakis (Ecole Polytechnique de Montreal)

An Efficient Factorization for the Group Inverse
Bernard F. Lamond (University of Arizona)

Thermal Properties of Contact (Homogenization and Singular Perturbations)
Helene Lanchon (Laboratoire d'Energetique & de Mecanique Theorique & Appliquee, Nancy, France)
Co-author: J. Saint-Jean Paulin

Spreadsheet Applications in Computer Aided Design
Thomas J. Langan (United States Naval Academy)

Largetime Behaviour in a Nonlinear Populations Dynamics Problem
Michel Langlais (Universite de Bordeaux 1)

Exact Solutions to Fredholm Integral Equations of the First Kind
Jesper Larsen (Math-Tech, Charlottenlund, Denmark)

On the Derivative Nonlinear Schrodinger Equation
Jyh-Hao Lee (Academia Sinica)

An Asymptotic Expansion for the Generalized Riemann Problem and Applications
Philippe Le Floch (Ecole Polytechnique, Centre de Mathematiques Appliquees, Palaiseau)
Co-author: P.A. Raviart

Stability of a Natural-Draught Fluid System
Bernhard J.G. Leidinger (M.B.B. - ERNO - Raumfahrttechnik GmbH, Bremen, F.R.G.)

Determining Consistent Initial Conditions for Differential-Algebraic Equations
Benedict J. Leimkuhler (University of Illinois)

Numerical Determination of Scattering Frequencies in Optics and Acoustics
Marc Lenoir (GHS-ENSTA, Centre de l'Yvette, Palaiseau, France)
Co-author: M. Vullierme-Ledard

On the Effectiveness of Numerical Techniques for Nonequilibrium Chemically Reacting Flows
D. Lentini (Universita di Roma "La Sapienza")
Co-author: M. Onofri

Statistics of the Photon Physical Significance of Temperature. Applications to a Unitary Conception of Particles
D. Lepretre (GARI/DGE/INSA, Toulouse, France)
Co-author: Pierre Lopez

On Optimal Control of Ordinary Volterra Integro - Differential Equations
Gunter Leugering (Technische Hochschule Darmstadt)

The Tunneling Algorithm for Global Optimization, Convergence Analysis and Test Results
A.V. Levy (The Nathan S. Kline Institute, New York)

An Annotation on the Definition of Least Upper Bound (Greatest Lower Bound) of a Poset
Weijian Li (Fudan University)
Co-author: Yongcai Liu

Methods for Combinations of the Ritz-Galerkin and Finite Element Methods
Z.C. Li (Concordia University)
Co-author: T.D. Bui

Unsteady Supersonic Aerodynamics for Arbitrary Time-Dependent Motions
Liviu Librescu (Virginia State University & Polytechnic Institute)

Two-Processor Runge-Kutta Methods for Ode's
I. Lie (Institute for Numerisk Mathematikk, Trondheim)
Co-author: S.P. Norsett

Instability of the Von Karman Vortex Trail and Boundedness of Disturbances
Chjan C. Lim (University of Michigan)

The Numerical Analysis of a Phase Field Model in Moving Boundary Problems
Jian-Tong Lin (Wright State University)

Mathematical Theory of Elastic-Orthotropic Plates in Plane Strain and Axi-Symmetric Deformations
Yihan Lin (University of British Columbia, Vancouver)

Numerical Evaluation of the Boundary of Periodic Regions of the O.D.E.
F.H. Ling (Shanghai Jiao Tong University)

Generality of Denavit-Hartenberg Formalism in the Homogeneous Modelling of the Joints in a Mechanical Articulated System
Pierre Lopez (GARI-DGE-INSA, Toulouse, France)
Co-author: R. Ferrier

On Convex Piecewise Cubic Interpolation
Jeronimo Lorente Pardo (Universidad de Granada)

Waiting Time Distribution in Some Queueing Systems with Server Vacations
Jacqueline Loris-Teghem (Universite de l'Etat a Mons)

Confinor and Anti-Confinor in Constrained "Lorenz" System
Rene Lozi (C.N.R.S., Universite de Nice)
Co-author: Shigehiro Ushiki

High-Order Accurate Characteristic Methods for Hyperbolic Conservation Laws
Bradley J. Lucier (Purdue University)

Graph Theoretical Methods for the Computation of Electrical Networks
Christoph Maas (University of Hamburg)

Modeling Optimization Test Problems with a Finite Number of Characterized Points
Nezam Mahdavi-Amiri (York University)

Modelling of the Two Phase Turbulent Flow in the Entrance Region
Hubert Mainardi (Universite d'Orleans)

A Finite Element Solution of the Stokes Problem with an Adaptive Multi Level Procedure
Jean-Francois Maitre (Ecole Centrale de Lyon)
Co-authors: E.M. Abdalass, F. Musy

Numerical Computation of the Scattering Frequencies for Acoustic Wave Equations
Georges Majda (Ohio State University)
Co-authors: Musheng Wei, Walter Strauss

Finite Element Analysis of Thermal Stresses in Steel Structures Subject to a Fire
Carmelo E. Majorana (University of Padua)
Co-author: R. Vitaliani

Statistical Electric Load Modeling: Analytical Results for a Non-Diffusion Hybrid-State Markovian Model
Roland Malhame (Ecole Polytechnique de Montreal)

Computing Maximum-Entropy Estimates
Francesco Malvestuto (ENEA, Roma, Italy)

Numerical Simulation of Supercooling During Solidification
Frans Mampaey (WTCM Gieterij, Zwijnaarde, Belgium)

On the Equivalence of Operators and the Implications to Preconditioned Iterative Methods
Thomas A. Manteuffel (Los Alamos, U.S.A.)
Co-author: Seymour V. Parter

Collapse of a Translating Bubble
Donatella Manzi (A.R.S. S.P.A., Milano, Italy)
Co-author: Andrea Prosperetti

Data Validation in Large-Scale Steady State Linear Data Systems
D. Maquin (Laboratoire C.N.R.S., LARA, Vandoeuvre, France)
Co-authors: M. Darouach, J. Fayolle, J. Ragot

Canonical Forms in Qualitative Control Theory for Time-Delay Systems
Vladimir M. Marchenko (Russian Institute of Technology)
Co-author: V.L. Merezsha

Partial Eigensolution of Large Banded Symmetric Matrix System Using the Woodbury Transformation
B. Marcos (Universite de Sherbrooke)
Co-author: J.G. Beliveau

Robust Numerical Procedures Which Reduce False Diffusion in the Solution of Convection-Diffusion Equation
N.G. Markatos (Thames Polytechnic)
Co-authors: M. Patel, M. Cross

Solving Stochastic Optimization Problems by Semi-Stochastic Approximation Procedures
Kurt Marti (Universitat der Bundeswehr Munchen)

Polar Complex Variables in Three Dimensions
E. Dale Martin (NASA Ames Research Center, U.S.A.)

A New Algorithm for Solving Sparse Nonlinear Least Square Problems
Jose Mario Martinez (IMECC-UNICAMP, Campinas, Brazil)

Computation of 2D Compressible Viscous Flow wih Upwind Implicit Schemes
Y. Marx (ENSM, Fluid Dynamics Group, Nantes, France)

Numerical Computation of Sensitivity in Shape Optimization Problems
Mohamed Masmoudi (Universite de Nice)

On the Uniform Convergence of a Quadrature Rule for PV Integrals
G. Mastroianni (C.N.R., Napoli, Italy)
Co-author: G. Criscuolo

Nonlinear-Wave Characteristics of Magnetomechanical Signal-Processing Devices
Gerard Maugin (Universite Pierre et Marie Curie)
Co-author: Abo-el-Nour Abd-Alla

Some Mathematical Aspects of the Conception and Realisation of Viscometers
G. Maurice (LEMTA, Institut National Polytechnique de Lorraine, Nancy, France)
Co-author: R. Kouitat

Enclosing Solutions of Nonlinear Systems of Equations
Gunter Mayer (Universitat Karlsruhe)

An Algorithm for Probabilistic Constrained Stochastic Programming
Janos Mayer (Institute of the Hungarian Academy of Sciences)

Coupling Particle and Finite Difference Methods on Inhomogenous Media
Anita Mayo (I.B.M.-T.J. Watson Research Center, U.S.A.)

Estimation of Physiological Non-Linear System in Medicine Using Dynamic Positron Emission Tomography (PET)
Bernard Mazoyer (Institut de Recherche Fondamentale, Orsay, France)
Co-authors: J. Delforge, A. Syrota

Symbolic Computation of Commutators
Giuseppe Mazzarella (University of Naples)

Mathematical Model for the EPR Rubber Crosslinking Optimization, in the Continuous Production of Power Cables
Salvatore Mazzullo (Centro Ricerche "G. Natta", Ferrara, Italy)
Co-authors: C. Cometto, L. Corbelli

An Economical Inversion Algorithm for Interval Velocities from Seismic Data
John McBain (Office of Naval Research, Arlington, U.S.A.)

Inversion and the Finitely Sampled Magnetotelluric Problem
John McBain (Office of Naval Research, Arlington, U.S.A.)

Asynchronous Multilevel Adaptive Methods for Large Scale Parallel Computation
S. McCormick (University of Colorado)

Behaviour of the Front Wheel of a Motorcycle During Heavy Braking
J.M. McGarry (University of Leeds)

Modelling Hydrodynamic Motion
Joseph F. McGrath (K.M.S. Fusion Inc., U.S.A.)
Co-authors: Darrell L. Hicks, Kenneth L. Kuttler

A Model for Oxygen Isotope Transport in Hydrothermal Systems
Robert McKibbin (University of Auckland)
Co-author: A. Absar

Thermal Convection in Layered Porous Media Heated from Below
Robert McKibbin (University of Auckland)

Local Switching Surface Structure for Closed-Loop Time-Optimal Control of Linear Systems
L. David Meeker (University of New Hampshire)

Control of Discrete Time Stochastic System with Stochastic Parameters
Driss Mehdi (Laboratoire d'Automatisme et de Recherche Appliquee, Vandoeuvre, France)
Co-author: C. Humbert

Random Variations in the Flow Path of Hydro Turbines - Analysis and Diagnostics
Z. Mehdi (Bahrat Heavy Electricals Ltd., Piplani, India)
Co-author: S.K. Khanna

Iterated Maps and the Newton-Raphson Algorithm
H. Meheryar (Institute for Reservoir Studies, Gujarat, India)

Defect Correction for Algebraic Riccati Equations
Volker Mehrmann (Universitat Bielefeld)
Co-author: Eng-Kan Tan

On Boundary Eigenvalue Operator Functions
R. Mennicken (Universitat Regensburg)
Co-author: M. Moller

Boundary Elements for Optimal Shape Design of Orthotropic Hollow Solids
R. Alsan Meric (Research Institute for Basic Sciences, Kocaeli, Turkey)

Sensitivity Analysis for Shape Inverse Problems Governed by Integro-Differential Equations
R. Alsan Meric (Research Institute for Basic Sciences, Kocaeli, Turkey)

ACRITH: Subroutine Library for Verified Results
Toni Merschen (IBM Germany, Boeblingen, F.R.G.)

Convection with Strongly Temperature-Dependent Viscosity in a Horizontal Layer
P. Metzener (Northwestern University)
Co-author: M. Matalon

Global Dynamical Behavior and Chaos of Driven Coupled Nonlinear Oscillators
Wolfgang Metzler (University of Kassel)

Sequential Algorithm for the Binary Skeleton
Fernand Meyer (Ecole des Mines, C.M.M., Fontainebleau)

Conjugate Residual Methods for Almost Symmetric Linear Systems
Juan C. Meza (Rice University)
Co-author: William W. Symes

Wave Propagation in Stratified Viscoelastic Media: A Fourier Approach
Nacer Mezouari (INRIA-Rocquencourt, France)

Incentive-Strategies in Stackelberg Games with Constraints
F. Mignanego (Univ. di Genova)
Co-author: G. Pieri

Modelling Some Engineering and Biomedical Systems by Iterative Methods of New Type
Miklos Mikolas (Technical University of Budapest)

Capillary Waves Produced by the Scattering of a Weak Shock from an Interface
Michael J. Miksis (Northwestern University)
Co-author: Lu Ting

On the Acceleration of Jacobi and Gauss-Seidel Iterations
J.P. Milaszewicz (Ciencias Exactas y Naturales, Buenos Aires)

A Fully Bayesian Approach to Nonlinear Parameter Estimation
Jeffrey A. Mills (Washington University)

Identification and Adaptive Control of a Blast Furnace: A Case Study
Chan Ming-Kam (Hong Kong Baptist College)
Co-authors: Li Men-Jan, Hu Shou-Jen, Ng Sze-Kui, Ngai Hung-Man

The Connection Matrix
Konstantin Mischaikow (Brown University)

Passive Systems: An Optimization of the Sensors Array Configuration
Stanko Mitrovic (Research and Development Department, Beograd, Yugoslavia)

The Kalman Filter in Dendroclimatology
J. Molenaar (University of Nijmegen)
Co-author: H. Visser

Formation of Spinless Water Spouts
John Molyneux (Widener University)
Co-author: Fred Daddi

Application To Robot Arm of Sampled Nonlinear Controller
Salvatore Monaco (Universita di L'Aquila)
Co-authors: D. Normand-Cyrot, S. Stornelli

Product Formulas for Fredholm Integral Equations with Rational Kernel Functions
Giovanni Monegato (Politecnico di Torino)

A Class of Convex Programs for Resource Allocation Problems on Directed Acyclic Graphs
Clyde L. Monma (Bell Communications Research, U.S.A.)
Co-authors: Victor K. Wei, Alexander Schrijver, Michael J. Todd

Some Isoperimetric Inequalities in Electrochemistry and Hele-Shaw Flows
Jacqueline Mossino (CNRS, Universite Paris Sud)
Co-author: B. Gustafsson

Some Numerical Schemes for Transonic Flow Problems
Marco M. Mostrel (University of California - Los Angeles)

Functional Modeling of Complex Integrated Circuits
G. Motet (DGE-INSA, Toulouse, France)
Co-author: M. Abid

Modelling of Human Instantaneous Memory
Simeon Jordanov Mrchev (Jordan Mishev, Jambol, Bulgaria)
Co-authors: A.B. Kogan, R.J. Iljuchenok, S. Vinogradova, A. Ivanov-Muromski

Limiting Lagrangians with a Primal Approach for a Nonconvex Program
R.N. Mukherjee (Banaras Hindu University)

Instrumental Variable Identification of Mechanical Systems
Peter C. Muller (Bergische Universitat)
Co-author: F. Roether

On Commutative One-Way Functions in GF(q) and Z/(n)
Winfried B. Muller (Universitat Klagenfurt)

Optimal Arma Parameters Estimation by Means of Non-Linear Filtering
Maria Pilar Munoz Garcia (Facultad Informatica de Barcelona)
Co-authors: M. Marti-Recober, J. Pages-Fita

A Numerical Model for Tsunami Generation and Propagation
A. Murli (University of Naples)
Co-author: M.A. Pirozzi

Versions of Nested Factorization When Faults and Local Grid Refinements are Present
Johannes Mykkelveit (Rogaland Research Institute)

A Filter-Algorithm for the Probabilistic Removal of Random Errors of Serial Data
Walter K. Nader (University of Alberta)

Morphological Instability and Imperfections
R. Narayanan (University of Florida)
Co-author: A. Nadarajah

Thermal Ignition in a Flow System
R. Narayanan (University of Florida)
Co-author: S. Pushpavanam

Numerical Methods in Nonparametric Data Fitting
Stephen G. Nash (John Hopkins University)

Efficient Vectorization on the Amdahl VP 1100 Supercomputer Using Standard FORTRAN
Jon Natvig (Rogaland University)

Decision Making in a System with Insufficient Input Data
Edward Nawarecki (Academy of Mining and Metallurgy, Krakow)
Co-author: S.K. Nawarecka

Optimal Observing and Replacement Strategies in a Model with Periods of Delayed Failure Detection
Ranko R. Nedeljkovic (Facility of Traffic & Transport Engineering, Beograd, Yugoslavia)

Asymptotic Behaviour for Distributed Parameters Circuits with Transistors
Pekka Neittaanmaki (Jyvaskyla University)
Co-author: C. Marinov

Block Picard-Lindelof Iteration
Olavi Nevanlinna (Helsinki University of Technology)

An Algorithm to Calculate the Lowest Eigensolutions of a Large Matrix
C.M.M. Nex (University of Cambridge)

On the Constitutive Distributed Parameter Description and Control of the Multipartial Electric Fields. Part I and Part II
Waclaw Niemiec (Silesian Technical University)

Convergence and Condition of Collocation Methods for Boundary Integral Equations Arising in Flow Calculations
Herbert Niessner (BBC Brown Boveri & Co., Baden, Switzerland)

Application of a Three-Dimensional Mathematical Model to the Study of Hydrodynamic Processes in the Adriatic Sea
Jacques C.J. Nihoul (Universite de Liege)
Co-authors: Francis Clement, Djenidi Salim Gher, G. Lebon

Adaptive Control of Parameter Polynomial Models
Markku T. Nihtila (Uppsala University, Sweden)

Notes on Nonlinear Stochastic and Deterministic Filtering
Markkut T. Nihtila (Uppsala University, Sweden)

Analysis of Quasi-Newton Updates for Unconstrained and Constrained Optimization
Jorge Nocedal (Northwestern Universtiy)

Solving Ordinary Differential Equations on a Hybrid Computation System
Matu-Tarow Noda (Ehime University)
Co-author: Hidetoshi Iwashita

Control of Distributed Parameter Systems with Diffusion and Transport
Erik J.L. Noldus (University of Ghent)

Random Vibration Analysis of A Nonlinear Oscillator - Non-Gaussian Closure Statistical Linearization
Mohammad N. Noori (Worcester Polytechnic Institute)
Co-authors: Hamid Davoodi, A. Saffar

Essential Mathematics for Software Engineers
A. Norcliffe (Sheffield City Polytechnic)
Co-author: Ian Huntley

Differential Equations and Interval-Analysis
Fernando Aleixo Oliveira (University of Coimbra)

A New Preconditioner for Linear Deconvolution Problems
Julia A. Olkin (Ferranti International Controls, U.S.A.)
Co-author: William W. Symes

Dynamic System Analysis of Water Pipe Networks
Kotaro Onizuka (Tokyo University of Agriculture & Technology)

On Optimal Recursive Multistep Prediction and Generalized Predictive Control: A State Space Formulation
Romero Ortega (Universidad Autonoma de Mexico)
Co-author: M. Espana

Applications of Classical and Zero-Total-Pressure-Loss Sets of Euler Equations to Delta Wings
A. Kandil Osama (Old Dominion University)
Co-author: Andrew H. Chuang

Necessary and Sufficient Conditions in Real and Complex Bifurcation Theory
Donal O'Shea (Clapp Laboratory, U.S.A.)

Higher Order Approximation of Three-Dimensional, Discrete Vortex Methods
Koichi Oshima (The Institute of Space and Astronautical Science, Tokyo, Japan)
Co-author: Y. Oshima

Automatic Selection of an Efficient Approximation to the Jacobian in Semi-Implicit Runge-Kutta Methods
Alexander Ostermann (Universitat Innsbruck)

Interactive Multicriterion Optimization System for Nonlinear Programming
Andrzej Osyczka (Politechnika Krakowska)

Energy Separation and Uniqueness of Motions of Elastic-Plastic Oscillators
R. Owen (Carnegie-Mellon University)

On-Line Estimation of Break-Points, Polynomial Orders as Well as Coefficients in Piecewise-Polynomial Approximation Function
Grazyna Anna Pajunen (Florida Atlantic University)

Stress Concentration in Composites - Delamination
E. Sanchez Palencia (C.N.R.S., Universite Paris VI)
Co-authors: D. Leguillon, H. Dumontet

Fast and Efficient Parallel Solution of Path Algebra Problems
V. Pan (Suny Albany)
Co-author: J. Reif

A Mathematical Model for BSP Kinetics
Endre Pap (Institute of Mathematics, Novi Sad, Yugoslavia)
Co-authors: B. Tosic, J. Malesevic, B. Milutinovic

PC-Generated Combinatorial Distributions
Lee Papayanopoulos (Columbia Business School, Rutgers GSM)

Subgradients in Optimization
Massimo Pappalardo (University of Pisa)

Indefinite Quadratic Programming
Panos M. Pardalos (Pennsylvania State University)

On a Model of General Economic Growth
Vesna Pasetta (Univerzitet U Beograddieboltu)

A Software Program for Dynamic Modelling of Flexible Mechanical Systems: Methodology and Implementation
L. Passeron (Aerospatiale, Cannes la Bocca, France)
Co-authors: Ch. Garnier, P. Rideau

Computer Control of Continuous Time Bilinear Systems Approximated by Discrete-Time State Affine Models
Darko Paszkiewicz (Institut National des Sciences Appliquees, Villeurbanne, France)
Co-authors: B. Neyran, D. Thomasset

A Mathematical Model in the Study of Marine Invertebrates
S.L. Paveri-Fontana (University of Roma)
Co-authors: S. Focardi, J.L. Deneuborg

Electrothermal Atomization in Chemistry
S.L. Paveri-Fontana (Universita di Roma)
Co-author: G. Tessari

Laplace to Navier-Stokes: A Constructive Approach
Fred R. Payne (University of Texas)
Co-authors: M. Mahmoudi, R. Mokkappat

A Numerical Method for the Calculation of Aircraft Afterbody Flows
A.J. Peace (Aircraft Research Assoc. Ltd., Bedford, U.K.)

Conditional Rewrite Rules: Theory and Applications to Automatic Program Generation
Alexandru A. Pelin (Florida International University)
Co-author: Robert Millar

On the Parabolic Models in Underwater Acoustics
Marie-Claude Pelissier (Universite de Toulon et du Var)
Co-author: B. Grandvuillemiri

Parallel Computation of Eulerian and Langrangian Turbulence
Richard B. Pelz (Rutgers University)
Co-author: R.L. Peskin

Estimation of Critical Properties of Gaseous Mixtures by a Computer Program
Joao Fernando Pereira Gomes (OPPI-SARL, Lisbon, Portugal)

Finite Element Analysis of Pulsatile Flow in Human Arteries
Karl Perktold (Technische Universitat Graz)

Conditional Rewrite Rules: Theory and Applications to Automatic Program Generation
Alexandru A. Perlin (Florida International University)

Shock-Temperature Spot Interaction
Bernadette Pernaud-Thomas (O.N.E.R.A., Chatillon, France)

Parallel Processing and Scientific Computing
R.H. Perrott (The Queen's University of Belfast)

Numerical Simulation of Non-Newtonian Blood Flow in Arteries
Reinfried O. Peter (Technische Universitat Graz)

Nonlinear Versus Linear Control in the Stabilization of a Class of Multi-Input Uncertain Systems
Ian R. Petersen (The University of New South Wales)

An Improved CVBEM for Plane Hydrodynamics
Titus Petrila (University of Cluj)

Imprecise Censoring Times
M.J. Phillips (Leicester University)

Efficient Polygon Handling in CAAD Applications
Caterina Pienovi (Consiglio Nazionale delle Ricerche, Genova, Italy)
Co-author: C. Gambaro

The Post-Buckling Behavior of a Nonlinearly Elastic Rod Subject to Symmetry-Breaking Perturbing Loads
John F. Pierce (West Virginia University)

Mode Localization and Eigenvalue Loci Veering Phenomena in Disordered Dynamical Systems
Christophe Pierre (The University of Michigan)

Computation of Viscous Flows Past Axisymmetric Bodies With and Without Propeller in Operation
J. Piquet (ENSM, Nantes, France)
Co-authors: P. Queutey, M. Visonneau

Symmetrization of Differential Equations in Nonlinear Elastodynamics
Adam Piskorek (University of Warsaw)

On Some Approximations of the Energy Variational Principles in the Finite Element Method
Apostol Poceski (University Kiril i Metodij, Yugoslavia)

Loss of Equilibrium Stability for a Gas-Filled Elastic Container
Paolo Podio-Guidugli (Universita di Roma)
Co-author: E. Virga

Periodic Orbits of Discrete-Time Adaptive Closed-Loop Systems
J.B. Pomet (Ecole Nationale Superieure Mines, Fontainebleau)
Co-authors: L. Praly, J.M. Coron

Generalized Time Finite Element Algorithm for Non-Linear Non-Stationary Mechanical Vibrations
Florin Poterasu (Polytechnical Institute of Jassy)

Shape Optimization of Elastic Uni-Dimensional Solids Subjected to Discrete System with Many Degrees of Freedom Action
Victor Florin Poterasu (Polytechnical Institute of Jassy)
Co-author: G. Ionescu

Parallel Iterative Algorithms for Solving Nonlinear Integral Equations
Florian A. Potra (University of Iowa)

Soliton Dynamics in Nonlinear Lattice Models of Elastic Solids
J. Pouget (Universite Pierre and Marie Curie)

Robust Adaptive Stabilization by High Gain Feedback and Switching
Dieter Pratzel-Wolters (Universitat Bremen)

Duals and Propagators: A New Canonical Formalism for General Nonlinear Equations
V. Protopopescu (Oak Ridge National Laboratory, U.S.A.)
Co-authors: R.B. Perez, D.G. Cacuci

Singular Perturbation Solutions for Solidifying Slabs, Cylinders and Spheres
Michael Prud'Homme (Ecole Polytechnique, Montreal)
Co-author: T. Hung Nguyen

Efficient Computational Methods for Nonlinear Convection-Diffusion Equations
Charles David Pruett (Virginia Commonwealth University)

Methode Multigrille Pour Modeles de Gisement (II) Notations Directionnelles et Applications; Resultats 2D et 3D
Pierre Puiseux (ELF-Aquitaine (Production), Pau, France)
Co-author: M. Khalil

Linear Multistep Methods for First-Order Boundary - Value Problems
Luigi Quartapelle (Politecnico di Milano)
Co-authors: S.C.R. Dennis, S. Rebay

Domain Decomposition Methods for Elliptic Partial Differential Equations
Alfio Quarteroni (Universita Cattolica di Brescia)

Fixed Point and Homomorphism Extension Are NP-Complete
A. Quilliot (Institut National Polytechnique, St.Martin d'Heres)

Richardson Extrapolation for Finite Element Approximation on Smooth Domain
Lin Quin (Academia Sinica, Institute of Systems Science)

Perturbed Bifurcation: An Asynthotic Method
Perregrina Quintela-Estevez (Universidad Santiago de Compostela)

Numerical Application of Shakedown Analysis in Soil Mechanics
L. Raad (American University of Beirut)
Co-authors: W. Najm, D. Weichert

Nonlinear Acoustics a Piezosemiconducting Medium: Parametric Interaction Approach
Pal Rabinkanti (K.K. Das College of Commerce, Calcutta)
Co-authors: Manojit Gupta, Dilip Kumar Sinha

New B-Splines (Polynomial, and Thin Plate Splines)
Christophe Rabut (I.N.S.A., Toulouse, France)

The Performance of the Preconditioned Conjugate Gradient Algorithm on the IBM 3090 VF Vector Processor
Guiseppe Radicati di Brozolo (IBM - E.C.S.E.C., Roma, Italy)
Co-authors: Y. Robert, M. Vitaletti

Perturbation Solutions Arising in Hydrodynamic Loading
M. Rahman (Technical University Nova Scotia)

A Cad-Cam System for Sculptured Surfaces in a Flow Field
P. Sundar Varada Raj (Bharat Heavy Electricals Ltd, Bhopal, India)
Co-author: Adarsh Swaroop

Eigenvalue Problem of Non-Linear Difference Equations
Jerzy Rakowski (Technical University of Poznan)

Uniformly Valid Analytical Solution to the Problem of a Decaying Shock Wave
Rishi Ram (Banaras Hindu University)
Co-authors: V.D. Sharma, P.L. Sachdev

Estimateurs a Retrecisseur du Parametre de Position de Lois a Symetrie Spherique
Jean-Pierre Raoult (Universite de Rouen)
Co-authors: D. Cellier, D. Fourdrinier, Ch. Robert

Oscillations and Nonlinear Hyperbolic Systems of Conservation Laws
Michel Rascle (Universite de Nice)

An Optimal Algorithm for Expansion of a Class of Binary Matrices
Tosic Ratko (Institute of Mathematics, Novi Sad, Yugoslavia)
Co-author: Paunic Djura

Unconstrained Optimization
Helmut Ratschek (University of Dusseldorf)

Mixed Numerical Methods in Viscoplasticity
M. Ravachol (Laboratoire des Ponts & Chaussees, Paris, France)

About Structural Controllability of Interconnected Dynamical Systems
Catherine Rech (E.N.S.I.E.G.-I.N.P.G., Laboratoire d'Automatique, Saint Martin d'Heres, France)

Approximation of Variational Inequalities of the Second Kind
B. Dayanand Reddy (University of Cape Town)

Finite Element Models of Fluid Flows
J.N. Reddy (Virginia Polytechnic Institute and State University)

A 2D Streamline Model with Periodical Automatic Regeneration of Streamlines
Gerard Renard (Institut Francais du Petrole)

Calcul Quasi Minimal du Modele Dynamique d'un Robot Manipulateur
M. Renaud (LAAS-CNRS, Toulouse, France)

Bifurcation and Stability in Viscoelasticity
D.W. Reynolds (N.I.H.E. Dublin, Ireland)

Increasing Mathematical Sophistication and Decreasing Effectiveness of Operations Research: How to Revert the Trend
D. Alonso Ribeiro (COPPE-UFRJ, Rio de Janeiro, Brazil)

Vector Methods for Computational Procedures for Fluid Flow
C.W. Richards (Thames Polytechnic)
Co-authors: C. Ierotheou, M. Cross

Group Analysis of the Helmholtz Equation in Ocean Acoustics
P. Childs Richards (Georgia Institute of Technology)
Co-author: W.F. Ames

Solution of Nonlinear Coupled Oscillators Differential Equation on SIMD Computers
Sandro Ridella (C.N.R., Istituto Circuiti Elettronici, Genova, Italy)
Co-authors: A. Corana, M. Muselli

Analytical Approximation for Offshore Pipelaying Problems
S.W. Rienstra (University of Nijmegen)

A Package for Interactive Data Representation
Manfred Ries (Universitat Trier)

The Nonlinear Dynamics of a Heated Channel
Rizwan-Uddin (University of Virginia)
Co-author: J.J. Dorning

A Comparative Analysis of Some Dense Matrix Factorization Routines on the IBM-3090 VF Vector Processor
Yves Robert (IBM - E.C.S.E.C., Roma, Italy)
Co-author: P. Sguazzero

A Fast Algorithm for Particle Simulations
Vladimir Rokhlin (Yale University)

New Development in Turbulence Modelling
G. Romberg (DFVLR-AVA, Gottingen, F.R.G.)

Motions of Fluids in an Electrolytic Cell for Production of Aluminium: A Mathematical Model and Its Numerical Computation
M. Romerio (Ecole Polytechnique Federale, Lausanne)
Co-author: J. Rappaz

Efficient Minimization of Transportation-Like Problems on L-Dimensional Arrays
David Romero (Universidad Nacional Autonoma, Mexico)

Exit in a Two-Dimensional Stochastic Model Arising in Population Dynamics
H. Roozen (Centre for Mathematics & Computer Science, Amsterdam, The Netherlands)

Nonlinear Mean Field-High Frequency Wave Interactions in the Induction Zone
Rodolfo R. Rosales (M.I.T.)
Co-author: Andrew Majda

Singular Approximation of Chaotic Slow-Fast Dynamical Systems
B. Rossetto (Universite de Toulon)
Co-author: M. Canalis-Durand

Deconvolution by an E-M Procedure
Carla Rossi (Universita di Roma "Tor Vergata")

A Non-Newtonian Approach to Mucocillary Transport
Marita Rozenson (Weizmann Institute of Science, Rehovot, Israel)
Co-author: Nadav Liron

Strict Band Matrices and Semiseparable Matrices
Pal Rozsa (Technical University of Budapest)

Enhanced Simulation of Pure 3D Nonlinear Incompressible Behavior Using Asymmetric Finite Elements
Vitoriano Ruas (Dept. d'Informatique, Rio de Janeiro, Brazil)

Multiple Steady States in One-Dimensional Electrodiffusion with Local Electroneutrality
Isaak Rubinstein (Stanford University)

Interactive Programming System for Scientific Computation
Siegfried M. Rump (IBM Germany, Boblingen, F.R.G.)

The Influence of Locomotive Tractive Force Transmission Upon the Dynamic Load in the Wheelset
Ladislav Rus (CKD Praha, Czechoslavakia)

Discontinuous Solutions of Hyperbolic Equations and Discrete Shock Waves
Viktor V. Rusanov (Academy of Sciences, Keldysh Institute of Applied Mathematics, Moscow)
Co-author: I.V. Bezmenov

A Variational Analysis of Amplitude Propagation in Trains of Internal Waves
John M. Russel (University of Oklahoma)

A Multidimensional Optimal Control Problem for the Harvesting of a Renewable Resource
Dennis Ryan (Wright State University)
Co-author: F. B. Hanson

Low Frequency Scattering of Scalar Waves By a Finite Rough Surface in a Half-Plane
Federico J. Sabina (University of Bath)
Co-author: V.M. Babich

On Non-Local Phenomena in a Harmonically Perturbed Self-Oscillator with a Set of Autonomous Limit Cycles
Y.A. Saet (Lamar University)
Co-author: G.L. Viviani

Global Elastic Behavior for Large Honeycomb and Reinforced Structures
Jeanine Saint-Jean-Paulin (C.N.R.S.-L.E.M.T.A., France)
Co-author: D. Cionarescu

Numerical Simulation of Filamentation in Laser-Plasma Interactions
S.G. Sajjadi (Conventry Lanchester Polytechnic)

An Algorithm for Classifying the Zeros of a Plane Vector Field
Takis Sakkalis (New Mexico State University)

The Euclidean Algorithm and the Sign of an Algebraic Number
Takis Sakkalis (New Mexico State University)

Waves in a Non-Linear Elastic Rod
Rogerio M. Saldanha Da Gama (Laboratorio Nacional de Computacao Cientifica, Brazil)
Co-author: Rubens Sampaio

Bifurcation of Solutions of Nerve Impulse Equations at a Triple Zero Eigenvalue
Isabel Salgado Labouriau (Universidade do Porto, Portugal)

Heat Transfer in Blood Flow Through an Artery
R. Ponnalagar Samy (Indian Institute of Technology, Bombay)

Pulsatile Flow of Viscoplastic Fluid Through Slowly Converging-Diverging Tubes
R. Ponnalagar Samy (Indian Institute of Technology, Bombay)

Spectral Properties of Stiff Problems
J. Sanchez-Hubert (Universite Pierre et Marie Curie)
Co-author: E. Sanchez-Palencia

Data Structure and Mesh Algorithm for a 3D Finite Elements Software
Orieta Santana (Laboratoire d'Electrotechnique, Saint-Martin-D'Heres, France)
Co-authors: J.L. Coulomb, J.C. Sabonnadiere

Augmented Order Robust Pole Placement Controllers
Rafael Santos Mendes (Laboratoire d'Automatique et d'Analyse de Systemes, Toulouse, France)
Co-author: Joseph Aguilar Martin

Finite Volume Procedure for Computing Inviscid Flows in Hydraulic Turbines
Andre Saxer (Ecole Polytechnie de Lausanne)
Co-authors: H. Felici, C. Neury, J.P. Therre

The Viscous Model Vorticity Equation
Ralph Saxton (University of New Orleans)

Convergence of Data Fitting Algorithms
R. Schaback (University of Gottingen)

Optimal Resource Extraction and Recycling
Martin Schafer (Federal Armed Forces University, Hamburg, F.R.G.)

Equilibrium of an Elastic Spherical Cap Pulled at the Rim
A. Schiaffino (University of Rome)
Co-authors: Paolo Podio Guidugli, M. Rosati, V. Valente

Symbolic Computation of Equations of Motion
Werner Schiehlen (University of Stuttgart)
Co-author: K.P. Schmoll

The Solution of Boundary Value Problems for Differential Inclusions by Simplicial Fixed Point Algorithms
Klaus Schilling (Dornier System GmbH, Friedrichshafen, F.R.G.)

Far-Field Boundary Conditions for Unsteady Transonic Flow
H. Schippers (National Aerospace Laboratory NRL, Amsterdam, The Netherlands)

Large Scale Constrained Nonlinear System Identification
Johannes P. Schloder (Universitat Bonn)

Pole-Variant Zero Assignment for Multivariable Systems with Two Inputs and Two Outputs
Joachim Schmidt (Ruhr Universitat)

An Overview of ESSL on the 3090 Vector Facility
S. Schmidt (IBM, U.S.A.)

Stability Analysis of a Dimension 4 Nonlinear System in Combustion Near a Degenerate Fixed Point
C. Schmidt-Laine (Ecole Centrale de Lyon)
Co-author: D. Serre

Postimproving Penalty Methods with Complementary Finite Elements in Elasticity
Uwe Schomburg (Universitat der Bundeswehr)

High Wave Number Modulations in Numerical Solutions of Some Dispersive Nonlinear Wave Equations
S.W. Schoombie (University of the Orange Free State)

On the Measurement of Integral Scales of Turbulence
Peter Schrader (Ruhr Universitat Bochum)

Block Reflectors: Computation and Applications
Robert S. Schreiber (Rensselaer Polytechnic Institute)
Co-author: Beresford N. Parlett

Smooth Spline Approximations to Functions
N.L. Schryer (AT&T Bell Laboratories, U.S.A.)

Easy Handling and Verified Solution of Algebraic Problems
Gunter Schumacher (Institut Fur Angewandte, Karlsruhe)

Nonstandard Scaling Matrices in Trust Region Gauss-Newton Methods
H. Schwetlick (Martin Luther Universitat Halle)
Co-author: V. Tiller

Global Existence for a Semiconductor System with Generation Terms
Thomas I. Seidman (University of Maryland)

Computer Simulation of a Parallel Machines Scheduling Algorithm
Roberto Semenzato (Universita di Padova)

Algebraic Design of Multivariable Controllers Via Rosenbrock's Theory for Linear Systems
Manuel de la Sen (Universidad del Pais Vasco)

On Computing an Equivalent Symmetric Matrix for a Nonsymmetric Matrix
S.K. Sen (Indian Institute of Science)
Co-author: V.Ch. Venkaiah

Modelisation of Fluid-Fluid Interfaces
Pierre Seppecher (Universite Pierre et Marie Curie)
Co-author: Renee Gatignol

Numerical Solution of a Multi-Mode Heat Transfer Within an Irradiated Nuclear Fuel Bundle Residing in Air
Paul Sermer (Central Nuclear Services, Toronto, Canada)

Parallelism in Seismic Computations: Three Cases Studies
P. Sguazzero (IBM - E.C.S.E.C., Roma, Italy)

Robust Optimal Control Via Mathematical Programming
Bahram Shafai (Northeastern University)
Co-author: Georgy Sotirov

Analysis and Synthesis of Hydrological Data Through Continuous Convolving
Hari Shankar (Indian Institute of Technology, New Delhi)

Loss of Strict Hyperbolicity in the Buckley-Leverett System for Three Phase Flow in a Porous Medium
Michael Shearer (North Carolina State University)

Solving Linear Partial Differential Equations by Exponential Splitting
Qin Sheng (University of Cambridge)

Identifying an Unknown Term in an Inverse Problem of Linear Diffusion Equations
A. Shidfar (Iran University of Science & Tech.)

On The Discrete Random Transform
Lin Shi-Ming (Northwestern Polytechnical University, China)
Co-author: W.M. Boerner

On Variable Modelling in Singularly Perturbed Nonlinear Optimal Control Problems
Josef Shinar (Technion - Israel Institute of Technology)

Solving Linear Partial Differential Equations by Exponential Splitting
Qin Shing (Technion - Israel Institute of Technology)

Minimization of Time for Solving the Initial Value Problem on a Parallel Computer
Yuri Shuraits (Institute of Control Sciences, Moscow)
Co-authors: E. Trakhtengerts, N.A. Egorov

Nonlinear Diffusion with Absorption
Xiao Shutie (Tsinghua University)
Co-author: Chen Changsheng

A Method for Image Representation Based on Bivariate Splines
Shoulamit C. Shwartz (Hasifa University)

Sparse Matrix Techniques for Resistivity Modelling of Geothermal Areas
Sven Sigurdsson (University of Iceland)
Co-author: R. Sigurdsson

Flow Paths Around Soil Penetrometers
V. Silvestri (Ecole Polytechnique, Montreal)
Co-author: C. Tabib

A New Algorithm for the Tridiagonal Reduction of Unsymmetric Matrices
Horst D. Simon (Boeing Computer Services, U.S.A.)

Sliding Mode Control of Rigid Manifolds of Elastic Robot Manipulators
Herbert Sira-Ramirez (Escuela de Ingenieria de Sistemas, Merida, Venezuela)

Parallelism in the Solution of Ordinary Differential Equations
Robert D. Skeel (University of Illinois)
Co-author: Hon-Wah Tam

On the Efficient Numerical Evaluation of the Hausdorff Distance
Andrzej M.J. Skulimowski (Institute of Automatic Control, Krakow, Poland)

Reimann Problem for a Van der Waals Fluid
Marshall Slemrod (Rensselaer Polytechnic Institute)

A Class of Efficient Nonlinear Interpolation Schemes
R.D. Small (University of New Brunswick)

Oscillations and Multiple Steady States in Cyclic Gene Models with Repression
Hal L. Smith (Arizona State University)

The Numerical Solution of the Plane Biharmonic Equation Using Harmonic and Harmonic-Related Polynomials
Kenneth C. Smith (University of Bradford)

The Development of Stewartson Layers in a Rotating Fluid
S.H. Smith (University of Toronto)

Solutions of Forced Wave Propagation Problems
William V. Smith (Brigham Young University)

Modelling and Filtering of Freeway Traffic Flow
S.A. Smulders (Centre For Mathematics and Computer Science, Amsterdam, The Netherlands)
Co-author: J.H. Van Schuppen

Homogeneity of Uniform Elastic Bodies
Jedrzej Sniatycki (University of Calgary)
Co-authors: M. Elzanowski, M. Epstein

Solution of an Integrodifferential Equation with B-Splines
I.M. Snyman (University of South Africa)
Co-author: S.A. Sofianos

Geometries for Fluid Mechanics and Aerodynamics
Helmut Sobieczky (Institute for Theoretical Fluid Mechanics, DFVLR, Gottingen, F.R.G.)

Fourier Transformation as a Tool for Generating Integral Equations from Differential Equations Defining Discontinuous Functions
Roman Solecki (University of Connecticut)
Co-author: Guoquan Zhao

Mooring by Unilateral Ropes
Jose Eduardo Souza de Cursi (Ecole Nationale Sup. de Mecanique)

Heat Conduction in Electronic Chips with Hot-Spots
Gunnar Spaar (Lund Institute of Technology, Sweden)

Reliability in Problems with Uncertain Inout Data
Herbert Sprouer (Universital Karlsruhe)
Co-author: E. Adams

Decomposition Algorithms to Count the Linear Extensions of Posets
George Steiner (McMaster University)

A Family of Mixed Finite Elements for the Elasticity Problem
Rolf Stenberg (Helsinki University of Technology)

Long Periodic Vorticity Waves in Channels and Lakes
T. Stocker (ETH-Eidgenossische Technische Hochschule, Zurich, Switzerland)
Co-author: K. Hutter

A Quadratic Spline Collocation Method for Singular Perturbation Two-Point Boundary Value Problems
Mirjana Stojanovic (University of Novi Sad)
Co-authors: K. Surla, M. Kulpinski

Conjugate Gradients and the Toeplitz-Circulant - Hankel Eigenvalue Problem
Gilbert Strang (M.I.T.)

Optimal Control Problems at the Starting-up of a Polymerization Reactor
Roland Strietzel (Technical University of Dresden)

Creating the Mathematical Models with Lumped and Distributed Parameters for Simulation of Glass Mass Flow in a Tank Furnace
Jan Studzinski (Polish Academy of Sciences)

On the State Estimation for Quasi-Stationary Stochastic Distributed Parameter Systems
Yoshifumi Sunahara (Kyoto Institute of Technology)
Co-authors: M. Ishikawa, S. Aihara

A Generic Computer Graphics Algorithm for the Surface Development and Manufacture of Complex Flow Ducts
Varada Raj Sundar (CAD Group, Bhopal, India)

The P-Version of Mixed Finite Element Methods for Elliptic Problems
Manil Suri (University of Maryland)

Implementation of an Explicit Solution for the Quadratic Dynamic Programming Problem
Wm. Richard S. Sutherland (Dalhousie University)

Trapping of Frontal-Vorticity Waves Due to Critical Layer Absorption
Gordon E. Swaters (University of Alberta)

A Vectorized Version of Fishpak
Roland Sweet (University of Colorado)

Spectral Analysis of an Adaptive Optical Reconstructor Operator
Roque K. Szeto (Hughes Aircraft Company, El Segundo, U.S.A.)

Ordering of the Variables in Oil Reservoir Simulation with Wells
Daniel B. Szyld (Duke University)
Co-authors: J.R.P. Rodrizuez, Flavio Dickstein

A Graph Valued Function for Postal Network
Yahya Tabesh (Isfahan University of Technology)

Applications de la Theorie des Matrices Booleennes a la Determination du Nombre Minimal de Tournois Reguliers
Claudette Tabib (College Edouard-Montpetit)

A Semiconductor Device Simulation Method Using the Voronoi Discretization on Boundary-Fitted Curvilinear Coordinates
Kazutami Tago (Energy Research Laboratory HITACHI, Ltd., Ibaraki, Japan)
Co-authors: Kazuyoshi Miki, Toru Kaga, Toru Toyabe

Estimation for the Solution of Partial Differential Equations
Djurdjica Takaci (Faculty of Sciences at University Nova Sad)

Global Stability of Generalized Volterra Delay-Diffusion Models
Yasuhiro Takeuchi (Shizuoka University)
Co-author: E. Beretta

Devices for Accelerating the Convergence of an Iterative Method for Derivatives Eigensystems
Roger C.E. Tan (La Trobe University)

On the Use of the Marquardt Method for the Finite Element Solution of Fluid Flow Problems at High Reynolds Number
Philippe A. Tanguy (Universite de Laval)
Co-authors: Daniel Fauchon, Robert E. Hayes

The Dynamics of Three Vortices Revisited
John Tavantzis (New Jersey Institute of Technology)
Co-author: Lu Ting

Efficiency's Measure, A Case of Study
Beatriz Regina Tavares Franciosi (Porto Alegre, Portugal)
Co-author: M.A. Oliveira Camargo

Dynamic Control for Ill Conditioned Linear Systems
Thierry Tedesco (Universite Paul Sabatier)

The $(T_{N(SV);N})$-Policy for the M/G/1 Queue
Jacques Teghem (Faculte Polytechnique de Mons, Belgium)
Co-author: Jacqueline Loris-Teghem

Stochastic Simulation of Grain Growth in 2 D-Polycrystals at the Microscopic Level
H. Telley (Swiss Federal Institute Technology)
Co-authors: A. Mocellin, Th. M. Liebling

Recent Problems in Uniform Asymptotic Expansions of Integrals
Nico M. Temme (Centre for Mathematics and Computer Science, Amsterdam, The Netherlands)

New FFT Algorithms for Large and Small Computers
Clive Temperton (Environmente Canada)

Solution of Riemann Problem and Convergence of Glimm Scheme for a Combustion Model System
Zhen-Huan Teng (Beijing University)

Proportional Integer Sharing Processes
Eric Terouanne (Universite Paul Valery)
Co-author: Jean-Luc Petit

Finite Element Formulations for Vorticity-Stream Function Computations
Tayfun E. Tezduyar (University of Houston)
Co-author: Roland Glowinski

The Free Surface in Cone and Plate and Parallel Plate Viscometers
R.W. Thatcher (U.M.I.S.T., Manchester, U.K.)
Co-authors: David Tidd, A. Kaye

Calculation of Two-Dimensional Turbulent Asymmetrical Wakes
Frank Thiele (Technische Universitat Berlin)
Co-author: F. Arnold

On the Convergence of Mixed Finite Element Methods
Jean-Marie Thomas (Universite de Pau)

A Model Reference Adaptive Control Scheme for Systems Having Measurable Disturbances and Multivariable Applications
Chai Tianyou (North-East University of Technology, Shenyagn)

A Stochastic Description of Copolymerisation and Network Formation in a Three-Stage Process
G.P.J.M. Tiemersma-Thoone (DSM Research BV, The Netherlands)
Co-authors: G.J.R. Scholtens, K. Dusek

Quick and Practical Approximations for Queueing Models
H.C. Tijms (Vrije University)

A K-Tree Generalisation Which Characterises Consistency of Dimensioned Engineering Drawings
Philip H. Todd (Tektronix, Inc., Computer Graphics Research, U.S.A.)

Partially Separable Nonlinear Network Optimization
Philippe L. Toint (Department of Mathematics, F.N.D.P., Namur, Belgium)
Co-author: D. Tuyttens

STAIRWAY: An Efficient Sparse Gaussian Solver for the Cyber 205 Supercomputer
W.S. Tortike (University of Alberta)
Co-author: S.M. Farouq Ali

Quadrature Rules for Stieltjes Integrals and Numerical Solution of Renewal-Type Integral Equations
M. Tortorella (AT&T Bell Laboratories, U.S.A.)
Co-author: M. Hamami

Piecewise C1 - Approximation, With Application to Water Steam Thermodynamic Functions
C.R. Traas (University of Twente)
Co-author: R.H.J. Gmelic Meyling

Numerical Experiments on the Eigenvalue Problem for Symmetric Toeplitz Matrices
William F. Trench (Trinity University)

An Investigation of the Mathematical Model of the Diesel Engine Cycle
Radivoje Trifunovic (University of Belgrade)

Nonlinear Stability Analysis of a Robot
Hans Troger (Institut fur Mechanik, Wien, Austria)
Co-authors: E. Lindtner, A. Steindl

A Nonconforming Finite Element Approximation of Linear Thin Shell Problems
Pascal Trouve (INRIA-Rocquencourt, France)

Shape Stabilization of Wave Equation
Christine Truchi (Ecole Nationale Sup. des Mines, Valbonne)
Co-author: J.P. Zolesio

Numerical Conformal Mapping via the Szego Kernel
Manfred R. Trummer (M.I.T.)

An Efficient Procedure for Estimating Mixed Spectrum with Recursive Least Squares on Amplified G-Harmonics
B. Truong-Van (Universite de Pau)

Vortex Blob Methods for Stratified Flow
Gretar Tryggvason (University of Michigan)

Some Practical Aspects of the Resolution of Large Linear Systems in Parallel
D. Trystram (Ecole Centrale de Paris)
Co-author: P. Laurent-Gengoux

On the Numerical Computation of the Derivatives of A B-Spline Series
N.K. Tsao (Wayne State University)
Co-authors: E.T.Y. Lee, T.C. Sun

Numerical Study of Convection in a Cylindrical Geometry: Breaking of Axisymmetry
Laurette S. Tuckerman (University of Texas - Austin)

Root-Squaring with SLI Arithmetic
P.R. Turner (University of Lancaster)
Co-author: C.W. Clenshaw

An Iterative Technique for Velocity Control of Redundant Robotic Manipulators
S.G. Tzafestas (National Technical University, Athens, Greece)
Co-author: A. Zagorianos

Expert Systems in Process Fault Detection and Supervisory Control
S.G. Tzafestas (National Technical University, Athens, Greece)

Path Following via Sensitivity Analysis Techniques
Klaus Ulrich (Universitat Hannover)

Geometric Foundations of Separation Engineering
Patrick Valentin (Societe Nationale ELF-Acquitaine, ELF-Solaize Research Center, Saint-Symphorien d'Ozon, France)

Free Oscillation Spectrum of a Self-Gravitating, Elastic, Uniformly Rotating Body
Bernard Valette (O.R.S.T.O.M. - I.P.G.P., Laboratoire de Sismologie, Paris, France)

Discretizations Conserving Energy and Other Constants of the Motion
F.P.H. Van Beckum (University of Twente)
Co-author: E. Van Groesen

Weir Flows
Jean-Marc Vanden-Broeck (University of Wisconsin-Madison)

Normal Forms in the Theory of Galloping for an Aeroelastic Oscillator with Two-Degrees-of-Freedom
C.G.A. Van der Beek (Delft University of Technology)
Co-author: A.H.P. Van Der Burgh

A Mathematical Model for a Falling Film Evaporator
R. Van Der Hout (AKZO, Corporate Research, CRS, Arnhem, The Netherlands)

The Implementation for a Block Relaxation Algorithm on the Parallel Processor
Ruud J. Van der Pas (State University of Utrecht)

On the Use of Digit Distributions in Pattern Recognition
J. Van der Pol (Amsterdam, The Netherlands)
Co-authors: H.P.M. Essink, R. de Jager

An Asymptotic Theory for a Class of Initial-Boundary Value Problems for Weakly Nonlinear Wave Equations
W.T. Van Horssen (Delft University of Technology)

An Efficient and Reliable Algorithm for Computing the Singular Subspace of a Matrix, Associated with its Smallest Singular Values
Sabine Van Huffel (Katholieke Universiteit Leuven)
Co-author: J. Vandewalle

A Rothe Galerkin Finite Element Method for the One Dimensional Heat Equation with Time Dependent Radiation Boundary Conditions
Roger Van Keer (University of Ghent)

On the Numerical Approximation of an Invariant Curve
M. Van Veldhuizen (Vrye Universiteit)

Symmetrizing a Matrix Exactly Using Floating-Point Modular Arithmetic
V. Ch. Venkaiah (Indian Institute of Science, Bangalore)
Co-author: S.K. Sen

Numerical Approximation of Singular Parabolic Problems
Claudio Verdi (Universita di Pavia)
Co-author: G. Sacchi

A New Levinson-Type Recursion for Structure and Parameter Estimation of Vector Arma Processes
S. Veres (Operations Research Department, Budapest, Hungary)
Co-author: J. Bokor

Heavy Gas Atmospheric Dispersion in Complex Environments
Emmanuel Vergison (Solvay S.A., Brussels, Belgium)
Co-author: J. Ch. Basler

Approximation of Nonintegrable Systems
F. Verhulst (Rijksuniversiteit Utrecht)

Trifou: A 3-D Eddy-Current Program
Jean-Claude Verite (Electricite de France)
Co-authors: G. Bunouff, P. Chaussecourte, C. Rose, G. Tanneau

Fingering in Displacement Processes with Magnetic Fluid in Porous Media
Alakh P. Verma (S.V.R. College of Eng. & Tech., India)
Co-author: A.K. Rajput

Dynamic Stochastic Programming: An Application to a Telecommunication Planning Problem
L. Verri (ITALTEL, Central Research Laboratories, Milano, Italy)
Co-authors: G. Gallassi, D. Terranova

Stationary Potential Flow of an Ideal Fluid Through a Resisting Boundary
Kresimir Veselic (Fern University)
Co-authors: Anton Suhadolc, Andro Mikelic

Modulef: A Biomedical Problem Solved by an Opened Finite Element Package
Marina Vidrascu (INRIA-Rocquencourt, France)

Bayesian Methods in Accelerated Life Testing
Reinhard Viertl (Technische Universitat Wien)

Optimization of Industrial Glass Furnaces
Pierre Villon (Universite de Technologie, Compiegne)
Co-authors: F. Mehl, J.P. Yvon

Optimal Control Versus PID
Gurvinder Singh Virk (University of Sheffield)

Raster Evaluation for Splines
Wolfgang Volk (Berlin, F.R.G.)

The Behaviour and Approximation of Solutions of a Quasilinear Second Order Differential Equation
Bozo Vrdoljak (University of Split)

Parallel Processing Experiment with the GMRES Method for Solving Large Nonsymmetric Linear Systems on the Cray XMP
Phuong Anh Vu (Cray Research, Inc., U.S.A.)

Numerical Integration of Totally Symmetric Functions Over the Rn: An Application to the Least Constants in the Sobolev Inequality
Jorg Waldvogel (Swiss Federal Institute of Technology)

On Finite Element Approximations of Problems in Duct Acoustics
Noel J. Walkington (The University of Texas at Austin)
Co-author: Thomas M. Wicks

Competition in the Gradostat
P. Waltman (Emory University)
Co-authors: W. Jager, J. So, B. Tang

A Note About the Generalized Newton Algorithm
Jia-Song Wang (Nanjing University)

The Model and Method of the Minimization of L1 Norm in Seismic Inverse Problem
Jia-Song Wang (Nanjing University)

Vibration Control in Elastic and Aeroelastic Systems with Time-Varying Spatial Domains
Paul K.C. Wang (University of California - Los Angeles)

Analysis of Two Special Cases of the Traveling Salesman Problem
Richard H. Warren (General Electric Company, U.S.A.)

Averaging, Fluctuation and Approximation of Parabolic Equation with Random Coefficients
Hisao Watanabe (Kyushu University)

The Moving Finite Element Method for Time-Dependent Partial Differential Equations
A.J. Wathen (University of Bristol)

Uniform Methods for Singular Boundary Point and Interior Turning Point
Abdul-Majid Wazwaz (College of Science and Technology, Jerusalem)
Co-author: F.B. Hanson

Hyperdimensional Data Analysis Using Parallel Coordinates
E.J. Wegman (George Mason University)

A Finite Element Method for the Shakedown Assessment of Elastic-Plastic Disks Under Variable Mechanical and Thermal Loads
D. Weichert (Institut fur Mechanik-Bochum)
Co-author: J. Gross-Weege

Sobolev Type Inequalities in the Limiting Cases and Its Applications to Singular Perturbation
Zhang Weitao (Institute of System Science, Beijing)

Error Analysis of Average and Variance Estimations for Finite Correlated Time Series
Zuyin Weng (Beijing Institute of Radio Measurements)

A Method for Computing the Intersection of Two Complex Matrices Under the Star Ordering
Hans Joachim Werner (Universitat Bonn)

Plasma Simulations in Technical Applications
Thomas Westermann (Nuclear Research Center Karlsruhe)

Factorization of the Wave Equation in a Stratified Medium
Vaughan H. Weston (Purdue University)

Stability-Border Continuation Method for Linear Parameter-Excited Systems
Bernhardt Weyh (University of Duisburg)
Co-author: Heinrich Kostyra

Stability Analysis of Variable Structure Control Systems
B.A. White (Royal Military College of Science, Shrivenham)

Superconvergent Finite Element Methods with Applications to Solid Mechanics
J.R. Whiteman (Brunel University)

Lagrange Approximation of Linear Collision Terms
Joachim Wick (Universitat Kaiserslautern)

Algebraic Theory of Variational Methods
C. Wielgosz (L.M.S. - E.N.S.M., Nantes, France)

Adaptive Tuning of Decentralized Controllers
P. Wiemer (Ruhr-University Bochum)

Interactive Sensitivity Analysis and Simulation in Capital Budgeting
Ph. Wieser (Ecole Polytechnique Federale, Switzerland)
Co-author: F.L. Perret

Vectorized High Order Finite Difference Methods for 2-D PDE's on General Domains
Helmut Wietschorke (Universitat Karlsruhe)
Co-authors: R. Weiss, W.S. Schonauer

On the Heat Transfer from a Horizontal Cylinder Inundated by a Thin, Vertical, Liquid Film
G. Wilks (University of Keele)
Co-author: R.J. Gribben

Local Gaussian Preconditioning of Symmetric and Non Symmetric Iterative Finite Element Equation Solvers
Sven Oivind Wille, (A.S. Veritas Research, Oslo, Norway)

The Application of Multigrid Methods to Jet Flow
M.J. Wilson (University of Leeds)
Co-author: S.A.E.G. Falle

Suspense - A Specification Tool for Numerical Problems and Algorithms
Guido Wirtz (Universitat Bonn)

On the Equation of Motion for Articulated Rigid-Body Systems
J. Wittenburg (Universitat Karlsruhe)

Direct Approach to Control Law Synthesis under Set-Membership Model of Uncertainty
Konrad Wojciechowski (University of Wageningen)
Co-author: Andrzej K. Krolikowski

Asymptotics of a Class of High Order Linear Ordinary Differential Equations and Applications
A.D. Wood (National Institute for Higher Education, Dublin, Ireland)
Co-author: P.B. Paris

An Unconditionally Stable Implementation of the Enthalpy Method
A.S. Wood (University of Bradford)
Co-author: G.E. Bell

Random Graph Theoretical and Statistical Mechanics Models for Monomolecular Films
Wojbor A. Woyczynski (Case Western Reserve University)

Inexact Methods for Nonsmooth Composite Functions
Stephen Wright (N.C. State University)

Cracks, Notches and the Method of Lines (Mol)
L.S. Xanthis (The Polytechnic of Central London)
Co-author: N.S. Lambrou

Application of System Science in Industry Outputs Price Reform in China
He Xiangwei (Beijing Institute of Information and Control)
Co-authors: Shao Ruyun, Chen Kai

Interacting Particle Models of Deposition
Bernard Ycart (Faculte des Sciences, Pau, France)

Convergence for Viscous Splitting of the Navier-Stokes Equations in Bounded Domains
Lung-An Ying (Peking University)

A Reliable Argument Principle Algorithm to Find the Number of Zeros of an Analytic Function in a Bounded Domain
Xingren Ying (Washington University)
Co-author: I. Norman Katz

Monotone Decomposition Theorems of Boolean Functions and Directed Methods for Monotone Decomposition
Liu Yongcai (Shanghai University Science & Technology)
Co-author: Li Weijian

Two-Sided Iterative Solution for a System of Nonlinear Numerical Equations
Zhao-Yong You (Xi'An Jiaotong University)
Co-author: Xiao-Jun Chen

The Iterative Solution of Partial Differential Equations with Vector, and Parallel Computers
David M. Young (University of Texas - Austin)

Combining Preconditioners for Krylov Subspace Methods
D.P. Young (Boeing Computer Services, U.S.A.)
Co-author: F.T. Johnson

Mean and Variance of Overflow Traffic in Loss Systems with Time-Varying Inputs
Naim Yunus (University Sains Malaysia)

Absolute Observability, Canonical Structure and Transmission Zeros of Linear Time-Invariant Systems
Yu Fan Zheng (East China Normal University)

The Application of Boltzmann Statistical Theory to Population Study
Zhou Zheng (Beijing Institute of Information & Control)
Co-authors: Wang Ling, Jing-Juan Yu

Superb: The Supremum Parallelizer
Hans Zima (Institut fur Informatik, Bonn, F.R.G.)

Optimization Models in Biology
Mariusz Ziolko (University of Mining and Metallurgy, Krakow)
Co-author: J. Kozlowski

Stability of the Method of Characteristics
Mariusz Ziolko (University of Mining and Metallurgy, Krakow, Poland)

Author Index